WADSWORTH
CENGAGE Learning

M&F 2010–2011 Edition
David Knox and Caroline Schacht

Editor-in-Chief: Michelle Julet

Sr. Publisher: Linda Schreiber

Director, 4LTR Press: Neil Marquardt

Developmental Editors: Laura Rush and Jamie Bryant, B-books, Ltd.

Assistant Editor: Melanie Cregger

Project Manager, 4LTR Press: Michelle Lockard

Sr. Editorial Assistant: Rachael Krapf

Executive Brand Marketing Manager: Robin Lucas

Executive Marketing Manager: Kimberly Russell

Marketing Manager: Andrew Keay

Marketing Communications Manager: Laura Localio

Production Director: Amy McGuire, B-books, Ltd.

Content Project Manager: Cheri Palmer

Media Editor: Lauren Keyes

Print Buyer: Paula Vang

Production Service: B-books, Ltd.

Sr. Art Director: Caryl Gorska

Internal Designer: Ke Design

Cover Design: Riezebos Holzbaur Design Group/Christopher Harris

Cover Image: photolibrary/Banana Stock

Photography Manager: Deanna Ettinger

Photo Researcher: Sam Marshall

Library of Congress Control Number: 2009934720

ISBN-13: 978-0-495-90545-5
ISBN-10: 0-495-90545-3

Wadsworth
20 Davis Drive
Belmont, CA 94002-3098
USA

Cengage Learning is a leading provider of customized learning solutions with office locations around the globe, including Singapore, the United Kingdom, Australia, Mexica, Brazil, and Japan. Locate your local office at **www.cengage.com/global**.

Cengage Learning products are represented in Canada by Nelson Education, Ltd.

To learn more about Wadsworth, visit **www.cengage.com/wadsworth**
Purchase any of our products at your local college store or at our preferred online store **www.ichapters.com**

Printed in the United States of America
1 2 3 4 5 6 7 12 11 10 09

Brief Contents

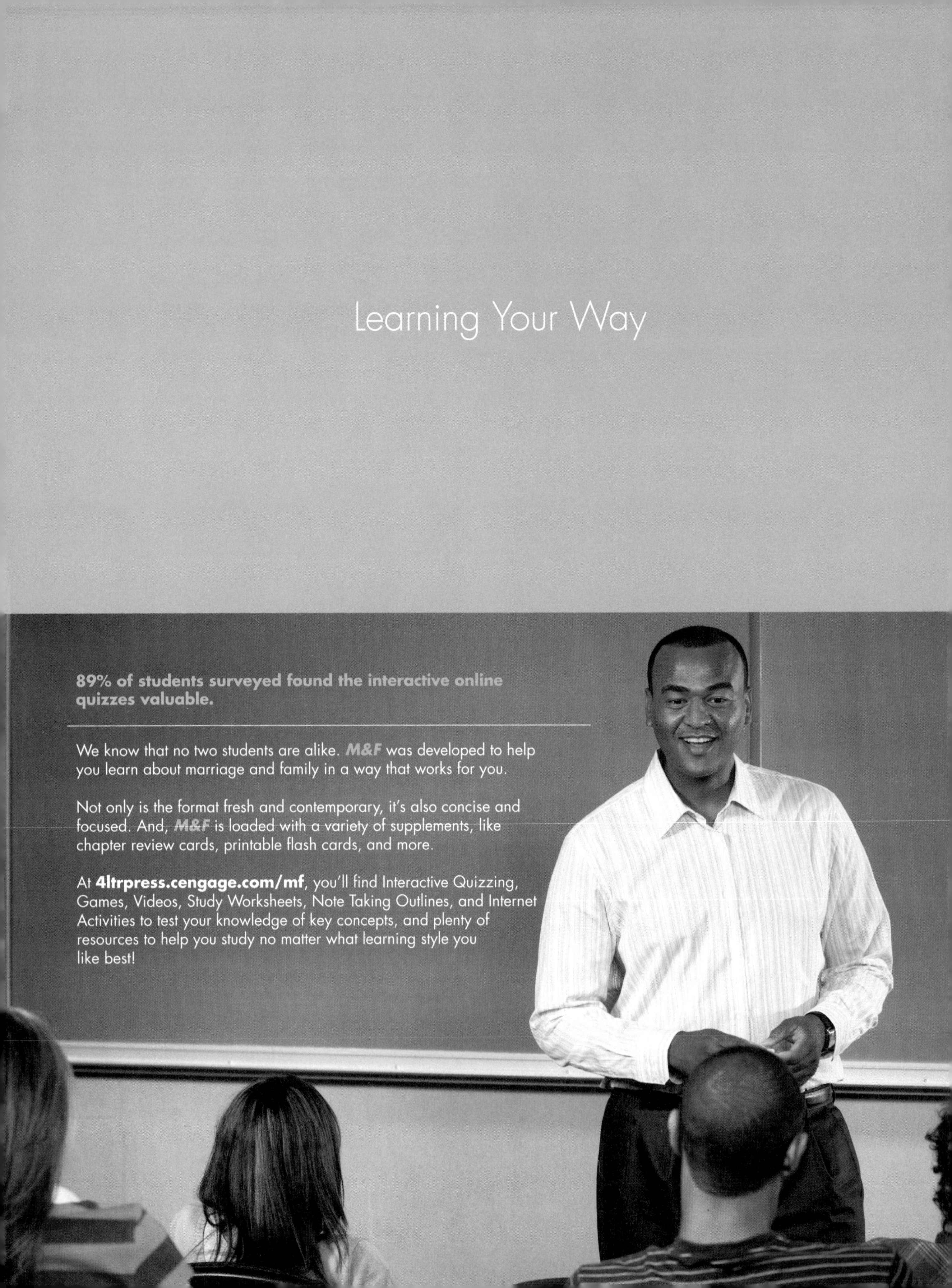
Learning Your Way
89% of students surveyed found the interactive online quizzes valuable.
We know that no two students are alike. M&F was developed to help you learn about marriage and family in a way that works for you.
Not only is the format fresh and contemporary, it's also concise and focused. And, M&F is loaded with a variety of supplements, like chapter review cards, printable flash cards, and more.
At 4ltrpress.cengage.com/mf, you'll find Interactive Quizzing, Games, Videos, Study Worksheets, Note Taking Outlines, and Internet Activities to test your knowledge of key concepts, and plenty of resources to help you study no matter what learning style you like best!

Contents

1 Relationships, Marriages, and Families Today 2

2 Gender 30

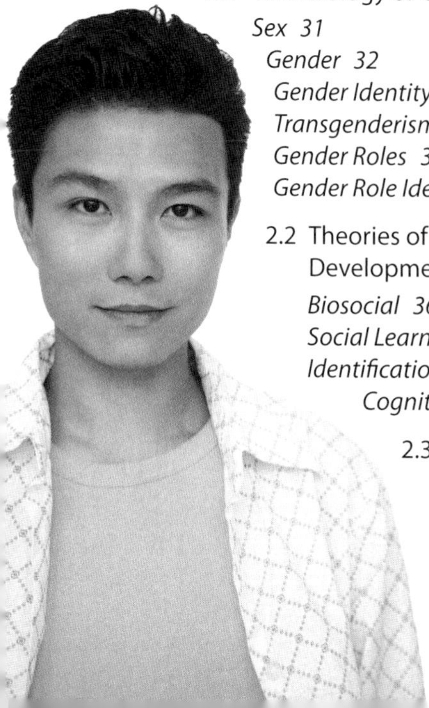

3 Singlehood, Hanging Out, Hooking Up, and Cohabitation 48

4 Communication 70

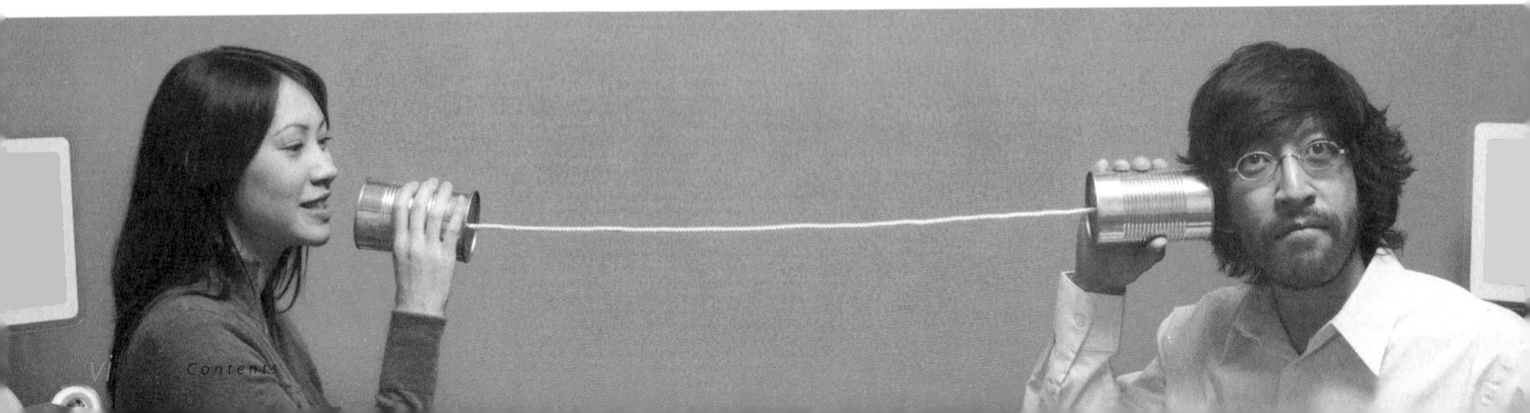

6 Marriage Relationships 116

7 Same-Sex Couples and Families 134

8 Sexuality in Relationships 154

9 Planning to Have Children 172

10 Parenting 188

11 Family and the Economy 206

12 Violence and Abuse in Relationships 220

13 Stress and Crisis in Relationships 236

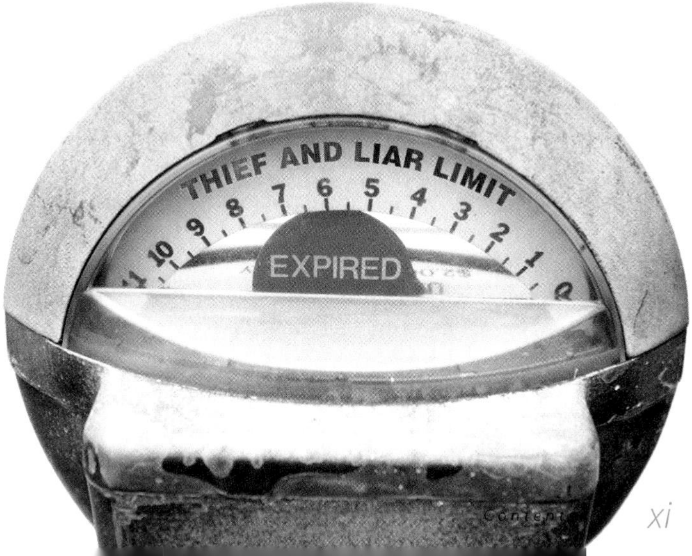

14 Divorce and Remarriage 254

Remember

The portable cards at the back of the book are designed to help you with the course!

Relationships, Marriages, and Families Today

SECTIONS

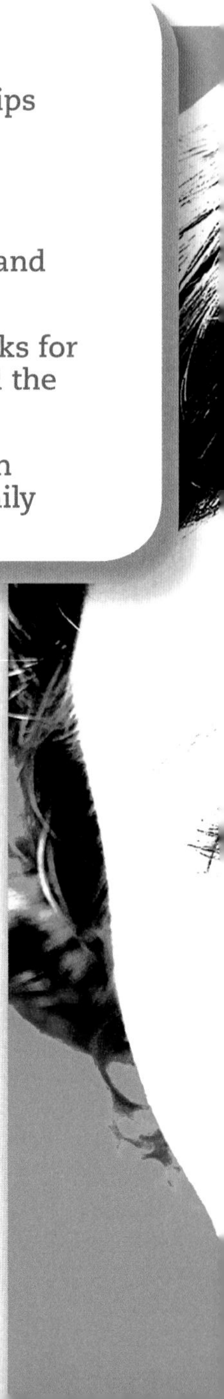

© Dynamic Graphics/Jupiterimages

1.1 Choices in Relationships

The central theme of this text is choices in relationships. Although we have many such choices to make, among the most important are whether to marry, who to marry, when to marry, whether to have children, whether to remain emotionally and sexually faithful to one's partner, and whether to use a condom. Though structural and cultural influences are operative, a choices framework emphasizes that individuals have some control over their relationship destiny by making deliberate choices to initiate, respond to, nurture, or terminate intimate relationships.

Facts about Choices in Relationships

The facts to keep in mind when making relationship choices include the following.

Not to Decide Is to Decide Not making a decision is a decision by default. If you are sexually active and decide not to use a condom, you have made a decision to increase your risk for contracting a sexually transmissible infection, including HIV. If you don't make a deliberate choice to end a relationship that is unfulfilling, abusive, or going nowhere, you have made a choice to continue in that relationship and have little chance of getting into a more positive and satisfying relationship. If you don't make a decision to be faithful to your partner, you have made a decision to be vulnerable to cheating.

Some Choices Require Correction Some of our choices, although appearing correct at the time that we make them, turn out to be disasters. Once we realize that a choice is having consistently negative consequences, we need to stop defending the old choice, reverse our position, make new choices, and move forward. Otherwise, one remains consistently locked into continued negative outcomes of "bad" choices. For example, choosing a partner who was loving and kind but who turns out to be abusive and dangerous requires correcting that choice. To stay in the abusive relationship will have predictable disastrous consequences—to make the decision to disengage and to move on opens the opportunity for a loving relationship with another partner. Living alone may also be a better alternative than living in a relationship in which you are abused or unhappy. Other examples of making corrections involve ending dead or loveless relationships (perhaps after investing time and effort to improve the relationship or love feelings), changing jobs or career, and changing friends.

Choices Involve Trade-Offs By making one choice, you relinquish others. Every relationship choice you make will have a downside and an upside. If you decide to stay in a

If these individuals do not decide to become committed to each other, they have decided to drift. Not to decide is to decide.

© Photolibrary/Alamy

relationship that becomes a long-distance relationship, you are continuing involvement in a relationship that is obviously important to you. However, you may spend a lot of time alone when you could be discovering new relationships. If you decide to marry, you will give up your freedom to pursue other emotional and/or sexual relationships, and you will also give up some of your control over how you spend your money—but you may also get a wonderful companion with whom to share life.

Choices Include Selecting a Positive or Negative View As Thomas Edison progressed toward inventing the lightbulb, he said, "I have not failed. I have found ten thousand ways that won't work." In spite of an unfortunate event in your life, you can choose to see the bright side. Regardless of your circumstances, you can choose to view a situation in positive terms. A breakup with a partner you have loved can be viewed as the end of your happiness or an opportunity to become involved in a new, more fulfilling relationship. The discovery of your partner cheating on you can be viewed as the end of the relationship or an opportunity to examine your relationship, to open up communication channels with your partner, and to develop a stronger relationship. Discovering that one is infertile can be viewed as a catastrophe or as a challenge to face adversity with one's partner. One's point of view does make a difference—it is the one thing we have control over.

sequential ambivalence the individual experiences one wish and then another.

simultaneous ambivalence the person experiences two conflicting wishes at the same time.

Choices Involve Different Decision-Making Styles Allen et al. (2008) identified four patterns in the decision-making process of 148 college students. These patterns and the percentage using each pattern included (1) "I am in control" (45 percent), (2) "I am experimenting and learning" (33 percent), (3) "I am struggling but growing" (14 percent), and (4) "I have been irresponsible" (3 percent). Of those who reported that they were in control, about a third (34 percent) said that they were "taking it slow," and about 11 percent reported that they were "waiting it out." Men were more likely to report that they were "in control." Hence, these college students could conceptualize their decision-making style; they knew what they were doing. Of note, only 3 percent labeled themselves as being irresponsible.

Choices Produce Ambivalence Choosing among options and trade-offs often creates ambivalence—conflicting feelings that produce uncertainty or indecisiveness as to a course of action to take. There are two forms of ambivalence: sequential and simultaneous. In **sequential ambivalence**, the individual experiences one wish and then another. For example, a person may vacillate between wanting to stay in a less-than-fulfilling relationship and wanting to end it. In **simultaneous ambivalence**, the person experiences two conflicting wishes at the same time. For example, the individual may feel both the desire to stay with the partner and the desire to break up at the same time. The latter dilemma is reflected in the saying, "You can't live with them, and you can't live without them." Some anxiety about choices is normative and should be embraced.

Most Choices Are Revocable; Some Are Not Most choices can be changed. For example, a person who has chosen to be sexually active with multiple partners can later decide to be monogamous or to abstain from sexual relations. Individuals who have in the past chosen to emphasize career, money, or advancement over marriage and family can choose to prioritize relationships over economic and career-climbing behaviors. People who have been unfaithful in the past can elect to be emotionally and sexually committed to a new partner.

Other choices are less revocable. For example, backing out of the role of spouse is much easier than backing out of the role of parent. Whereas the law permits disengagement from the role of spouse, as with a formal divorce decree, the law ties parents to dependent offspring (through child support, for example). Hence, the decision to have a child is usually irrevocable. Choosing to have unprotected sex can also result in a lifetime of coping with sexually transmitted infections.

Choices of Generation Y Those in **Generation Y** (typically born between 1979 and 1984) are the children of the baby boomers. About 40 million of them, these Generation Yers (also known as the Millennial or Internet Generation) have been the focus of their parents' attention. They have been nurtured, coddled, and scheduled into day-care centers for getting ahead. The result is a generation of high self-esteem, self-absorbed individuals who believe they "are the best." Unlike their parents, who believe in paying one's dues, getting credentials, and sacrifice through hard work to achieve economic stability, Generation Yers focus on fun, enjoyment, and flexibility. They might choose a summer job at the beach if it buys a burger and a room with six friends over an internship at IBM that smacks of the corporate America sellout. Generation Yers, knowing college graduates who work at McDonald's, may wonder, "*why bother attending college?*" and instead seek innovative ways of drifting through life; they may continue to live with their parents, live communally, or get food by "dumpster diving." In effect, they are the generation of immediate gratification; they focus only on the here and now. The need for Social Security is too far off, and health care is available, so they stay "for free at the local hospital emergency room."

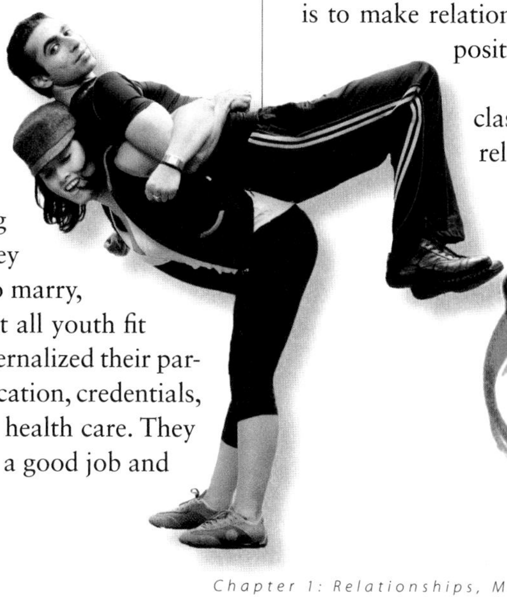

Generation Yers are also relaxed about relationship choices. Rather than pair-bond, they hang out, hook up, and live together. They are in no hurry to find "the one," to marry, or to begin a family. To be sure, not all youth fit this characterization. Some have internalized their parents' values, and are focused on education, credentials, a stable job, a retirement plan, and health care. They may view education as the ticket to a good job and expect their college to provide a credential they can market. However, increasingly, Generation Yers are taking their time getting to the altar and focusing on education, career, and enjoying their freedom in the meantime (Generation Y Data 2007).

Generation Y children of the baby boomers, typically born between 1979 and 1984. Also known as the Millennial or Internet Generation.

Choices Are Influenced by the Stage in the Family Life Cycle The choices a person makes tend to be individualistic or familistic, depending on the stage of the family life cycle that the person is in. Before marriage, individualism characterizes the thinking and choices of most people because individuals need only be concerned with their own needs. Most people delay marriage in favor of completing school, becoming established in a career, and enjoying the freedom of singlehood.

Once married, and particularly after having children, the person's familistic values and choices change as the needs of a spouse and children begin to influence. Evidence of familistic choices is reflected in the fact that spouses with children are less likely to divorce than spouses without children.

Making Wise Choices Is Facilitated by Learning Decision-Making Skills Choices occur at the individual, couple, and family levels. Deciding to transfer to another school or take a job out of state may involve all three levels, whereas the decision to lose weight is more likely to be an individual decision. Regardless of the level, the steps in decision making include setting aside enough time to evaluate the issues involved in making a choice, identifying alternative courses of action, carefully weighing the consequences for each choice, and being attentive to your own inner voice ("Listen to your senses"). The goal of most people is to make relationship choices that result in the most positive and least negative consequences.

Students in my marriage and family class identify their "best" and "worst" relationship choices (see Table 1.1).

The goal of most people is to make relationship choices that result in the most positive and least negative consequences.

Table 1.1
"Best" and "Worst" Choices Identified by University Students

Best Choice	Worst Choice
Waiting to have sex until I was older and involved.	Cheating on my partner.
Ending a relationship with someone I did not love.	Getting involved with someone on the rebound.
Insisting on using a condom with a new partner.	Making decisions about sex when drunk.
Ending a relationship with an abusive partner.	Staying in a relationship I knew was dead.
Forgiving my partner and getting over cheating.	Changing schools to be near my partner.
Getting out of a relationship with an alcoholic.	Not going after someone I really wanted.

Global, Structural/Cultural, and Media Influences on Choices

Choices in relationships are influenced by global, structural/cultural, and media factors. Although a major theme of this book is the importance of taking active control of your life in making relationship choices, it is important to be aware that the social world in which you live restricts and channels such choices. For example, enormous social disapproval for marrying someone of another race is part of the reason that 95 percent of all individuals in the United States marry someone of the same race.

Globalization Families exist in the context of world globalization. Economic, political, and religious happenings throughout the world affect what happens in your marriage and family in the United States. When the price of oil per barrel increases in the Middle East, gasoline costs more, leaving fewer dollars to spend on other items. When the stock market in Hong Kong drops 500 points, Wall Street reacts, and U.S. stocks drop. The politics of the Middle East (for example, terrorist or nuclear threats from Iran) impact Homeland Security measures so that getting through airport security to board a plane may take longer.

institution established and enduring patterns of social relationships.

The country in which you live also affects your happiness and well-being. For example, in one study, citizens of thirteen countries were asked to indicate their level of life satisfaction on a scale from 1 (dissatisfied) to 10 (satisfied): citizens in Switzerland averaged 8.3, those in Zimbabwe averaged 3.3, and those in the United States averaged 7.4 (Veenhoven 2007). The Internet, CNN, and mass communications provide global awareness so that families are no longer isolated units.

Social Structure The social structure of a society consists of institutions, social groups, statuses, and roles.

1. **Institutions.** The largest elements of society are social **institutions**, which may be defined as established and enduring patterns of social relationships. In addition to the family, major institutions of society include the economy, education, and religion. Institutions affect individual decision making. For example, you live in a capitalistic society where economic security is valued—the number-one value held by college students (Pryor et al. 2008). In effect, the more time you spend focused on obtaining money, the less time you have for relationships. You are now involved in the educational institution that will impact your choice of a mate (college-educated people tend to select and marry one another). Religion also affects sexual and relationship choices (for example, religion may result in delaying first intercourse, not using a condom, or marrying someone of the same faith). The family is a universal institution. Spouses who "believe in the institution of the family" are highly committed to maintaining their marriage and do not regard divorce as an option.

2. **Social groups.** Institutions are made up of social groups, defined as two or more people who share a common identity, interact, and form a social relationship. Most individuals spend their day going between social groups. You may awaken in the context (social group) of a roommate, partner, or spouse. From there you go to class with other students, lunch with friends, work with the boss, and talk on the phone to your parents. So, within twenty-four hours you have been in at least five social groups. These social groups have varying influences on your choices. Your roommate influences who you have in your room for how long, your friends may want to eat at a particular place, your boss will assign you certain duties, and your

parents may want you to come home for the weekend.

Your interpersonal choices are influenced mostly by your partner and peers (for example, your sexual values, use of condoms, and the amount of alcohol you consume). Thus, selecting a partner and peers carefully is important.

The age of the partner you select to become romantically involved with is influenced by social context. If you are a woman, your parents and peers will probably approve of you dating and marrying "someone a little older." Likewise, if you are a man, your parents and peers will probably approve of you dating and marrying someone "a little younger." This pattern of women dating older men and men dating younger women makes up the **mating gradient**, conceptualized in Figure 1.1. As a woman progresses through college, her acceptable dating pool shrinks because there are fewer men her age or older, whereas a man experiences a higher number of younger women acceptable to date as he ages.

Social groups may be categorized as primary or secondary. **Primary groups**, which tend to involve small numbers of individuals, are characterized by interaction that is intimate and informal. Parents are members of one's primary group and may exercise enormous influence over one's mate choice. Lehmiller and Agnew (2007) found that romantically involved couples that perceived that their parents and friends did not approve of their relationship were more likely to break up than couples that viewed approval from these social networks.

Although parents may register direct disapproval, they may also be less direct in how they influence the mate choices of their children; by living in the "right" neighborhood, joining a particular church, and enrolling their children in a college or university, parents influence the context in which their children are likely to meet and select a "suitable" marriage partner.

mating gradient the tendency for husbands to marry wives who are younger and have less education and less occupational success.

primary group small, intimate, informal group.

Figure 1.1
Effect of Mating Gradient on Potential Partners for Students at East Carolina University

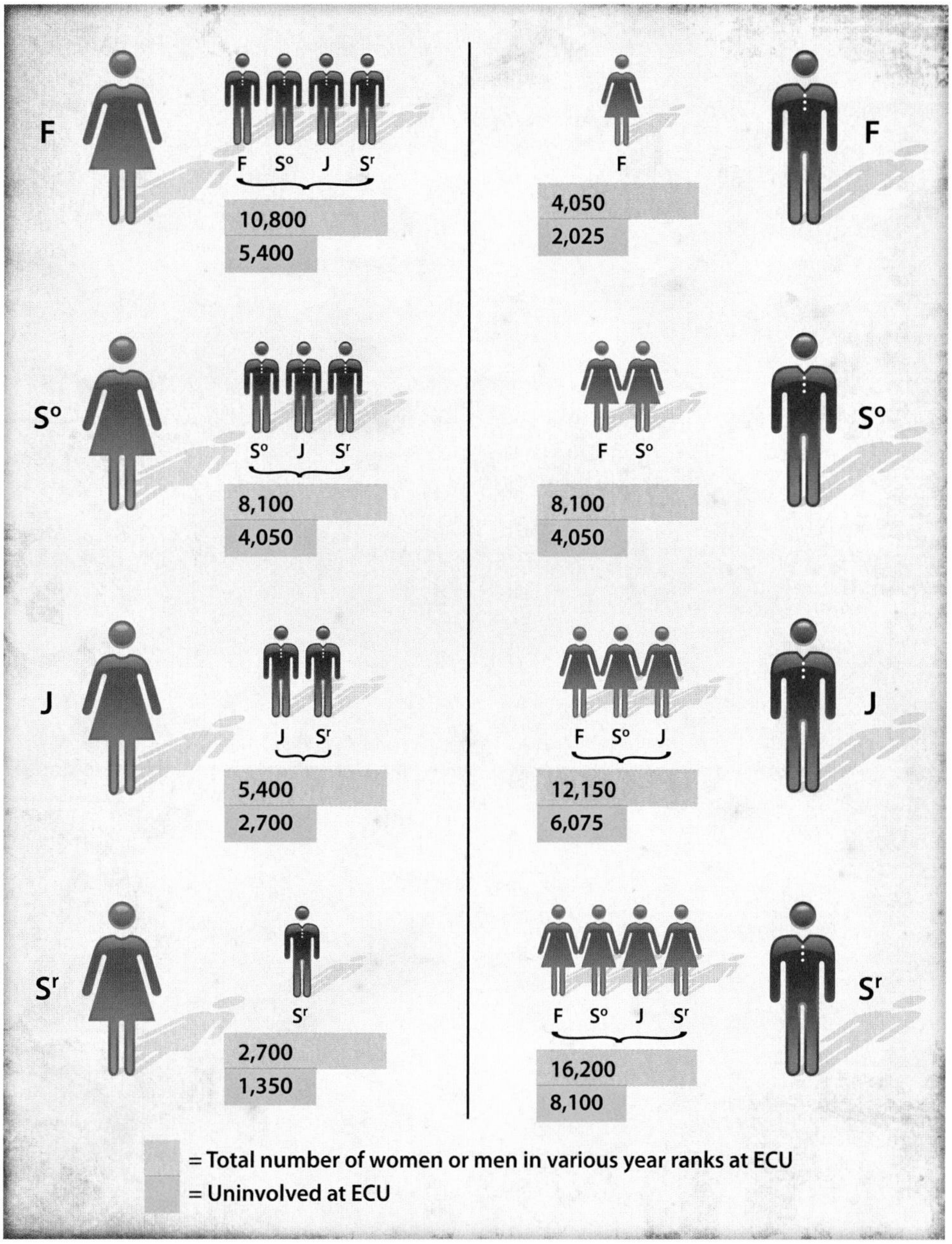

secondary group large or small group characterized by impersonal and formal interaction.

status a social position a person occupies within a social group.

role the behavior with which individuals in certain status positions are expected to engage.

beliefs definitions and explanations about what is thought to be true.

values standards regarding what is good and bad, right and wrong, desirable and undesirable.

individualism philosophy in which decisions are made on the basis of what is best for the individual.

familism philosophy in which decisions are made in reference to what is best for the family as a collective unit.

collectivism pattern that one regards group values and goals as more important than one's own values and goals.

In contrast to primary groups, **secondary groups** may involve small or large numbers of individuals and are characterized by interaction that is impersonal and formal. Being in a context of classmates, coworkers, or fellow students in the library are examples of secondary groups. Members of secondary groups have much less influence over one's relationship choices than members of one's primary groups.

Most people regard primary groups as crucial for their personal happiness and feel adrift if they have only secondary relationships. Indeed, in the absence of close primary ties, they may seek meaning in secondary group relationships. The late comedian George Carlin said that his "fans were his family" because he was in a different town performing more than half the weekends a year, implying he had no "real" family.

3. **Statuses.** Just as institutions consist of social groups, social groups consist of statuses. A **status** is a position a person occupies within a social group. The statuses we occupy largely define our social identity. The statuses in a family may consist of mother, father, child, sibling, stepparent, and so on. In discussing family issues, we refer to statuses such as teenager, cohabitant, and spouse. Statuses are relevant to choices in that many choices can significantly change one's status. Making decisions that change one's status from single person to spouse to divorced person can influence how people feel about themselves and how others treat them.

4. **Roles.** Every status is associated with many **roles**, or sets of rights, obligations, and expectations associated with a status. Our social statuses identify who we are; our roles identify what we are expected to do. Roles guide our behavior and allow us to predict the behavior of others. Spouses adopt a set of obligations and expectations associated with their status. By doing so, they are better able to influence and predict each other's behavior.

Because individuals occupy a number of statuses and roles simultaneously, they may experience role conflict. For example, the role of the parent may conflict with the role of the spouse, employee, or student. If your child needs to be driven to the math tutor, your spouse needs to be picked up at the airport, your employer wants you to work late, and you have a final exam all at the same time, you are experiencing role conflict.

Culture Just as social structure refers to the parts of society, culture refers to the meanings and ways of living that characterize people in a society. Two central elements of culture are beliefs and values.

1. **Beliefs. Beliefs** refer to definitions and explanations about what is true. The beliefs of an individual or couple influence the choices they make. Dual-earner couples that believe that young children flourish best with a full-time parent in the home make different child-care decisions than do couples who believe that day care offers opportunities for enrichment. If a person believes that children are best served by being reared with two parents, the decision regarding what to do about a premarital pregnancy or an unhappy marriage will differ from a decision made by a person who believes that single-parent families can provide an enriching context for rearing children.

2. **Values. Values** are standards regarding what is good and bad, right and wrong, desirable and undesirable. Values influence choices. **Individualism** involves making decisions that are more often based on what serves the individual's rather than the family's interests (**familism**). Americans are characteristically individualistic, whereas Hispanics are characteristically familistic. **Collectivism** emphasizes doing what is best for the group (not specific to the family group); this is characteristic of Asian and Asian American families.

Those who remain single, who live together, who seek a childfree lifestyle, and who divorce are more likely to be operating from an individualistic philosophical perspective than those who marry, do not live together before marriage, rear children, and stay married (a familistic value). Collectivistic values would be illustrated by an Asian child on a swim team who would work for the good of the team, not for personal acclaim.

These elements of social structure and culture play a central role in making interpersonal choices and decisions. One of the goals of this text is to encourage awareness of how powerful social structure and culture are in influencing decision making. Sociologists refer to this awareness as the **sociological imagination** (or sociological mindfulness). For example, though most people in the United States assume that they are free to select their own sex partner, this choice (or lack of it) is, in fact, heavily influenced by structural and cultural factors. Most people date, have sex with, and marry a person of the same racial background. Structural forces influencing race relations include segregation in housing, religion, and education. The fact that African Americans and European Americans live in different neighborhoods, worship in different churches, and often attend different schools makes meeting a person of a different race unlikely. When such encounters occur, prejudice and bias may influence these interactions so that individuals are hardly "free" to act as they choose. Hence, cultural values (transmitted by and through parents and peers) generally do not support or promote mixed racial interaction, relationship formation, and marriage. In a study of college students, DeCuzzi et al. (2006) found that both European Americans and African Americans tended to view their respective groups more positively. Consider the last three relationships in which you were involved, the level of racial similarity, and the structural and cultural influences on those relationships.

sociological imagination the perspective of how powerful social structure and culture are in influencing personal decision making.

© Michal Rozanski/iStockphoto.com / © Tamera Rees/iStockphoto.com

Media Pescosolido et al. (2008) emphasized that "individuals do not come to social interaction devoid of affect and motivation and that all social interactions take place in a context in which organizations, media and larger cultures structure normative expectations which create the possibility of marking 'difference'" (p. 431). Media in all of its forms (television, Internet, movies, print) influences how we think about and make our relationship choices. Media exposure colors the "acceptability" of cohabitation, abortion, same-sex relationships, divorce, single-parent families, and so on. For example, Vogel et al. (2008) noted that the greater the television exposure, the lower the willingness to seek therapy.

Other Influences on Relationship Choices

Aside from structural and cultural influences on relationship choices, other influences include family of origin (the family in which you were reared), unconscious motivations, habit patterns, individual personality, and previous experiences.

1. **Family of origin.** Your family of origin is a major influence on your subsequent choices and relationships. Over half (55.9 percent) of 658 undergraduates at a southeastern university (in random sample phone interview) reported that they were "very close" to their family, and almost another third (31.2 percent) reported that they were "somewhat close" (Bristol and Farmer 2005). Such closeness

may translate into the desire for parental approval for one's choice of partner. In addition, Busby et al. (2005) analyzed data on 6,744 individuals in regard to the impact of their family of orientation on their current relationship functioning; they found that the quality of the parents' relationship was strongly related to the quality of their own relationship. The happiness of one's family of origin may vary by sex. Dotson-Blake et al. (2009) analyzed data from 288 undergraduate and graduate students and found that 57 percent of males compared to 45 percent of females viewed their parents' marriage as "excellent."

2. **Unconscious motivations.** Unconscious processes are operative in our choices. A person reared in a lower-class home without adequate food and shelter may become overly concerned about the accumulation of money and may make all decisions in reference to obtaining, holding, or hoarding economic resources.

3. **Habits.** Habit patterns also influence choices. People who are accustomed to and enjoy spending a great deal of time alone may be reluctant to make a commitment to live with people who make demands on their time. A person who has workaholic tendencies is unlikely to allocate enough time to a relationship to make it flourish. Alcohol abuse is associated with a higher number of sexual partners and not using condoms to avoid pregnancy and contraction of sexually transmitted infections.

4. **Personality.** One's personality (for example, introverted, extroverted; passive, assertive) also influences choices. For example, people who are assertive are more likely than those who are passive to initiate conversations with someone they are attracted to at a party. People who are very quiet and withdrawn may never choose to initiate a conversation even though they are attracted to someone. People with a bipolar disorder who are manic one part of the semester (or relationship) and depressed the other part are likely to make different choices when each process is operative.

5. **Relationships and life experiences.** Current and past relationship experiences also influence one's perceptions and choices. Individuals currently in a relationship are more likely to hold relativistic sexual values (choose intercourse over abstinence). Similarly, people who have been cheated on in a previous relationship are vulnerable to not trusting a new partner.

The life of Ann Landers illustrates how life experience changes one's views and choices. After Landers's death, her only child, Margo Howard (2003), noted that her mother was once against premarital intercourse, divorce, and involvement with a married man. However, when Margo told her that unmarried youth were having intercourse, Landers shifted her focus to the use of contraception. When Margo divorced, Landers began to say that ending an unfulfilling marriage is an option. When Landers was divorced and fell in love with a married man, she said that you can't control who you fall in love with.

TO DARE IS TO LOSE ONE'S FOOTING MOMENTARILY. NOT TO DARE IS TO LOSE ONESELF.

—*Søren Kierkegaard, philosopher*

The effect of one relationship on another is illustrated by the life of Chet Baker, legendary trumpet player. His biographer details Baker's innocent love for a woman who dumped him to marry his buddy. ". . . [T]hat woman to whom he had opened his heart and who turned out to be a liar and a phony, permanently marred his attitude toward love. In future relationships he would veer between a need for mothering and paranoid mistrust" (Gavin 2003, 28).

In spite of the numerous influences on choices, the theme of this text is to not let destiny direct you but to make deliberate choices.

Having emphasized that making choices in relationships is the theme of this text and having reviewed the mechanisms operative in those choices, we later discuss other frameworks used to examine marriage and family relationships. Next, we look at the meaning of marriage.

1.2 Marriage

All of us were born into a family and will end up in a family (however one defines this concept) of our own. "Raising a family" remains one of the top values in life (superseded only by "being well off financially") for undergraduates. In a nationwide study of 240,000 undergraduates in 340 colleges and universities, 76 percent identified this as an essential objective (77 percent for financial success) (Pryor et al. 2008). In a smaller study of 1,319 undergraduates, a "happy marriage" was the top value in life (over financial security and "having a career I love"; Knox and Zusman 2009). Clearly, relationships and family life are important values.

Traditionally, **marriage** has been viewed as a legal relationship that binds a man and a woman together for reproduction and the subsequent physical and emotional care and socialization of children. Each society works out its own details of what marriage is. In the United States, marriage is a legal contract between a heterosexual couple and the state in which they reside that specifies economic cooperation and encourages sexual fidelity (we discuss gay marriage later in the chapter). The fine print and implied factors implicit during a wedding ceremony include the following elements.

Elements of Marriage

Several elements comprise the meaning of marriage in the United States.

Legal Contract Marriage in our society is a legal contract into which two people of different sexes and legal age (usually eighteen or older) may enter when they are not already married to someone else. The marriage license certifies that a legally empowered representative of the state married the individuals, often with two witnesses present.

Under the laws of the state, the license means that spouses will jointly own all future property acquired and that each will share in the estate of the other. In most states, whatever the deceased spouse owns is legally transferred to the surviving spouse at the time of death. In the event of divorce and unless the couple had a prenuptial agreement, the property is usually divided equally regardless of the contribution of each partner. The license also implies the expectation of sexual fidelity in the marriage. Though less frequent because of no-fault divorce, infidelity is a legal ground for both divorce and alimony in some states.

The marriage license is also an economic license that entitles a spouse to receive payment for medical bills by a health insurance company if the partner is insured, to collect Social Security benefits at the death of the other spouse, and to inherit from the estate of the deceased. One of the goals of gay rights advocates who seek the legalization of marriage between homosexuals is that the couple will have the same rights and benefits as heterosexuals.

Though the courts are reconsidering the definition of what constitutes a "family," the law is currently designed to protect spouses, not lovers or cohabitants. An exception is **common-law marriage**, in which a heterosexual couple cohabit and present themselves as married; they will be regarded as legally married in those states that recognize such marriages. Common-law marriages exist in ten states and the District of Columbia.

Emotional Relationship Most people in the United States regard being in love with the person they marry as an important reason for staying married. Almost half (47.9 percent) of 1,319 undergraduates reported that they would divorce if they no longer loved their spouse (Knox and Zusman 2009). This emphasis on love is not shared throughout the world. Individuals in other cultures (for example, India and Iran) do not require love feelings to marry—love is expected to follow, not precede, marriage. In these countries, parental approval and similarity of religion, culture, and education are considered more important criteria for marriage than love.

Sexual Monogamy Marital partners expect sexual fidelity. Over two-thirds (68 percent) of 1,319 undergraduates agreed, "I would divorce a spouse who had an affair," and a similar percentage (69 percent) agreed that they would end a relationship with a partner who cheated on them (Knox and Zusman 2009).

marriage a legal contract signed by a couple with the state in which they reside that regulates their economic and sexual relationship.

common-law marriage a marriage by mutual agreement between cohabitants without a marriage license or ceremony (recognized in some, but not all, states).

polygamy a generic term referring to a marriage involving more than two spouses.

polygyny a form of polygamy in which one husband has two or more wives.

Legal Responsibility for Children Although individuals marry for love and companionship, one of the most important reasons for the existence of marriage from the viewpoint of society is to legally bond a male and a female for the nurture and support of any children they may have. In our society, child rearing is the primary responsibility of the family, not the state.

Marriage is a relatively stable relationship that helps to ensure that children will have adequate care and protection, will be socialized for productive roles in society, and will not become the burden of those who did not conceive them. Even at divorce, the legal obligation of the father and mother to the child is theoretically maintained through child-support payments.

Announcement/Ceremony The legal bonding of a couple is often preceded by an announcement in the local newspaper and a formal ceremony, usually in a place of worship.

Such a ceremony reflects the cultural importance of the event. Telling parents, siblings, and friends about wedding plans helps to verify the commitment of the partners and also helps to marshal the social and economic support to launch the couple into marital orbit. Most people in our society decide to marry, and the benefits of doing so are enormous. When married people are compared with singles, the differences are strikingly in favor of the married (see Table 1.2).

Types of Marriage

Although we think of marriage in the United States as involving one man and one woman, other societies view marriage differently. **Polygamy** is a form of marriage involving more than two spouses. Polygamy occurs "throughout the world . . . and is found on all continents and among adherents of all world religions" (Zeitzen 2008, 8). There are three forms of polygamy: polygyny, polyandry, and pantagamy.

Polygyny **Polygyny** involves one husband and two or more wives and is practiced illegally in the United States by some religious fundamentalist groups. These groups are primarily in Arizona, New Mexico, and Utah (as well as Canada), and have splintered off from the Church of Jesus Christ of Latter-day Saints (commonly known as the Mormon Church). To be clear, the Mormon Church does not practice or condone polygyny. The group that split off represents only about 5 percent of Mormons in Utah and is called the Fundamentalist Church of Jesus Christ of the Latter-day Saints (FLDS). Members of the group feel that the practice of polygyny is God's will. Although illegal, polygynous individuals are rarely prosecuted for several reasons:

1. **One legal marriage; many religious marriages.** In effect, a man will legally marry one wife and bring other wives into the unit via a religious ceremony. There is no law against a husband living with several women and their children in the same house as long as he is only legally married to one of them.
2. **Population.** Mormons outnumber non-Mormons, representing 70 percent of the population in Utah (there is limited public momentum for prosecuting Mormons).
3. **Prosecution witness.** The absence of finding someone willing to testify for the prosecution makes mounting a case difficult.
4. **Prosecution priorities.** Local prosecutors elect to spend available resources on organized crime and drug trafficking rather than on illegal civil relationships.

© Jose Luis Pelaez Inc./Blend Images/Jupiterimages

Table 1.2

Benefits of Marriage and the Liabilities of Singlehood

	Benefits of Marriage	Liabilities of Singlehood
Health	Spouses have fewer hospital admissions, see a physician more regularly, and are "sick" less often.	Single people are hospitalized more often, have fewer medical checkups, and are "sick" more often.
Longevity	Spouses live longer than single people.	Single people die sooner than married people.
Happiness	Spouses report being happier than single people.	Single people report less happiness than married people.
Sexual satisfaction	Spouses report being more satisfied with their sex lives, both physically and emotionally.	Single people report being less satisfied with their sex lives, both physically and emotionally.
Money	Spouses have more economic resources than single people.	Single people have fewer economic resources than married people.
Lower expenses	Two can live more cheaply together than separately.	Cost is greater for two singles than one couple.
Drug use	Spouses have lower rates of drug use and abuse.	Single people have higher rates of drug use and abuse.
Connectedness	Spouses are connected to more individuals who provide a support system—partner, in-laws, etc.	Single people have fewer individuals upon whom they can rely for help.
Children	Rates of high school dropouts, teen pregnancies, and poverty are lower among children reared in two-parent homes.	Rates of high school dropouts, teen pregnancies, and poverty are higher among children reared by single parents.
History	Spouses develop a shared history across time with significant others.	Single people may lack continuity and commitment across time with significant others.
Crime	Spouses are less likely to be involved in crime.	Single people are more likely to be involved in crime.
Loneliness	Spouses are less likely to report loneliness.	Single people are more likely to report being lonely.

5. **Jail space.** There is no jail space for housing the hundreds of individuals if convicted.

In spite of the tolerance toward polygyny in Utah, there are some abuses. Some former wives in plural marriages report (a) living in poverty (it is difficult for one man to financially support fifteen wives), (b) nonconsent (some existing wives resent their husbands taking new wives), and (c) child sex abuse (some children as young as 14 are made "new wives"). Warren Steed Jeffs, a former president of FLDS, was charged in Arizona with child sex abuse for arranging the marriage of a teenager to an older married man. In 2007, he was found guilty of two counts of rape as an accomplice and sentenced to ten years to life, and has begun serving his sentence at the Utah State Prison.

Although the items on the previous list may be factors in explaining why polygynous individuals are rarely prosecuted, Hall (2009) has provided an alternative explanation:

> It is important to note that in many, if not almost all, of these unions only the first wife is legally married to the husband and the other wives are comparable to cohabiting women (in their minds "spiritual wives"). Hence, there is technically no bigamy to prosecute. In addition, pockets of these groups are known to exist in several locations across the country, yet are not being aggressively prosecuted by the local governments. It seems that religion may be less of a factor in the lack of prosecution than other considerations. A primary one is that people in general don't have a stomach for taking kids away from parents or taking parents away from their children and locking them up—unless there is clear danger involved. This is particularly true in that our society is becoming increasingly open to "nontraditional" (kind of an ironic term in this sense) relationship forms among adults. There have been a couple of raids in the past (1953 in Arizona and Texas in 2008), and there was clear national backlash against what was perceived as overly-aggressive breaking up of families.

There is some concern that, because the wives in the FLDS community have limited education, job skills, and contact with the outside community, they are disadvantaged in extricating themselves if they wanted to. Tapestry Against Polygamy is an organization that has helped women break free from bigamous marriages.

Polygyny among fundamentalist Mormons serves a religious function in that large earthly families are believed to result in large heavenly families. Notice that

polyandry a form of polygamy in which one wife has two or more husbands.

polyamory a term meaning "many loves," whereby three or more men and women have a committed emotional and sexual relationship.

open relationship a stable relationship in which the partners regard their own relationship as primary but agree that each may have emotional and physical relationships with others.

pantagamy a group marriage in which each member of the group is "married" to the others.

family a group of two or more people related by blood, marriage, or adoption.

polygynous sex is a means to accomplish another goal—large families. See http://www.absalom.com/mormon/polygamy/faq.htm for more information.

It is often assumed that polygyny exists to satisfy the sexual desires of the man, that the women are treated like slaves, and that jealousy among the wives is common. In most polygynous societies, however, polygyny has a political and economic rather than a sexual function. Polygyny is a means of providing many male heirs to continue the family line. In addition, a man with many wives can produce a greater number of children for domestic or farm labor. Wives are not treated like slaves (although women have less status than men in general), all household work is evenly distributed among the wives, and each wife is given her own house or own sleeping quarters. Jealousy is minimal because the husband often has a rotational system for conjugal visits, which ensures that each wife has equal access to sexual encounters.

Polyandry The Buddhist Tibetans foster yet another brand of polygamy, referred to as **polyandry**, in which one wife has two or more (up to five) husbands. These husbands, who may be brothers, pool their resources to support one wife. Polyandry is a much less common form of polygamy than polygyny. The major reason for polyandry is economic. A family that cannot afford wives or marriages for each of its sons may find a wife for the eldest son only. Polyandry allows the younger brothers to also have sexual access to the one wife or marriage that the family is able to afford.

Polyamory Polyamory (also known as **open relationships**) is a lifestyle in which two lovers do not forbid one another from having other lovers. By agreement, each partner may have numerous other emotional and sexual relationships. Some (about 20 percent) of the 100 members of Twin Oaks Intentional Community in Louisa, Virginia, are polyamorous in that each partner may have several emotional or physical relationships with others at the same time. Although not legally married, these adults view themselves as emotionally bonded to each other and may even rear children together. Polyamory is not swinging, as polyamorous lovers are concerned about enduring, intimate relationships that include sex (Tupelo and Freeman, 2008). We discuss polyamory in greater detail in Chapter 5.

Pantagamy Pantagamy describes a group marriage in which each member of the group is "married" to the others. Pantagamy is a more formal arrangement than polyamory and is reflected in communes (for example, Oneida) of the nineteenth and twentieth centuries. Pantagamy is illegal in the United States.

Our culture emphasizes monogamous marriage and values stable marriages.

1.3 Family

Most people who marry choose to have children and become a family. However, the definition of what constitutes a family is sometimes unclear. This section examines how families are defined, their numerous types, and how marriages and families have changed in the past fifty years.

Definitions of Family

The U.S. Census Bureau defines **family** as a group of two or more people related by blood, marriage, or adoption. This definition has been challenged because it does not include foster families or long-term couples (heterosexual or homosexual) that live together. The answer to "who is family?" is important because access to resources such as health care, Social Security, and retirement benefits is involved. Cohabitants are typically not viewed as "family" and are not accorded health benefits, Social Security, and retirement benefits of the partner. Indeed, the "live-in partner" or a partner who is gay (although a long-term significant other) may not be allowed to see the beloved in the hospital, which limits visitation to "family only." Nevertheless, the definition of who counts as family is being challenged. In some cases, families are being defined by function rather than by structure—what is the level of emotional and financial commitment and interdependence? How long have they lived together? Do the partners view themselves as a family?

Friends sometimes become family. Due to mobility, spouses may live several states away from their respective families. Although they may visit their families for holidays, they often develop close friendships with others on whom they rely locally for emotional and physical support on a daily basis.

Sociologically, a family is defined as a kinship system of all relatives living together or recognized as a social unit, including adopted people. The family is regarded as the basic social institution because of its important functions of procreation and socialization, and because it is found in some form in all societies. Henley-Walters et al. (2002) studied family life in different cultures and noted that in many respects,

> the experience of living in a family is the same in all cultures. For example, relationships between spouses and parents and children are negotiated; most relationships within families are hierarchical; the work of the home is primarily the responsibility of the wife; . . . destructive conflict between spouses or parents and children is damaging to children. . . . (p. 449)

Same-sex couples (for example, Rosie O'Donnell, her partner, and their children, as well as Ellen DeGeneres and her partner) certainly define themselves as family. Maine, Iowa, New Hampshire, Vermont, Massachusetts, and Connecticut recognize marriages between same-sex individuals. Short of marriage, New Jersey recognizes committed gay relationships as **civil unions**.

Other states typically do not recognize same-sex marriages or civil unions (and thus people moving from these states to another state lose the privileges associated with marriage), but domestic partner rights are a hot topic of debates in states and cities around the country. In addition, some corporations, such as Disney, are recognizing the legitimacy of such relationships by providing medical coverage for partners of employees. **Domestic partnerships** are considered an alternative to marriage by some individuals and tend to reflect more egalitarian relationships than those between traditional husbands and wives.

civil union a pair-bonded relationship given legal significance in terms of rights and privileges.

domestic partnership a relationship in which individuals who live together are emotionally and financially interdependent and are given some kind of official recognition by a city or corporation so as to receive partner benefits.

Some individuals view their pets as part of their family. In a Harris Poll survey of 2,455 adults, 88 percent regarded their pets as family members—more women (93 percent) than men (84 percent), and more dog owners (93 percent) than cat owners (89 percent). Two-thirds of pet owners buy holiday presents for

Pets are routinely included in family photos.

family of orientation the family of origin into which a person is born.

family of origin the family into which an individual is born or reared, usually including a mother, father, and children.

family of procreation the family a person begins by getting married and having children.

nuclear family family consisting of an individual, his or her spouse, and his or her children, or of an individual and his or her parents and siblings.

their pets (Harris Poll 2007). Some pet owners buy accident insurance—Progressive© car insurance covers pets. Custody battles have been fought and financial trusts have been set up for pets, and lawsuits against veterinarians have been filed over pets that did not make it out of surgery (Parker 2005).

Types of Families

There are various types of families.

Family of Origin Also referred to as the **family of orientation**, this is the family into which you were born or the family in which you were reared. It involves you, your parents, and your siblings. When you go to your parents' home for the holidays, you return to your **family of origin**. Your experiences in your family of origin have an impact on subsequent outcome behavior. For example, if you are born into a family on welfare, you are less likely to attend college than if your parents are both educated and affluent.

Novilla et al. (2006) emphasized the important role the family of orientation plays in health promotion. Children learn the importance of a healthy diet, exercise, moderate alcohol use, and so on, from their family of orientation and may duplicate these patterns in their own families.

Siblings represent an important part of one's family of origin. Indeed, a team of researchers (Meinhold et al. 2006) noted that the relationship with one's siblings, particularly sister–sister relationships, represent the most enduring relationship in a person's lifetime. Sisters who lived near one another and who did not have children reported the greatest amount of intimacy and contact.

Family of Procreation The **family of procreation** represents the family that you will begin when you marry and have children. Of U.S. citizens living in the United States, 96.3 percent marry and establish their own family of procreation (*Statistical Abstract of the United States, 2009*, Table 33). Across the life cycle, individuals move from the family of orientation to the family of procreation.

Nuclear Family The **nuclear family** refers to either a family of origin or a family of procreation. In practice, this means that your nuclear family consists of you, your parents, and your siblings; or you, your spouse, and your children. Generally, one-parent households are not referred to as nuclear families. They are binuclear families if both parents are involved in the child's life, or single-parent families if one parent is involved in

the child's life and the other parent is totally out of the picture.

Sociologist George Peter Murdock (1949) emphasized that the nuclear family is a "universal social grouping" found in all of the 250 societies he studied. Not only does it channel sexual energy between two adult partners who reproduce, but also their cooperation in the care for and socialization of offspring to be productive members of society. "This universal social structure, produced through cultural evolution in every human society, as presumably the only feasible adjustment to a series of basic needs, forms a crucial part of the environment in which every individual grows to maturity" (p. 11). Neyer and Lang (2003) emphasized that closeness to one's kinship members continues throughout one's life and found some evidence for "blood being thicker than water" (p. 310).

Traditional, Modern, and Postmodern Family Silverstein and Auerbach (2005) distinguished between three central concepts of the family. The **traditional family** is the two-parent nuclear family, with the husband as breadwinner and wife as homemaker. The **modern family** is the dual-earner family, where both spouses work outside the home. **Postmodern families** represent a departure from these models, such as lesbian or gay couples and mothers who are single by choice, which emphasizes that a healthy family need not be heterosexual or include two parents.

Binuclear Family A **binuclear family** is a family in which the members live in two separate households. This family type is created when the parents of the children divorce and live separately, setting up two separate units, with the children remaining a part of each unit. Each of these units may also change again when the parents remarry and bring additional children into the respective units (**blended family** or **stepfamily**). Hence, the children

Marriage can be thought of as a set of social processes that lead to the establishment of family. Every society or culture has mechanisms of guiding their youth into permanent emotionally, legally, or socially bonded heterosexual relationships that are designed to result in procreation and care of offspring. Some differences between marriage and family, as identified by sociologist Dr. Lee Axelson, are listed below.

Marriage	Family
Usually initiated by a formal ceremony	Formal ceremony not essential
Involves two people	Usually involves more than two people
Ages of the individuals tend to be similar	Individuals represent more than one generation
Individuals usually choose each other	Members are born or adopted into the family
Ends when spouse dies or is divorced	Continues beyond the life of the individual
Sex between spouses is expected and approved	Sex between near kin is neither expected nor approved
Requires a license	No license needed to become a parent
Procreation expected	Consequence of procreation
Spouses are focused on each other	Focus changes with addition of children
Spouses can voluntarily withdraw from marriage with obligations to children	Spouses/parents cannot easily withdraw voluntarily with approval of state
Money in unit is spent on the couple	Money is used for the needs of children
Recreation revolves around adults	Recreation revolves around children

traditional family the two-parent nuclear family with the husband as breadwinner and wife as homemaker.

modern family the dual-earner family, in which both spouses work outside the home.

postmodern family non-traditional families emphasizing that a healthy family need not be heterosexual or have two parents.

binuclear family family in which the members live in two households.

blended family (stepfamily) a family created when two individuals marry and at least one of them brings a child or children from a previous relationship or marriage.

extended family the nuclear family or parts of it plus other relatives.

may go from a nuclear family with both parents, to a binuclear unit with parents living in separate homes, to a blended family when parents remarry and bring additional children into the respective units.

Extended Family The **extended family** includes not only the nuclear family but other relatives as well. These relatives include grandparents, aunts, uncles, and cousins. An example of an extended family living together would be a husband and wife, their children, and the husband's parents (the children's grandparents).

1.4 Changes in Marriage and the Family

Whatever family we experience today was different previously and will change yet again. A look back at some changes in marriage and the family follow.

The Industrial Revolution and Family Change

The Industrial Revolution refers to the social and economic changes that occurred when machines and factories, rather than human labor, became the dominant mode for the production of goods. Industrialization occurred in the United States during the early- and mid-1800s and represents one of the most profound influences on the family.

Before industrialization, families functioned as an economic unit that produced goods and services for its own consumption. Parents and children worked together in or near the home to meet the survival needs of the family. As the United States became industrialized, more men and women left the home to sell their labor for wages. The family was no longer a self-sufficient unit that determined its work hours. Rather, employers determined where and when family members would work. Whereas children in preindustrialized America worked on farms and contributed to the economic survival of the family, children in industrialized America became economic liabilities rather than assets. Child labor laws and mandatory education removed children from the labor force and lengthened their dependence on parental support. Eventually, both parents had to work away from the home to support their children. The dual-income family had begun.

During the Industrial Revolution, urbanization occurred as cities were built around factories and families moved to the city to work in the factories. Living space in cities was crowded and expensive, which contributed to a decline in the birthrate and thus smaller families. The development of transportation systems during the Industrial Revolution made it possible for family members to travel to work sites away from the home and to move away from extended kin. With increased mobility, many extended families became separated into smaller nuclear family units consisting of parents and their children. As a result of parents leaving the home to earn wages and the absence of extended kin in or near the family household, children had less adult supervision and moral guidance. Unsupervised children roamed the streets, increasing the potential for crime and delinquency.

Industrialization also affected the role of the father in the family. Employment outside the home removed men from playing a primary role in child care and in other domestic activities. The contribution men made to the household became primarily economic.

Finally, the advent of industrialization, urbanization, and mobility is associated with the demise of familism and the rise of individualism. When family members functioned together as an economic unit, they were dependent on one another for survival and were concerned about what was good for the family. This familistic focus on the needs of the family has since shifted to a focus on self-fulfillment—individualism. Families from familistic cultures such as China who immigrate to the United States soon discover that their norms, roles, and values begin to alter in reference to the industrialized, urbanized, individualistic patterns and thinking. Individualism and the quest for personal fulfillment are thought to have contributed to high divorce rates, absent fathers, and parents spending less time with their children.

Hence, although the family is sometimes blamed for juvenile delinquency, violence, and divorce, it is more accurate to emphasize changing social norms and conditions of which the family is a part. When industrialization takes parents out of the home so that they can no longer be constant nurturers and supervisors, the likelihood of aberrant acts by children and adolescents increases. One explanation for school violence is that absent, career-focused parents have failed to provide close supervision for their children.

Changes in the Last Half-Century

Enormous changes have occurred in marriage and the family in the last fifty years. Among changes, divorce has replaced death as the endpoint for the majority of marriages, marriage and intimate relations have become legitimate objects of scientific study, feminism and changes in gender roles in marriage have risen, and remarriages have declined (Amato et al. 2007). Other changes include a delay in age at marriage, increased acceptance of singlehood, cohabitation, and childfree marriages. Even the definition of what constitutes a family is being revised, with some emphasizing that durable emotional bonds between individuals is the core of "family" whereas others insist on a more legalistic view, emphasizing connections by blood, marriage, or adoption mechanisms. Table 1.3 on the next page contrasts marriage and family in the fifties with marriages and families in 2010.

In spite of the persistent and dramatic changes in marriage and the family, marriage and the family continue to be resilient. Using this **marriage-resilience perspective**, changes in the institution of marriage are not viewed negatively nor are they indicative that marriage is in a state of decline. Indeed, these changes are thought to have "few negative consequences for adults, children, or the wider society" (Amato et al. 2007, 6).

Families amid a Context of Terrorism

Prior to September 11, 2001, with the exception of Timothy McVeigh bombing the federal building in Oklahoma in 1995, terrorism was thought to exist mostly in foreign countries. Today, husbands, wives, parents, and children live under a new cloud of terror anxiety with "threat levels," "airport security measures," and "new terrorist bombing reports" as part of the evening news. Richman et al. (2008) assessed the effects of the terrorist attacks of September 11, 2001, on subsequent mental health among participants in a Midwestern town study. The researchers revealed higher stress and alcohol consumption levels two and four years after the attacks, even after controlling for sociodemographic characteristics and pre-September 11 distress and drinking patterns.

In spite of changes, marriage and the family continue to be resilient.

1.5 Theoretical Frameworks for Viewing Marriage and the Family

Although we emphasize choices in relationships as the framework for viewing marriage and the family, other conceptual theoretical frameworks are helpful in understanding the context of relationship decisions. All **theoretical frameworks** are the same in that they provide a set of interrelated principles designed to explain a particular phenomenon and provide a point of view. In essence, theories are explanations.

Social Exchange Framework

In a review of 673 empirical articles, exchange theory was the most common among those using a theoretical perspective (Taylor and Bagd 2005). The **social exchange framework** views interaction in terms of cost and profit.

The social exchange framework also operates from a premise of **utilitarianism**—that individuals rationally weigh the rewards and costs associated with behavioral choices. Each interaction between spouses, parents, and children can be understood in terms of each individual's seeking the most benefits at the least cost so as to have the highest "profit" and avoid a "loss" (White and Klein 2002). Both men and women marry because they perceive more benefits than costs for doing so. Similarly, those who remain single or who divorce perceive fewer benefits and

marriage-resilience perspective the view that changes in the institution of marriage are not indicative of a decline and do not have negative effects.

theoretical framework a set of interrelated principles designed to explain a particular phenomenon and to provide a point of view.

social exchange framework spouses exchange resources, and decisions are made on the basis of perceived profit and loss.

utilitarianism the doctrine holding that individuals rationally weigh the rewards and costs associated with behavioral choices.

TABLE 1.3
MARRIAGES AND FAMILIES: 1950 AND 2010

	1950	2010
Family Relationship Values	Strong values for marriage and the family. Individuals who wanted to remain single or childless were considered deviant, even pathological. Husband and wife should not be separated by jobs or careers.	Individuals who remain single or childfree experience social understanding and sometimes encouragement. Single and childfree people are no longer considered deviant or pathological but are seen as self-actuating individuals with strong job or career commitments. Husbands and wives can be separated by jobs or career reasons and live in a commuter marriage. Married women in large numbers have left the role of full-time mother and housewife to join the labor market.
Gender Roles	Rigid gender roles, with men earning income and wives staying at home, to take care of children.	Fluid gender roles, with most wives in workforce, even after birth of children. Part-time work is the modal choice for married woman.
Sexual Values	Marriage was regarded as the only appropriate context for intercourse in middle-class America. Living together was unacceptable, and a child born out of wedlock was stigmatized. Virginity was sometimes exchanged for marital commitment.	For many, concerns about safer sex have taken precedence over the marital context for sex. Virginity is no longer exchanged for anything. Living together is regarded as not only acceptable but sometimes preferable to marriage. For some, unmarried single parenthood is regarded as a lifestyle option. It is certainly less stigmatized.
Homogamous Mating	Strong social pressure existed to date and marry within one's own racial, ethnic, religious, and social class group. Emotional and legal attachments were heavily influenced by obligation to parents and kin.	Dating and mating have become more heterogamous, with more freedom to select a partner outside one's own racial, ethnic, religious, and social class group. Attachments are more often by choice.
Cultural Silence on Intimate Relationships	Intimate relationships were not an appropriate subject for the media.	Talk shows, interviews, and magazine surveys are open about sexuality and relationships behind closed doors.
Divorce	Society strongly disapproved of divorce. Familistic values encouraged spouses to stay married for the children. Strong legal constraints kept couples together. Marriage was forever.	Divorce has replaced death as the endpoint of a majority of marriages. Less stigma is associated with divorce. Individualistic values lead spouses to seek personal happiness. No-fault divorce allows for easy divorce. Marriage is tenuous. Increasing numbers of children are being reared in single-parent households apart from other relatives.
Familism versus Individualism	Families were focused on the needs of children. Mothers stayed home to ensure that the needs of their children were met. Adult concerns were less important.	Adult agenda of work and recreation has taken on increased importance, with less attention being given to children. Children are viewed as more sophisticated and capable of thinking as adults, which frees adults to pursue their own interests. Day care is used regularly.
Homosexuality	Same-sex emotional and sexual relationships were a culturally hidden phenomenon. Gay relationships were not socially recognized.	Gay relationships are increasingly a culturally open phenomenon. Some definitions of the family include same-sex partners. Domestic partnerships are increasingly given legal status in some states. Same-sex marriage is a hot social and political issue.
Scientific Scrutiny	Aside from Kinsey, few studies were conducted on intimate relationships.	Acceptance of scientific study of marriage and intimate relationships.
Family Housing	Husbands and wives lived in the same house.	Husbands and wives may "live apart together" (LAT), which means that, although they are emotionally and economically connected, they (by choice) maintain two households, houses, condos, or apartments. They may be separated for reasons of career, or mutually desire the freedom and independence of having a separate domicile.

more costs for marriage. Foster (2008) confirmed that narcissists require a high profit margin in relationships they stay in. We examine how the social exchange framework is operative in mate selection later in the text.

A social exchange view of marital roles emphasizes that spouses negotiate the division of labor on the basis of exchange. For example, a man participates in child care in exchange for his wife earning an income, which relieves him of the total financial responsibility. Social exchange theorists also emphasize that power in relationships is the ability to influence, and avoid being influenced by, the partner.

The various bases of power, such as money, the need for a partner, and brute force, may be expressed in various ways, including withholding resources, decreasing investment in the relationship, and violence.

Family Life Course Development Framework

The **family life course development** framework is the second most frequently used theory in empirical family studies (Taylor and Bagd 2005), and emphasizes the important role transitions that occur in different periods of life and in different social contexts. For example, young married couples become parents, which changes the interaction between the couple. As spouses age and retire, their new roles impact not only them but also their partners.

The family life course developmental framework has a basis in sociology (for example, role transitions), whereas the **family life cycle** has a basis in psychology, with its emphasis on developmental tasks at different ages and stages. If developmental tasks at one stage are not accomplished, functioning in subsequent stages will be impaired. For example, one of the developmental tasks of early marriage is to emotionally and financially separate from one's family of origin. If such separation does not take place, independence as individuals and as a couple is impaired.

The family life course development framework may help to identify the choices with which many individuals are confronted throughout life. Each family stage presents choices. For example, never-married people are choosing partners, newly married people are making choices about careers and when to begin a family, soon-to-be-divorced people are making decisions about custody, child support, and division of property, and remarried people are making choices with regard to stepchildren and ex-spouses. Grandparents are making choices about how much child care to which they want to commit, and widows or widowers are concerned with where to live (with children, a friend, alone, or in a retirement home).

Both optimists and pessimists contribute to our society. The optimist invents the airplane and the pessimist the parachute.
— G. B. Stern, British novelist

© Mike Kemp/Rubberball/Jupiterimages

family life course development the stages and process of how families change over time.

family life cycle stages which identify the various challenges faced by members of a family across time.

structural-function framework emphasizes how marriage and family contribute to the larger society.

functionalists structural functionalist theorists who view the family as an institution with values, norms, and activities meant to provide stability for the larger society.

Structural-Function Framework

The **structural-function framework** emphasizes how marriage and family contribute to society. Just as the human body is made up of different parts that work together for the good of the individual, society is made up of different institutions (family, education, economics, and so on) that work together for the good of society. **Functionalists** view the family as an institution with values, norms, and activities meant to provide stability for the larger society. Such stability is dependent on families performing various functions for society.

First, families serve to replenish society with socialized members. Because our society cannot continue to exist without new members, we must have some way of ensuring a continuing supply. However, just having new members is not enough. We need socialized members—those who

feral children wild, undomesticated children who are thought to have been reared by animals.

conflict framework view that individuals in relationships compete for valuable resources.

can speak our language and know the norms and roles of our society. So-called **feral** (meaning wild, not domesticated) **children** are those who are thought to have been reared by animals. Newton (2002) details nine such children, the most famous of which was Peter the Wild Boy found in the Germanic woods at the age of 12 and brought to London in 1726. He could not speak; growling and howling were his modes of expression. He lived until the age of 70 and never learned to talk. Feral children emphasize that social interaction and family context make us human.

Real-life Genie is a young girl who was discovered in the 1970s who had been kept in isolation in one room in her California home for twelve years by her abusive father (James 2008). She could barely walk and could not talk. Although provided intensive therapy at UCLA and the object of thousands of dollars of funded research, Genie progressed only briefly. Today, she is in her early fifties, institutionalized, and speechless. Her story illustrates the need for socialization; the legal bond of marriage and the obligation to nurture and socialize offspring help to ensure that this socialization will occur.

Second, marriage and the family promote the emotional stability of the respective spouses. Society cannot provide enough counselors to help us whenever we have problems. Marriage ideally provides in-residence counselors who are loving and caring partners with whom people share their most difficult experiences.

Children also need people to love them and to give them a sense of belonging. This need can be fulfilled in a variety of family contexts (two-parent families, single-parent families, extended families). The affective function of the family is one of its major offerings. No other institution focuses so completely on meeting the emotional needs of its members as marriage and the family.

Third, families provide economic support for their members. Although modern families are no longer self-sufficient economic units, they provide food, shelter, and clothing for their members. One need only consider the homeless in our society to be reminded of this important function of the family.

In addition to the primary functions of replacement, emotional stability, and economic support, other functions of the family include the following:

- Physical care—families provide the primary care for their infants, children, and aging parents. Other agencies (neonatal units, day care centers, assisted-living residences) may help, but the family remains the primary and recurring caretaker. Spouses are also concerned about the physical health of one another by encouraging the partner to take medications and to see the doctor.
- Regulation of sexual behavior—spouses are expected to confine their sexual behavior to each other, which reduces the risk of having children who do not have socially and legally bonded parents, and of contracting or spreading sexually transmitted infections.
- Status placement—being born in a family provides social placement of the individual in society. One's family of origin largely determines one's social class, religious affiliation, and future occupation. Prince William, the son of Prince Charles and the late Diana, was automatically in the upper class and destined to be in politics by virtue of being born into a political family.
- Social control—spouses in high-quality, durable marriages provide social control for each other that results in less criminal behavior. Parole boards often note that the best guarantee that a person released from prison will not return to prison is a spouse who expects the partner to get a job and avoid criminal behavior and who reinforces these goals.

Conflict Framework

Conflict framework views individuals in relationships as competing for valuable resources. Conflict theorists recognize that family members have different goals and values that result in conflict. Conflict is inevitable between social groups (such as parents and children). Conflict theory provides a lens through which to view these differences. Whereas functionalists look at family practices as good for the whole, conflict theorists recognize that not all family decisions are good for every member of the family. Indeed, some activities that are good for one member are not good for others. For example, a woman who has devoted her life to staying home and taking care of the children may decide to return to school or to seek full-time employment. This may be a good decision for her personally, but her husband and children may not like it. Similarly, divorce may have a positive outcome for spouses in turmoil but a negative outcome for children, whose standard of living and access to the noncustodial parent are likely to decrease.

Conflict theorists also view conflict not as good or bad but as a natural and normal part of relationships. They regard conflict as necessary for change and growth of individuals, marriages, and families. Cohabitation relationships, marriages, and families all have the potential for conflict. Cohabitants are in conflict about commitment to marry, spouses are in conflict about the division of labor, and parents are in conflict with their children over rules such as curfew, chores, and homework. These three units may also be in conflict with other systems. For example, cohabitants are in conflict with the economic institution for health benefits for their partners. Similarly, employed parents are in conflict with their employers for flexible work hours, maternity or paternity benefits, and day-care or eldercare facilities.

Karl Marx emphasized that conflict emanates from struggles over scarce resources and for power. Though Marxist theorists viewed these sources in terms of the conflict between the owners of the means of production (bourgeoisie) and the workers (proletariat), they are also relevant to conflicts within relationships. The first of these concepts, conflict over scarce resources, reflects the fact that spouses, parents, and children compete for scarce resources such as time, affection, and space. Spouses may fight with each other over how much time should be allocated to one's job, friends, or hobbies. Parents are sometimes in conflict with each other over who will do what housework or child care. Children are in conflict with their parents over time, affection, what programs to watch on television, and money.

Conflict theory is also helpful in understanding choices in relationships with regard to mate selection and jealousy. Unmarried individuals in search of a partner are in competition with other unmarried individuals for the scarce resources of a desirable mate. Such conflict is particularly evident in the case of older women in competition for men. At age 85 and older, there are twice as many women (3.9 million) as there are men (1.9 million) (*Statistical Abstract of the United States, 2009*, Table 10). Jealousy is also sometimes about scarce resources. People fear that their "one and only" will be stolen by someone else who has no partner.

Conflict theorists also emphasize conflict over power in relationships. Premarital partners, spouses, parents, and teenagers also use power to control one another. The reluctance of some courtship partners to make a marital commitment is an expression of wanting to maintain their autonomy, because marriage implies a relinquishment of power from each partner to the other. Spouse abuse is sometimes the expression of one partner trying to control the other through fear, intimidation, or force. Divorce may also illustrate control. The person who executes the divorce is often the person with the least interest in the relationship. Having the least interest gives that person power in the relationship. Parents and adolescents are also in a continuous struggle over power. Parents attempt to use the withholding of privileges and resources as power tactics to bring compliance in their adolescent. However, adolescents may use the threat of suicide as their ultimate power ploy to bring their parents under control.

Symbolic Interaction Framework

Symbolic interaction framework views marriages and families as symbolic worlds in which the various members give meaning to each other's behavior. Human behavior can be understood only by the meaning attributed to behavior (White and Klein 2002). Herbert Blumer (1969) used the term *symbolic interaction* to refer to the process of interpersonal interaction. Concepts inherent in this framework include

symbolic interaction framework views marriage and families as symbolic worlds in which the various members give meaning to each other's behavior.

the definition of the situation, the looking-glass self, and the self-fulfilling prophecy.

Definition of the Situation Two people who have just spotted each other at a party are constantly defining the situation and responding to those definitions. Is the glance from the other person (1) an invitation to approach, (2) an approach, or (3) a misinterpretation—the other person was looking at someone behind the person? The definition a person arrives at will affect subsequent interaction.

Looking-Glass Self The image people have of themselves is a reflection of what other people tell them about themselves (Cooley 1964). People may develop an idea of who they are by the way others act toward them. If no one looks at or speaks to them, they will begin to feel unsettled, according to Cooley. Similarly, family members constantly hold up social mirrors for one another into which the respective members look for definitions of self.

The importance of the family (and other caregivers) as an influence on the development and maintenance of a positive self-concept cannot be overemphasized (Brown et al. 2009). Orson Welles, known especially for his film *Citizen Kane*, once said that he was taught that he was wonderful and that everything he did was perfect. He never suffered from a negative self-concept. Cole Porter, known for creating such memorable songs as "I Get a Kick out of You," had a mother who held up social mirrors offering nothing but praise and adoration. Because children spend their formative years surrounded by their family, the self-concept they develop in that setting is important to their feelings about themselves and their positive interaction with others.

G. H. Mead (1934) believed that people are not passive sponges but evaluate the perceived appraisals of others, accepting some opinions and not others. Although some parents teach their children that they are worthless, they may eventually overcome the definition.

Self-Fulfilling Prophecy Once people define situations and the behaviors in which they are expected to engage, they are able to behave toward one another in predictable ways. Such predictability of behavior also tends to exert influence on subsequent behavior. If you feel that your partner expects you to be faithful, your behavior is likely to conform to these expectations. The expectations thus create a self-fulfilling prophecy.

family systems framework views each member of the family as part of a system and the family as a unit that develops norms of interaction.

Symbolic interactionism as a theoretical framework helps to explain various choices in relationships. Individuals who decide to marry have defined their situation as a committed reciprocal love relationship. This choice is supported by the belief that the partners will view each other positively (looking-glass self) and be faithful spouses and cooperative parents (self-fulfilling prophecies).

Later we will discuss the negative emotion of violence. Turner (2007) noted how symbolic interactionism may be used to explain the dynamics of intense emotions (such as violence). He suggested that extreme violence (which depends on intense negative emotions) has its genesis when negative emotions about the self and identity are repressed. All of this happens through the manipulation of symbols (negative self-reference statements) inside one's head.

Family Systems Framework

Systems theory is the most recent of all the theories for understanding family interaction (White and Klein 2002). The **family systems framework** views each member of the family as part of a system and the family as a unit that develops norms of interacting, which may be explicit (for example, parents specify chores for the children) or implied (for example, spouses expect fidelity from each other). These rules serve various functions, such as allocating the resources (money for vacation), specifying the division of power (who decides how money is spent), and defining closeness and distance between systems (seeing or avoiding parents or grandparents).

Rules are most efficient if they are flexible. For example, they should be adjusted over time in response to children's growing competence. A rule about not leaving the yard when playing may be appropriate for a 4-year-old but inappropriate for a 15-year-old. The rules and individuals can be understood only by recognizing that "all parts of the system are interconnected" (White and Klein 2002, 122).

The image people have of themselves is a reflection of what other people tell them about themselves.

Family members also develop boundaries that define the individual and the group and separate one system or subsystem from another. A boundary is a "border between the system and its environment that affects the flow of information and energy between the environment and the system" (White and Klein 2002, 124). A boundary may be physical, such as a closed bedroom door, or social, such as expectations that family problems will not be aired in public. Boundaries may also be emotional, such as communication, which maintains closeness or distance in a relationship. Some family systems are cold and abusive; others are warm and nurturing.

In addition to rules and boundaries, family systems have roles (leader, follower, scapegoat) for the respective family members. These roles may be shared by more than one person or may shift from person to person during an interaction or across time. In healthy families, individuals are allowed to alternate roles rather than being locked into one role. In problem families, one family member is often allocated the role of scapegoat, or the cause of all the family's problems (for example, an alcoholic spouse).

Family systems may be open, in that they are open to information and interaction with the outside world, or closed, in that they feel threatened by such contact. The Amish have closed family systems and minimize contact with the outside world. Some communes also encourage minimal outside exposure. Twin Oaks Intentional Community of Louisa, Virginia, does not permit any of its almost 100 members to own or keep televisions in their rooms. Exposure to the negative drumbeat of the evening news is seen as harmful.

Feminist Framework

Although a **feminist framework** views marriage and family as contexts of inequality and oppression, there are eleven feminist perspectives, including lesbian feminism (oppressive heterosexuality and men's domination of social spaces), psychoanalytic feminism (cultural domination of men's phallic-oriented ideas and repressed emotions), and standpoint feminism (neglect of women's perspective and experiences in the production of knowledge) (Lorber 1998). Regardless of which feminist framework is being discussed, all feminist frameworks have the themes of inequality and oppression. According to feminist theory, gender structures our experiences (for example, women and men will experience life differently because there are different expectations for the respective genders) (White and Klein 2002). Feminists seek equality in their relationships with their partners.

© Amanda Rohde/iStockphoto.com

feminist framework views marriage and the family as contexts for inequality and oppression.

1.6 Evaluating Research in Marriage and the Family

New "Research Study" is a frequent headline in popular magazines (such as *Cosmopolitan, Glamour, Redbook*) promising accurate information about "hooking up," "what women want," "what men want," or other relationship, marriage, and family issues. As you read such articles, as well as the research in such texts as this, be alert to their potential flaws. Following a list of steps in the research process, some specific issues to keep in mind when evaluating research follow:

Steps in the Research Process

Several steps are used in conducting research.

1. Identify the topic or focus of research. Select a focus about which you are passionate. For example, are you interested in studying cohabitation of college students? Give your projected study a title in the form of a question—"Do People Who Cohabit before Marriage Have Happier Marriages Than Those Who Do Not?"

2. Review the literature. Go online to the various databases of your college or university and

read research that has already been published on cohabitation. Not only will this prevent you from "reinventing the wheel" (you might find a research study has already been conducted on exactly what you want to study), but it will give you ideas for your study.

3. Develop hypotheses. A **hypothesis** is a suggested explanation for a phenomenon. For example, you might suggest that cohabitation results in greater marital happiness and less divorce because the partners have a chance to "test-drive" each other.

4. Decide on a method of data collection. To test your hypothesis, do you want to interview college students, give them a questionnaire, or ask them to complete an online questionnaire? Of course, you will also need to develop a list of questions for your interview or survey questionnaire.

5. Get **IRB approval.** To ensure the protection of people who agree to be interviewed or who complete questionnaires, researchers must submit a summary of their proposed research to the Institutional Review Board (IRB) of their institution. The IRB reviews the research to ensure that the project is consistent with research ethics and poses no undue harm to participants. Important considerations include collecting data from individuals who are told that their participation is completely voluntary, that maintains their anonymity, and that is confidential.

6. Collect and analyze data. There are various statistical packages designed to analyze data to discover if your hypotheses are true or false.

7. Write up and publish results. Writing up and submitting your findings for publication is important so that your study becomes part of the literature in cohabitation.

hypothesis a suggested explanation for a phenomenon.

IRB approval Institutional Review Board approval is the OK by one's college, university, or institution that the proposed research is consistent with research ethics standards and poses no undo harm to participants.

random sample sample in which each person in the population being studied has an equal chance of being included in the sample.

Evaluating Research Quality

The following issues need to be considered in evaluating research.

Sample Some of the research on marriage and the family is based on random samples. In a **random sample,** each individual in the population has an equal chance of being included in the sample. Random sampling involves randomly selecting individuals from an identified population. That population often does not include the homeless or people living on military bases. Studies that use random samples are based on the assumption that the individuals studied are similar to and therefore representative of the population that the researcher is interested in.

Because of the trouble and expense of obtaining random samples, most researchers study subjects to whom they have convenient access. This often means students in the researchers' classes. The result is an overabundance of research on "convenience" samples consisting of white, Protestant, middle-class college students. Because college students cannot be assumed to be similar in their attitudes, feelings, and behaviors to their noncollege peers or older adults, research based on college students cannot be generalized beyond the base population.

In addition to having a random sample, having a large sample is important. The American Council on

Education and the University of California (Pryor et al. 2008) collected a national sample of 240,580 first-semester undergraduates at 340 colleges and universities throughout the United States; the sample was designed to reflect the responses of 1.4 million first-time, full-time students entering four-year colleges and universities. If only fifty college students had been in the sample, the results would have been less valuable in terms of generalizing beyond that sample. Similarly, Corra et al. (2009) analyzed national data collected over a thirty-year period from the 1972–2002 General Social Survey (GSS) to discover the influence of sex (male or female) and race (white or black) on the level of reported marital happiness. This is an enormous set of random samples and gives validity to the findings.

Be alert to the sample size of the research you read. Most studies are based on small samples. In addition, considerable research has been conducted on college students, leaving us to wonder about the attitudes, values, and behaviors of noncollege students.

Control Groups In an example of a study that concludes that an abortion (or any independent variable) is associated with negative outcomes (or any dependent variable), the study must necessarily include two groups: (1) women who have had an abortion, and (2) women who have not had an abortion. The latter would serve as a **control group**—the group not exposed to the independent variable you are studying (**experimental group**). Hence, if you find that women in both groups in your study develop negative attitudes toward sex, you know that abortion cannot be the cause. Be alert to the existence of a control group, which is usually not included in research studies.

Age and Cohort Effects In some research designs, different cohorts or age groups are observed and/or tested at one point in time. One problem that plagues such research is the difficulty—even impossibility—of discerning whether observed differences between the subjects studied are due to the research variable of interest, cohort differences, or some variable associated with the passage of time (for example, biological aging).

A good illustration of this problem is found in research on changes in marital satisfaction over the course of the family life cycle. In such studies, researchers may compare the levels of marital happiness reported by couples that have been married for different lengths of time. For example, a researcher may compare the marital happiness of two groups of people—those who have been married for fifty years and those who have been married for five years. However, differences between these two groups may be due to (1) differences in age (age effect), (2) differences in the historical time period that the two groups have lived through (cohort effect), or (3) differences in the lengths of time the couples have been married (research variable). Keeping these issues in mind when you read studies on marital satisfaction over time is helpful.

Terminology In addition to being alert to potential shortcomings in sampling and control groups, you should consider how the phenomenon being researched is defined. For example, if you are conducting research on living together, how would you define this term? How many people, of what sex, spending what amount of time, in what place, engaging in what behaviors will constitute your definition. Indeed, researchers have used more than twenty definitions of what constitutes living together.

What about other terms? Considerable research has been conducted on marital success, but how is the term to be defined? What is meant by marital satisfaction, commitment, interpersonal violence, and sexual fulfillment? Before reading too far in a research study, be alert to the definitions of the terms being used. Exactly what is the researcher trying to measure?

Researcher Bias Although one of the goals of scientific studies is to gather data objectively, it may be impossible for researchers to be totally objective. Researchers are human and have values, attitudes, and beliefs that may influence their research methods and findings. It may be important to know what the researcher's bias is in order to evaluate the findings. For example, a researcher who does not support abortion rights may conduct research that focuses only on the negative effects of abortion. Similarly, researchers funded by corporations who have a vested interest in having their products endorsed are also suspect. Thomas (2007) investigated the toy industry and noted that some researchers are funded by corporations that make them celebrities when their research supports outcomes conducive to selling products to infants and toddlers. In 2008, Dr. Eric Westman of Duke University compared the Atkins diet with two other diets and found the Atkins diet superior in terms of patients keeping weight off. However, the Robert C. and Veronica Atkins Foundation funded the study. Similarly,

control group group used to compare with the experimental group that is not exposed to the independent variable being studied.

experimental group the group exposed to the independent variable.

in 2008, the makers of Crestor (AstraZeneca) funded a study that showed the cholesterol-lowering statin drug resulted in big reductions in heart attacks, strokes, and deaths in people with so-called healthy cholesterol levels. Should we be suspicious?

Time Lag Typically, a two-year lag exists between the time a study is completed and the study's appearance in a professional journal. Because textbook production takes even longer than getting an article printed in a professional journal, they do not always present the most current research, especially on topics in flux. In addition, even though a study may have been published recently, the data on which the study was based may be old. Be aware that the research you read in this or any other text may not reflect the most current, cutting-edge research.

Distortion and Deception Our society is no stranger to distortion and deception—"weapons of mass destruction"? Writers at the prestigious *New York Times* have been fired after they were discovered to have fabricated articles. Chemists at the University of Utah reported that they had discovered "cold fusion." "They hadn't, it turned out. . . ." (Lemonick 2006, 43).

Distortion and deception, deliberate or not, also exist in marriage and family research. Marriage is a very private relationship that happens behind closed doors; individual interviewees and respondents to questionnaires have been socialized not to reveal the intimate details of their lives to strangers. Hence, they are prone to distort, omit, or exaggerate information, perhaps unconsciously, to cover up what they may feel is no one else's business. Thus, researchers sometimes obtain inaccurate information.

Marriage and family researchers know more about what people say they do than about what they actually do. An unintentional and probably more common form of distortion is inaccurate recall. Sometimes researchers ask respondents to recall details of their relationships that occurred years ago. Time tends to blur some memories, and respondents may relate not what actually happened, but rather what they remember to have happened, or, worse, what they wish had happened.

© Dynamic Graphics/Jupiterimages

Other Research Problems Nonresponse on surveys and the discrepancy between attitudes and behaviors are other research problems. With regard to nonresponse, not all individuals who complete questionnaires or agree to participate in an interview are willing to provide information about such personal issues as date rape and partner abuse. Such individuals leave the questionnaire blank or tell the interviewer they would rather not respond. Others respond but give only socially desirable answers. The implications for research are that data gatherers do not know the nature or extent to which something may be a problem because people are reluctant to provide accurate information. Computer-administered self-interviewing (CASI) has been suggested as a way of collecting sensitive information (for example, sexual). However, a study by Testa et al. (2005) comparing CASI and self-administered mailed questionnaires found that the latter had a higher response rate with greater disclosure of sensitive information.

The discrepancy between people's attitudes and their behavior is another cause for concern about the validity of research data. It is sometimes assumed that if people have a certain attitude (for example, a belief that extramarital sex is wrong), then their behavior will be consistent with that attitude (avoidance of extramarital sex). However, this assumption is not always accurate. People do indeed say one thing and do another. This potential discrepancy should be kept in mind when reading research on various attitudes.

Finally, most research reflects information that volunteers provide. However, volunteers may not represent nonvolunteers when they are completing surveys. In view of the research cautions identified here, you might ask, "Why bother to report the findings?" The quality of some family science research is excellent. For example, articles published in *Journal of Marriage and the Family* (among other journals) reflect the high level of methodologically sound articles that are being published. There are even some less sophisticated journals that provide useful information on marital, family, and other relationship data. Table 1.4 summarizes potential inadequacies of any research study.

Table 1.4

Potential Research Problems in Marriage and Family

Weakness	Consequences	Example
Sample not random	Cannot generalize findings	Opinions of college students do not reflect opinions of other adults.
No control group	Inaccurate conclusions	Study on the effect of divorce on children needs control group of children whose parents are still together.
Age differences between groups of respondents	Inaccurate conclusions	Effect may be due to passage of time or to cohort differences.
Unclear terminology	Inability to measure what is not clearly defined	What is living together, marital happiness, sexual fulfillment, good communication, quality time?
Researcher bias	Slanted conclusions	Male researcher may assume that, because men usually ejaculate each time they have intercourse, women should have an orgasm each time they have intercourse.
Time lag	Outdated conclusions	Often-quoted Kinsey sex research is more than seventy years old.
Distortion	Invalid conclusions	Research subjects exaggerate, omit information, and/or recall facts or events inaccurately. Respondents may remember what they wish had happened.

Diversity in Relationships

The "one-size-fits-all" model of relationships and marriage is nonexistent. Individuals may be described as existing on a continuum from heterosexuality to homosexuality, from rural to urban dwellers, and from being single and living alone to being married and living in communes. Emotional relationships range from being close and loving to being distant and violent. Family diversity includes two parents (other or same-sex), single-parent families, blended families, families with adopted children, multigenerational families, extended families, and families representing different racial, religious, and ethnic backgrounds. *Diversity* is the term that accurately describes marriage and family relationships today.

Gender

SECTIONS

Mike May was blinded at age 3 due to a chemical explosion. Years later, after he had married and had children, he was offered the possibility of sight again (through stem cell technology). It worked. However, Mike had to learn to "see"; he could not distinguish between women and men. His wife, Jennifer, went with him to a coffee shop and taught him what to look for to discover whether a person was a woman or a man. For a woman, she taught him to look for:

> *Swinging hips* ("Women walk with a bounce; men don't.")
> *Purses* ("Men don't carry things over their shoulders, at least in the United States.")
> *Tight pants* ("Women sometimes paint them on; men's are more baggy.")
> *Bellies* ("Women show them a lot these days. Men almost never do.")
> *Jewelry* ("Some men might wear necklaces, but very few wear shiny bracelets.")
>
> (Kurson 2007, 191–92)

Although Jennifer detailed a current cultural list of symbolic behavior that helps to differentiate the sexes, sociologists note that one of the defining moments in an individual's life is when the sex of a fetus (in the case of an ultrasound) or infant (in the case of a birth) is announced. "It's a boy" or "It's a girl" immediately summons an onslaught of cultural programming affecting the color of the nursery (for example, blue for a boy and pink for a girl), name of the baby (there are few gender-free names such as Chris), occupational choices (in spite of Nancy Pelosi and Hillary Clinton, few women are in politics), and Jennifer's list of symbolic behaviors. In this chapter, we examine variations in gender roles and the way they express themselves in various relationships. We begin by looking at the terms used to discuss gender issues.

2.1 Terminology of Gender Roles

In common usage, the terms *sex* and *gender* are often interchangeable, but sociologists, family or consumer science educators, human development specialists, and health educators do not find these terms synonymous. After clarifying the distinction between *sex* and *gender*, we discuss other relevant terminology, including *gender identity, gender role,* and *gender role ideology.*

Sex

Sex refers to the biological distinction between females and males. Hence, to be assigned as a female or male, several factors are used to determine the biological sex of an individual:

sex the biological distinction between being female and being male.

- *Chromosomes*: XX for females; XY for males
- *Gonads*: Ovaries for females; testes for males
- *Hormones*: Greater proportion of estrogen and progesterone than testosterone in females; greater proportion of testosterone than estrogen and progesterone in males
- *Internal sex organs*: Fallopian tubes, uterus, and vagina for females; epididymis, vas deferens, and seminal vesicles for males
- *External genitals*: Vulva for females; penis and scrotum for males

hermaphrodites (intersexed individuals) people with mixed or ambiguous genitals.

intersex development refers to congenital variations in the reproductive system, sometimes resulting in ambiguous genitals.

true hermaphroditism an extremely rare condition in which individuals are born with both ovarian and testicular tissue.

pseudohermaphroditism refers to a condition in which an individual is born with gonads matching the sex chromosomes, but with genitals either ambiguous or resembling those of the opposite sex.

gender the social and psychological behaviors associated with being female or male.

Even though we commonly think of biological sex as consisting of two dichotomous categories (female and male), biological sex exists on a continuum. Sometimes not all of the items identified are found neatly in one person (who would be labeled as a female or a male). Rather, items typically associated with females or males might be found together in one person, resulting in mixed or ambiguous genitals; such persons are called **hermaphrodites** or **intersexed individuals**. Indeed, the genitals in these intersexed (or middlesexed) individuals (about 2 percent of all births) are not clearly male or female (Crawley et al. 2008). **Intersex development** refers to congenital variations in the reproductive system, sometimes resulting in ambiguous genitals. Even if chromosomal makeup is XX or XY, too much or too little of the wrong kind of hormone during gestation can also cause variations in sex development. Although intersexed people currently prefer the term *intersexed* or *middlesexed*, intersex conditions that may result from hormonal abnormalities are referred to as *hermaphroditism* and *pseudohermaphroditism* in the medical literature. **True hermaphroditism** is an extremely rare condition in which individuals are born with both ovarian and testicular tissue. More common than hermaphroditism is **pseudohermaphroditism**, which refers to a condition in which an individual is born with gonads matching the sex chromosomes, but with genitals either ambiguous or resembling those of the other sex. Meyer-Bahlburg (2005) suggested that the genesis of the ambiguity may be the brain anatomy, whereby the wiring is different for those with gender identity disorder (GID). Genetics, hormones, and brain mechanisms might all underlie neuroanatomic changes inducing intersexuality.

Gender

Gender refers to the social and psychological characteristics associated with being female or male. For example, women see themselves (and men agree) as moody and easily embarrassed; men see themselves (and women agree) as competitive, sarcastic, and sexual (Knox et al. 2004). In popular usage, gender is dichotomized as an either/or concept (feminine or masculine). Each gender has some characteristics of the other. However, gender may also be viewed as existing along a continuum of femininity and masculinity.

There is an ongoing controversy about whether gender differences are innate as opposed to learned or socially determined. Just as sexual orientation may be best explained as an interaction of biological and social or psychological variables, gender differences may also be seen as a consequence of both biological and social or psychological influences. For example, Irvolino et al. (2005) studied the genetic and environmental effects on the sex-typed behavior of 3,999 3- to 4-year-old twin and nontwin sibling pairs and concluded that their gender role behavior was a function of both genetic inheritance (for example, chromosomes and hormones) and social factors (for example, male/female models such as parents, siblings, peers).

Whereas some researchers emphasize an interaction of the biological and social, others emphasize a biological imperative as the basis of gender role behavior. As evidence for the latter, the late John Money, psychologist and former director of the now-defunct Gender Identity Clinic at Johns Hopkins University School of Medicine, encouraged the parents of a boy (Bruce) to rear him as a girl (Brenda) because of a botched

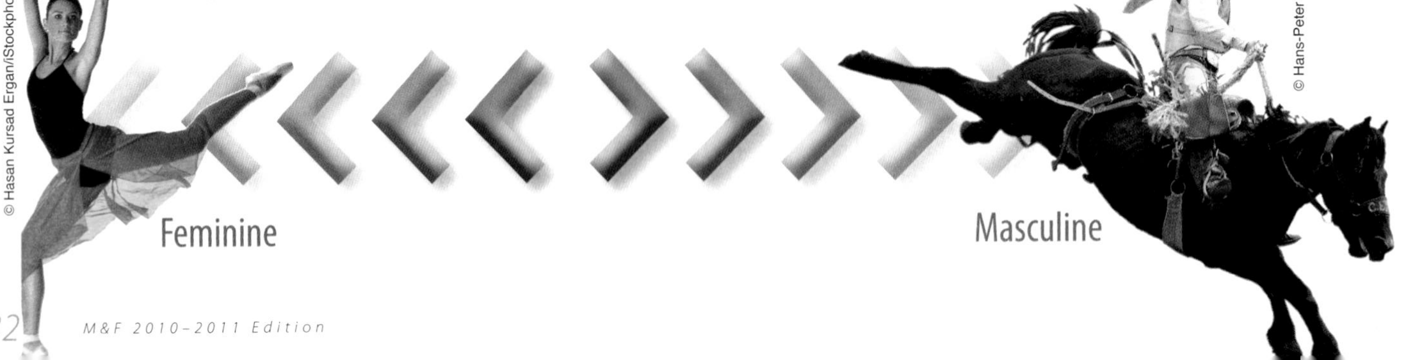

© Laura Neal/iStockphoto.com

circumcision that rendered the infant without a penis. Money argued that social mirrors dictate one's gender identity, and thus, if the parents treated the child as a girl (e.g. name, dress, toys), the child would adopt the role of a girl and later that of a woman. The child was castrated and sex reassignment began.

However, the experiment failed miserably; the child as an adult (David Reimer—his real name) reported that he never felt comfortable in the role of a girl and had always viewed himself as a boy. He later married and adopted his wife's two children. In the book *As Nature Made Him: The Boy Who Was Raised as a Girl* (Colapinto 2000), David worked with a writer to tell his story. His courageous decision to make his poignant personal story public has shed light on scientific debate on the "nature/nurture" question. In the past, David's situation was used as a textbook example of how "nurture" is the more important influence in gender identity, if a reassignment is done early enough. Today, his case makes the point that one's biological wiring dictates gender outcome (ibid.). Indeed, David Reimer noted in a television interview, "I was scammed," referring to the absurdity of trying to rear him as a girl. Distraught with the ordeal of his upbringing and beset with financial difficulties, he committed suicide in May 2004 via a gunshot to the head.

The story of David Reimer is a landmark in terms of the power of biology in determining gender identity and other research supports the critical role of biology. Cohen-Kettenis (2005) emphasized that biological influences in the form of androgens in the prenatal brain are very much at work in creating one's gender identity.

Nevertheless, **socialization** (the process through which we learn attitudes, values, beliefs, and behaviors appropriate to the social positions we occupy) does impact gender role behaviors, and social scientists tend to emphasize the role of social influences in gender differences. Although her research is controversial, Margaret Mead (1935) focused on the role of social learning in the development of gender roles in her study of three cultures.

She visited three New Guinea tribes in the early 1930s and observed that the Arapesh socialized both men and women to be feminine, by Western standards. The Arapesh people were taught to be cooperative and responsive to the needs of others. In contrast, the Tchambuli were known for dominant women and submissive men—just the opposite of our society. Both of these societies were unlike the Mundugumor, which socialized only ruthless, aggressive, "masculine" personalities. The inescapable conclusion of this cross-cultural study is that human beings are products of their social and cultural environment and that gender roles are learned. As Peoples observed, "cultures construct gender in different ways" (2001, 18). Indeed, the very terms we use to describe various body parts of women and men carry notions of power and use. A penis is a "probe" (active) whereas a vagina is a "hole" (to be filled) (Crawley et al. 2008).

socialization the process through which we learn attitudes, values, beliefs, and behaviors appropriate to the social positions we occupy.

gender identity the psychological state of viewing oneself as a girl or a boy, and later as a woman or a man.

gender dysphoria the condition in which one's gender identity does not match one's biological sex.

transgender a generic term for a person of one biological sex who displays characteristics of the opposite sex.

Gender Identity

© Dmitriy Tereschenko/iStockphoto.com

Gender identity is the psychological state of viewing oneself as a girl or a boy, and later as a woman or a man. Such identity is largely learned and is a reflection of society's conceptions of femininity and masculinity.

Some individuals experience **gender dysphoria**, a condition in which one's gender identity does not match one's biological sex. An example of gender dysphoria is transsexualism (discussed in the next section).

Transgenderism

The word **transgender** is a generic term for a person of one biological sex who displays characteristics of the other sex. **Cross-dresser** is a broad term for individuals who may dress or present themselves in the gender of the opposite sex. Some cross-dressers are heterosexual adult males who enjoy dressing and presenting themselves as women.

Cross-dressers may also be women who dress as men and present themselves as men. Some cross-dressers are bisexual or homosexual. Another term for cross-dresser is **transvestite**, although the latter term is commonly associated with homosexual men who dress provocatively as women to attract men—sometimes as sexual customers.

Transsexuals are people with the biological and anatomical sex of one gender (for example, male) but the self-concept of the opposite sex (that is, female). "I am a woman trapped in a man's body" reflects the feelings of the male-to-female transsexual, who may take hormones to develop breasts and reduce facial hair and may have surgery to artificially construct a vagina. Such a person lives full-time as a woman.

The female-to-male transsexual is one who is a biological and anatomical female but feels "I am a man trapped in a woman's body." This person may take male hormones to grow facial hair and deepen her voice and may have surgery to create an artificial penis. This person lives full-time as a man. Thomas Beatie, born a biological woman, viewed himself as a man and transitioned from living as a woman to living as a man. His wife could not have children so Tom agreed to be artificially inseminated (because he had ovaries), became pregnant, and delivered a child in the summer of 2008. The media referred to him as "The Pregnant Man"; they asked him, if he gave birth to the child, could he be the father? Technically, Oregon law defines birth as an expulsion or extraction from the mother, so Tom is the technical mother. However, the new parents could petition the courts and have him declared as the father and his wife as the mother (Heller 2008).

Individuals need not take hormones or have surgery to be regarded as transsexuals. The distinguishing variable is living full-time in the role of the gender opposite one's biological sex. A man or woman who presents full-time as the opposite gender is a transsexual by definition. Table 2.1 may help to keep the categories clear.

Some transsexuals prefer the term **transgenderist**, which refers to individuals who live in a gender role that does not match their biological sex but who have no desire to surgically alter their genitalia (as do people who are transsexual). Another variation is the she-male, a person who looks like and has the breasts of a woman yet has the genitalia and reproductive system of a male.

cross-dresser a generic term for individuals who may dress or present themselves in the gender of the opposite sex.

transvestite term commonly associated with homosexual men who dress provocatively as women to attract men.

transsexual an individual who has the anatomical and genetic characteristics of one sex but the self-concept of the other.

transgenderist an individual who lives in a gender role that does not match his or her biological sex, but has no desire to surgically alter his or her genitalia.

gender roles behaviors assigned to women and men in a society.

Gender Roles

Gender roles are the social norms that dictate what is socially regarded as appropriate female and male behavior. All societies have expectations of how boys and girls, men and women "should" behave. Gender roles influence women and men in virtually every sphere of life, including family and occupation. One prevalent norm in family life is that women end up devoting more time to child rearing and child care. Lareau and Weininger (2008) studied the division of labor of parents getting their children to various leisure activities and found that traditional gender roles were the norm and that mothers "are the ones who must satisfy these demands" (p. 450). The sheer number of activities forces some mothers to cut back. One mother said:

> I know that all these things are good for my child and they develop all those things but time-wise, I just don't have time for all that stuff. I mean I have to live up to my label as the meanest Mom in the world and I try to

Table 2.1
Transgender Categories

Category	Biological Sex	Sexual Orientation	Most Usual Case
Cross-dresser	Either	Either	Male heterosexual dresses as woman
Transvestite	Male	Homosexual	Homosexual male dresses as woman
Transsexual	Either	Either	Heterosexual male in woman's body who wants surgery

Gender Differences in Viewing Romantic Relationships

Abowitz et al. (2009) assessed the respective gender views of romantic relationships of 326 undergraduates and found that men were significantly more likely to believe that bars are good places to meet a potential partner, that cohabitation improves marriage, that men control relationships, and that people will "cheat" if they feel they will not be caught. In contrast, women were significantly more likely to believe that love is more important than factors like age and race in choosing a mate, that couples stop "trying" after they marry, and that women know when their men are lying.

> do a real good job of it [laughs]. I told her she could be either on soccer or softball. She'll be on one of them. (p. 450)

Walzer (2008) noted that marriage is a place where men and women "do gender" in the sense that roles tend to be identified as breadwinning, housework, parenting, and emotional expression and are gender-differentiated. She also noted that divorce generates "redoing" gender in the sense that it changes the expectations for masculine and feminine behavior in families (for example, women become breadwinners, men become single parents, and so on).

The term **sex roles** is often confused with and used interchangeably with the term *gender roles*. However, whereas gender roles are socially defined and can be enacted by either women or men, sex roles are defined by biological constraints and can be enacted by members of one biological sex only—for example, wet nurse, sperm donor, and childbearer.

Gender Role Ideology

sex roles behaviors defined by biological constraints.

gender role ideology the proper role relationships between women and men in a society.

Gender role ideology refers to beliefs about the proper role relationships between women and men in any given society. Where there is gender equality, there is enhanced relationship satisfaction (Walker and Luszcz, 2009). Egalitarian wives are also most happy with their marriages if their husbands share both the work and the emotions of managing the home and caring for the children (Wilcox and Nock 2006).

In spite of the rhetoric regarding the entrenchment of egalitarian interaction between women and men in the United States, there is evidence of traditional gender roles in mate selection with men in the role of initiating relationships. When 692 undergraduate females at a large southeastern university were asked if they had ever asked a new guy to go out, 60.1 percent responded, "no" (Ross et al., forthcoming). In another study, 30 percent of the female respondents reported a preference of marrying a traditional man (one who saw his role as provider and who was supportive of his wife staying home to rear children) (McGinty et al. 2006). Some undergraduate men also prefer a traditional wife.

Traditional American gender role ideology has perpetuated and reflected patriarchal male dominance and male bias in almost every sphere of life. Even our language reflects this male bias. For example, the words *man* and *mankind* have traditionally been used to refer to all humans. There has been a growing trend away from using male-biased language. Dictionaries have begun to replace *chairman* with *chairperson* and *mankind* with *humankind*.

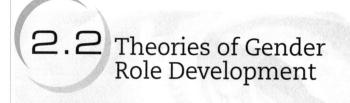

2.2 Theories of Gender Role Development

Various theories attempt to explain why women and men exhibit different characteristics and behaviors.

Biosocial

In the discussion of gender at the beginning of the chapter, we noted the profound influence of biology on one's gender. **Biosocial theory** (also referred to as **sociobiology**) emphasizes that social behaviors (for example, gender roles) are biologically based and have an evolutionary survival function. For example, women tend to select and mate with men whom they deem will provide the maximum parental investment in their offspring. The term **parental investment** refers to any investment by a parent that increases the offspring's chance of surviving and thus increases reproductive success. Parental investments require time and energy. Women have a great deal of parental investment in their offspring (including nine months of gestation), and they tend to mate with men who have high status, economic resources, and a willingness to share those economic resources.

The biosocial explanation for mate selection is extremely controversial. Critics argue that women may show concern for the earning capacity of a potential mate because they have been systematically denied access to similar economic resources, and selecting a mate with these resources is one of their remaining options. In addition, it is argued that both women and men, when selecting a mate, think more about their partners as companions than as future parents of their offspring.

Social Learning

Derived from the school of behavioral psychology, the social learning theory emphasizes the roles of reward and punishment in explaining how a child learns gender role behavior. This is in contrast to the biological explanation for gender roles. For example, consider two young brothers who enjoy playing "lady"; each of them puts on a dress, wears high-heeled shoes, and carries a pocketbook. Their father came home early one day and angrily demanded, "Take those clothes off and never put them on again. Those things are for women." The boys were punished for "playing lady" but rewarded with their father's approval for playing cowboys, with plastic guns and "Bang! You're dead!" dialogue.

Reward and punishment alone are not sufficient to account for the way in which children learn gender roles. Another way children learn is when parents or peers offer direct instruction (for example, "girls wear dresses" or "a man stands up and shakes hands"). In addition, many of society's gender rules are learned through modeling. In modeling, children observe and imitate another's behavior. Gender role models include parents, peers, siblings, and characters portrayed in the media.

The impact of modeling on the development of gender role behavior is controversial. For example, a modeling perspective implies that children will tend to imitate the parent of the same sex, but children in all cultures are usually reared mainly by women. Yet this persistent female model does not seem to interfere with the male's development of the behavior that is considered appropriate for his gender. One explanation suggests that boys learn early that our society generally grants boys and men more status and privileges than girls and women. Therefore, boys devalue the feminine and emphasize the masculine aspects of themselves.

Identification

Freud was one of the first researchers to study gender role development. He suggested that children acquire the characteristics and behaviors of their same-sex parent through a process of identification. Girls identify with their mothers; boys identify with their fathers. For example, girls are more likely to become involved in taking care of children because they see women as the primary caregivers

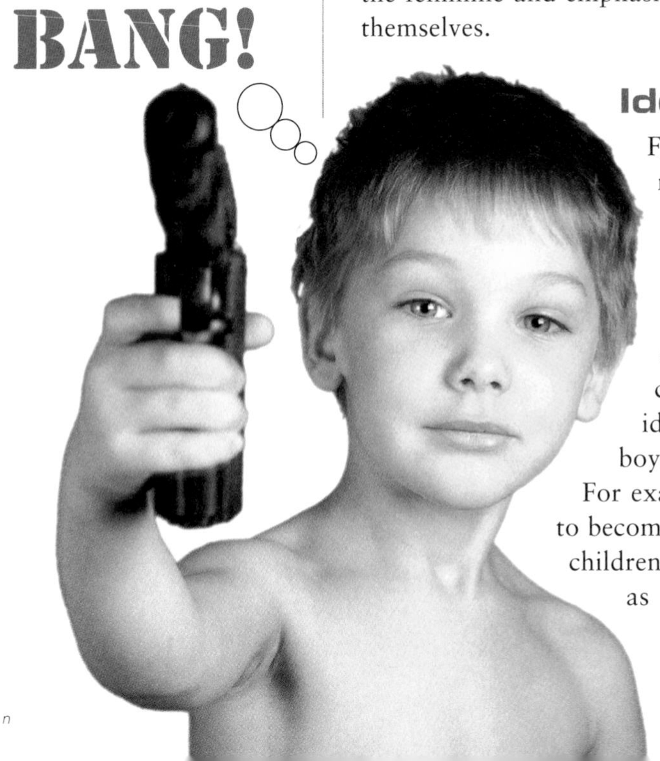

© Iva Barmina/iStockphoto.com

biosocial theory (sociobiology) emphasizes the interaction of one's biological or genetic inheritance with one's social environment to explain and predict human behavior.

parental investment any investment by a parent that increases the chance that the offspring will survive and thrive.

> ALL WOMEN BECOME LIKE THEIR MOTHERS. THAT IS THEIR TRAGEDY. NO MAN DOES. THAT'S HIS.
>
> — *Oscar Wilde, playwright*

of young children. In effect, they identify with their mothers and will see their own primary identity and role as those of a mother. Likewise, boys will observe their fathers and engage in similar behaviors to lock in their own gender identity. The classic example is the son who observes his father shaving and wants to do likewise (be a man too).

Cognitive-Developmental

The cognitive-developmental theory of gender role development reflects a blend of biological and social learning views. According to this theory, the biological readiness of the child, in terms of cognitive development, influences how the child responds to gender cues in the environment (Kohlberg 1966). For example, gender discrimination (the ability to identify social and psychological characteristics associated with being female or male) begins at about age 30 months. However, at this age, children do not view gender as a permanent characteristic. Thus, even though young children may define people who wear long hair as girls and those who never wear dresses as boys, they also believe they can change their gender by altering their hair or changing clothes.

Not until age 6 or 7 do children view gender as permanent (Kohlberg 1966; 1969). In Kohlberg's view, this cognitive understanding involves the development of a specific mental ability to grasp the idea that certain basic characteristics of people do not change. Once children learn the concept of gender permanence, they seek to become competent and proper members of their gender group. For example, a child standing on the edge of a school playground may observe one group of children jumping rope while another group is playing football. That child's gender identity as either a girl or a boy connects with the observed gender-typed behavior, and the child joins one of the two groups. Once in the group, the child seeks to develop behaviors that are socially defined as gender-appropriate.

2.3 Agents of Socialization

Three of the four theories discussed in the preceding section emphasize that gender roles are learned through interaction with the environment. Indeed, though biology may provide a basis for one's gender identity, cultural influences in the form of various socialization agents (parents, peers, religion, and the media) shape the individual toward various gender roles. These powerful influences in large part dictate what people think, feel, and do in their roles as man or woman. In the next section, we look at the different sources influencing gender socialization.

Family

The family is a gendered institution with female and male roles highly structured by gender. The names parents assign to their children, the clothes they dress them in, and the toys they buy them all reflect gender. Parents may also be stricter on female children—determining the age they are allowed to leave the house at night, the time of curfew and using directives such as "call your mamma when you get to the party."

The importance of the father in the family was noted in Pollack's (2001) study of adolescent boys. "America's boys are crying out for a new gender revolution that does for them what the last forty years of feminism has tried to do for girls and women," he stated (p. 18). This new revolution will depend on fathers who teach their sons that feelings and relationships are important. How equipped do you feel today's fathers are to provide these new models for their sons?

Siblings also influence gender role learning. As noted in Chapter 1, the relationship with one's sibling

© Eliza Snow/iStockphoto.com / © Wendy Idele/Nonstock/Jupiterimages

(particularly in sister-sister relationships) is likely to be the most enduring of all relationships (Meinhold et al. 2006). Also, growing up in a family of all sisters or all brothers intensifies social learning experiences toward femininity or masculinity. A male reared with five sisters and a single-parent mother is likely to reflect more feminine characteristics than a male reared in a home with six brothers and a stay-at-home dad.

Race/Ethnicity

The race and ethnicity of one's family also influence gender roles. Although African American families are often stereotyped as being matriarchal, the more common pattern of authority in these families is egalitarian (Taylor 2002). Both President Barack and First Lady Michelle Obama have law degrees from Harvard, and their relationship appears to be very egalitarian.

The fact that African American women have increased economic independence provides a powerful role model for young African American women. A similar situation exists among Hispanics, who represent the fastest-growing segment of the U.S. population. Mexican American marriages have great variability, but where the Hispanic woman works outside the home, her power may increase inside the home. However, because Hispanic men are much more likely to be in the labor force than Hispanic women, traditional role relationships in the family are more likely to be the norm.

Peers

Though parents are usually the first socializing agents that influence a child's gender role development, peers become increasingly important during the school years. Haynie and Osgood (2005) analyzed data from the National Longitudinal Study of Adolescents reflecting responses from adolescents in grades 7 through 12 at 132 schools over an eleven-year period and confirmed the influence of peers in delinquent behavior. If friends drank, smoked cigarettes, skipped school without an excuse, and became involved in serious fights, then the adolescents had an increased likelihood that they would also engage in delinquent acts. Regarding gender, the gender role messages from adolescent peers are primarily traditional. Boys are expected to play sports and be career-oriented. Female adolescents are under tremendous pressure to be physically attractive and thin, popular, and achievement-oriented. Female achievements may be traditional (cheerleading) or nontraditional (sports or academics). Adolescent females are sometimes in conflict because high academic success may be viewed as being less than feminine.

Peers also influence gender roles throughout the family life cycle. In Chapter 1, we discussed the family life cycle and noted the various developmental tasks throughout the cycle. With each new stage, role changes are made and one's peers influence those role changes. For example, when a couple moves from being childfree to being parents, peers who are also parents quickly socialize them into the role of parent and the attendant responsibilities.

> Women in general *want to be loved for what they are and men for what they accomplish.*
>
> —*Theodor Reik*

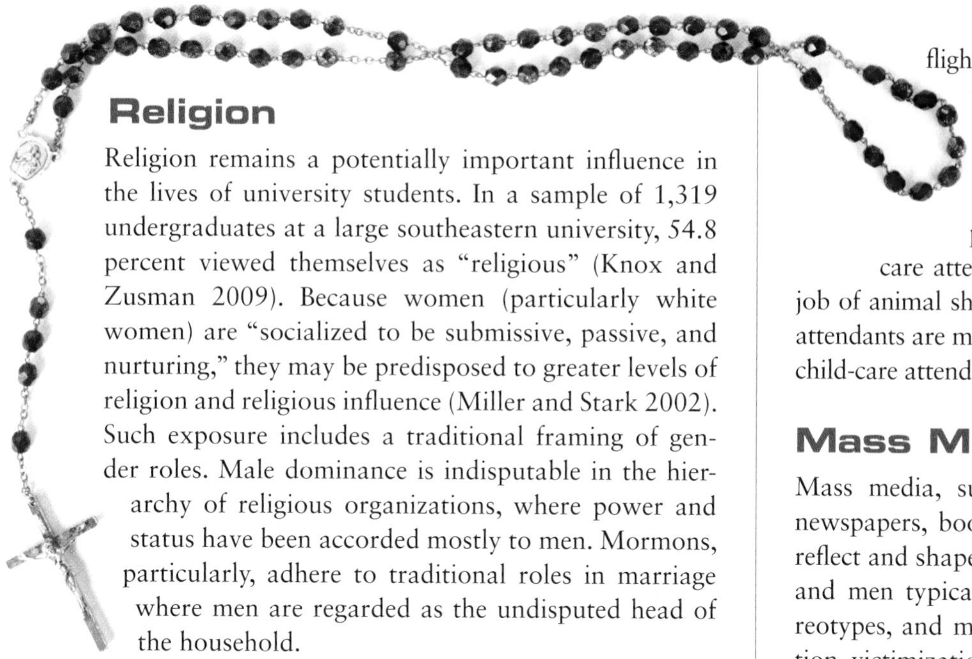

© iStockphoto.com

Religion

Religion remains a potentially important influence in the lives of university students. In a sample of 1,319 undergraduates at a large southeastern university, 54.8 percent viewed themselves as "religious" (Knox and Zusman 2009). Because women (particularly white women) are "socialized to be submissive, passive, and nurturing," they may be predisposed to greater levels of religion and religious influence (Miller and Stark 2002). Such exposure includes a traditional framing of gender roles. Male dominance is indisputable in the hierarchy of religious organizations, where power and status have been accorded mostly to men. Mormons, particularly, adhere to traditional roles in marriage where men are regarded as the undisputed head of the household.

The Roman Catholic Church does not have female clergy, and men dominate the nineteen top positions in the U.S. dioceses. Popular books marketed to the Christian right also emphasize traditional gender roles. Denton (2004) found that conservative Protestants are committed to the ideology that the husband is the head of the family. However, this implies "taking spiritual leadership" and may not imply dominance in marital decision making (p. 1174).

Education

The educational institution serves as an additional socialization agent for gender role ideology. However, such an effect must be considered in the context of the society or culture in which the "school" exists and of the school itself. Schools are basic cultures of transmission in that they make deliberate efforts to reproduce the culture from one generation to the next. Sumsion (2005) noted that even having male teachers in the lower grades does not seem to disrupt traditional gender stereotypes that young children have.

Economy

The economy of the society influences the roles of the individuals in the society. The economy is a very gendered institution. **Occupational sex segregation** denotes the fact that women and men are employed in gender-segregated occupations, that is, occupations in which workers are either primarily male or female (for example, men may work as mechanical or electrical engineers and women as flight attendants). Female-dominated occupations tend to require less education, have lower status, and pay lower salaries than male-dominated occupations. If men typically occupy a role, it tends to pay more. For example, the job of child-care attendant requires more education than the job of animal shelter attendant. However, animal shelter attendants are more likely to be male and earn more than child-care attendants, who are more likely to be female.

Mass Media

Mass media, such as movies, television, magazines, newspapers, books, music, and computer games, both reflect and shape gender roles. Media images of women and men typically conform to traditional gender stereotypes, and media portrayals depicting the exploitation, victimization, and sexual objectification of women are common. Observe the array of magazine covers at any newsstand. Compare the number of provocative females with males. Similarly, notice the traditional gender role scripting in popular television programs. Kim et al. (2007) identified the gender or sexual scripting of twenty-five prime-time television shows and found evidence of the traditional scripting (for example, men are dominant and "need sex"; women are passive and valued for their bodies).

Rivadeneyraa and Lebob (2008) studied ninth grade students and found that watching "romantic" television (for example, soaps, Lifetime movies) was associated with having more traditional gender role attitudes in dating situations. However, watching nonromantic television dramas and educational television was related to having less traditional dating role attitudes.

Self-help parenting books are also biased toward traditional gender roles. A team of researchers conducted a content analysis of six of the best-selling self-help books for parents and found that 82 percent were wrought with stereotypical gender role messages (Krafchick et al. 2005).

The cumulative effects of family, peers, religion, education, the economy, and mass media perpetuate gender stereotypes. Each agent of socialization reinforces gender roles that are learned from other agents of socialization, thereby creating a gender role system that is deeply embedded in our culture.

occupational sex segregation the concentration of women in certain occupations and men in other occupations.

2.4 Consequences of Traditional Gender Role Socialization

This section discusses different consequences, both negative and positive, for women and men, of traditional female and male socialization in the United States.

Consequences of Traditional Female Role Socialization

Table 2.2 summarizes some of the negative and positive consequences of being socialized as a woman in U.S. society. Each consequence may or may not be true for a specific woman. For example, although women in general have less education and income, a particular woman may have more education and a higher income than a particular man.

Table 2.2

Consequences of Traditional Female Role Socialization

Negative Consequences	Positive Consequences
Less education/income (more dependent)	Longer life
Feminization of poverty	Stronger relationship focus
Higher STD/HIV infection risk	Keeps relationships on track
Negative body image	Bonding with children
Less marital satisfaction	Identity not tied to job

Negative Consequences of Traditional Female Role Socialization There are several negative consequences of being socialized as a woman in our society.

1. Less income. Although women now earn 46 percent of PhDs (Welch 2008), they have lower academic rank (Probert 2005) and earn less money. The lower academic rank is because women give priority to the care of their children and family (Aissen and Houvouras 2006). In addition, women tend to be more concerned about the nonmonetary aspects of work. In a study of 102 seniors and 504 alumni from a mid-sized Midwestern public university that rated forty-eight job characteristics, women gave significantly higher ratings to family life accommodations, pleasant working conditions, travel, and interpersonal relationships. Women still earn about two-thirds of what men earn, even when the level of educational achievement is identical (see Table 2.3). Their visibility in the ranks of high corporate America is also still low. Of *Fortune* 500 companies, women run only thirteen as CEO. Angela Braly is one of them. She heads WellPoint and earns $9.1 million annually (*Fortune* 500 2008).

feminization of poverty the idea that women disproportionately experience poverty.

Table 2.3

Women's and Men's Median Income with Similar Education

	Bachelor's	Master's	Doctoral Degree
Men	$54,403	$67,425	$90,511
Women	$35,094	$46,250	$61,091

Source: *Statistical Abstract of the United States, 2009*. 128th ed. Washington, DC: U.S. Bureau of the Census, Table 680.

With divorce being a nearly 45 percent probability for marriages begun in the 2000s, the likelihood of being a widow for five or more years, and the almost certain loss of her parenting role midway through her life, a woman without education and employment skills is often left high and dry. As one widowed mother of four said, "The shock of realizing you have children to support and no skills to do it is a worse shock than learning that your husband is dead." In the words of a divorced, 40-year-old mother of three, "If young women think it can't happen to them, they are foolish."

2. Feminization of poverty. Another reason many women are relegated to a lower income status is the **feminization of poverty**. This term refers to the disproportionate percentage of poverty experienced by women living alone or with their children. Single mothers are particularly associated with poverty.

When head-of-household women are compared with married-couple households, the median income is $28,829 versus $69,404 (*Statistical Abstract of the United States, 2009*, Table 677). The process is cyclical—poverty contributes to

© Zak Waters/Alamy

teenage pregnancy—because teens have limited supervision and few alternatives to parenthood.

Such early childbearing interferes with educational advancement and restricts women's earning capacity, which keeps them in poverty. Their offspring are born into poverty, and the cycle begins anew.

Even if they get a job, women tend to be employed fewer hours than men, and they earn less money, even when they work full-time. Not only is discrimination in the labor force operating against women, but women also usually make their families a priority over their employment, which translates into less income. Such prioritization is based on the patriarchal family, which ensures that women stay economically dependent on men and are relegated to domestic roles. Such dependence limits the choices of many women.

Low pay for women is also related to the fact that they tend to work in occupations that pay relatively low incomes. Indeed, women's lack of economic power stems from the relative dispensability of women's labor (it is easy to replace) and how work is organized (men control positions of power). Women also live longer than men, and poverty is associated with being elderly (Lipsitz 2005).

When women move into certain occupations, such as teaching, there is a tendency in the marketplace to segregate these occupations from men's, and the result is a concentration of women in lower-paid occupations. The salaries of women in these occupational roles increase at slower rates. For example, salaries in the elementary and secondary teaching profession, which is predominately female, have not kept pace with inflation.

sexism an attitude, action, or institutional structure that subordinates or discriminates against an individual or group because of their sex.

Conflict theorists assert that men are in more powerful roles than women and use this power to dictate incomes and salaries of women and "female professions." Functionalists also note that keeping salaries low for women keeps women dependent and in child-care roles so as to keep equilibrium in the family. Hence, for both conflict and structural reasons, poverty is primarily a feminine issue. One of the consequences of being a woman is to have an increased chance of feeling economic strain throughout life.

3. **Higher risk for sexually transmitted infections.** Gender roles influence a woman's vulnerability to sexually transmitted infection and HIV, not only because women receive more bodily fluids from men, who have a greater number of partners (and are therefore more likely to be infected), but also because some women feel limited power to influence their partners to wear condoms.

4. **Negative body image.** There are more than 3,800 beauty pageants annually in the United States. The effect for many women who do not match the cultural ideal is to have a negative body image. Although women are becoming less likely to view themselves as overweight, the obsession of even females of normal weight to mirror the cultural ideal is strong (Neighbors et al. 2008).

Women also live in a society that devalues them in a larger sense. Their lives and experiences are not taken as seriously. **Sexism** is defined as an attitude, action, or institutional structure that subordinates or discriminates against individuals or groups because of their biological sex. Sexism against women reflects the tradition of male dominance and presumed male superiority in American society. It is reflected in the fact that women are rarely found in power positions in our society. In the 111th Congress, only 75 of the 435 members of the House of Representatives are women. Seventeen women and eighty-three men are serving as senators. A signal of change emerged when Democrat Nancy Pelosi became Speaker of the House in 2007. Hillary Clinton is also a strong

© Kristian Dowling/Getty Images

female role model—and has strong support, which was in evidence when she was a close second to Barack Obama in becoming the Democratic candidate for the presidency.

5. **Less marital satisfaction.** Corra et al. (2006) analyzed General Social Survey data over a thirty-year period (1972–2002), controlled for socioeconomic factors such as income and education, and found that women reported less marital satisfaction than men. Similarly, twice as many husbands as wives among 105 late-life couples (average age, 69 years) reported that they had "no disappointments in the marriage" (15 percent versus 7 percent), suggesting greater dissatisfaction among wives (Henry et al. 2005). The lower marital satisfaction of wives is attributed to power differentials in the marriage. Traditional husbands expect to be dominant, which translates into their earning an income and the expectation that the wife not only will earn an income but also will take care of the house and children. The latter expectation results in a feeling of unfairness. Analysis of other large national samples has yielded the same finding of lower marital satisfaction among wives (Faulkner et al. 2005).

Positive Consequences of Traditional Female Role Socialization We have discussed the negative consequences of being born and socialized as a woman. However, there are also decided benefits.

1. **Longer life expectancy.** Women have a longer life expectancy than men. It is not clear if their greater longevity is related to biological or to social factors.

2. **Stronger relationship focus.** Women continue to prioritize family relationships over work relationships (Stone 2007). Female family members, in contrast to male family members, are viewed as more nurturing and responsive (Monin et al. 2008).

3. **Keep relationships on track.** Because women evidence more concern for relationships, they are more likely to be motivated to keep them on track and to initiate conversation when there is a problem. In a study of 203 undergraduates, two-thirds of the women, in contrast to 60 percent of the men, reported that they were likely to start a discussion about a problem in their relationship (Knox et al. 1998). Ingram et al. (2008) also noted that, of 300,000 crisis calls to a national hotline over a five-year period, women were more likely than men to call by a two-to-one margin.

4. **Bonding with children.** Another advantage of being socialized as a woman is the potential to have a closer bond with children. In general, women tend to be more emotionally bonded with their children than men. Although the new cultural image of the father is of one who engages emotionally with his children, many fathers continue to be content for their wives to take care of their children, with the result that mothers, not fathers, become more emotionally bonded with their children.

© Uwe Umstatter/Photolibrary

Consequences of Traditional Male Role Socialization

Male socialization in American society is associated with its own set of consequences. Both the negative and positive consequences are summarized in Table 2.4. As with women, each consequence may or may not be true for a specific man.

Negative Consequences of Traditional Male Role Socialization There are several negative consequences associated with being socialized as a man in U.S. society.

1. **Identity synonymous with occupation.** Ask men who they are, and many will tell you what they do. Society tends to equate a man's identity with his occupational role. Male socialization toward greater involvement in the labor force is evident in governmental statistics.

   Maume (2006) analyzed national data on taking vacation time and found that men were much less likely to do so. They cited fear that doing so would affect their job or career performance evaluation. Women, on the other hand, were much more likely to use all of their vacation time. However, the "work equals identity" equation for men may be changing. Increasingly, as women are more present in the labor force and become co-providers, men become co-nurturers and co-homemakers. In addition, more stay-at-home dads and fathers are seeking full custody in divorce litigation. These changes challenge cultural notions of masculinity.

   That men work more and play less may translate into fewer friendships and relationships. In a study of 377 university students, 25.9 percent of the men compared to 16.7 percent of the women reported feeling a "deep sense of loneliness" (Vail-Smith et al. 2007). Similarly, Grief (2006) reported that a quarter of 386 adult men reported that they did not have enough friends. Grief also suggested some possible reasons for men having few friends—homophobia, lack of role models, fear of being vulnerable, and competition between men. McPherson et al. (2006) also found that men reported fewer confidantes than women.

2. **Limited expression of emotions.** Some men feel caught between society's expectations that they be competitive, aggressive, and unemotional and their own desire to be more cooperative, passive, and emotional. Indeed, men see themselves as less emotional and loving than women (Hill 2007), and are pressured to disavow any expression that could be interpreted as feminine (for example, be emotional). Indeed, 55 percent of the soldiers serving in Iraq or Afghanistan reported that they feared they would appear "weak" if they expressed feelings of fear or symptoms of post-traumatic stress disorder (Thompson 2008). In addition, Cordova et al. (2005) studied a sample of husbands and wives, and confirmed that the men were less able to express their emotions than the women. Notice that men are repeatedly told to "prove their manhood" (which implies not being emotional), whereas women in our culture have no dictum "to 'prove their womanhood'—the phrase itself sounds ridiculous" (Kimmel 2001, 33). Indeed, men today should realize that by shedding their traditional masculinity, they will "live longer, happier, and healthier lives, lives characterized by close and caring relationships with children, with women, and with other men" (Kimmel 2006, 187).

3. **Fear of intimacy.** Men may be socialized to withhold information about

Table 2.4
Consequences of Traditional Male Role Socialization

Negative Consequences	Positive Consequences
Identity tied to work role	Higher income and occupational status
Limited emotionality	More positive self-concept
Fear of intimacy; more lonely	Less job discrimination
Disadvantaged in getting custody	Freedom of movement; more partners to select from; more normative to initiate relationships
Shorter life	Happier marriage

themselves that encourages the development of intimacy. Giordano et al. (2005) analyzed data from the National Longitudinal Study of Adolescent Health, consisting of more than 9,000 interviews, and found that adolescent boys reported less willingness to disclose than adolescent girls.

4. **Custody disadvantages.** Courts are sometimes biased against divorced men who want custody of their children. Because divorced fathers are typically regarded as career-focused and uninvolved in child care, some are relegated to seeing their children on a limited basis, such as every other weekend or four evenings a month.

5. **Shorter life expectancy.** Men typically die five years sooner (at age 76) than women (*Statistical Abstract of the United States, 2009,* Table 100). One explanation is that the traditional male role emphasizes achievement, competition, and suppression of feelings, all of which may produce stress. Not only is stress itself harmful to physical health, but it may lead to compensatory behaviors such as smoking, alcohol and other drug abuse, and dangerous risk-taking behavior (all of which is higher in males).

In sum, the traditional male gender role is hazardous to men's physical health. However, as women have begun to experience many of the same stresses and behaviors as men, their susceptibility to stress-related diseases has increased. For example, since the 1950s, male smoking has declined whereas female smoking has increased, resulting in an increased incidence of lung cancer in women.

Benefits of Traditional Male Socialization

As a result of higher status and power in society, men tend to have a more positive self-concept and greater confidence in themselves. In a sample of 288, 48 percent of undergraduate/graduate men, in contrast to 30 percent of undergraduate women agreed that "we determine whatever happens to us, and nothing is predestined" (Dotson-Blake et al. 2008). Men also enjoy higher incomes and an easier climb up the good-old-boy corporate ladder; they are rarely stalked or targets of sexual harassment. Other benefits are the following:

1. **Freedom of movement.** Men typically have no fear of going anywhere, anytime. Their freedom of movement is unlimited. Unlike women, who are taught to fear rape and to be aware of their surroundings, walk in well-lit places, and not walk alone after dark, men are less attuned to these fears and perceptions. They can be alone in public and be anxiety-free about something ominous happening to them.

2. **Greater available pool of potential partners.** Because of the mating gradient (men marry "down" in age and education whereas women marry "up"), men tend to marry younger women so that a 35-year-old man may view women from 20 years to 40 years as possible mates. However, a woman of age 35 is more likely to view men her same age or older as potential mates. As she ages, fewer men are available; this is not so for men.

3. **Norm of initiating a relationship.** Men are advantaged because traditional norms allow men to be aggressive in initiating relationships with women. In contrast, women are less often aggressive in initiating a relationship. In a study of 1,027 undergraduates, 61.1 percent of the female respondents reported that they had not "asked a guy to go out" (Ross et al., forthcoming).

We have been discussing the respective ways in which traditional gender role socialization affects women and men. Table 2.5 summarizes twelve implications that traditional gender role socialization has for the relationships of women and men.

2.5 Changing Gender Roles

Imagine a society in which women and men each develop characteristics, lifestyles, and values that are independent of gender role stereotypes. Characteristics such as strength, independence, logical thinking, and aggressiveness are no longer associated with maleness, just as passivity, dependence, emotions, intuitiveness, and nurturance are no longer associated with femaleness. Both sexes are considered equal, and women and men may pursue the same occupational, political, and domestic roles. Some gender scholars have suggested that people in such a society would be neither feminine nor masculine but would be described as androgynous. The next subsections dis-

© Fancy/Veer/Corbis

Table 2.5
Effects of Gender Role Socialization on Relationship Choices

Women
1. A woman who is not socialized to pursue advanced education may feel pressure to stay in an unhappy relationship with someone on whom she is economically dependent.
2. Women who are socialized to play a passive role and not initiate relationships are limiting interactions that could develop into valued relationships.
3. Women who are socialized to accept that they are less valuable and important than men are less likely to seek or achieve egalitarian relationships with men.
4. Women who internalize society's standards of beauty and view their worth in terms of their age and appearance are likely to feel bad about themselves as they age. Their negative self-concept, more than their age or appearance, may interfere with their relationships.
5. Women who are socialized to accept that they are solely responsible for taking care of their parents, children, and husband are likely to experience role overload. Potentially, this could result in feelings of resentment in their relationships.
6. Women who are socialized to emphasize the importance of relationships in their lives will continue to seek relationships that are emotionally satisfying.
Men
1. Men who are socialized to define themselves more in terms of their occupational success and income and less in terms of positive individual qualities leave their self-esteem and masculinity vulnerable should they become unemployed or work in a low-status job.
2. Men who are socialized to restrict their experience and expression of emotions are denied the opportunity to discover the rewards of emotional interpersonal sharing.
3. Men who are socialized to believe it is not their role to participate in domestic activities (child rearing, food preparation, house cleaning) will not develop competencies in these life skills. Potential partners often view domestic skills as desirable qualities.
4. Heterosexual men who focus on cultural definitions of female beauty overlook potential partners who might not fit the cultural beauty ideal but who would nevertheless be good life companions.
5. Men who are socialized to view women who initiate relationships in negative ways are restricted in their relationship opportunities.
6. Men who are socialized to be in control of relationship encounters may alienate their partners, who may desire equal influence in relationships.

cuss androgyny, gender role transcendence, and gender postmodernism.

Androgyny

Androgyny typically refers to being neither male nor female but a blend of both traits. Two forms of androgyny are described here:

1. Physiological androgyny refers to intersexed individuals, discussed earlier in the chapter. The genitals are neither clearly male nor female, and there is a mixing of "female" and "male" chromosomes and hormones.
2. Behavioral androgyny refers to the blending or reversal of traditional male and female behavior, so that a biological male may be very passive, gentle, and nurturing and a biological female may be very assertive, rough, and selfish. Identifying an androgynous individual as male or female may be difficult.

Androgyny may also imply flexibility of traits; for example, an androgynous individual may be emotional in one situation, logical in another, assertive in another, and so forth. Ward (2001) classified 311 (159 male, 152 female) undergraduates at the National University of Singapore as androgynous (33.8 percent men and 16.0 percent women), feminine (11.0 percent men and 39.6 percent women), masculine (35.7 percent men and 13.9 percent women), and undifferentiated (19.5 percent men and 30.6 percent women). Peters (2005) emphasized that the blending of genders is inevitable.

Cheng (2005) found that androgynous individuals have a broad coping repertoire and are much more able to cope with stress. As evidence, Moore et al. (2005) found that androgynous individuals with Parkinson's disease were not only better able to cope with their disease but also reported having a better quality of life

androgyny a blend of traits that are stereotypically associated with masculinity and femininity.

than those with the same disease who expressed the characteristics of one gender only. Similarly, androgynous individuals reported much less likelihood of having an eating disorder (Hepp et al. 2005).

Woodhill and Samuels (2003) emphasized the need to differentiate between positive and negative androgyny. **Positive androgyny** is devoid of the negative traits associated with masculinity and femininity. Antisocial behavior has been associated with masculinity (Ma 2005) as well as traits such as aggression, hard-heartedness, indifference, selfishness, showing off, and vindictiveness. Negative aspects of femininity include being passive, submissive, temperamental, and fragile. The researchers also found that positive androgyny is associated with psychological health and well-being.

Gender Role Transcendence

Beyond the concept of androgyny is that of gender role transcendence. We associate many aspects of our world, including colors, foods, social or occupational roles, and personality traits, with either masculinity or femininity. The concept of **gender role transcendence** involves abandoning gender schema (for example, becoming "gender aschematic" [Bem 1983]) so that personality traits, social or occupational roles, and other aspects of our lives become divorced from gender categories. However, such transcendence is not equal for women and men. Although females are becoming more masculine, in part because our society values whatever is masculine, men are not becoming more feminine. Indeed, adolescent boys may be described as very gender-entrenched.

Beyond gender role transcendence is gender postmodernism.

positive androgyny a view of androgyny that is devoid of the negative traits associated with masculinity and femininity.

gender role transcendence abandoning gender frameworks and looking at phenomena independent of traditional gender categories.

Heaven help *the American-born boy with a talent for ballet.*

—*Camille Paglia*

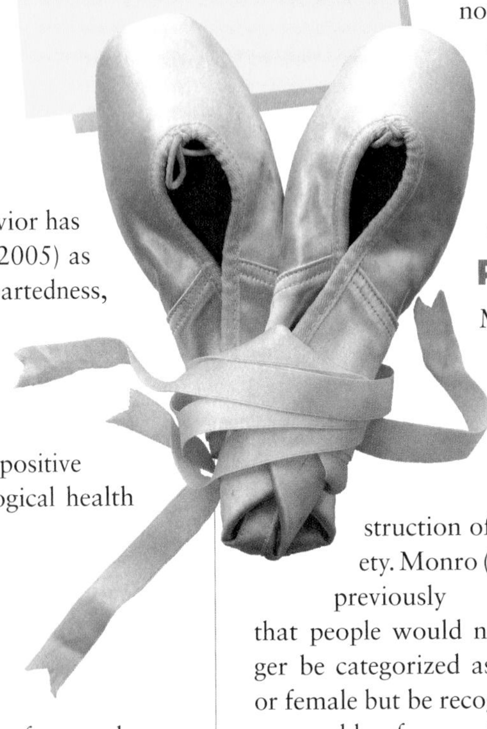

Gender Postmodernism

Mirchandani (2005) emphasized that empirical postmodernism can help us see into the future. Such a view would abandon the notion that the genders are natural and focus on the social construction of individuals in a gender-fluid society. Monro (2000) previously noted that people would no longer be categorized as male or female but be recognized as capable of many identities—"a third sex" (p. 37). A new conceptualization of "trans" people calls for new social structures, "based on the principles of equality, diversity and the right to self determination" (p. 42). No longer would our society telegraph transphobia but embrace pluralization "as an indication of social evolution, allowing greater choice and means of self-expression concerning gender" (p. 42).

© Image 100/Jupiterimages / © Photodisc/Getty Images

PEOPLE WOULD NO LONGER BE CATEGORIZED AS MALE OR FEMALE BUT BE RECOGNIZED AS CAPABLE OF MANY IDENTITIES.

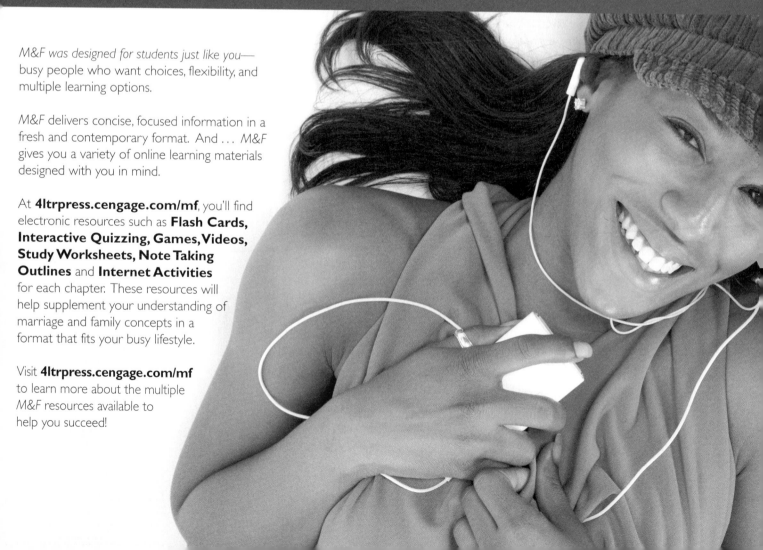

Singlehood, Hanging Out, Hooking Up, and Cohabitation

SECTIONS

Singlehood and the years before marriage are thought of as some of "the best years of our lives." To keep the "best years of life" (singlehood) going, individuals are delaying marriage. Concerns about launching one's career, paying off debts, and enjoying the freedom of singlehood (which implies avoiding expectations of either a spouse or a child) have propelled a pattern adopted by today's youth to put off marriage in a ten-year float from the late teens to the late twenties.

Young American adults are not alone: individuals in France, Germany, and Italy are engaging in a similar pattern of delaying marriage. In the meantime, the process of courtship has evolved, with various labels and patterns, including "hanging out" (undergraduates rarely use the term *dating*), "hooking up" (the new term for "one-night stand"), and "pairing off," which may include cohabitation as a prelude to marriage. We begin with examining singlehood versus marriage.

3.1 Singlehood

In this section, we discuss how social movements have increased the acceptance of singlehood, the various categories of single people, the choice to be permanently unmarried, the human immunodeficiency virus (HIV) infection risk associated with this choice, and the fact that more people are delaying marriage.

Individuals Are Delaying Marriage Longer

American adults are more likely to live alone today than in the past. The proportion of households consisting of one person living alone increased from 30.1 percent in 2005 to 31.1 percent in 2007 (*Statistical Abstract of the United States: 2009*. Table 61). In part, this is due to the fact that American women and men are staying single longer (see Figure 3.1).

Figure 3.1
Median Age at First Marriage in America in Selected Years, by Sex

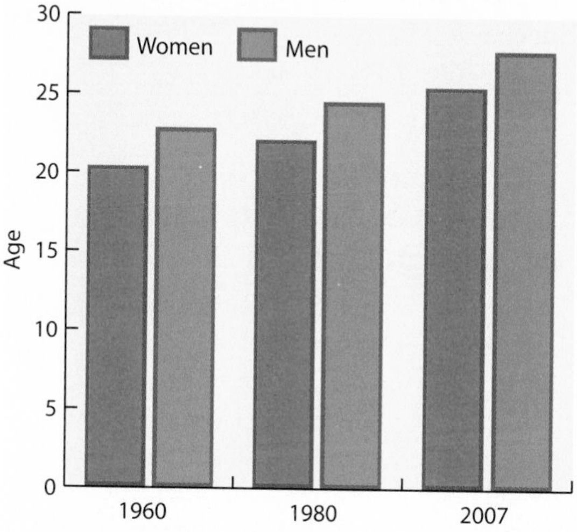

Source: U.S. Census Bureau. 2006. *America's Families and Living Arrangements: 2006.* Table MS-2 Estimated Median Age at First Marriage, by Sex (updated for 2007). http://www.census.gov.

Whether people today are embracing singlehood forever or just delaying marriage is unknown. We won't know until they reach their seventies, which is the age by which more than 95 percent typically marry. We do know, however, that people are opting for singlehood longer than in the past (Vanderkam 2006). Table 3.1 lists the standard reasons people give for remaining single. The primary advantage of remaining single is freedom and control over one's life. Once a decision has been made to involve another in one's life, one's choices become vulnerable to the influence of

Table 3.1
Reasons to Remain Single

Benefits of Singlehood	Limitations of Marriage
Freedom to do as one wishes	Restricted by spouse or children
Variety of lovers	One sexual partner
Spontaneous lifestyle	Routine, predictable lifestyle
Close friends of both sexes	Pressure to avoid close other-sex friendships
Responsible for one person only	Responsible for spouse and children
Spend money as one wishes	Expenditures influenced by needs of spouse and children
Freedom to move as career dictates	Restrictions on career mobility
Avoid being controlled by spouse	Potential to be controlled by spouse
Avoid emotional and financial stress of divorce	Possibility of divorce

that other person. The person who chooses to remain single may view the needs and influence of another person as things to avoid.

Some people do not set out to be single but simply remain in singlehood longer than they anticipated—and discover that they like it. Meredith Kennedy (2008) is a never-married veterinarian who has found that singlehood works very well for her. She writes:

> As a little girl, I always assumed I'd grow up to be swept off my feet, get married, and live happily ever after. Then I hit my 20s, became acquainted with reality, and discovered that I had a lot of growing up to do. I'm still working on it now, in my late 30s.
>
> I've gotten a lot out of my relationships with men over the years, some serious and some not so serious, and they've all left their impression on me. But gradually I've moved away from considering myself "between boyfriends" to getting very comfortable with being alone, and finding myself good company. The thought of remaining single for the rest of my life doesn't bother me, and the freedom that comes with it is very precious. I've worked and traveled all over the world, and my schedule is my own. I don't think this could have come about with the responsibilities of marriage and a family, and the time and space I have as a single woman have allowed me to really explore who I am in this life.
>
> It's not always easy to explain why I'm single in a culture that expects women to get married and to have children, but the freedom and independence I have allow me to lead a unique and interesting life.

Social Movements and the Acceptance of Singlehood

Though more than 95 percent of American adults eventually marry (*Statistical Abstract of the United States, 2009*, Table 56) and only 5 percent of a sample of 1,293 adolescents predicted that they would never marry (Manning et al. 2007), more people are delaying marriage and enjoying singlehood. The acceptance of singlehood as a lifestyle can be attributed to social movements—the sexual revolution, the women's movement, and the gay liberation movement.

The sexual revolution involved openness about sexuality and permitted intercourse outside the context of marriage. No longer did people feel compelled to wait until marriage for involvement in a sexual relationship. Hence, the sequence changed from dating, love, maybe intercourse with a future spouse, and then marriage and parenthood to "hanging out," "hooking up" with numerous partners, maybe living together (in one or more relationships), marriage, and children.

The women's movement emphasized equality in education, employment, and income for women. As a result, rather than get married and depend on a husband for income, women earned higher degrees, sought career opportunities, and earned their own income. This economic independence brought with it independence of choice. Women could afford to remain single or to leave an unfulfilling or abusive relationship.

The gay liberation movement, with its push for recognition of same-sex marriage, has increased the visibility of gay people and relationships. Though some gay people still marry heterosexuals to provide a traditional social front, the gay liberation movement has provided support for a lifestyle consistent with one's sexual orientation. This includes rejecting traditional heterosexual marriage. Today, some gay pair-bonded couples regard themselves as married even though they are not legally wed. Some gay couples have formal wedding ceremonies in which they exchange rings and vows of love and commitment.

In effect, there is a new wave of youth who feel that their commitment is to themselves in early adulthood and to marriage in their late twenties and thirties, if at all. The increased acceptance of singlehood translates into staying in school or getting a job, establishing oneself in a career, and becoming economically and emotionally independent from one's parents. The old pattern was to leap from high school into marriage. The new pattern of these Generaton Yers (discussed in Chapter 1) is to wait until after college, become established in a career, and enjoy themselves. A few (less than 5 percent of American adults) opt for remaining single forever.

The value young adults attach to singlehood may vary by race. Brewster (2006) interviewed forty African American men 18 to 25 years of age in New York City. These men dated numerous women and some maintained serious relationships with multiple women simultaneously. These men "have redefined family for themselves, and marriage is not included in the definition" (p. 32).

" *Marriage is only one of many acceptable family forms.*

© Creatas/Photolibrary

Alternatives to Marriage Project

According to the mission statement identified on the website of the Alternatives to Marriage Project (http://www.unmarried.org/aboutus.php), the emphasis of the Alternatives to Marriage Project (ATMP) is to advocate "for equality and fairness for unmarried people, including people who are single, choose not to marry, cannot marry, or live together before marriage." The nonprofit organization is not against marriage but provides support and information for the unmarried and "fights discrimination on the basis of marital status. . . . We believe that marriage is only one of many acceptable family forms, and that society should recognize and support healthy relationships in all their diversity."

The Alternatives to Marriage Project is open to everyone, "including singles, couples, married people, individuals in relationships with more than two people, and people of all genders and sexual orientations. We welcome our married supporters, who are among the many friends, relatives, and allies of unmarried people."

Legal Blurring of the Married and Unmarried

The legal distinction between married and unmarried couples is blurring. Whether it is called the deregulation of marriage or the deinstitutionalization of marriage, the result is the same—more of the privileges previously reserved for the married are now available to unmarried and/or same-sex couples. As noted earlier, domestic partnership conveys rights and privileges (for example, health benefits for a partner) previously available only to married people.

MY STRONG OBJECTION IS TO THE NOTION THAT THERE'S ONE KIND OF RELATIONSHIP THAT'S BEST FOR EVERYONE.

— *Judith Stacey*

3.2 Categories of Singles

The term **singlehood** is most often associated with young unmarried individuals. However, there are three categories of single people: the never-married, the divorced, and the widowed. Figure 3.2 shows the distribution of the American adult population by relationship status.

Never-Married Singles

Kevin Eubanks (former *Tonight Show* music director), Oprah Winfrey, Diane Keaton, and Drew Carey are examples of heterosexuals who have never married. Nevertheless, it is rare for people to remain single their entire life. One reason is stigma. DePaulo (2006) asked 950 undergraduate college students to describe single people. In contrast to married people, who were described as "happy, loving, and stable," single people were described as "lonely, unhappy, and insecure."

singlehood being unmarried.

Though the never-married singles consist mostly of those who want to marry someday, these individuals are increasingly comfortable delaying marriage to pursue educational and career opportunities. Others, such as African American women who have never married, note a lack of potential marriage partners. Educated black women report a particularly difficult time finding eligible men from which to choose.

There is a great racial divide in terms of remaining single. In her article, "Marriage Is for White People," Jones (2006, B5) notes:

> Sex, love, and childbearing have become *a la carte* choices rather than a package deal that comes with marriage. Moreover, in an era of brothers on the "down low," the spread of sexually transmitted diseases and the decline of the stable blue-collar jobs that black men used to hold, linking one's fate to a man makes marriage a risky business for a black woman.

In spite of the viability of singlehood as a lifestyle, stereotypes remain, as never-married people are viewed as desperate or swingers. Indeed, a cultural norm still disapproves of singlehood as a lifelong lifestyle.

Jessica Donn (2005) emphasized that, because mature "adulthood" implies that one is married, single people are left to negotiate a positive identity outside of marriage. She studied the subjective well-being of 171 self-identified heterosexual, never-married singles (40 men and 131 women), aged 35 to 45 years, who were not currently living with a romantic partner. Results revealed that women participants reported higher life satisfaction and positive affect than men. Donn hypothesized that having social connections and close friendships (greater frequency of social contact, more close friends, someone to turn to in times of dis-

Figure 3.2
U.S. Adult Population by Relationship Status
N=222,557,000

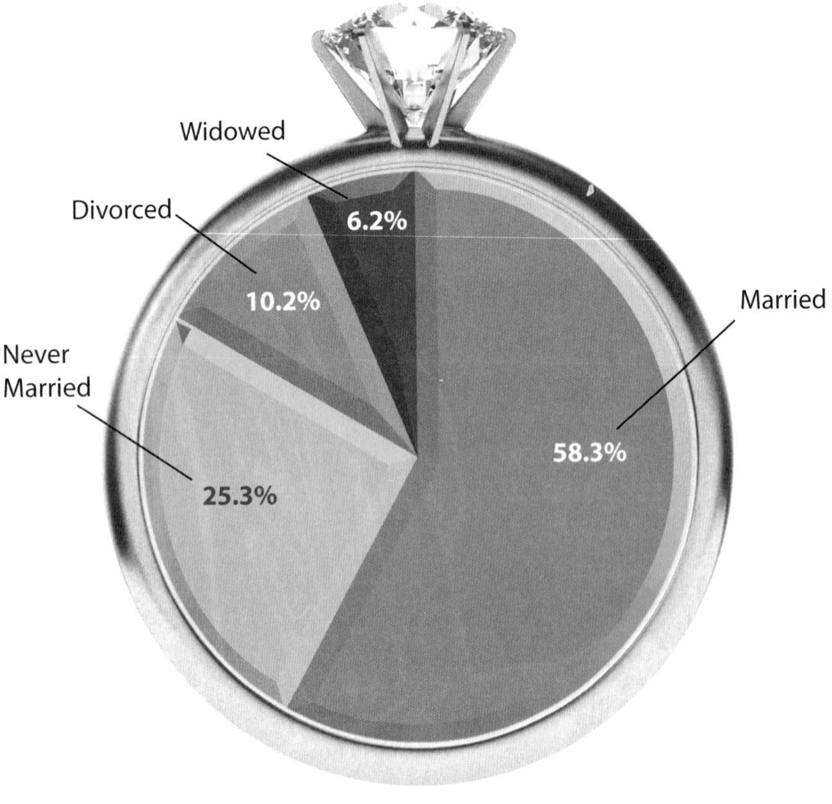

Source: *Statistical Abstract of the United States, 2009.* 128th ed. Washington, DC: U.S. Bureau of the Census, Table 56.

tress, and greater reciprocity with a confidant) was the variable associated with higher subjective well-being for both women and men (but women evidenced greater connectedness). Other findings included that relationships, career, older age, and avoiding thoughts of old age contributed to a sense of well-being for never-married men. For women, financial security, relationships, achievement, and control over their environment contributed to a sense of well-being.

Sharp and Ganong (2007) interviewed thirty-two white, never-married college-educated women ages 28 to 34 who revealed a sense of uncertainty about their lives. One reported:

> Like all or nothing, it is either—you assume [your life] is either going to be great or horrible. You just have to get better at accepting the fact that you don't know, it is probably somewhere in between and you are just going to have to wait and see.

Although some were despondent that they would never meet a man and have children, others (particularly when they became older) reminded themselves of the advantages of being single: freedom, financial independence, ability to travel, and so on. However, the overriding theme of these respondents was that they were "running out of time to marry and to have children." The researchers emphasized the enormous cultural expectation to follow age-graded life transitions and to stay on time and on course. People outside the norm struggle with managing their difference, with varying degrees of success. The chart on the next page details some issues involved with being a single woman.

When single and married people are compared, married people report being happier. Lucas et al. (2003) analyzed data from a fifteen-year longitudinal study of more than 24,000 individuals and found that married people were happier than single people and hypothesized that marriage may draw people who are already more satisfied than average (for example, unhappy people may be less in demand as marriage partners). Wienke and Hill (2009) also compared singles with marrieds and cohabitants (both heterosexual and homosexual) and found that singles were less happy regardless of sexual orientation.

Divorced Singles

Divorced people are also regarded as single. For many divorced people, the return to singlehood is not an easy transition. Knox and Corte (2007) studied a sample of people going through divorce, many of whom reported their unhappiness and a desire to reunite with their partner.

Most divorced individuals have children. Most of these are single mothers, but increasingly, single fathers have sole custody of their children. Most single parents prioritize their roles of "single parent" as a parent first and as a single adult second. One newly divorced single

A NEVER-MARRIED SINGLE WOMAN'S VIEW OF SINGLEHOOD

A never-married woman, 40 years of age, spoke to the marriage and family class about her experience as a single woman. The following is from the outline she developed and the points she made about each topic.

Stereotypes about Never-Married Women

Various assumptions are made about the never-married woman and why she is single. These include the following:

Unattractive—She's either overweight or homely, or else she would have a man.

Lesbian—She has no real interest in men and marriage because she is homosexual.

Workaholic—She's career-driven and doesn't make time for relationships.

Poor interpersonal skills—She has no social skills, and she embarrasses men.

History of abuse—She has been turned off to men by the sexual abuse by, for example, her father, a relative, or a date.

Negative previous relationships—She's been rejected again and again and can't hold a man.

Man-hater—Deep down, she hates men.

Frigid—She hates sex and avoids men and intimacy.

Promiscuous—She is indiscriminate in her sexuality so that no man respects or wants her.

Too picky—She always finds something wrong with each partner and is never satisfied.

Too weird—She would win the Miss Weird contest, and no man wants her.

Positive Aspects of Being Single

1. Freedom to define self in reference to own accomplishments, not in terms of attachments (for example, spouse).
2. Freedom to pursue own personal and career goals and advance without the time restrictions posed by a spouse and children.
3. Freedom to come and go as you please and to do what you want, when you want.
4. Freedom to establish relationships with members of both sexes at desired level of intensity.
5. Freedom to travel and explore new cultures, ideas, values.

Negative Aspects of Being Single

1. Increased extended-family responsibilities. The unmarried sibling is assumed to have the time to care for elderly parents.
2. Increased job expectations. The single employee does not have marital or family obligations and consequently can be expected to work at night, on weekends, and on holidays.
3. Isolation. Too much time alone does not allow others to give feedback such as "Are you drinking too much?" "Have you had a checkup lately?" or "Are you working too much?"
4. Decreased privacy. Others assume the single person is always at home and always available. They may call late at night or drop in whenever they feel like it. They tend to ask personal questions freely.
5. Less safety. A single woman living alone is more vulnerable than a married woman with a man in the house.
6. Feeling different. Many work-related events are for couples, husbands, and wives. A single woman sticks out.
7. Lower income. Single women have much lower incomes than married couples.
8. Less psychological intimacy. The single woman does not have an emotionally intimate partner at the end of the day.
9. Negotiation skills lie dormant. Because single people do not negotiate issues with someone on a regular basis, they may become deficient in compromise and negotiation skills.
10. Patterns become entrenched. Because no other person is around to express preferences, the single person may establish a very repetitive lifestyle.

Maximizing One's Life as a Single Person

1. Frank discussion. Talk with parents about your commitment to and enjoyment of the single lifestyle and request that they drop marriage references. Talk with siblings about joint responsibility for aging parents and your willingness to do your part. Talk with employers about spreading workload among all workers, not just those who are unmarried and child-free.
2. Relationships. Develop and nurture close relationships with parents, siblings, extended family, and friends to have a strong and continuing support system.
3. Participate in social activities. Go to social events with or without a friend. Avoid becoming a social isolate.
4. Be cautious. Be selective in sharing personal information such as your name, address, and phone number.
5. Money. Pursue education to maximize income; set up a retirement plan.
6. Health. Exercise, have regular checkups, and eat healthy food. Take care of yourself.

THERE WERE 13.2 MILLION DIVORCED FEMALES AND 9.6 MILLION DIVORCED MALES IN THE UNITED STATES IN 2007.

—*Statistical Abstract of the United States, 2009,* Table 55

parent said, "My kids come first. I don't have time for anything else now." We discuss the topic of single parenthood in greater detail in Chapter 10.

Divorced people tend to die earlier than married people. On the basis of a study of 44,000 deaths, Hemstrom (1996) observed that, "on the whole, marriage protects both men and women from the higher mortality rates experienced by unmarried groups" (p. 376). One explanation is the protective aspect of marriage. "The protection against diseases and mortality that marriage provides may take the form of easier access to social support, social control, and integration, which leads to risk avoidance, healthier lifestyles, and reduced vulnerability" (p. 375). Married people also look out for the health of the other. Spouses often prod each other to go to the doctor and help each other remember things, like taking their medicine. Single people often have no one in their life to nudge them toward regular health maintenance.

Widowed Singles

Although divorced people often choose to return to singlehood, widowed people are forced into singlehood. The stereotype of the widow and widower is utter loneliness, but Ha (2008) compared widowed people with married couples and though the former are less likely to have a confidant, they received greater support from children, friends, and relatives. Hence, widowed people may have a broader array of relationships than married people.

Nevertheless, widowhood is a difficult time for most individuals. Widowed men, in contrast to widowed women, may feel particularly disconnected from their children (Kalmijn 2007).

In a study conducted by the American Association of Retired Persons (AARP), most widows were between the ages of 40 and 69. Almost a third (31 percent) were in an exclusive relationship and another 32 percent were dating nonexclusively. Of the remaining 37 percent, only 13 percent reported that they were actively looking; 10 percent said that they had no desire to date; and the rest said they were open to meeting someone but not obsessive about it (Mahoney 2006).

3.3 Ways of Finding a Partner

One of the unique qualities of the college or university environment is that it provides a context in which to meet thousands of potential partners of similar age, education, and social class. This context will likely never recur following graduation. Although people often meet through friends or on their own in school, work, or recreation contexts, an increasing number are open to a range of alternatives, from hanging out to the Internet.

Hanging Out

The term *hanging out* has made its way into the professional literature (Harcourt 2005; Thomas 2005). **Hanging out**, also referred to as getting together, refers to going out in groups where the agenda is to meet others and have fun. The individuals may watch television, rent a video, go to a club or party, and/or eat out. Hanging out may be considered "testing the waters," as a possible prelude to more serious sexual involvement and commitment (Luff and Hoffman 2006). Of 1,319 undergraduates, 93.4 percent reported that "hanging out for me is basically about meeting people and having fun" (Knox and Zusman 2009). Hanging out may occur in group settings such as at a bar, a sorority or fraternity party, or a small gathering of friends that keeps expanding. Friends may introduce individuals, or they may meet someone "cold," as in initiating a conversation. There is usually no agenda beyond meeting and having fun. Of the 1,319 respondents, only 4 percent said that hanging out was about beginning a relationship that may lead to marriage (Knox and Zusman 2009).

hanging out refers to going out in groups where the agenda is to meet others and have fun.

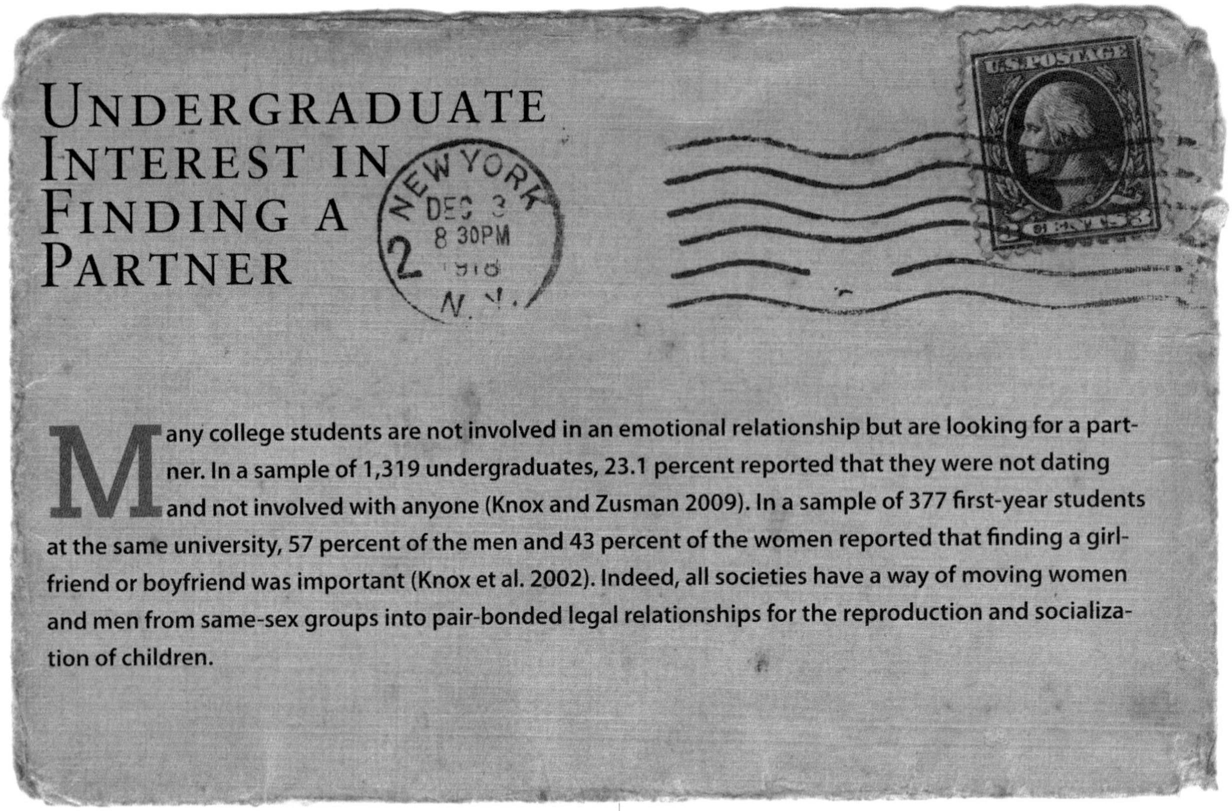

Undergraduate Interest in Finding a Partner

Many college students are not involved in an emotional relationship but are looking for a partner. In a sample of 1,319 undergraduates, 23.1 percent reported that they were not dating and not involved with anyone (Knox and Zusman 2009). In a sample of 377 first-year students at the same university, 57 percent of the men and 43 percent of the women reported that finding a girlfriend or boyfriend was important (Knox et al. 2002). Indeed, all societies have a way of moving women and men from same-sex groups into pair-bonded legal relationships for the reproduction and socialization of children.

Hooking Up

Hooking up is also a term that has entered the social science literature and has become the focus of research. **Hooking up** is defined as a one-time sexual encounter in which there are generally no expectations of seeing one another again. The nature of the sexual expression may be making out, oral sex, and/or sexual intercourse. The term is also used to denote getting together periodically for a sexual encounter, with no strings attached (Luff and Hoffman 2006). Renshaw (2005) wrote his dissertation on hooking up and noted that hooking up involves playful or nonserious interaction where one's body and the context of alcohol move the encounter to a sexual ending.

In a sample of 1,319 undergraduates, 29 percent reported that they had "hooked up" (had oral sex or sexual intercourse) the first time they met someone. As a hypothetical question, when respondents were asked if they were to "hook up" with the right person, and felt good about the interaction, would they have oral or sexual intercourse with a person they just met, 32.4 percent responded "yes" (Knox and Zusman 2009).

hooking up a one-time sexual encounter in which there are generally no expectations of seeing one another again.

Bogle (2008) interviewed fifty-seven college students and alumni at two universities in the eastern United States about their experiences with dating and sex. She found that hooking up had become the *primary* means for heterosexuals to get together on campus. About half (47 percent) of hookups start at a party and involve alcohol, with men averaging five drinks and women three drinks (England and Thomas 2006). The sexual behaviors that were reported to occur during a hookup included kissing and nongenital touching (34 percent), hand stimulation of genitals (19 percent), oral sex (22 percent), and intercourse (23 percent) (England and Thomas 2006).

Researchers (Bogle 2008; Eshbaugh and Gute 2008) note that, although hooking up may be an exciting sexual adventure, it is fraught with feelings of regret. Some of the women in their studies were particularly disheartened to discover that hooking up usually did not result in the development of a relationship that went beyond a one-night encounter. Eshbaugh and Gute (2008) examined hooking up as a predictor of sexual regret in 152 sexually active college women and identified two sexual behaviors that were particularly predictive of participants' regret: (a) engaging in sexual intercourse with someone once and only once, and (b) engaging in intercourse with someone known for less than twenty-four hours. Noncoital hookups (performing and receiving oral sex) were not significantly related to regret. Indeed,

Bogle (2002; 2008) noted three outcomes of hooking up. In the first, previously described, nothing results from the first-night sexual encounter. In the second, the college students will repeatedly hook up with each other on subsequent occasions of "hanging out." However, a low level of commitment characterizes this type of relationship, in that each person is still open to hooking up with someone else. A third outcome of hooking up, and the least likely, is that the two people begin going out or spending time together in an exclusive relationship. Hence, hooking up is most often a sexual adventure that rarely results in the development of a relationship. Stepp (2007) interviewed two groups of high school and one group of college students and discovered the same outcome in most relationships involving hooking up—they rarely end with the couple becoming a monogamous pair. However, there are exceptions. When 1,319 undergraduates were presented with the statement "People who 'hook up' and have sex the first night don't end up in a stable relationship," 16.7 percent disagreed (Knox and Zusman 2009).

Renshaw (2005) noted that hooking up is becoming normative and replacing contemporary patterns of dating. He suggested that it is "shrouded in deception," "contains individual health risks associated with sex," and "may also threaten traditional conceptions of marriage and family."

The Internet—Meeting Online: Upside and Downside

"We could never let anyone know how we really met," remarked a couple who had met on the Internet. "People who meet online are stigmatized as desperate losers." As more individuals use the Internet, such stigmatization is changing. Almost three-fourths (74 percent) of single Americans have used the Internet to find a romantic partner, and 15 percent report that they know someone who met their spouse or significant other online (Madden and Lenhart 2006). Yahoo.com claims over 375 million visits to their online dating site each month (Whitty 2007).

Online meetings will continue to increase as people delay getting married and move beyond contexts where hundreds or thousands of potential partners are available (the undergraduate coed classroom filled with same-age potential mates is rarely equaled in the workplace after college). The profiles that individuals construct or provide for others to view reflect impression management, presenting an image that is perceived to be what the target audience wants. In this regard, men tend to emphasize their status characteristics (for example, income, education, career), whereas women tend to emphasize their youth, trim body, and beauty (Spitzberg and Cupach 2007).

The attraction of online dating is its efficiency. It takes time and effort to meet someone at a coffee shop for an hour, only to discover that the person has habits (for example, does or doesn't smoke) or values (too religious or too agnostic) that would eliminate them as a potential partner. Match.com features 8 million profiles that can be scanned at one's leisure. For noncollege people who are busy in their job or career, the Internet offers the chance to meet someone outside their immediate social circle. "There are only six guys in my office," noted one Internet user. "Four are married and the other two are alcoholics. I don't go to church and don't like bars so the Internet has become my guy store."

More than 1,000 Internet sites are designed for the purpose of meeting a partner online (Jerin and Dolinsky 2007). More than $500 million are spent on these sites (Stringfield 2008). Right Mate at Heartchoice.com offers not only a way to meet others but a free "Right Mate Checkup" to evaluate whether the person is right for you. Some websites exist to target specific interests such as black singles (BlackPlanet.com), Jewish singles (Jdate.com), and gay people (Gay.com). In one study on online dating, women received an average of fifty-five replies compared to men who reported receiving thirty-nine replies. Younger women (average age of 35), attractive women, and those who wrote longer profiles were more successful in generating replies (Whitty 2007).

Although hooking up may be an exciting sexual adventure, it is fraught with feelings of regret.

Pros:
• Highly efficient
• Develop a relationship without visual distraction
• Avoid crowded, loud, uncomfortable locations, like bars
• The opportunity to try on new identities
ONLINE MEETING
Cons:
• Ease of deception
• Relationship involvement escalates too quickly.
• Inability to assess "chemistry" through the computer
• A lot of competition

© Comstock Images/Jupiterimages

In addition to the efficiency of meeting someone online, another advantage is the opportunity to try on new identities. Defining identity as an understanding of who one is, Yurchisin et al. (2005) interviewed individuals who had used an Internet dating service in the past year and found that some posted a profile of who they wanted to be rather than who they were. For example, a respondent reported wanting to be more athletic so she checked various recreational activities she didn't currently engage in but wanted to. The Internet is also a place for people to "try out" a gay identity if they are very uncomfortable doing so in real life, around people they know. Of the respondents in the previously mentioned Internet study, 40 percent reported that they had lied online. Men tend to lie about their economic status, and women tend to lie about their physical appearance, weight, or age. Other lies include marital status (Gibbs et al. 2006). Kassem Saleh was married yet maintained fifty simultaneous online relationships with other women. He allegedly wrote intoxicating love letters, many of which were cut and paste jobs to various women. He made marriage proposals to several and some bought wedding gowns in anticipation of the wedding (Albright 2007). Although Saleh is an "Internet guy," it is important to keep in mind that people not on the Internet may also be very deceptive and cunning. To suggest that the Internet is the only place where deceivers lurk is to turn a blind eye to those who meet through traditional channels.

McGinty (2006) noted the importance of using Internet dating sites safely, including not giving out home or business phone numbers or addresses, always meeting the person in one's own town with a friend, and not posting photos that are "too revealing," as these can be copied and posted elsewhere. She recommends being open and honest: "Let them know who you are and who you are looking for," she suggests.

The Internet may also be used to find out information about a partner. Argali.com can be used to find out where the Internet mystery person lives, Zabasearch.com to learn how long the person has lived there, and Zoominfo.com to discover where the person works. The person's birth date can be found at Birthdatabase.com. Women might want to see if any red flags have been posted on the Internet at Dontdatehimgirl.com.

For individuals who learn about each other online, what is it like to finally meet? First, they tend to meet each other relatively quickly; often, after three e-mail exchanges, they will move to a phone call and set up a time to meet during that phone call (McKenna 2007). When they do meet, Baker (2007) is clear: "If they have presented themselves accurately and honestly online, they encounter few or minor surprises at the first meeting offline or later on in further encounters" (p. 108). About 7 percent end up marrying someone they met online (Albright 2007).

Internet Partners: The Downside Although most Internet exchanges or relationships are positive, it is important to be cautious of meeting someone online. See the website WildXAngel.com for horror stories of online dating. Although the Internet is a good place to meet new people, it also allows someone you rejected or an old lover to monitor your online behavior. Most sites note when you have been online last, so if you reject someone online by saying, "I'm really not ready for a relationship," that same person can log on and see that you are still looking. Some individuals become obsessed with a person they meet online and turn into a cyberstalker when rejected. This will be discussed in Chapter 12.

Video Chatting

Video chatting moves beyond communicating by typewritten words and allows potential partners to see each other while chatting online. One of the largest online

© Sven Hagolani/zefa/Corbis

communities with downloadable software is iSpQ ("Eye Speak," http://www.ispq.com/), which enables people to visually meet with others all over the world. Hodge (2003) noted, "Unlike conventional chat rooms, iSpQ does not have a running dialogue or conversation for anyone to view. Video chatting allows users to have personal or private conversations with another user. Hence an individual cannot only write information but see the person with whom he or she is interacting online." Half of the respondents in Hodge's study of video chat users reported "meeting people and having fun" as their motivation for video chatting.

Speed-Dating: The Eight-Minute Date

Dating innovations that involve the concept of speed include the eight-minute date. The website www.8minutedating.com/ identifies these "Eight-Minute Dating Events" throughout the country, where a person has eight one-on-one dates that last eight minutes each. If both parties are interested in seeing each other again, the organizer provides contact information so that the individuals can set up another date. Speed-dating is cost-effective because it allows daters to meet face-to-face without burning up a whole evening. Adams et al. (2008) interviewed participants who had experienced speed-dating to assess how they conceptualized the event. They found that women were more likely to view speed-dating as an investment of time and energy to find someone (58 percent versus 25 percent), whereas men were more likely to see the event as one of exploration (for example, see how flexible a person was) (75 percent versus 17 percent). Wilson et al. (2006) collected data on nineteen young men who had three-minute social exchanges with nineteen young women and found that those partners who wanted to see each other again had more in common than those who did not want to see each other again. Common interests were assessed using the compatibility quotient (CQ).

International Dating

Go to google.com and type in "international brides," and you will see an array of sites dedicated to finding foreign women for Americans. Not listed is Ivan Thompson, who specializes in finding Mexican women for his American clients. As documented in the movie *Cowboy del Amor* (Ohayon 2005), Ivan takes males (one at a time) to Mexico (Torreón is his favorite place). For $3,000, Ivan places an ad in a local newspaper for a young (age 20 to 35), trim (less than 130 pounds), single woman "interested in meeting an American male for romance and eventual marriage" and waits in a hotel for the phone to ring. They then meet and interview "candidates" in the hotel lobby. Ivan says his work is done when his client finds a woman he likes.

Advertising for a partner is not unusual. Jagger (2005) conducted a content analysis of 1,094 advertisements and found that young men and older women were the most likely to advertise for a partner. The researcher also noted a trend in women seeking younger men.

3.4 Functions of Involvement with a Partner

Meeting and becoming involved with someone has at least seven functions:

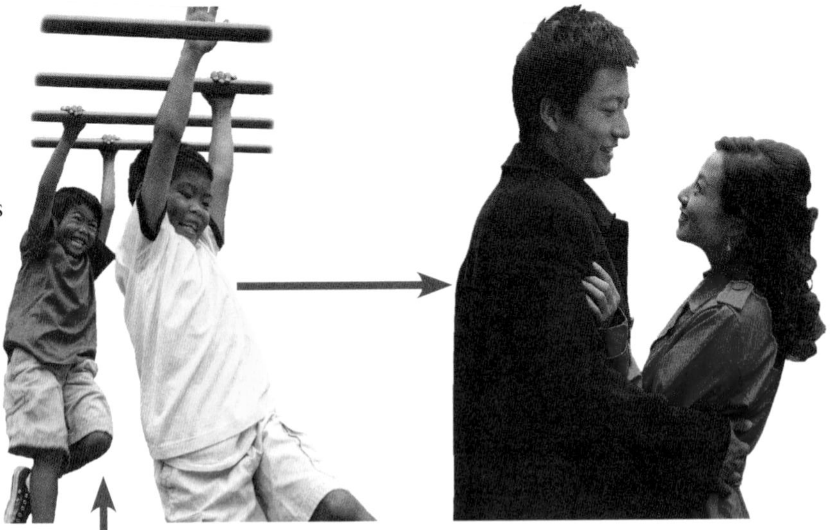

1. **Confirmation of a social self.** In Chapter 1, we noted that symbolic interactionists emphasize the development of the self. Parents are usually the first social mirrors in which we see ourselves and receive feedback about who we are; new partners continue the process. When we are hanging out with a person, we are continually trying to assess how that person sees us: Does the person like me? Will the person want to be with me again? When the person gives us positive feedback through speech and gesture, we feel good and tend to have positive self-perception. Hanging out provides a context for the confirmation of a strong self-concept in terms of how we perceive our effect on other people.

2. **Recreation.** The focus of hanging out and pairing off is fun. Reality television programs such as *Next, Blind Date, Elimidate,* and *The Bachelor* always use recreational activities as a context to help participants interact; being a fun person seems to be a criterion for being selected. The couples may make only small talk and learn very little about each other—what seems important is not that they have common interests, values, or goals but that they "have fun."

3. **Companionship, intimacy, and sex.** Beyond fun, major motivations for finding a new person and pairing off are companionship, intimacy, and sex. The impersonal environment of a large university makes a secure relationship very appealing. "My last two years have been the happiest ever," remarked a senior in interior design. "But it's because of the involvement with my partner. During my freshman and sophomore years, I felt alone. Now I feel loved, needed, and secure."

4. **Anticipatory socialization.** Before puberty, boys and girls interact primarily with same-sex peers. A fifth grader may be laughed at if showing an interest in someone of the other sex. Even when boy-girl interaction becomes the norm at puberty, neither sex may know what is expected of the other. Meeting a new partner and hanging out provides the first opportunity for individuals to learn how to interact with other-sex partners. Though the manifest function of hanging out is to teach partners how to negotiate differences (for example, how much sex and how soon), the latent function is to help them learn the skills necessary to maintain long-term relationships (empathy, communication, and negotiation, for example). In effect, pairing off involves a form of socialization that anticipates a more permanent union in one's life. Individuals may also try out different role patterns, like dominance or passivity, and try to assess the feel and comfort level of each.

5. **Status achievement.** Being involved with someone is usually associated with more status than being unattached and alone. Some may seek such involvement because of the associated higher status. Others may become involved for peer acceptance and conformity to gender roles, not for emotional reasons. Though the practice is becoming less common, some gay people may pair off with someone of the other sex so as to provide a heterosexual cover for their sexual orientation.

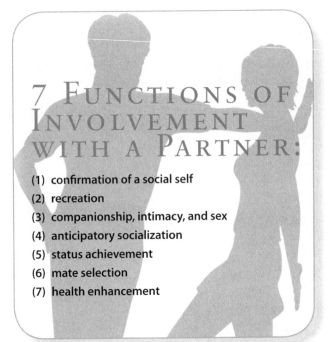

7 Functions of Involvement with a Partner:

(1) confirmation of a social self
(2) recreation
(3) companionship, intimacy, and sex
(4) anticipatory socialization
(5) status achievement
(6) mate selection
(7) health enhancement

© Blend Images/Jupiterimages / © Alex Mares-Manton/Photolibrary

© Robert Adrian Hillman/iStockphoto.com

6. **Mate selection.** Finally, pairing off may eventually lead to marriage, which remains a major goal in our society (Pryor et al. 2008). Selecting a mate has become big business. B. Dalton, one of the largest bookstore chains in the United States, carries about 200 titles on relationships, about 50 of which are specifically geared toward finding a mate.
7. **Health enhancement.** In Chapter 1, we reviewed the benefits of marriage. Not the least of marital benefits is health. Specifically, there is a direct relationship between getting married for the first time and the cessation of smoking (Weden and Kimbro 2007).

3.5 Cohabitation

Cohabitation, also known as **living together**, is becoming more normative. In a sample of 1,293 adolescents, 75 percent regarded living together as an option before marriage (Manning et al. 2007). Of 1,319 undergraduates at a large southeastern university, 71.9 percent reported that they would live with a partner they were not married to, and 16 percent were or had already done so (Knox and Zusman 2009). Reasons for the increase in cohabitation include career or educational commitments; increased tolerance of society, parents, and peers; improved birth control technology; desire for a stable emotional and sexual relationship without legal ties; and greater disregard for convention. Twenge (2006) surveyed university students and found that 62 percent paid little attention to social conventions. Cohabitants also regard living together as a vaccination against divorce. Later, we will review studies emphasizing that this hope is more often an illusion.

Almost 60 percent of women (59 percent) have lived with a partner before age 24. Most of these relationships were short-lived, with 20 percent resulting in marriage (Schoen et al. 2007). People who live together before marriage are more likely to be high school dropouts than college graduates (60 percent versus 37 percent), to have been married, to be less religious or traditional, and to be supportive of egalitarian gender roles (Baxter 2005). Indeed, compared to married people, cohabitant couples are more likely to include men who perform household chores (Davis et al. 2007). Cohabitants are also less likely to be getting economic support from their parents (Eggebeen 2005).

cohabitation (living together) two unrelated adults (by blood or by law) involved in an emotional and sexual relationship who sleep in the same residence at least four nights a week.

Definitions of Cohabitation

Research has used more than twenty definitions of cohabitation. These various definitions involve variables such as duration of the relationship, frequency of overnight visits, emotional or sexual nature of the relationship, and sex of the partners.

Cohabitation without children or marriage needs to be viewed not only as a legitimate end-state in itself, but also as a legitimate form of premarriage.

—*William Pinsof, family psychologist*

POSSLQ an acronym used by the U.S. Census Bureau that stands for "people of the opposite sex sharing living quarters."

Most research on cohabitation has been conducted on heterosexual live-in couples. Even partners in relationships may view the meaning of their cohabitation differently, with women viewing it more as a sign of a committed relationship moving toward marriage and men viewing it as an alternative to marriage or as a test to see whether future commitment is something to pursue. We define cohabitation as two unrelated adults involved in an emotional and sexual relationship who sleep overnight in the same residence on a regular basis. The terms used to describe live-in couples include cohabitants and **POSSLQs** (people of the opposite sex sharing living quarters), the latter term used by the U.S. Census Bureau.

Eight Types of Cohabitation Relationships

There are various types of cohabitation:

1. **Here and now.** These new partners have an affectionate relationship and are focused on the here and now, not the future of the relationship. Only a small proportion of people living together report that the "here and now" type characterizes their relationship (Jamieson et al. 2002). Rhoades et al. (2009) studied a sample of 240 cohabitating heterosexual couples and found that wanting to spend more time together was one of the top motivations for living together.

2. **Testers.** These couples are involved in a relationship and want to assess whether they have a future together. As in the case of here-and-now cohabitants, only a small proportion of cohabitants characterize themselves as "testers" (Jamieson et al. 2002). Rhoades et al. (2009) found that those who were motivated to live together to test their relationship (in contrast to spending more time together) reported more negative couple communication, more aggression, and lower relationship adjustment.

3. **Engaged.** These couples are in love and are planning to marry. Although not all cohabitants consider marriage their goal, most view themselves as committed to each other (Jamieson et al. 2002). Oppenheimer (2003) studied a national sample of cohabitants and found that cohabiting European Americans were much more likely to marry their partner than cohabiting African Americans—51 percent versus 22 percent. Dush et al. (2005) compared people in marriage; people in cohabiting, steady dating, and casual dating relationships; and people who dated infrequently or not at all. They found that individuals in a happy relationship, independent of the nature of the relationship, reported higher subjective well-being. In addition, the more committed the relationship, the higher the subjective well-being. Hence, for cohabiting people who define their relationships as involved, committed, or engaged, we would expect higher levels of subjective well-being.

4. **Money savers.** These couples live together primarily out of economic convenience. They are open to the possibility of a future together but regard such a possibility as unlikely.

5. **Pension partners.** This type is a variation of the money savers category. These individuals are older, have been married before, still derive benefits from their previous relationships, and are living with someone new. Getting married would mean giving up their pension benefits from the previous marriage. An example is a widow from the war in Afghanistan who was given military benefits due to a spouse's death. If remarried, the widow forfeits both health and pension benefits, but now lives with a new partner and continues to get benefits from the previous marriage.

© iStockphoto.com / © Thinkstock Images/Jupiterimages

6. **Security blanket cohabiters.** Some of the individuals in these cohabitation relationships are drawn to each other out of a need for security rather than mutual attraction.

7. **Rebellious cohabiters.** Some couples use cohabitation as a way of making a statement to their parents that they are independent and can make their own choices. Their cohabitation is more about rebelling from parents than being drawn to each other.

8. **Marriage never (cohabitants forever).** These couples feel that a real relationship is a commitment of the heart, not a legal document. Living together provides both companionship and sex without the responsibilities of marriage. Skinner et al. (2002) found that individuals in long-term cohabiting relationships scored the lowest in terms of relationship satisfaction when compared with married and remarried couples. The "marriage never" couples are rare (celebrities Johnny Depp and Vanessa Paradis, Goldie Hawn and Kurt Russell, and Susan Sarandon and Tim Robbins are examples of couples who live together, have children, and have opted not to marry).

There are various reasons and motivations for living together as a permanent alternative to marriage. Some may have been married before and don't want the entanglements of another marriage. Others feel that the real bond between two people is (or should be) emotional. They contend that many couples stay together because of the legal contract, even though they do not love each other any longer. "If you're staying married because of the contract," said one partner, "you're staying for the wrong reason." Some couples feel that they are "married" in their hearts and souls and don't need or want the law to interfere with what they feel is a private act of commitment. For most couples, living together is a short-lived experience. About 55 percent will marry and 40 percent will break up within five years of beginning cohabitation (Smock 2000). Some couples who view their living together as "permanent" seek to have it defined as a **domestic partnership**.

Same-Sex Cohabitation and Race

Although U.S. Census Bureau surveys do not ask about sexual orientation or gender identity, same-sex cohabiting couples may identify themselves as "unmarried partners." Those couples in which both partners are men or both are women are considered to be same-sex couples or households for purposes of research. Of the six million unmarried partner households, seven percent consist of two males; six percent, two females (*Statistical Abstract of the United States: 2009*, Table 62). Of these almost 800,000 (779,867) households, we might estimate that about 13 percent or about 100,000 (101,382) are black couples. These couples must cope with both heterosexism and racism.

domestic partnership a relationship in which individuals who live together are emotionally and financially interdependent and are given some kind of official recognition by a city or corporation so as to receive partner benefits.

Consequences of Cohabitation

Although living together before marriage does not ensure a happy, stable marriage, it has some potential advantages.

Advantages of Cohabitation Many unmarried couples who live together report that it is an enjoyable, maturing experience. Other potential benefits of living together include the following:

1. **Sense of well-being.** Compared to uninvolved individuals or those involved but not living together, cohabitants are likely to report a sense of well-being (particularly if the partners see a future together). They are in love, the relationship is new, and the disenchantment that frequently occurs in long-term relationships has not had time to surface. One student reported, "We have had to make some adjustments in terms of moving all our stuff into one place, but we very much enjoy our life together." Although young cohabitants report high levels of enjoyment compared to single people, cohabitants in midlife who have never married when compared to married spouses, report lower levels of relational and subjective well-being. However, those who have been married before and are currently cohabitating do not evidence lower levels of relational or personal well-being when compared to married people (Hansen et al. 2007).

2. **Delayed marriage.** Another advantage of living together is remaining unmarried—and the longer one waits to marry, the better. Being older at the time of marriage is predictive of marital happiness

and stability, just as being young (particularly 18 years and younger) is associated with marital unhappiness and divorce. Hence, if a young couple who have known each other for a short time is faced with the choice of living together or getting married, their delaying marriage while they live together seems to be the better choice. Also, if they break up, the split will not go on their "record" as would a divorce.

3. **Knowledge about self and partner.** Living with an intimate partner provides couples with an opportunity for learning more about themselves and their partner. For example, individuals in a cohabiting relationship may find that their role expectations are more (or less) traditional than they had previously thought. Learning more about one's partner is a major advantage of living together. A person's values (calling parents daily), habits (leaving the lights on), and relationship expectations (how emotionally close or distant) are sometimes more fully revealed when living together than in a traditional dating context.

4. **Safety.** Particularly for females, living together provides a higher level of safety not enjoyed by single females who live alone. Of course, living with a roommate or group of friends would provide a similar margin of safety.

Disadvantages of Cohabitation There is a downside for individuals and couples who live together.

1. **Feeling used or tricked.** We have mentioned that women are more prone than men to view cohabitation as reflective of a more committed relationship. When expectations differ, the more invested partner may feel used or tricked if the relationship does not progress toward marriage. One partner said, "I always felt we would be getting married, but it turns out that he never saw a future for us."

2. **Problems with parents.** Some cohabiting couples must contend with parents who disapprove of or do not fully accept their living arrangement. For example, cohabitants sometimes report that, when visiting their parents' homes, they are required to sleep in separate beds in separate rooms. Some cohabitants who have parents with traditional values respect these values, and sleeping in separate rooms is not a problem. Other cohabitants feel resentful of parents who require them to sleep separately. Some parents express their disapproval of their child's cohabiting by cutting off communication, as well as economic support, from their child. Other parents display lack of acceptance of cohabitation in more subtle ways. One woman who had lived with her partner for two years said that her partner's parents would not include her in the family's annual photo portrait. Emotionally, she felt very much a part of her partner's family and was deeply hurt that she was not included in the family portrait. Still other parents are completely supportive of their children's cohabiting and support their doing so. "I'd rather my kid live together than get married and, besides, it is safer for her and she's happier," said one father.

3. **Economic disadvantages.** Some economic liabilities exist for those who live together instead of getting married. Cohabitants typically do not benefit from their partner's health insurance, Social Security, or retirement benefits. In most cases, only spouses qualify for such payoffs.

Given that most relationships in which people live together are not long-term and that breaking up is not uncommon, cohabitants might develop a written and signed legal agreement should they

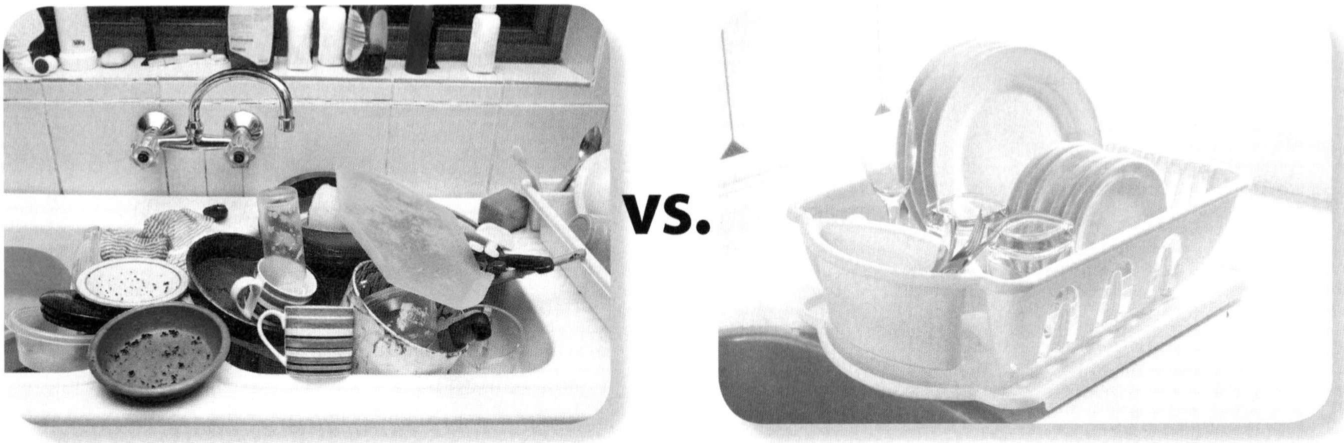

© Don Bayley/iStockphoto.com / © iStockphoto.com

Cohabitation is one way to learn about differences in housekeeping styles, such as when it's time to do the dishes.

purchase a house, car, or other costly items together. The written agreement should include a description of the item, to whom it belongs, how it will be paid for, and what will happen to the item if the relationship terminates. Purchasing real estate together may require a separate agreement, which should include how the mortgage, property taxes, and repairs will be shared. The agreement should also specify who gets the house if the partners break up and how the value of the departing partner's share will be determined.

If the couple have children, another agreement may be helpful in defining custody, visitation, and support issues in the event the couple terminates the relationship. Such an arrangement may take some of the romance out of the cohabitation relationship, but it can save a great deal of frustration should the partners decide to go their separate ways.

In addition, couples who live together instead of marrying can protect themselves from some of the economic disadvantages of living together by specifying their wishes in wills; otherwise, their belongings will go to next of kin or to the state. They should also own property through joint tenancy with rights of survivorship.

This means that ownership of the entire property will revert to one partner if the other partner dies. In addition, the couple should save for retirement, because live-in companions may not access Social Security benefits, and some company pension plans bar employees from naming anyone other than a spouse as the beneficiary.

4. **Effects on children.** About 40 percent of children will spend some time in a home where the adults are cohabitating. In addition to being disadvantaged in terms of parental income and education, they are likely to experience more disruptions in family structure. Raley et al. (2005) analyzed data on children who lived with cohabiting mothers (from the National Survey of Families and Households) and found that these children fared exceptionally poorly and sometimes were significantly worse off than were children who lived with divorced or remarried mothers. The researchers reasoned that the instability associated with cohabitation may account for why these children do less well.

5. **Other issues.** More than a million cohabitants are over the age of 50. When compared to married people, they report more depressive symptoms independent of their economic resources, social support, and physical health (Brown et al. 2005). These middle-aged individuals may possibly prefer to be married and their unhappiness reflects that preference.

Having Children while Cohabitating

Sassler and Cunningham (2008) interviewed twenty-five never-married women who were cohabiting with their heterosexual partners. Most (two-thirds) reported that they wanted to be married before having a child. Indeed, none of the respondents planned on having a child in the near future and none were actively trying to conceive. However, some noted that marriage made no difference. One respondent noted:

> I don't see a reason to get married if you don't want to get married. Everyone is so traditional, which is really silly because it's not a traditional world anymore. Like when my sister had her baby. I mean, she was concerned about . . . "I wonder what they are going to think if I don't have a husband?" I mean, who cares? Who cares if you don't have a husband? What does marriage have to do with anything? (p. 13)

What is the effect of cohabitation on one's future marital happiness and stability? For women who have only one cohabitation experience with the man they marry, there is no increased risk of divorce (Teachman 2003). However, people commonly have more than one cohabitation experience, and those who have multiple cohabitation experiences prior to marriage are more likely to end up in marriages characterized by violence, depression, divorce, and lower levels of happiness

common-law marriage a marriage by mutual agreement between a cohabiting man and woman without a marriage license or ceremony.

palimony referring to the amount of money one "pal" who lives with another "pal" may have to pay if the partners terminate their relationship.

and positive communication (Cohan and Kleinbaum 2002; Booth et al. 2008).

Legal Aspects of Living Together

In recent years, the courts and legal system have become increasingly involved in relationships in which couples live together. Some of the legal issues concerning cohabiting partners include common-law marriage, palimony, child support, and child inheritance. Lesbian and gay couples also confront legal issues when they live together.

Technically, cohabitation is against the law in some states. For example, in North Carolina, cohabitation is a misdemeanor punishable by a fine not to exceed $500, imprisonment for not more than six months, or both. Most law enforcement officials view cohabitation as a victimless crime and feel that the general public can be better served by concentrating upon the crimes that do real damage to citizens and their property.

Common-Law Marriage The concept of **common-law marriage** dates to a time when couples who wanted to be married did not have easy or convenient access to legal authorities (who could formally sanction their relationship so that they would have the benefits of legal marriage). Thus, if the couple lived together, defined themselves as husband and wife, and wanted other people to view them as a married couple, they would be considered married in the eyes of the law.

Despite the assumption by some that heterosexual couples who live together a long time have a common-law marriage, only eleven jurisdictions recognize such marriages. In ten states (Alabama, Colorado, Idaho, Iowa, Kansas, Rhode Island, South Carolina, Montana, Pennsylvania, and Texas) and the District of Columbia, a heterosexual couple may be considered married if they are legally competent to marry, if the partners agree that they are married, and if they present themselves to the public as a married couple. A ceremony or compliance with legal formalities is not required.

In common-law states, individuals who live together and who prove that they were married "by common law" may inherit from each other or receive alimony and property in the case of relationship termination. They may also receive health and Social Security benefits, as would other spouses who have a marriage license. In states not recognizing common-law marriages, the individuals who live together are not entitled to benefits traditionally afforded married individuals. More than three-quarters of the states have passed laws prohibiting the recognition of common-law marriages within their borders.

Palimony A takeoff on the word *alimony*, **palimony** refers to the amount of money one "pal" who lives with another "pal" may have to pay if the partners terminate their relationship. In 2005, for instance, comedian Bill Maher was the target of a $9 million palimony suit by ex-girlfriend Coco Johnsen. In 2007, Candace Cabbil sued NBA star Latrell Sprewell for $200 million, claiming that he broke their long-term cohabitation agreement. Cabbil also alleged that Sprewell had fathered four of her children.

Avellar and Smock (2005) compared the economic well-being of men and women who ended their relationship. Whereas the economic standing of the cohabiting man declined moderately, that of the former cohabiting woman declined steeply, leaving a substantial proportion of women in poverty (and even more so for African American and Hispanic women).

Child Support Heterosexual individuals who conceive children are responsible for those children whether they are living together or married. In most cases, the custody of young children will be given to the mother, and the father will be required to pay child support. In effect, living together is irrelevant with regard to parental obligations. However, a woman who agrees to have a child with her lesbian partner cannot be forced to pay child support if the couple breaks up. In 2005, the Massachusetts Supreme Judicial Court ruled that their informal agreement to have a child together did not constitute an enforceable contract.

Couples who live together or who have children together should be aware that laws traditionally applying only to married couples are now being applied to many unwed relationships. Palimony, distribution of property, and child support payments are all possibilities once two people cohabit or parent a child.

Child Inheritance Children born to cohabitants who view themselves as spouses and who live in common-law states are regarded as legitimate and can inherit from their parents. However, children born to cohabitants who do not present themselves as married or who do not live in common-law states are also able to inherit. A biological link between the parent and the offspring is all that needs to be established.

3.6 Intentional Communities

Although single people often live in apartments, condominiums, or single-family houses, living in an intentional community (previously called a commune) is an alternative. Less than 10 percent (7 percent) of 1,319 undergraduates at a large southeastern university agreed, "Someday, I would like to live in an intentional community [a commune]" (Knox and Zusman 2009). Intentional communities are not just for single or young people. Married people and seniors also embrace communal living. Blue Heron Farm, for example, is home to a collective of individuals, ages 20 to 70, who live in ten houses (Yeoman 2006). The range of intentional community alternatives may be explored via the Cohousing Association of the United States (www.cohousing.org) and the Federation of Egalitarian Communities (www.thefec.org). In the next section, we describe one such intentional community.

> *Home life is no more natural to us than a cage to a cockatoo.*
>
> —*George B. Shaw*

Twin Oaks: An Alternative Context for Living

Twin Oaks is a community of ninety adults and fifteen children living together on 450 acres of land in Louisa, Virginia (about forty-five minutes east of Charlottesville and one hour west of Richmond). The **commune** (an older term), now known as an **intentional community** (comprising a group of people who choose to live together on the basis of shared values and worldview), was founded in 1967, and is one of the oldest nonreligious intentional communities in the United States.

The membership is 55 percent male and 45 percent female. Most of the members are white, but there is a wide range of racial, ethnic, and social class backgrounds. Members include gay, straight, bisexual, and transgender people; most are single, never-married adults, but there are married couples and families. Sexual values range from celibate to monogamous to polyamorous (about 25 percent of the community are involved in more than one emotional or sexual relationship at the same time). The age range of the members is newborn to 78, with an average age of 40.

intentional community (commune) group of people living together on the basis of shared values and worldview.

The average length of stay for current members is over seven years. There are no officially sanctioned religious beliefs at Twin Oaks—it is not a "spiritual" community, although its members represent various religious values (Jewish, Christian, pagan, atheist, and so on). The core values of the community are nonviolence (no guns, low tolerance for violence of any kind in the community, no parental use of violence against children), egalitarianism (no leader, with everyone having equal political power and the same access to resources), and environmental sustainability (the community endeavors to live off the land, growing its own food and heating its buildings with wood from the forest).

Another core value is income sharing. All the members work together to support everyone in the community instead of working individually to support themselves. Money earned from the three community businesses (weaving hammocks, making tofu, and writing indexes for books) is used to provide food and other basic needs for all members. No one needs to work outside the community. Each member works in the community in a combination of income-producing and domestic jobs. An hour of cooking or gardening receives the same credit as an hour of fixing computers or business management. No matter what work one chooses to do, each member's commitment to the community is forty-two hours of work a week. This includes preparing meals, taking care of children, cleaning bathrooms, and maintaining buildings—activities not considered in the typical mainstream forty-hour workweek.

The community places high value on actively creating a "homegrown" culture. Members provide a large amount of their own entertainment, products, and services. They create homemade furniture, present theatrical and musical performances, and enjoy innovative community holidays such as Validation Day (a distant relative of Valentine's Day, minus the commercialism). The community also encourages participation in activism outside the community. Many members are activists for peace and justice, feminism, and ecological organizations. (This section is based on information provided by Kate Adamson and Ezra Freeman, members of Twin Oaks, and is used with their permission.)

3.7 Living Apart Together

The premise of the current, most common relationship paradigm is that individuals are socialized to believe that "more is better". . . that the more time they spend together, including moving in together, the better. In effect, loving and committed couples automatically assume that they will marry or live together in one residence and that to do otherwise or to have "spaces in their togetherness," to quote Gibran, would suggest that they do not "really" love each other and aren't "really" committed to each other.

living apart together (LAT) a committed couple who does not live in the same home.

However, a new lifestyle and family form has emerged called living apart together (Hess 2009). The definition of **living apart together (LAT)** is a committed couple who does not live in the same home (and others such as children or elderly parents may live in those respective homes). Three criteria must be met for a couple to be defined as an LAT couple: (1) they must define themselves as a committed couple; (2) others must define the partners as a couple; and (3) they must live in separate domiciles. The lifestyle of living apart together involves partners in loving and committed relationships (married or unmarried) identifying their independent needs in terms of the degree to which they want time and space away from each other. People living apart together exist on a continuum from partners who have separate bedrooms and baths in the same house to those who live in a separate place (apartment, condo, house) in the same or different cities. LAT couples are not those couples who are forced by their career or military assignment to live separately. Rather, LAT partners choose to live in separate domiciles.

This new lifestyle or family form has been identified as a new social phenomenon in several Western European countries (for example, France, Sweden, Norway), as well as in the United States (Lara 2005).

Disadvantages of LAT

1. **Stigma or disapproval.** Because the norm that married couples move in together is firmly entrenched, couples who do not do so are suspect.
2. **Cost.** Certainly, maintaining two separate living arrangements can be more expensive than sharing a house. But there are ways LAT couples manage their lifestyle: living in two condominiums that are cheaper than two houses or setting up housing out of high-priced real estate areas.
3. **Inconvenience.** Unless the partners live in a duplex or two condominiums in the same building, going between the two places to share meals or be together can be inconvenient.
4. **Lack of shared history.** Because the adults are living in separate quarters, a lot of what goes on in each house does not become a part of the life history of the other. For example, children in one place don't benefit as much from the other adult who lives in another domicile most of the time.
5. **Waking up alone.** Although some LAT partners sleep together overnight, others say goodnight and sleep in separate beds or houses. A potential disadvantage is waking up and beginning the day alone without the early-morning connection with one's beloved.

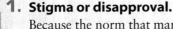

Advantages of LAT

1. **Space and privacy.** Having two places enables each partner to have a separate space, which provides a measure of privacy for individual activities, especially if one partner works at home. Couples also benefit from two residences if they have guests, giving both the guest and the couple ample space for privacy.

2. **Variable sleep needs.** Although some partners enjoy going to bed at the same time and sleeping in the same bed, others like to go to bed at radically different times and to sleep in separate beds or rooms. The LAT arrangement allows for partners to have their own sleep needs or schedules met without interfering with a partner. A frequent comment from LAT partners is, "My partner thrashes throughout the night and kicks me, not to speak of the wheezing and teeth grinding, so to get a good night's sleep, I need to sleep somewhere else."

3. **Allergies.** Individuals who have cat or dog allergies may need to live in a separate antiseptic environment from their partner who loves animals and won't live without them. "He likes his dog on the couch," said one woman.

4. **Variable social needs.** Partners differ in terms of their need for social contact with friends, siblings, and parents. The LAT arrangement allows for the partner who enjoys frequent time with others to satisfy that need without subjecting the other to the presence of a lot of people in one's life space. One wife could enjoy having her siblings and parents around without expense to her husband, who lived upstairs in a separate condo. This arrangement also works well for remarried families with children from previous marriages. This way, a married couple may live in a duplex with their own kids living with them.

5. **Keeping the relationship exciting.** Time apart from our beloved can make time together feel more precious. The term **satiation** means that a stimulus loses its value with repeated exposure. Just as we tire of eating the same food, listening to the same music, or watching the same movie, so satiation is relevant to relationships. An LAT relationship can help a couple avoid their interest in each other becoming satiated, and maintain some of the excitement in seeing or being with each other.

6. **Self-expression and comfort.** Partners often have very different tastes in home decor, music, temperature, and levels of cleanliness or orderliness. With two separate places, each can decorate to their respective tastes, listen to their own music, heat and cool to their preference, and maintain order and cleanliness without bothering their partner.

7. **Elder care.** One partner may be taking care of an elderly parent in the parents' house or in his or her own house. Either way, the partners may have a preference not to live in the same house as a parent. An LAT relationship allows for the partner taking care of the elderly parent to do so and a place for the couple to be alone.

8. **Maintaining one's lifetime residence.** Some retirees, widows, and widowers meet, fall in love, and want to enjoy each other's companionship. However, they don't want to move out of their own house. The LAT arrangement allows each partner to maintain a separate residence but to enjoy the new relationship.

9. **Leaving inheritances to children from previous marriages.** Having separate residences allows respective partners to leave their family home or residential property to their children from their first marriage without displacing their surviving spouse.

satiation the state in which a stimulus loses its value with repeated exposure.

Communication

SECTIONS

communication the process of exchanging information and feelings between two or more people.

nonverbal communication the "message about the message," using gestures, eye contact, body posture, tone, volume, and rapidity of speech.

Philosopher Arthur Schopenhauer noted the delicate movements of two porcupines huddling together on a cold winter night. They need each other for warmth but must avoid the pain of quills pricking their delicate skin. They continually move and adjust so as to achieve the maximum amount of warmth with the least amount of sticking. So it is in relationships, individuals are constantly seeking the warmth of the emotional relationship yet must be careful to avoid painful conflict.

"Good communication" is regarded as the primary factor responsible for a strong relationship. It provides a way for individuals to find that balance between warmth and pain. Individuals report that communication confirms the quality of their relationship ("We can talk all night about anything and everything") or condemns their relationship ("We have nothing to say to each other; we are getting a divorce"). Couples before marriage also have different content themes than couples who are married. Knobloch (2008) noted that people before marriage are focused on individual issues (for example, "Why don't you spend more leisure time with me?") in contrast to married people who couch issues in the context of "we" (for example, "How will we rear our children?"). In this chapter, we examine various issues related to communication and identify some communication principles and skills. We begin by looking at the nature of interpersonal communication.

4.1 The Nature of Interpersonal Communication

Communication can be defined as the process of exchanging information and feelings between two or more people. Communication is both verbal and nonverbal. Although most communication is focused on verbal content, most (estimated to be as high as 80 percent) interpersonal communication is nonverbal. **Nonverbal communication** is the "message about the message," the gestures, eye contact, body posture, tone, volume, and rapidity of speech. Even though a person says, "I love

© Blend Images/Jupiterimages

you and am faithful to you," crossed arms and lack of eye contact will convey a very different meaning than the same words accompanied by a tender embrace and sustained eye-to-eye contact. Bos et al. (2007) found that the greater the congruence between verbal and nonverbal communication, the better. Indeed, when the two are congruent, a person experiences fewer stressful events and lowered depression.

We tend to assign more importance to nonverbal cues than verbal cues (Preston 2005). In effect, we like to hear sweet words but we feel more confident when we see behavior that supports the words. "Show me the money, honey" is a phrase that reflects a partner's focus on behavior rather than words.

One of the by-products of communication is the feeling of intimacy. Individuals differ in their capacity for intimacy. Jane Fonda noted in her autobiography that, "danger lies in intimacy and that far away is safe" (Fonda 2005, 34). She noted that if she tried to be close with her father, he would erupt into an angry rage so that she leaned to smile, act happy, and never try to connect or become emotionally intimate with her father for fear of him turning his rage on her. Having a supportive communicative partner—one who listens and is engaged—is a plus for any relationship.

4.2 Conflicts in Relationships

Conflict can be defined as the process of interaction that results when the behavior or desires of one person interfere with the behavior or desires of another. A professor in a marriage and family class said, "If you haven't had a conflict with your partner, you haven't been going together long enough." This section explores the inevitability, desirability, sources, and styles of conflict in relationships.

Inevitability of Conflict

If you are alone this Saturday evening from six o'clock until midnight, you are assured of six conflict-free hours. However, if you plan to be with your partner, roommate, or spouse during that time, the potential for conflict exists. Whether you eat out, where you eat, where you go after dinner, and how long you stay must be negotiated. Although it may be relatively easy for you and your companion to agree on the evening agenda, marriage involves the meshing of desires on an array of issues for potentially sixty years or more. Indeed, conflict is inevitable in any intimate relationship. DeMaria (2005) studied 129 married couples who signed up for a communication and marriage education workshop presumably designed for couples who were already functioning well and who were there for a "relationship tune-up." Analysis of the data revealed that the couples were highly distressed, conflicted, devitalized, and lacked communication skills. Hence, one need not be on the verge of divorce in the office of a marriage counselor to profit from learning effective communication skills.

Benefits of Conflict

Conflict can be healthy and productive for a couple's relationship. Ignoring an issue may result in the partners becoming increasingly resentful and dissatisfied with the relationship. Indeed, not talking about a concern can do more damage to a relationship than bringing up the issue and discussing it (Campbell 2005). Couples in trouble are not those who disagree but those who never discuss their disagreements.

conflict the interaction that occurs when the behavior or desires of one person interfere with the behavior or desires of another.

Sources of Conflict

Conflict has numerous sources, some of which are easily recognized, whereas others are hidden inside the web of marital interaction.

1. **Behavior.** Stanley et al. (2002) noted that money was the issue over which a national sample of couples reported that they argued the most. The behavioral expression of a money issue might include how the partner spends money (for example, excessively), the lack of communication about spending (for example, does not consult the partner), and the target (for example, items considered unnecessary by the partner). However, marital conflict is not limited to behavioral money issues. Stanley et al. (2002) found that remarried couples argued most about the children (for example, rules for and discipline of). In a sample of 105 older married couples (average age, 69), the most often reported behavior problem was related to leisure activities (Henry et al. 2005). One 67-year-old wife noted, "My spouse watches too much football and after 48 years, I get upset" (p. 249).
2. **Cognitions and perceptions.** Aside from your partner's actual behavior, your cognitions and perceptions of a behavior can be a source of satisfaction or dissatisfaction. One husband complained that his wife "had boxes of coupons everywhere and always kept the house a wreck." The wife made the husband aware that she saved $100 on their grocery bill every week and asked him to view the boxes and the mess as "saving money." He changed his view and the clutter ceased to be a problem.

© Aldo Murillo/iStockphoto.com

© Image Source Black/Jupiterimages /© Tony Sanchez-Espinosa/iStockphoto.com / © iStockphoto.com / © Jose Luis Pelaez, Inc./Image Source/Pnk/Jupiterimages

3. **Value differences.** Because you and your partner have had different socialization experiences, you may also have different values—about religion (one feels religion is a central part of life; the other does not), money (one feels uncomfortable being in debt; the other has the buy-now-pay-later philosophy), in-laws (one feels responsible for parents when they are old; the other does not), and children (number, timing, discipline). The effect of value differences depends less on the degree of the difference than on the degree of rigidity with which each partner holds values. Dogmatic and rigid thinkers, feeling threatened by value disagreement, may try to eliminate alternative views and thus produce more conflict. Partners who recognize the inevitability of difference may consider the positives of an alternative view and move toward acceptance. When both partners do this, the relationship takes priority and the value differences suddenly become less important.

4. **Inconsistent rules.** Partners in all relationships develop a set of rules to help them function smoothly. These unwritten but mutually understood rules include what time you are supposed to be home after work, whether you should call if you are going to be late, how often you can see friends alone, and when and how to make love. Conflict results when the partners disagree on the rules or when inconsistent rules develop in the relationship. For example, one wife expected her husband to take a second job so they could afford a new car, but she also expected him to spend more time at home with the family.

5. **Leadership ambiguity.** Unless a couple has an understanding about which partner will make decisions in which area (for example, the wife may make decisions about money management, and the husband may make decisions about rearing the children), unnecessary conflict may result. Whereas some couples may want to discuss certain issues, others may want to develop a clear specification of roles.

Styles of Conflict

Spouses develop various styles of conflict. If you were watching a videotape of various spouses disagreeing over the same issue, you would notice at least six styles of conflict. These styles have been described by Greeff and De Bruyne (2000) as the following:

Competing Style

Partners who use a **competing style of conflict** are both assertive and uncooperative. Both try to force their way on the other so that there is a winner and a loser.

competing style of conflict conflict style in which partners are both assertive and uncooperative. Each tries to force a way on the other so that there is a winner and a loser.

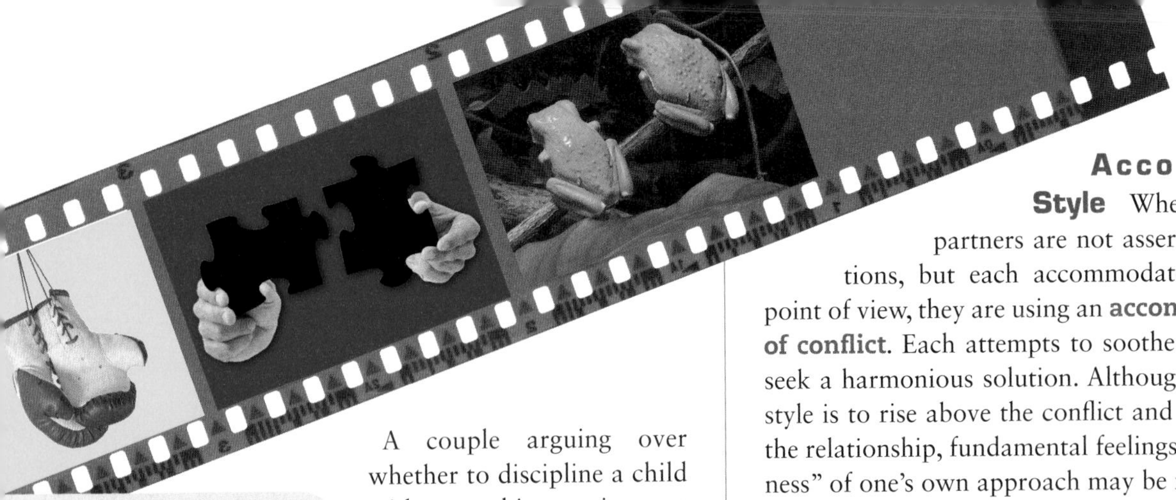

collaborating style of conflict conflict style in which partners are both assertive and cooperative. Each expresses views and cooperates to find a solution.

compromising style of conflict conflict style in which there is an intermediate solution in which both partners find a middle ground they can live with.

avoiding style of conflict conflict style in which partners are neither assertive nor cooperative, so as to avoid confrontation.

accommodating style of conflict conflict style in which the respective partners are not assertive in their positions but are cooperative. Each attempts to soothe the other and to seek a harmonious solution.

parallel style of conflict style of conflict whereby both partners deny, ignore, and retreat from addressing a problem issue.

A couple arguing over whether to discipline a child with a spanking or time-out would resolve the argument with the dominant partner's forcing a decision.

Collaborating Style When partners use a **collaborating style of conflict**, both assertive and cooperative. Both partners express their views and cooperate to find a solution. A spanking, time-out, or just talking to the child might resolve the previous issue, but both partners would be satisfied with the resolution.

Compromising Style With the **compromising style**, partners generally find an intermediate solution: both partners would find a middle ground they could live with—perhaps spanking the child for serious infractions such as playing with matches in the house and imposing a time-out for talking back.

Avoiding Style Partners who use an **avoiding style** are neither assertive nor cooperative. They would avoid a confrontation and let either parent control the disciplining of the child. Thus the child might be both spanked and put in a time-out. Marchand and Hock (2000) noted that depressed spouses were particularly likely to use avoidance as a conflict-resolution strategy.

Accommodating Style When the respective partners are not assertive in their positions, but each accommodates to the other's point of view, they are using an **accommodating style of conflict**. Each attempts to soothe the other and to seek a harmonious solution. Although the goal of this style is to rise above the conflict and keep harmony in the relationship, fundamental feelings about the "rightness" of one's own approach may be maintained.

Parallel Style Finally, with a **parallel style**, both partners deny, ignore, and retreat from addressing a problem issue. "Don't talk about it, and it will go away" is the theme of this conflict style. Gaps begin to develop in the relationship; neither partner feels free to talk, and both believe that they are misunderstood. They eventually become involved in separate activities rather than spending time together.

Greeff and De Bruyne (2000) studied fifty-seven couples who had been married at least ten years and found that the collaborating style was associated with the highest level of marital and spousal satisfaction. The competitive style, used by either partner, was associated with the lowest level of marital satisfaction. Regardless of the style of conflict, partners who say positive things to each other at a ratio of 5:1 (positives to negatives) seem to stay together (Gottman 1994).

4.3 Principles and Techniques of Effective Communication

People who want effective communication in their relationship follow various principles and techniques, including the following:

1. **Make communication a priority.** Communicating effectively implies making communication an important priority in a couple's relationship. When communication is a priority, partners make time for it to occur in a setting without interruptions: they are alone; they do not answer the phone; and they turn the television off. Making communication a priority results in the exchange of more information between partners, which increases the knowledge each partner has about the other.

2. **Establish and maintain eye contact.** Partners who look at each other when they are talking not only communicate an interest in each other but also are able to gain information about the partner's feelings and responses to what is being said. Not looking at your partner may be interpreted as lack of interest and prevents you from observing nonverbal cues.

3. **Ask open-ended questions.** When your goal is to find out your partner's thoughts and feelings about an issue, using **open-ended questions** is best. Such questions (for example, "How do you feel about me?") encourage answers that contain a lot of information. **Closed-ended questions** (for example, "Do you love me?"), which allow for a one-word answer such as yes or no, do not provide the opportunity for the partner to express a range of thoughts and feelings.

4. **Use reflective listening.** Effective communication requires being a good listener. One of the skills of a good listener is the ability to use the technique of **reflective listening**, which involves paraphrasing or restating what the person has said to indicate that the listener understands. For example, suppose you ask your partner, "How was your day?" and your partner responds, "I felt exploited today at work because I went in early and stayed late and a memo from my new boss said that future bonuses would be eliminated because of a company takeover." Listening to what your partner is both saying and feeling, you might respond, "You feel frustrated because you really worked hard and felt unappreciated . . . and it's going to get worse."

   Reflective listening serves the following functions:

   a. it creates the feeling for speakers that they are being listened to and are being understood

   b. it increases the accuracy of the listener's understanding of what the speaker is saying.

   If a reflective statement does not accurately reflect what a speaker thinks and feels, the speaker can correct the inaccuracy by restating the thoughts and feelings.

   An important quality of reflective statements is that they are nonjudgmental. For example, suppose two lovers are arguing about spending time with their respective friends and one says, "I'd like to spend one night each week with my friends and not feel guilty about it." The partner may respond by making a statement that is judgmental (critical or evaluative), such as those exemplified in Table 4.1.

open-ended question question that encourages answers that contain a great deal of information.

closed-ended question question that allows for a one-word answer and does not elicit much information.

reflective listening paraphrasing or restating what a person has said to indicate that the listener understands.

Table 4.1
Judgmental and Nonjudgmental Responses to a Partner's Saying, "I'd Like to Spend One Evening a Week with My Friends"

Nonjudgmental, Reflective Statements	Judgmental Statements
You value your friends and want to maintain good relationships with them.	You only think about what you want.
You think it is healthy for us to be with our friends some of the time.	Your friends are more important to you than I am.
You really enjoy your friends and want to spend some time with them.	You just want a night out so that you can meet someone new.
You think it is important that we not abandon our friends just because we are involved.	You just want to get away so you can drink.
You think that our being apart one night each week will make us even closer.	You are selfish.

"I" statements statements that focus on the feelings and thoughts of the communicator without making a judgment on others.

"you" statements statements that blame or criticize the listener and often result in increasing negative feelings and behavior in the relationship.

Judgmental responses serve to punish or criticize people for what they think, feel, or want and often result in frustration and resentment. Table 4.1 also provides several examples of nonjudgmental reflective statements.

5. **Use "I" statements.** **"I" statements** focus on the feelings and thoughts of the communicator without making a judgment on others. Because "I" statements are a clear and nonthreatening way of expressing what you want and how you feel, they are likely to result in a positive change in the listener's behavior.

   In contrast, **"you" statements** blame or criticize the listener and often result in increasing negative feelings and behavior in the relationship. For example, suppose you are angry with your partner for being late. Rather than say, "You are always late and irresponsible" (which is a "you" statement), you might respond with, "I get upset when you are late and will feel better if you call and let me know." The latter focuses on your feelings and a desirable future behavior rather than blaming the partner for being late.

6. **Touch.** Hertenstein et al. (2007) identified the various meanings of touch, such as conveying emotion, attachment, bonding, compliance, power, and intimacy. The researchers also emphasized the importance of using touch as a mechanism of nonverbal communication to emphasize one's point or meaning.

7. **Use "soft" emotions.** Sanford (2007) identified "hard" emotions (for example angry or aggravated) or "soft" emotions (sad or hurt) displayed during conflict. The use of hard emotions resulted in an escalation of negative communication, whereas the display of "soft" emotions resulted in more benign communication and an increased focus on the importance of resolving interpersonal conflict.

8. **Avoid negative expressivity.** Rayer and Volling (2005) studied the levels of emotional expressivity (both positive and negative) and found that negative expressivity had a strong impact on marital love and conflict. Because intimate partners are capable of hurting each other so intensely, be careful how you criticize or communicate disapproval to your partner.

9. **Say positive things about your partner.** A team of researchers found that emotional expressiveness was strongly related to marital adjustment, particularly when coupled with the suppression of negative statements (Ingoldsby et al. 2005).

   People like to hear others say positive things about them. These positive statements may be in the form of compliments (for example, "You look terrific!") or appreciation ("Thanks for putting gas in the car"). Gable et al. (2003) asked fifty-eight heterosexual dating couples to monitor their interaction with one another. The respondents observed that they were overwhelmingly positive, at a five-to-one ratio.

10. **Tell your partner what you want.** Focus on what you want rather than on what you don't want. Rather than say, "You always leave the bathroom a wreck," an alternative might be "Please hang up your towel after you take a shower." Rather than say, "You never call me when you are going to be late," say "Please call me when you are going to be late."

I wish you'd tell me when you buy something . . .

© Brand X Pictures/Jupiterimages

11. Stay focused on the issue. **Branching** refers to going out on different limbs of an issue rather than staying focused on the issue. If you are discussing the overdrawn checkbook, stay focused on the checkbook. To remind your partner that he or she is equally irresponsible when it comes to getting things repaired or doing housework is to get off the issue of the checkbook. Stay focused.

12. Make specific resolutions to disagreements. To prevent the same issues or problems from recurring, agreeing on what each partner will do in similar circumstances in the future is important. For example, if going to a party together results in one partner's drinking too much and drifting off with someone else, what needs to be done in the future to ensure an enjoyable evening together? In this example, a specific resolution would be to decide how many drinks the partner will have within a given time period.

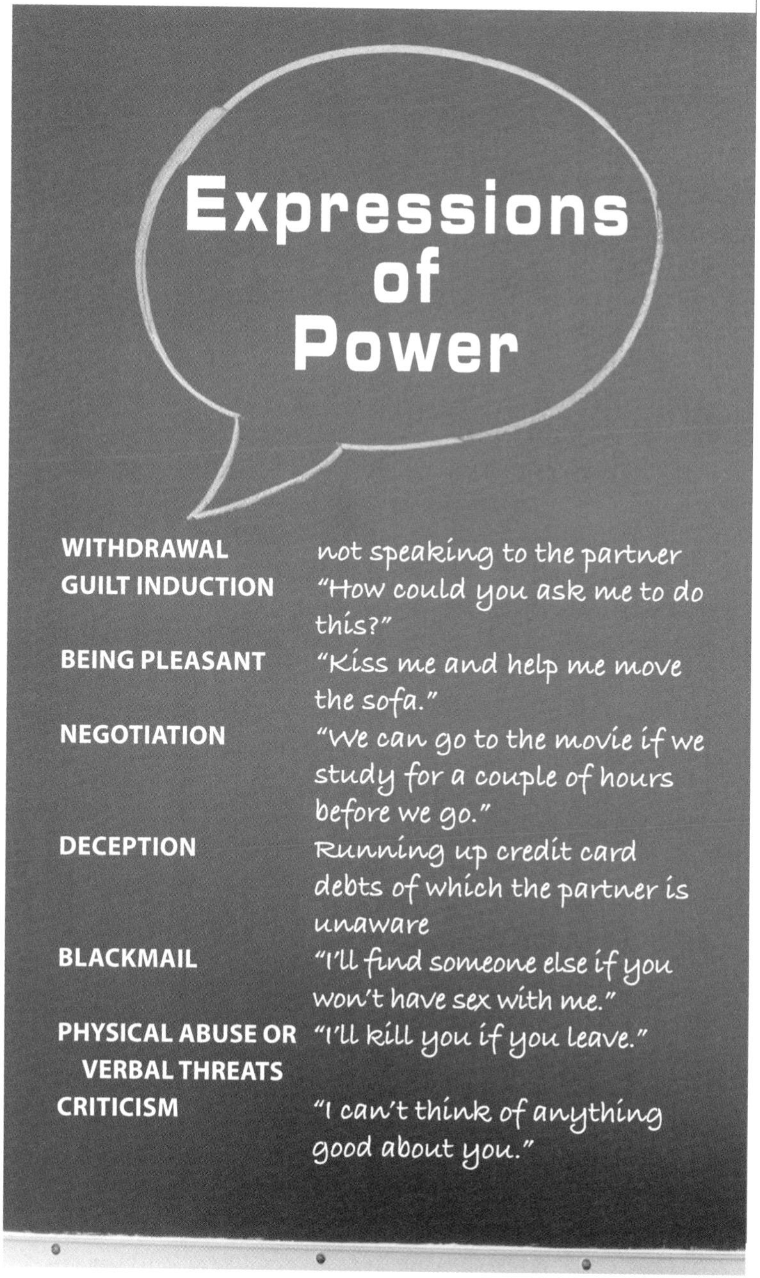
Expressions of Power
WITHDRAWAL
not speaking to the partner
GUILT INDUCTION
"How could you ask me to do this?"
BEING PLEASANT
"Kiss me and help me move the sofa."
NEGOTIATION
"We can go to the movie if we study for a couple of hours before we go."
DECEPTION
Running up credit card debts of which the partner is unaware
BLACKMAIL
"I'll find someone else if you won't have sex with me."
PHYSICAL ABUSE OR VERBAL THREATS
"I'll kill you if you leave."
CRITICISM
"I can't think of anything good about you."

branching in communication, going out on different limbs of an issue rather than staying focused on the issue.

congruent message one in which verbal and nonverbal behaviors match.

power the ability to impose one's will on one's partner and to avoid being influenced by the partner.

13. Give congruent messages. **Congruent messages** are those in which the verbal and nonverbal behaviors match. A person who says, "Okay, you're right" and smiles while embracing the partner is communicating a congruent message. In contrast, the same words accompanied by leaving the room and slamming the door communicate a very different message.

14. Share power. One of the greatest sources of dissatisfaction in a relationship is a power imbalance and conflict over power (Kurdek 1994). **Power** is the ability to impose one's will on one's partner and to avoid being influenced by the partner.

In general, the spouse with the more prestigious occupation, higher income, and more education exerts the greater influence on family decisions. Indeed, Dunbar and Burgoon (2005) noted that the greater the perception of one's own power, the more dominant one was in conversation with the partner.

However, power may also take the form of love and sex. The person in the relationship who loves less and who needs sex less has enormous power over the partner who is very much in love and who is dependent on the partner for sex. This pattern reflects the principle of least interest we discussed earlier in the text.

15. Keep the process of communication going. Communication includes both content (verbal and nonverbal information) and process (interaction). It is important not to allow difficult content to shut down the communication process (Turner 2005). To ensure that the process continues, the partners should focus on the fact that sharing information is essential and reinforce each other for keeping the process alive. For example, if your partner tells you something that you do that bothers him or her, it is important to thank him or her for telling you that rather than becoming defensive. In this way, your partner's feelings about you stay out in the open rather than hidden behind a wall of resentment. Otherwise, if you punish such disclosure because you don't like the content, subsequent disclosure will stop.

Although effective communication skills can be learned, Robbins (2005) noted that physiological capacities may enhance or impede the acquisition of these skills. She noted that people with attention-deficit/hyperactivity disorder (ADHD) might have deficiencies in basic communication and social skills. Being able to communicate effectively is valuable. Rosof (2005) found that positive couple communication is related to feelings of individual fulfillment in an intimate relationship.

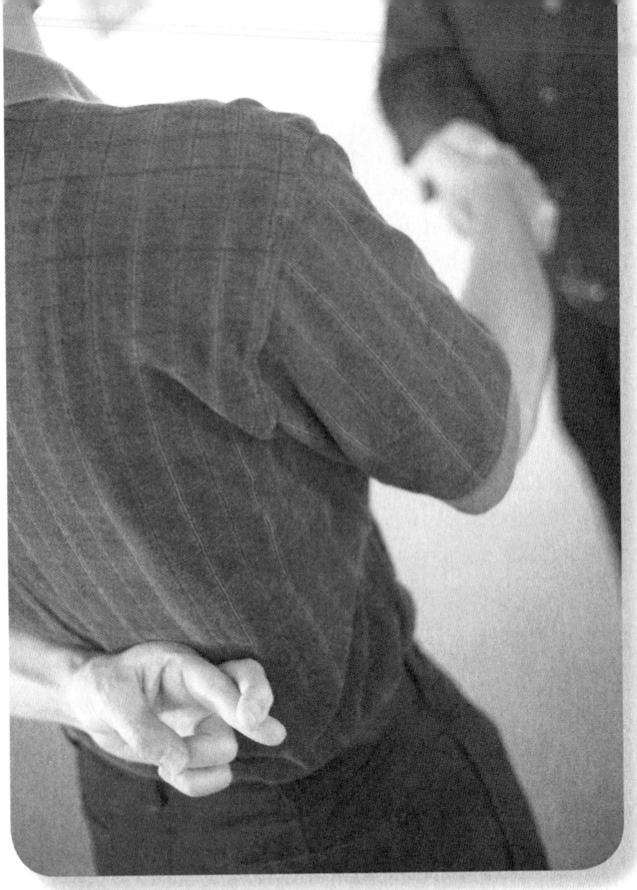
© SW Productions/Getty Images

4.4 Self-Disclosure, Honesty, and Lying

Shakespeare noted in Macbeth that "the false face must hide what the false heart doth know," suggesting that withholding and dishonesty may affect the way one feels about one's self and relationships with others. All of us make choices, consciously or unconsciously, about the degree to which we disclose, are honest, and/or lie.

Self-Disclosure in Intimate Relationships

One aspect of intimacy in relationships is self-disclosure, which involves revealing personal information and feelings about oneself to another person. McKenna et al. (2002) found that a positive function of meeting online is that people were better able to express themselves and disclose on the Internet than in person. Gibbs et al. (2006) also noted that people seeking a partner on the Internet were more honest about their disclosures if they had the goal of a long-term relationship.

Relationships become more stable when individuals disclose themselves—their formative years, previous relationships (positive and negative), experiences of elation and sadness or depression, and goals (achieved and thwarted). We will note in the discussion of love in Chapter 5 that self-disclosure is a psychological condition necessary for the development of love. To the degree that you disclose yourself to another, you invest yourself in and feel closer to that person. People who disclose nothing are investing nothing and remain aloof. One way to encourage disclosure in one's partner is to make disclosures about one's own life and then ask about the partner's life. Patford (2000) found that the higher the level of disclosure, the more committed the spouses were to each other.

Honesty in Intimate Relationships

Lying is pervasive in American society. Presidential candidate John Edwards repeatedly lied about his affair with Rielle Hunter, a member of his campaign staff, until he was caught meeting her at a hotel. Investment consultant Bernie Madoff lied to 4,800 clients over 25 years and stole over fifty billion dollars from them. Baseball hitter Alex Rodriguez ("A-Rod") lied to investigators about steroid use. Politicians routinely lie to citizens ("Lobbyists can't buy my vote"), and citizens lie to the government (via cheating on taxes). Teachers lie to students

("The test will be easy"), and students lie to teachers ("I studied all night"). Parents lie to their children ("It won't hurt"), and children lie to their parents ("I was at a friend's house"). Dating partners lie to each other ("I've had a couple of previous sex partners"), women lie to men ("I had an orgasm"), and men lie to women ("I'll call"). The price of lying is high—distrust and alienation. A student in class wrote:

> At this moment in my life I do not have any love relationship. I find college dating to be very hard. The guys here lie to you about anything and you wouldn't know the truth. I find it's mostly about sex here and having a good time before you really have to get serious. That is fine, but that is just not what I am all about.

In addition to lying to gain sexual access is the behavior of cheating—having sex with someone else while involved in a relationship with a romantic partner. When 1,319 undergraduates were asked if they had cheated on a partner they were involved with, 37.4 percent responded that they had done so (Knox and Zusman 2009). McAlister et al. (2005) noted that extradyadic activity (defined as kissing or "sexual activity") among young adults who were dating could be predicted. Those young adults who had a high number of previous sexual partners, who were impulsive, who were not satisfied in their current relationship, and who had attractive alternatives were more vulnerable to being unfaithful.

Forms of Dishonesty and Deception

Dishonesty and deception take various forms. In addition to telling an outright lie, people may exaggerate the truth, pretend, conceal the truth, or withhold information. Regarding the latter, in virtually every relationship, partners may not share things with each other about themselves or their past. We often withhold information or keep secrets in our intimate relationships for what we believe are good reasons—we believe that we are protecting our partners from anxiety or hurt feelings, protecting ourselves from criticism and rejection, and protecting our relationships from conflict and disintegration. Finkenauer and Hazam (2000) found that happy relationships depend on withholding information. The researchers contend, "Nobody wants to be criticized (for example, 'You're really fat') or talk about topics that are known to be conflictive (for example, 'You should not have spent that much money')." Ennis et al. (2008) noted three types of lies: (1) self-centered to protect one's self ("I didn't do it."); (2) oriented to protect another ("Your hair looks good today."); or (3) altruistic to protect a third party ("She didn't do it.").

Lying in College Student Relationships

Lying is epidemic in college student relationships. In response to the statement, "I have lied to a person I was involved with," 77 percent of 1,319 undergraduates reported "yes" (Knox and Zusman 2009). Almost one in four (23.9 percent) reported having lied to a partner about their previous number of sexual partners (ibid.).

Even in "monogamous" relationships, there is considerable lying. Vail-Smith et al. (2010) found that 27.2 percent of the males and 19.8 percent of the females of 1,341 undergraduates reported having oral, vaginal, or anal sex outside of a relationship that their partner considered monogamous. People most likely to cheat in these "monogamous" relationships were men over the age of 20, those who were binge drinkers, members of a fraternity, male NCAA athletes, and those who reported that they were "nonreligious." The data suggest a need for people in "committed" relationships to reconsider their risk of sexually transmitted infections and to protect themselves via condom usage.

In addition, one of the ways in which college students deceive their partners is by failing to disclose that they have an STI. Approximately 25 percent of college students will contract an STI while they are in college (Purkett 2009).

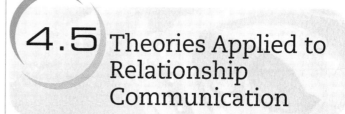

4.5 Theories Applied to Relationship Communication

Symbolic interactionism and social exchange are theories that help to explain the communication process.

Symbolic Interactionism

Interactionists examine the process of communication between two actors in terms of the meanings each attaches to the actions of the other. Definition of the situation, the looking-glass self, and taking the role of the other (discussed in Chapter 1) are all relevant to

GENDER DIFFERENCES IN COMMUNICATION

Women and men differ in their approach to and patterns of communication. Women are more communicative about relationship issues, view a situation emotionally, and initiate discussions about relationship problems. Deborah Tannen (1990; 2006) is a specialist in communication. She observed that, to women, conversations are negotiations for closeness in which they try "to seek and give confirmations and support, and to reach consensus" (1990, 25). A woman's goal is to preserve intimacy and avoid isolation. To men, conversations are about winning and achieving the upper hand.

Women also tend to approach a situation emotionally. A husband might react to a seriously ill child by putting pressure on the wife to be mature about the situation (for example, stop crying) and by encouraging stoicism (asking her not to feel sorry for herself). Wives, on the other hand, want their husbands to be more emotional (by asking them to cry to show that they really care that their child is ill). Mothers and fathers also speak differently to their children. Shinn and O'Brien (2008) observed the interactions between parents and their third grade children and found that mothers used more affiliative (relationship) speech than fathers, and fathers used more assertive speech than mothers. No sex differences in children's speech were found, suggesting that these differences do not emerge until later.

Women disclose more in their relationships than men do, according to Gallmeier et al. (1997). In this study of 360 undergraduates, women were more likely to disclose information about previous love relationships, previous sexual relationships, their love feelings for the partner, and what they wanted for the future of the relationship. They also wanted their partners to reciprocate their (the women's) disclosure, but such disclosure was not forthcoming. Punyanunt-Carter (2006) confirmed that female college students are more likely to disclose than are male college students.

Behringer (2005) found that in spite of the fact that women and men may have different communication foci, they both value openness, honesty, respect, humor, and resolution as principal components of good communication. They also each endeavor to create a common reality. Hence, although spouses may be on different pages, they are reading the same book.

understanding how partners communicate. With regard to resolving a conflict over how to spend the semester break (for example, vacation alone or go to see parents), the respective partners must negotiate their definitions of the situation (is it about their time together as a couple or their loyalty to their parents?). The looking-glass self involves looking at each other and seeing the reflected image of someone who is loved and cared for and someone with whom a productive resolution is sought. Taking the role of the other involves each partner's understanding the other's logic and feelings about how to spend the break.

Social Exchange

Exchange theorists suggest that the partners' communication can be described as a ratio of rewards to costs. Rewards are positive exchanges, such as compliments, compromises, and agreements. Costs refer to negative exchanges, such as critical remarks, complaints, and attacks. When the rewards are high and the costs are low, the outcome is likely to be positive for both partners (profit). When the costs are high and the rewards low, neither may be satisfied with the outcome (loss).

When discussing how to spend the semester break, the partners are continually in the process of exchange—not only in the words they use but also in the way they use them. If the communication is to continue, both partners need to feel acknowledged for their points of view and to feel a sense of legitimacy and respect. Communication in abusive relationships is characterized by the parties criticizing and denigrating each other, which usually results in a shutdown of the communication process.

4.6 Fighting Fair: Seven Steps in Conflict Resolution

When a disagreement ensues, it is important to establish rules for fighting that will leave the partners and their relationship undamaged after the disagreement. Such guidelines for fair fighting include not calling each other names, not bringing up past misdeeds, not attacking each other, and not beginning a heated discussion late at night. In some cases, a good night's sleep has a way of altering how a situation is viewed and may even result in the problem no longer being an issue.

Fighting fairly also involves keeping the interaction focused, respective, and moving toward a win-win outcome. If recurring issues are not discussed and resolved, conflict may create tension and distance in the relationship, with the result that the partners stop talking, stop spending time together, and stop being intimate. A conflictual, unsatisfactory marriage is similar to divorce in terms of its impact on the diminished psychological, social, and physical well-being of the partners (Hetherington 2003). Developing and using skills for fair fighting and conflict resolution are critical for the maintenance of a good relationship. Resolving issues via communication is not easy. Rhoades et al. (2009) noted that relationship partners who decided to live together to test their relationship indicated that communication was difficult/negative.

The Seven Steps

The seven steps for fair fighting and resolution of interpersonal conflict:

1. Address recurring, disturbing issues
2. Identify new desired behaviors
3. Identify perceptions to change
4. Summarize you partner's perspective
5. Generate alternative win-win solutions
6. Forgive
7. Be alert to defense mechanisms

Howard Markman is head of the Center for Marital and Family Studies at the University of Denver. He and his colleagues have been studying 150 couples at yearly intervals (beginning before marriage) to determine those factors most responsible for marital success. They have found that communication skills that reflect the ability to handle conflict, which they call "constructive arguing," are the single biggest predictor of marital success over time (Marano 1992). According to Markman, "Many people believe that the causes of marital problems are the differences between people and problem areas such as money, sex, children. However, our findings indicate it is not the differences that are important, but how these differences and problems are handled, particularly early in marriage" (Marano 1992, 53).

Address Recurring, Disturbing Issues

Addressing issues in a relationship is important. As noted earlier, couples who stack resentments rather than discuss conflictual issues do no service to their relationship. Indeed, the healthiest response to feeling upset about a partner's behavior is to engage the partner in a discussion about the behavior. Not to do so is to let the negative feelings fester, which will result in emotional and physical withdrawal from the relationship. For example, Pam is jealous that Mark spends more time with other people at parties than with her. "When we go someplace together," she blurts out, "he drops me to disappear with someone else for two hours." Her jealousy is spreading to other areas of their relationship. "When we are walking down the street and he turns his head to look at another woman, I get furious." If Pam and Mark don't discuss her feelings about Mark's behavior, their relationship may deteriorate as a result of a negative response cycle: He looks at another woman and she gets angry; he gets angry at her getting angry and finds that he is even more attracted to other women; she gets angrier because he escalates his looking at other women, and so on.

To bring the matter up, Pam might say, "I feel jealous when you spend more time with other women at parties than with me. I need some help in dealing with these feelings." By expressing her concern in this way, she has identified the problem from her perspective and asked her partner's cooperation in handling it.

When discussing difficult relationship issues, it is important to avoid attacking, blaming, or being negative. Such reactions reduce the motivation of the partner to talk about an issue and thus reduce the probability of a positive outcome.

Using good timing in discussing difficult issues with your partner is also important. In general, it is best to discuss issues or conflicts when:

(1) you and your partner are in private

(2) you and your partner have ample time to talk

(3) you and your partner are rested and feeling generally good (avoid discussing conflict issues when one of you is tired, upset, or under unusual stress).

Identify New Desired Behaviors

Dealing with conflict is more likely to result in resolution if the partners focus on what they want rather than what they don't want. For example, rather than tell Mark she doesn't want him to spend so much time with other women at parties, Pam might tell him that she wants him to spend more time with her at parties.

Identify Perceptions to Change

Rather than change behavior, changing one's perception of a behavior may be easier and quicker. Rather than expect one's partner to always be "on time," it may be easier to drop the expectation that one's partner be on time and to stop being mad about something that doesn't matter. Pam might also decide that it does not matter that Mark looks at and talks to other women. If she feels secure in his love for her, the behavior is inconsequential.

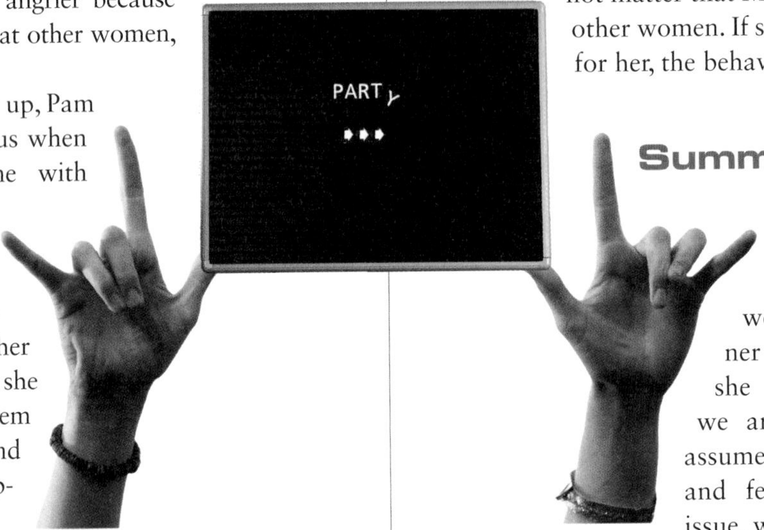

© Workbook Stock/Jupiterimages / © Tristan Fewings/PYMCA/Jupiterimages

Summarize Your Partner's Perspective

We often assume that we know what our partner thinks and why he or she does things. Sometimes we are wrong. Rather than assume how our partner thinks and feels about a particular issue, we might ask open-ended questions in an effort to learn our partner's thoughts and feelings about a particular situation.

Pam's words to Mark might be, "What is it like for you when we go to parties?" and "How do you feel about my jealousy?" Once your partner has shared thoughts about an issue with you, summarizing your partner's perspective in a nonjudgmental way is important. After Mark has told Pam how he feels about their being at parties together, she can summarize his perspective by saying, "You feel that I cling to you more than I should, and you would like me to let you wander around without feeling like you're making me angry." (She may not agree with his view, but she knows exactly what it is—and Mark knows that she knows.) In addition, Mark should summarize Pam's view—"You enjoy our being together and prefer that we hang relatively close to each other when we go to parties. You do not want me off in a corner talking to another girl or dancing."

© Brand X Pictures/Jupiterimages

Generate Alternative Win-Win Solutions

Looking for win-win solutions to conflicts is imperative. Solutions in which one person wins means that one person's needs are not met. As a result, the person who loses may develop feelings of resentment, anger, hurt, and hostility toward the winner and may even look for ways to get even. In this way, the winner is also a loser. In intimate relationships, one winner really means two losers.

Generating win-win solutions to interpersonal conflict often requires **brainstorming**. The technique of brainstorming involves suggesting as many alternatives as possible without evaluating them. Brainstorming is crucial to conflict resolution because it shifts the partners' focus from criticizing each other's perspective to working together to develop alternative solutions.

Knox et al (1995) reported research on the degree to which 200 college students who were involved in ongoing relationships were involved in win-win, win-lose, and lose-lose relationships. Descriptions of the various relationships follow:

Win-win relationships are those in which conflict is resolved so that each partner derives benefits from the resolution. For example, suppose a couple have a limited amount of money and disagree on whether to spend it on eating out or on seeing a current movie. One possible win-win solution might be for the couple to eat a relatively inexpensive dinner and rent a movie.

An example of a **win-lose solution** occurs when one partner benefits at the expense of the other. One of the partners would get what he or she wanted (eat out or go to a movie), while the other partner got nothing of what he or she wanted. Caughlin and Ramey (2005) studied demand-and-withdraw patterns in parent–adolescent dyads and found that the demand on the part of one of them was usually met by withdrawal on the part of the other partner. Such demand may reflect a win-lose interaction.

A **lose-lose solution** is one in which neither partner benefits. Both partners get nothing that they want; in the scenario presented, the partners would neither go out to eat nor see a movie and would be mad at each other.

More than three-quarters (77.1 percent) of the 200 students reported being involved in a win-win relationship, with men and women reporting similar percentages. Of the respondents, 20 percent were involved in win-lose relationships. Only 2 percent reported that they were involved in lose-lose relationships. Of the students in win-win relationships, 85 percent reported that they expected to continue their relationship, in contrast to only 15 percent of

brainstorming suggesting as many alternatives as possible without evaluating them.

win-win relationship a relationship in which conflict is resolved so that each partner derives benefits from the resolution.

win-lose solution a solution to a conflict in which one partner benefits at the expense of the other.

lose-lose solution a solution to a conflict in which neither partner benefits.

students in win-lose relationships. No student in a lose-lose relationship expected the relationship to last.

After a number of solutions are generated, each solution should be evaluated and the best one selected. In evaluating solutions to conflicts, it may be helpful to ask the following questions:

1. Does the solution satisfy both individuals? Is it a win-win solution?
2. Is the solution specific? Does it specify exactly who is to do what, how, and when?
3. Is the solution sensible? Can both parties realistically follow through with what they have agreed to do?
4. Does the solution prevent the problem from recurring?
5. Does the solution specify what is to happen if the problem recurs?

Kurdek (1995) emphasized that conflict-resolution styles that stress agreement, compromise, and humor are associated with marital satisfaction, whereas conflict engagement, withdrawal, and defensive styles are associated with lower marital satisfaction. In Kurdek's study of 155 married couples, the style where the wife engaged the husband in conflict and the husband withdrew was particularly associated with low marital satisfaction for both spouses.

Communicating effectively and creating a win-win context in one's relationship contributes to a high quality marital relationship, which research indicates is good for one's emotional and physical health.

Forgive

Too little emphasis is placed on forgiveness as an emotional behavior that can move a couple from a deadlock to resolution. Forgiveness requires acknowledging to one's self and the partner that either of them can make a mistake and that the focus should be on moving beyond the incident . . . to "let it go." It takes more energy to hold on to resentment than to move beyond it. One reason some people do not forgive a partner for a transgression is that one can use the fault to control the relationship. "I wasn't going to let him forget," said one woman of her husband who had an affair with his coworker.

Gordon et al. (2005) emphasized that forgiveness is an important factor in a couple's recovery from infidelity on the part of one or both partners. Toussaint and Webb (2005) noted that women and men are equally forgiving. Day and Maltby (2005) found that individuals who are not capable of forgiveness tend to withdraw from social relationships and to become more lonely and/or socially isolated.

"Forgiveness ought to be like a cancelled note—torn in two, and burned up, so that it never can be shown against one.
—Henry Ward Beecher

© Jaggat Images/Alamy

Be Alert to Defense Mechanisms

Effective conflict resolution is sometimes blocked by **defense mechanisms**—unconscious techniques that function to protect individuals from anxiety and to minimize emotional hurt. The following paragraphs discuss some common defense mechanisms.

Escapism is the simultaneous denial of and withdrawal from a problem. The usual form of escape is avoidance. The spouse becomes "busy" and "doesn't have time" to think about or deal with the problem, or the partner may escape into recreation, sleep, alcohol, drugs, or work. Denying and withdrawing from problems in relationships offer no possibility for confronting and resolving the problems.

Rationalization is the cognitive justification for one's own behavior that unconsciously conceals one's true motives. For example, one wife complained that her husband spent too much time at the health club in the evenings. The underlying reason for the husband's going to the health club was to escape an unsatisfying home life. However, the idea that he was in a dead marriage was too painful and difficult for the husband to face, so he rationalized to himself and his wife that he spent so much time at the health club because he made a lot of important business contacts there. Thus, the husband concealed his own true motives from himself (and his wife).

Projection occurs when one spouse unconsciously attributes individual feelings, attitudes, or desires to the partner while avoiding recognition that these are his or her own thoughts, feelings, and desires. For example, the wife who desires to have an affair may accuse her husband of being unfaithful to her. Projection may be seen in such statements as "You spend too much money" (projection for "I spend too much money") and "You want to break up" (projection for "I want to break up"). Projection interferes with conflict resolution by creating a mood of hostility and defensiveness in both partners. The issues to be resolved in the relationship remain unchanged and become more difficult to discuss.

Displacement involves shifting one's feelings, thoughts, or behaviors from the person who evokes them onto someone else. The wife who is turned down for a promotion and the husband who is driven to exhaustion by his boss may direct their hostilities (displace them) onto each other rather than toward their respective employers. Similarly, spouses who are angry at each other may displace this anger onto someone else, such as their children.

By knowing about defense mechanisms and their negative impact on resolving conflict, you can be alert to them in your own relationships.

defense mechanisms unconscious techniques that function to protect individuals from anxiety and minimize emotional hurt.

escapism the simultaneous denial of and withdrawal from a problem.

rationalization the cognitive justification for one's own behavior that unconsciously conceals one's true motives.

projection attributing one's own feelings, attitudes, or desires to one's partner while avoiding recognition that these are one's own thoughts, feelings, and desires.

displacement shifting one's feelings, thoughts, or behaviors from the person who evokes them onto someone else.

Truth is better than friction.
—Charles Herguth

© Polka Dot Images/Jupiterimages

When Silence Is Golden

Even in the midst of a heated quarrel, some words should never be spoken. To do so is to destroy the relationship forever. William Berle, adopted son of comedian Milton Berle, recalled being in an argument with his dad who lashed out at him, "Oh yeah? Well, I did make one mistake and that was twenty-seven years ago when we adopted you . . . how do you like that you little prick!" (Berle and Lewis 1999, 188). The son recalled being devastated and walking into the next room to take out a pistol to kill himself. Luckily, a knock on the door interrupted his plan.

Chapter 5

Love and Finding a Partner

SECTIONS

© Flirt/Jupiterimages

J. H. Newman, cardinal and philosopher, noted that we tend to fear less that life will end but, rather, that life will never begin. His point targets the importance of love in one's life that provides an unparalleled richness, meaning, and happiness. Demir (2008) emphasized that involvement in a romantic relationship moves one to a new level of happiness independent of one's personality. In other words, although some individuals have personalities that tend to be happy anyway, love moves them to an even higher level.

We are also reminded that young lovers note that nothing is like being in love. The late trumpet player Chet Baker also noted the value of love in one's life—"I don't think life is really worth all the pain and effort and struggling if you don't have somebody that you love very much" (Gavin 2003, 349). When asked what word best characterizes her relationship with her husband Jay Leno, Mavis Leno said, "Joy. I don't just love Jay—I'm madly in love. And I say this as someone who didn't think the state could persist" (Burford 2005, 174).

For many, being in love is a prerequisite for remaining married. Falling out of love paves the way for divorce. Almost half (47.9 percent) of 1,319 university students surveyed reported that they would divorce their spouse if they fell out of love (Knox and Zusman 2009). Lovers affect each other. Schoebi (2008) identified hard (angry) and soft (depressed) emotions and the degree to which the emotions of one spouse affected another. When one spouse was experiencing hard emotions, the partner tended to mirror those, particularly when feeling interpersonal insecurity.

Love is very much a part of student life. More than half (57.8 percent) of 1,319 undergraduates surveyed reported that they were emotionally involved with one person, engaged, or married (Knox and Zusman 2009).

This chapter is concerned with the nature of love (both ancient and modern views), various theories of the origin of love, how love develops in a new relationship, and problems associated with love. Because jealousy in love relationships is common, we also examine its causes and consequences.

CHAPTER 5 LOVE

5.1 Ways of Conceptualizing Love

Love is elusive and impossible to define by those caught in its spell. Love is often confused with lust and infatuation (Jefson 2006). Love is about deep, abiding feelings; **lust** is about sexual desire; and **infatuation** is about emotional feelings based on little actual exposure to the love object. In the following section, we look at the various ways of conceptualizing love.

lust sexual desire.

infatuation emotional feelings based on little actual exposure to the love object.

Love Styles

Theorist John Lee (1973; 1988) identified a number of styles of love that describe the way lovers relate to each other. Keep in mind that the same individual may view love in more than one way at a time or may view love in different ways at different times. These love styles are also independent of one's sexual orientation—no one love style is characteristic of heterosexuals or homosexuals.

1. **Ludic.** In the **ludic love style**, the ludic lover views love as a game, refuses to become dependent on any one person, and does not encourage another's intimacy. Two essential skills of the ludic lover are to juggle several partners at the same time and to manage each relationship so that no one partner is seen too often. These strategies help to ensure that the relationship does not deepen into an all-consuming love. Don Juan represented the classic ludic lover. "Love 'em and leave 'em" is the motto of the ludic lover. Tzeng et al. (2003) found that, whereas men were more likely than women to be ludic lovers, ludic love characterized the love style of college students the least.

   In a study (Paul et al. 2000) of "hookups" between college students, certain love styles were characteristic of students who hooked up. Distinguishing features of those who had noncoital hookups were a ludic love style and high concern for personal safety. These individuals may have been participating in collegiate cultural expectations by engaging in "playful" sexual exploration but refraining from intercourse out of their concern for personal safety. Those who engaged in coital hookups were also characterized by ludic love styles that included heavy drinking. The researchers worried that the combination of ludic orientation (motivated by the thrill of the game) and alcohol intoxication might be a precursor to sexual experiences that were forced or unwanted by a partner.

   The ludic lover is sometimes characterized as manipulative and uncaring. However, ludic lovers may also be compassionate and very protective of another's feelings. For example, some uninvolved, soon-to-graduate seniors avoid involvement with anyone new and become ludic lovers so as not to encourage anyone.

ludic love style love style in which love is viewed as a game whereby the love interest is one of several partners, is never seen too often, and is kept at an emotional distance.

pragma love style love style that is logical and rational. The love partner is evaluated in terms of assets and liabilities.

eros love style love style characterized by passion and romance.

manic love style an out-of-control love whereby the person "must have" the love object. Obsessive jealousy and controlling behavior are symptoms of manic love.

© iStockphoto.com

2. **Pragma.** The **pragma love style** is the love of the pragmatic—that which is logical and rational. Pragma lovers assess their partners on the basis of assets and liabilities. Economic security may be regarded as very important. Pragma lovers do not become involved with interracial, long-distance, or age discrepant partners, because logic argues against doing so. Bulcroft et al. (2000) noted that, increasingly, individuals are becoming more pragmatic about their love choices.

3. **Eros.** Just the opposite of the pragmatic love style, the **eros love style** is one of passion and romance. Intensity of both emotional and sexual feelings dictates one's love involvements. Research varies on the degree to which passion characterizes most relationships. Tzeng et al. (2003) assessed the love styles of more than 700 college students and found that eros was the most common love style of women and men. Similarly, a *Redbook* (2005) survey found that 6 percent of 700 readers reported that "passionate" best described their relationship. Hendrick et al. (1988) found that couples who were more romantically and passionately in love were more likely to remain together than couples who avoided intimacy by playing games with each other.

4. **Mania.** The person with a **manic love style** feels intense emotion and sexual passion but is out

of control. The person is possessive, dependent, and "must have" the beloved. People who are extremely jealous and controlling reflect manic love. "If I can't have you, no one else will" is sometimes the mantra of the manic lover. Stalking is an expression of love gone wild. O. J. Simpson once said, "If I killed her, it would be because I loved her . . . right?"

5. **Storge.** The **storge love style** is a calm, soothing, nonsexual love devoid of intense passion. Respect, friendship, commitment, and familiarity are characteristics that help to define the relationship. The partners care deeply about each other but not in a romantic or lustful sense. Their love is also more likely to endure than fleeting romance. One's grandparents who have been married fifty years and who still love and enjoy each other are likely to have a storge type of love.
6. **Agape.** One of the forms of love identified by the ancient Greeks, the **agape love style** is characterized by a focus on the well-being of the beloved, with little regard for reciprocation. These nurturing and caring partners are concerned only about the welfare and growth of each other. The love parents have for their children is often described as agape love.

Romantic versus Realistic Love

Love may also be described as being on a continuum from romanticism to realism. For some people, love is romantic; for others, it is realistic. **Romantic love** is characterized by such beliefs as "love at first sight," and "If I were really in love, I would marry someone I had known for only a short time." Regarding these beliefs, 26.2 percent of 1,319 undergraduates reported that they had experienced love at first sight; a similar percentage (26 percent) reported that they would marry quickly if they were in love (Knox and Zusman 2009). Men were significantly more likely than women to believe in love at first sight (32.3 percent versus 24.3 percent). One explanation is that men must be visually attracted to young, healthy females to inseminate them. This biologically based reproductive attraction is interpreted as a love attraction so that the male feels immediately drawn to the female, but he may actually see an egg needing fertilization. Further evidence that males are more romantic than females is from Dotson-Blake et al. (2008) who found that men were significantly more likely (85 percent versus 73 percent) than women to believe that they could solve any relationship problem as long as they were in love.

In regard to love at first sight, Barelds and Barelds-Dijkstra (2007) studied the relationships of 137 married couples or cohabitants (together for an average of twenty-five years) and found that those who fell in love at first sight had similar relationship quality to those couples who came to know each other more gradually. Huston et al. (2001) found that, after two years of marriage, the couples who had fallen in love more slowly were just as happy as couples who fell in love at first sight.

The symptoms of romantic love include drastic mood swings, palpitations of the heart, and intrusive thoughts about the partner. F. Scott Fitzgerald immortalized the concept of romantic obsession in *The Great Gatsby*. Of Daisy Buchanan, he wrote, "She was the first girl I ever loved and I have faithfully avoided seeing her . . . to keep that illusion perfect." He actually was writing about a real-life true love, Ginevra King, whom he had met when she was 16; she eventually married another man (West 2005).

Infatuation is sometimes regarded as synonymous with romantic love. Infatuation comes from the same root word as *fatuous*, meaning "silly" or "foolish," and refers to a state of passion or attraction that is not based on reason. Infatuation is characterized by the tendency to idealize the love partner. People who are infatuated magnify their lovers' positive qualities ("My partner is always happy") and overlook or minimize their negative qualities ("My partner doesn't have a problem with alcohol; he just likes to have a good time").

> Jealousy *is all the fun you think they had.*
>
> —Erica Jong, American writer

storge love style a love consisting of friendship that is calm and nonsexual.

agape love style love style characterized by a focus on the well-being of the love object, with little regard for reciprocation. The love of parents for their children is agape love.

romantic love an intense love whereby the lover believes in love at first sight, only one true love, and that love conquers all.

conjugal (married) love the love between married people characterized by companionship, calmness, comfort, and security.

In contrast to romantic love is realistic love. Realistic love is also known as conjugal love. **Conjugal (married) love** is less emotional, passionate, and exciting than romantic love and is characterized by companionship, calmness, and security.

Triangular View of Love

Sternberg (1986) developed the "triangular" view of love, consisting of three basic elements: intimacy, passion, and commitment. The presence or absence of these three elements creates various types of love experienced between individuals, regardless of their sexual orientation. These various types include:

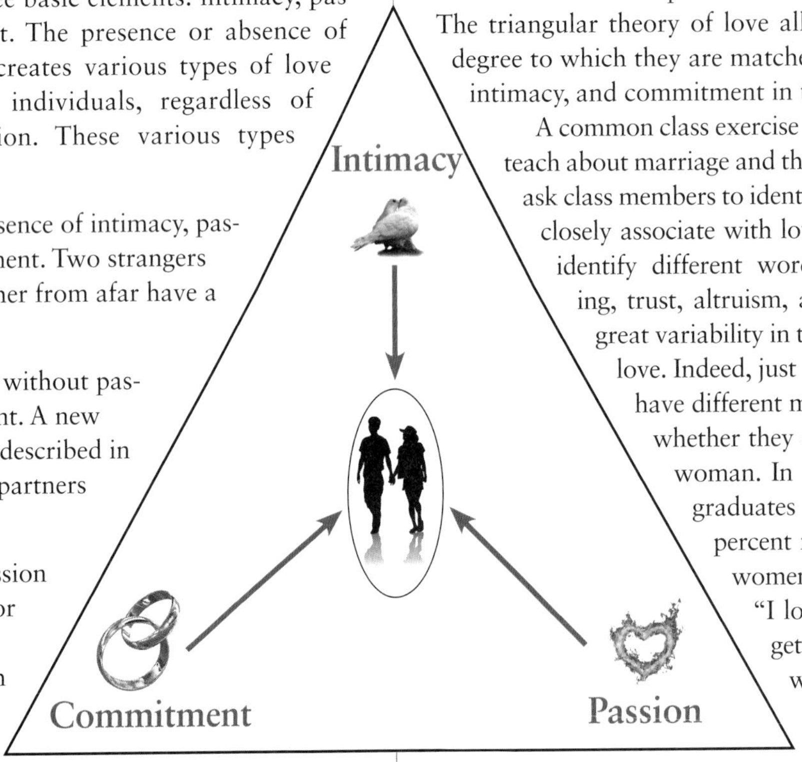

1. **Nonlove**—the absence of intimacy, passion, and commitment. Two strangers looking at each other from afar have a nonlove.
2. **Liking**—intimacy without passion or commitment. A new friendship may be described in these terms of the partners liking each other.
3. **Infatuation**—passion without intimacy or commitment. Two people flirting with each other in a bar may be infatuated with each other.
4. **Romantic love**—intimacy and passion without commitment. Love at first sight reflects this type of love.
5. **Conjugal love** (also known as companionate love)—intimacy and commitment without passion. A couple who has been married for fifty years is said to illustrate conjugal love.
6. **Fatuous love**—passion and commitment without intimacy. Couples who are passionately wild about each other and talk of the future but do not have an intimate connection with each other have a fatuous love.
7. **Empty love**—commitment without passion or intimacy. Couples who stay together for social and legal reasons but who have no spark or emotional sharing between them have an empty love.
8. **Consummate love**—combination of intimacy, passion, and commitment; Sternberg's view of the ultimate, all-consuming love.

Individuals bring different combinations of the elements of intimacy, passion, and commitment (the triangle) to the table of love. One lover may bring a predominance of passion, with some intimacy but no commitment (romantic love), whereas the other person brings commitment but no passion or intimacy (empty love). The triangular theory of love allows lovers to see the degree to which they are matched in terms of passion, intimacy, and commitment in their relationship.

A common class exercise among professors who teach about marriage and the family is to randomly ask class members to identify one word they most closely associate with love. Invariably, students identify different words (commitment, feeling, trust, altruism, and so on), suggesting great variability in the way we think about love. Indeed, just the words "I love you" have different meanings, depending on whether they are said by a man or a woman. In a study of 147 undergraduates (72 percent female, 28 percent male), men (more than women) reported that saying "I love you" was a ploy to get a partner to have sex, whereas women (more than men) reported that saying "I love you" was a reflection of their feelings, independent of a specific motive (Brantley et al. 2002).

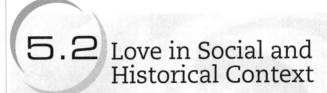

5.2 Love in Social and Historical Context

Though we think of love as an individual experience, the society in which we live exercises considerable control over our love object of choice and conceptualizes it in various ways.

WHEN LOVERS ARE AGREED, NOT EVEN THEIR PARENTS CAN CONTROL THEM.

—*Bia Xingqian, Tang Dynasty*
The Little Book of Chinese Proverbs

Social Control of Love

The ultimate social control of love is **arranged marriage.** The parents select the mate for their child in an effort to prevent any potential love relationship from forming with the "wrong" person and to ensure that the child marries the "right" person. Such a person must belong to the desired social class and have the economic resources that the parents desire. Marriage is regarded as the linking of two families; the love feelings of the respective partners are irrelevant. Love is expected to follow marriage, not precede it. Parents arrange 80 percent of marriages in China, India, and Indonesia (three countries representing 40 percent of the world's population).

In an arranged marriage, the couple may get a fifteen-minute meeting, followed in a few months by a wedding. However, love marriages—where the individuals meet, fall in love, and then convince and cajole their parents to approve of a wedding—are slowly becoming more common in Eastern societies (Jones 2006). Similarly, to accommodate the needs of traditional parents in a small village in Western Turkey who want to arrange the marriage of their children but appear "modern," anthropologist Hart (2007) observed that they now allow a period of time for the couple to develop romantic love feelings for each other.

America is a country that prides itself on the value of individualism. We proclaim that we are free to make our own choices. Not so fast, however. Love may be blind, but it knows what color a person's skin is. The data are clear—potential spouses seem to see and select people of similar color, as more than 95 percent of people marry someone of their own racial background (*Statistical Abstract of the United States, 2009*, Table 59). Hence, parents and peers may approve of their offsprings' and friends' love choice when the partner is of the same race and disapprove when the choice is not. These approval and disapproval mechanisms illustrate social control of love.

Another example of the social control of love is that individuals attracted to someone of the same sex quickly feel the social and cultural disapproval of this attraction. (We discuss same-sex relationships later in the text.) These relationships are challenged by the lack of institutional support. Even though Maine, Iowa, New Hampshire, Vermont, Massachusetts, and Connecticut permit same-sex marriage, these couples are no longer "married" once they cross the border into another state. Nevertheless, regardless of the law, same-sex love is common. Diamond (2003) emphasized that individuals are biologically wired and capable of falling in love and establishing intense emotional bonds with members of their own or the opposite sex (hence, one's partners for love desire and for sexual desire can be different).

Because romantic love is such a powerful emotion and marriage such an important relationship, mate selection is not left to chance in connecting an outsider into an existing family. Parents inadvertently influence the mate choice of their children by moving to certain neighborhoods, joining certain churches, and enrolling their children in certain schools. Doing so increases the chance that their offspring will "hang out" with, fall in love with, and marry people who are similar in race, education, and social class. Peers exert a similar influence on homogenous mating by approving or disapproving certain partners. Their motive is similar to that of parents—they want to feel comfortable around the people their peers bring with them to social encounters. Both parents and peers are influential, as most offspring and friends end up falling in love with and marrying people of the same race, education, and social class.

Social approval of one's partner is normally important for a love relationship to proceed on course. Even the engagement of Prince Charles to Camilla Parker Bowles received the "blessing" of Queen Elizabeth, and 70 percent of Britons either approved of it or did not care (Soriano 2005).

Partners also use love to control each other. Fehr and Harasymchuk (2005) noted that a comment expressing dissatisfaction by a romantic partner has considerably more negative emotional

arranged marriage mate selection pattern whereby parents select the spouse of their offspring.

impact than a similar comment by a friend. In effect, we give considerable credence to what our love partner thinks of us, and we get upset when they criticize us.

Love in Medieval Europe—from Economics to Romance

Love in the 1100s was a concept influenced by economic, political, and family structure. In medieval Europe, land and wealth were owned by kings controlling geographical regions—kingdoms. When so much wealth and power were at stake, love was not to be trusted as the mechanism for choosing spouses for royal offspring. Rather, marriages of the sons and daughters of the aristocracy were arranged with the heirs of other states with whom an alliance was sought. Love was not tied to marriage but was conceptualized as an adoration of physical beauty (often between a knight and his beloved) and as spiritual and romantic, even between people not married or of the same sex. Hence, romantic love had its origin in extramarital love and was not expected between spouses (Trachman and Bluestone 2005).

The presence of kingdoms and estates and the patrimonial households declined with the English revolutions of 1642 and 1688 and the French Revolution of 1789. No longer did aristocratic families hold power; the power was transferred to individuals through parliaments or other national bodies. Even today, English monarchs are figureheads, with parliament handling the real business of international diplomacy. Because wealth and power were no longer in the hands of individual aristocrats, the need to control mate selection decreased and the role of love changed. Marriage became less of a political and business arrangement and more of a mutually desired emotional union. Just as bureaucratic structure held partners together in medieval society, a new mechanism—love—would now provide the emotional and social bonding.

Hence, love in medieval times changed from a feeling irrelevant to marriage—because individuals (representing aristocratic families) were to marry even though they were not in love—to a feeling that bonded a woman and a man together for marriage.

dowry (trousseau) the amount of money or valuables a woman's father pays a man's father for the man to marry his daughter. It functioned to entice the man to marry the woman, because an unmarried daughter stigmatized the family of the woman's father.

Love in Colonial America

Love in colonial America was similar to that in medieval times. Marriage was regarded as a business arrangement between the fathers of the respective families (Dugan 2005). An interested suitor would approach the father of a girl to express his desire to court his daughter. The fathers would generally confer on the amount of the **dowry** (also known as the **trousseau**), which included the money and/or valuables the girl's father would pay the boy's father. Because unmarried women were stigmatized, marrying them off was desirable; thus, the dowry was an added inducement for a boy to marry the girl. Fathers could deny their daughters a dowry if the daughters were unwilling to marry the man their father chose. Love was not totally absent, however; sometimes a girl could persuade her father to tell the suitor she was not interested.

5.3 How Love Develops in a New Relationship

Various social, physical, psychological, physiological, and cognitive conditions affect the development of love relationships.

Americans have radar for love.

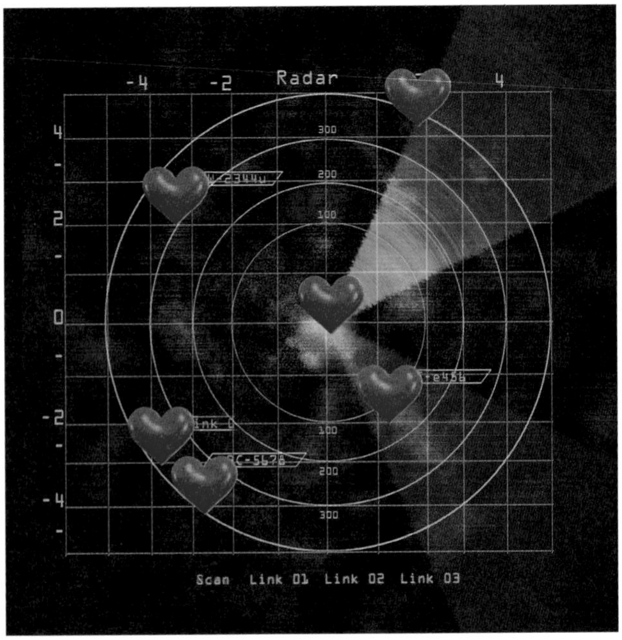

Social Conditions for Love

Love is a social label given to an internal feeling. Our society promotes love through popular music, movies, and novels. These media convey the message that love is an experience to pursue, enjoy, and maintain. People who fall out of love are encouraged to try again: "love is lovelier the second time you fall." Unlike people reared in Eastern cultures, Americans grow up in a context to turn on their radar for love.

Body Type Condition for Love

The probability of being involved in a love relationship is influenced by approximating the cultural ideal of physical appearance. Halpern et al. (2005) analyzed data on a nationally representative sample of 5,487 African American, white, and Hispanic adolescent females and found that, for each one-point increase in body mass index (BMI), the probability of involvement in a romantic relationship dropped by 6 percent. Hence, to the degree that a woman approximates the cultural ideal of being trim and "not being fat," she increases the chance of attracting a partner and becoming involved in a romantic love relationship. One of our former students dropped from 225 pounds to 125 pounds and noted, "You wouldn't believe the dramatic difference in the way guys noticed and talked to me [between] when I was beefed up and when I was trim. I was engaged within three months of getting the weight off and am now married."

Ambwani and Strauss (2007) found that body image has an effect on sexual relations and that relationships affect their self-image. Hence, women who felt positive about their body were more likely to report having sexual relations with a partner. The fact that they were in a relationship was associated with positive feelings about themselves.

Psychological Conditions for Love

Two psychological conditions associated with the development of healthy love relationships are high self-esteem and self-disclosure.

Self-Esteem High self-esteem is important for defining success (Bianchi and Povilavicius 2006). High self-esteem is also important for developing healthy love relationships because it enables individuals to feel worthy of being loved. Feeling good about yourself allows you to believe that others are capable of loving you. Individuals with low self-esteem doubt that someone else can love and accept them (DeHart et al. 2002).

Benefits of Self-Esteem

1. **It allows one to be open and honest with others about both strengths and weaknesses.**
2. **It allows one to feel generally equal to others.**
3. **It allows one to take responsibility for one's own feelings, ideas, mistakes, and failings.**
4. **It allows for the acceptance of both strengths and weaknesses in one's self and others.**
5. **It allows one to validate one's self and not to expect the partner to do this.**
6. **It permits one to feel empathy—a very important skill in relationships.**
7. **It allows separateness and interdependence, as opposed to fusion and dependence.**

Positive physiological outcomes also follow from high self-esteem. People who feel good about themselves are less likely to develop ulcers and are likely to cope with anxiety better than those who don't. In contrast, low self-esteem has devastating consequences for individuals and the relationships in which they become involved. Not feeling loved as a child and, worse, feeling rejected and abandoned creates the context for the development of a negative self-concept and mistrust of others. People who have never felt loved and wanted may require constant affirmation from a partner as to their worth, and may cling desperately to that person out of fear of being abandoned. Such dependence (the modern term is *codependency*) may also encourage staying in unhealthy relationships (for example, abusive and alcoholic relationships), because the person may feel "this is all I deserve." Fuller and Warner (2000) studied 257 college students and observed that women had higher codependency scores than men. Codependency was also associated with being reared in families that were stressful and alcoholic.

One characteristic of individuals with low self-esteem is that they may love too much and be addicted to unhealthy love relationships. Petrie et al. (1992) studied fifty-two women who reported that they were involved in unhealthy love relationships in which they had selected men with problems (such as alcohol or other drug addiction) that they attempted to solve at the expense of neglecting themselves. "Their preoccupation with correcting the problems of others may be an attempt to achieve self-esteem," the researchers noted (p. 17). "I know I can help this man" is the motif of these women.

Although having positive feelings about one's self when entering into a love relationship is helpful, sometimes these develop after one becomes involved in the relationship. "I've always felt like an ugly duckling," said one woman. "But once I fell in love with him and him with me, I felt very different. I felt very good about myself then because I knew that I was somebody that someone else loved." High self-esteem, then, is not necessarily a prerequisite for falling in love. People who have low self-esteem may fall in love with someone else as a result of feeling deficient. The love they perceive the other person has for them may compensate for the perceived deficiency and improve their self-esteem. This phenomenon can happen with two individuals with low self-esteem—love can elevate the self concepts of both individuals.

Self-Disclosure Disclosing one's self is necessary if one is to love—to feel invested in another (Radmacher and Azmitia 2006). Disclosed feelings about the partner included "how much I like the partner," "my feelings about our sexual relationship," "how much I trust my partner," "things I dislike about my partner," and "my thoughts about the future of our relationship"—all of which were associated with relationship satisfaction. Of interest in Ross's (2006) findings is that disclosing one's tastes and interests was negatively associated with relationship satisfaction. By telling a partner too much detail about what one likes, partners may discover something that turns them off and lowers relationship satisfaction.

Kito (2005) examined the self-disclosure patterns of 145 college students (both American and Japanese), and found that self-disclosure was higher in romantic relationships than in friendships and that Americans were more disclosing than the Japanese. The researcher also found that disclosure was higher in same-sex friendships than in cross-sex friendships.

It is not easy for some people to let others know who they are, what they feel, or what they think. They may fear that, if others really know them, they will be rejected as a friend or lover. To guard against this possibility, they may protect themselves and their relationships by allowing only limited information about their past behaviors and present thoughts and feelings. Some people keep others at a distance—they do not want psychological intimacy.

Trust is the condition under which people are most willing to disclose themselves. When people trust someone, they tend to feel that whatever feelings or information they share will not be judged and will be kept safe with that person. If trust is betrayed, people may become bitterly resentful and vow never to disclose themselves again. One woman said, "After I told my partner that I had had an abortion, he told me that I was a murderer and he never wanted to see me again. I was devastated and felt I had made a mistake telling him about my past. You can bet I'll be careful before I disclose myself to someone else" (personal communication).

Gallmeier et al. (1997) studied the communication patterns of 360 undergraduates at two universities and found that women were significantly more likely to disclose information about themselves. Specific areas of disclosure included previous love relationships, what they wanted for the future of the relationship, and what their partners did that they did not like.

Eight Dimensions of Self-Disclosure:

1. background and history
2. feelings toward the partner
3. feelings toward self
4. feelings about one's body
5. attitudes toward social issues
6. tastes and interests
7. money and work
8. feelings about friends

Physiological and Cognitive Conditions for Love

Physiological and cognitive variables are also operative in the development of love. The individual must be physiologically aroused and interpret this stirred-up state as love (Walster and Walster 1978).

Suppose, for example, that Dan is afraid of flying, but his fear is not particularly extreme and he doesn't like to admit it to himself. This fear, however, does cause him to be physiologically aroused. Suppose that Dan takes a flight and finds himself sitting next to Judy on the plane. With heart racing, palms sweating, and breathing labored, Dan chats with Judy as the plane takes off. Suddenly, Dan discovers that he finds Judy terribly attractive, and he begins to try to figure out ways that he can continue seeing her after the flight is over. What accounts for Dan's sudden surge of interest in Judy? Is

Love as a Context for Problems

Though love may bring great joy, it also creates a context for problems. Four such problems are simultaneous loves, involvement in an abusive relationship, making risky or dangerous choices, and the emergence of stalking.

© Stefan Klein/iStockphoto.com / © Pixland/Jupiterimages

Destruction of Existing Relationships

Sometimes the development of one love relationship is at the expense of another. A student in our classes noted that when she was 16, she fell in love with a person at work who was 25. Her parents were adamant in their disapproval and threatened to terminate the relationship with their daughter if she continued to see this man. The student noted that she initially continued to see her lover and to keep their relationship hidden. However, eventually she decided that giving up her family was not worth the relationship, so she stopped seeing him permanently. Others in the same situation would end the relationship with their parents and continue the relationship with their beloved. Choosing to end a relationship with one's parents is a downside of love.

Simultaneous Loves

For all the wonder of love, awareness that one's partner is in love with or having sex with someone else can create heartbreak. As we will discuss in the section on jealousy later in the chapter, multiple involvements are not a problem for some individuals or couples (for example, in compersion or polyamorous relationships). However, most people are not comfortable knowing their partner has other emotional or sexual relationships. Only 2.2 percent of 1,319 undergraduates agreed: "I can feel good about my partner having an emotional/sexual relationship with someone else" (Knox and Zusman 2009). In a study on "undesirable marriage forms," 91.2 percent of 111 undergraduates reported that they would "never participate" in a "group marriage" (Billingham et al. 2005). Hence, for most individuals, simultaneous lovers are viewed as a problem.

Unrequited or Unfulfilling Love Relationships

It is not unusual for lovers to vary in the intensity of their love for each other. The interesting question is whether being the person who loves more in a relationship is better than the person who loves less. The person who loves more may suffer more anguish. Such was the case of Jack Twist (in the now-classic *Brokeback Mountain*), who was hurt that his love interest, Ennis Del Mar, would not make time for them to continue their clandestine meetings on Brokeback Mountain.

Love as a Context for Risky, Dangerous, or Questionable Choices

Plato said that "love is a grave mental illness," and some research suggests that individuals in love make risky, dangerous, or questionable decisions. In a study on "what I did for love," college students reported that "driving drunk," "dropping out of school to be with my partner," and "having sex without protection" were among the more dubious choices they had made while they were under the spell of love (Knox et al. 1998). Similarly, a team of researchers examined the relationship between having a romantic love partner and engaging in minor acts of delinquency (for example, smoking cigarettes, getting drunk, skipping school); they found that females were particularly influenced by their "delinquent" boyfriends (Haynie et al. 2005). Their data source was the National Longitudinal Study of Adolescent Health. Furthermore, researchers have found that women who are "romantically in love" are less likely to use condoms with their partners. Doing so isn't regarded as very romantic, and they elect not to inject realism into a love context (East et al. 2007).

jealousy an emotional response to a perceived or real threat to an important or valued relationship.

reactive jealousy feelings that the partner may be straying.

anxious jealousy obsessive ruminations about the partner's alleged infidelity make one's life a miserable emotional torment.

possessive jealousy attacking the partner who is perceived as being unfaithful.

Judy really that appealing to him, or has he taken the physiological arousal of fear and mislabeled it as attraction? (Brehm 1992, 44).

Although most people who develop love feelings are not aroused in this way, they may be aroused or anxious about other issues (being excited at a party or feeling apprehensive about meeting someone), and may mislabel these feelings as those of attraction when they meet someone.

In the absence of one's cognitive functioning, love feelings are impossible. Individuals with brain cancer who have had the front part of their brain (between the eyebrows) removed are incapable of love. Indeed, emotions are not present in them at all (Ackerman 1994). The social, physical, psychological, physiological, and cognitive conditions are not the only factors important for the development of love feelings. The timing must also be right. There are only certain times in life (for example, when educational and career goals are met or within sight) when people seek a love relationship. When those times occur, a person is likely to fall in love with another person who is there and who is also seeking a love relationship. Hence, many love pairings exist because each of the individuals is available to the other at the right time—not because they are particularly suited for each other.

5.4 Jealousy in Relationships

Jealousy can be defined as an emotional response to a perceived or real threat to an important or valued relationship. People experiencing jealousy fear being abandoned and feel anger toward the partner or the perceived competition (Guerrero et al. 2005). As Buss (2000) emphasized, "Jealousy is an adaptive emotion, forged over millions of years. . . . It evolved as a primary defense against threats of infidelity and abandonment" (p. 56). People become jealous when they fear replacement. Although jealousy does not occur in all cultures (polyandrous societies value cooperation, not sexual exclusivity; Cassidy and Lee 1989), it does occur in our society and among both heterosexuals and homosexuals.

Of 1,319 university students, 41.7 percent reported, "I am a jealous person" (Knox and Zusman 2009). In another study, 185 students gave information about their experience with jealousy (Knox et al. 1999). On a continuum of 0 ("no jealousy") to 10 ("extreme jealousy"), with 5 representing "average jealousy," these students reported feeling jealous at a mean level of 5.3 in their current or last relationship. Students who had been dating a partner for a year or less were significantly more likely to report higher levels of jealousy (mean = 4.7) than those who had dated 13 months or more (mean = 3.3). Hence, jealously is more likely to occur early in a couple's relationship.

Types of Jealousy

Barelds-Dijkstra and Barelds (2007) identified three types of jealousy as reactive jealousy, anxious jealousy, and possessive jealousy. **Reactive jealousy** occurs as a reaction to something the partner is doing (for example, coming home late every night). Reactive jealousy consists of feelings that the partner may be straying. **Anxious jealousy** is obsessive ruminations about the partner's alleged infidelity that make one's life a miserable emotional torment. **Possessive jealousy** involves an attack on the partner or the alleged person to whom the partner is showing attention.

Causes of Jealousy

Jealousy can be triggered by external or internal factors.

External Causes External factors refer to behaviors a partner engages in that are interpreted as (1) an emotional and/or sexual interest in someone (or something) else, or (2) a lack of emotional and/or sexual interest in the primary partner. In the study of 185 students previously referred to, the respondents identified "actually talking to a previous partner" (34 percent) and "talking about a previous partner" (19 percent) as the most common sources of their jealousy. Also, men were more likely than women to report feeling jealous when their partner talked to a previous partner, whereas women were more likely than men to report feeling jealous when their partner danced with someone else.

Internal Causes Jealousy may also exist even when no external behavior indicates the partner is involved or interested in an **extradyadic relationship**—an emotional or sexual involvement between a member of a couple and someone other than the partner. Internal causes of jealousy refer to characteristics of individuals that predispose them to jealous feelings, independent of their partner's behavior, such as:

1. **Mistrust.** If an individual has been deceived or cheated on in a previous relationship, that individual may learn to be mistrustful in subsequent relationships. Such mistrust may manifest itself in jealousy. Mistrust and jealousy may be intertwined. Tilley and Brackley (2005) examined the factors involved for men convicted of assaulting a female and suggested that both jealousy and mistrust may have been involved in aggression against a female.
2. **Low self-esteem.** Individuals who have low self-esteem tend to be jealous because they lack a sense of self-worth and hence find it difficult to believe anyone can value and love them (Khanchandani 2005). Feelings of worthlessness may contribute to suspicions that someone else is valued more.
3. **Anxiety.** In general, individuals who experience higher levels of anxiety also display more jealousy (Khanchandani 2005).
4. **Lack of perceived alternatives.** Individuals who have no alternative person or who feel inadequate in attracting others may be particularly vulnerable to jealousy. They feel that, if they do not keep the person they have, they will be alone.
5. **Insecurity.** Individuals who feel insecure in a relationship with their partner may experience higher levels of jealousy. Khanchandani (2005) found that individuals who had been in relationships for a shorter time, who were in less committed relationships, and who were less satisfied with their relationships were more likely to be jealous (Pines 1992).

Consequences of Jealousy

Jealousy can have both desirable and undesirable consequences.

extradyadic relationship emotional or sexual involvement between a member of a couple and someone other than the partner.

Desirable Outcomes

Barelds-Dijkstra and Barelds (2007) studied 961 couples and found that reactive jealousy is associated with a positive effect on the relationship. Not only may reactive jealousy signify that the partner is cared for (the implied message is "I love you and don't want to lose you to someone else"), but also the partner may learn that the development of other romantic and sexual relationships is unacceptable.

The researchers noted that making the partner jealous may also have the positive function of assessing the partner's commitment and of alerting the partner that one could leave for greener mating pastures. Hence, one partner may deliberately evoke jealousy to solidify commitment and ward off being taken for granted. In addition, sexual passion may be reignited if one partner perceives that another would take the love object away. That people want what others want is an adage that may underlie the evocation of jealousy.

Undesirable Outcomes Shakespeare referred to jealousy as the "green-eyed monster," suggesting that it sometimes leads to undesirable outcomes for relationships. Anxious jealousy with its obsessive ruminations about the partner's alleged infidelity can make an individual miserable, and such jealousy spills over into one's evaluation or experience of the relationship as negative. If the anxious jealousy results in repeated unwarranted accusations, a partner can tire of such attacks and end the relationship.

In its extreme form, jealousy may have devastating consequences. In the name of love, people have stalked or shot the beloved and killed themselves in reaction to rejected love. Barelds-Dijkstra and Barelds (2007) noted that possessive jealousy involves an attack on a partner or an alleged person to whom the partner is showing attention. Possessive jealousy is likely to have negative consequences for a relationship.

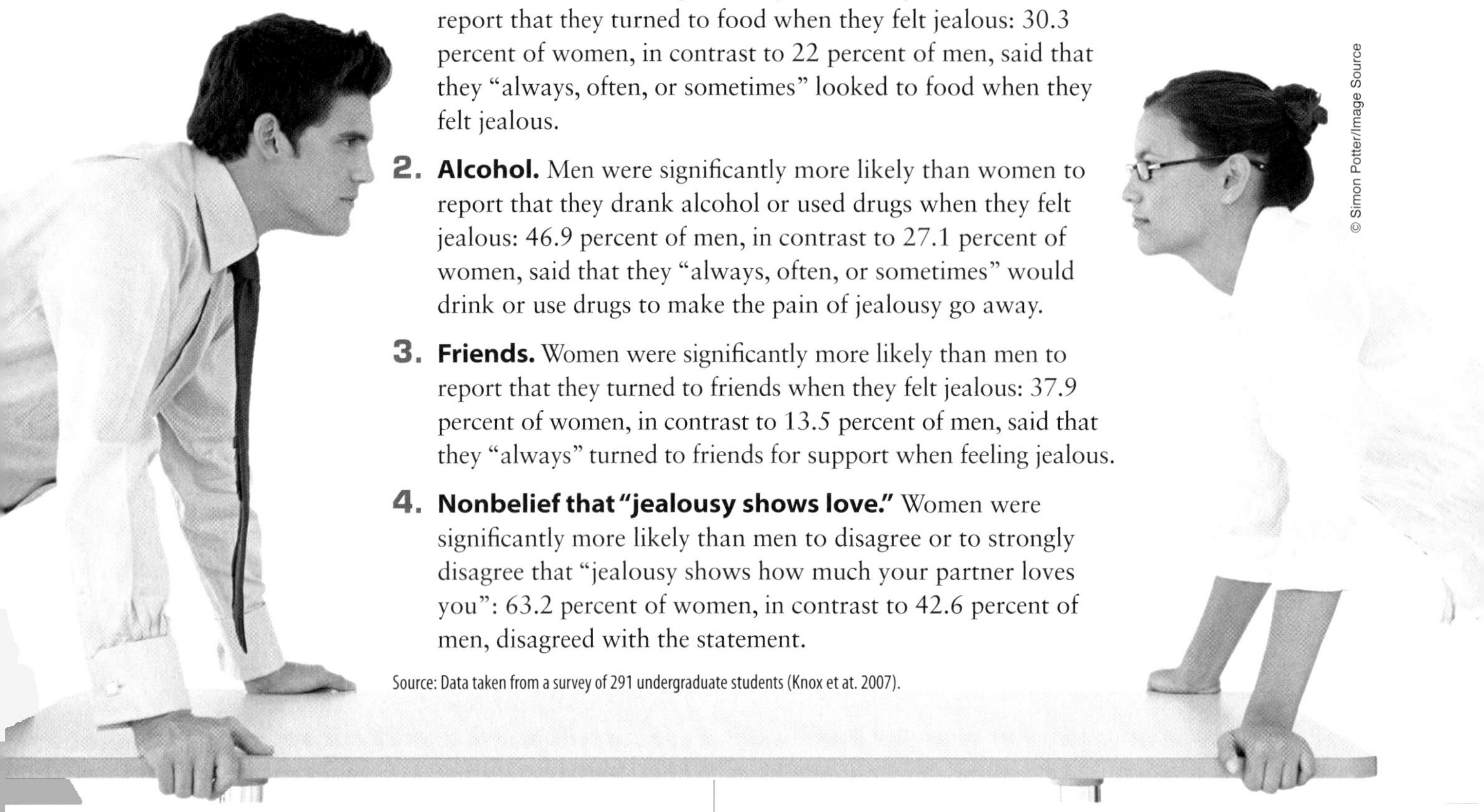

Gender Differences in Coping with Jealousy

1. **Food.** Women were significantly more likely than men to report that they turned to food when they felt jealous: 30.3 percent of women, in contrast to 22 percent of men, said that they "always, often, or sometimes" looked to food when they felt jealous.
2. **Alcohol.** Men were significantly more likely than women to report that they drank alcohol or used drugs when they felt jealous: 46.9 percent of men, in contrast to 27.1 percent of women, said that they "always, often, or sometimes" would drink or use drugs to make the pain of jealousy go away.
3. **Friends.** Women were significantly more likely than men to report that they turned to friends when they felt jealous: 37.9 percent of women, in contrast to 13.5 percent of men, said that they "always" turned to friends for support when feeling jealous.
4. **Nonbelief that "jealousy shows love."** Women were significantly more likely than men to disagree or to strongly disagree that "jealousy shows how much your partner loves you": 63.2 percent of women, in contrast to 42.6 percent of men, disagreed with the statement.

Source: Data taken from a survey of 291 undergraduate students (Knox et at. 2007).

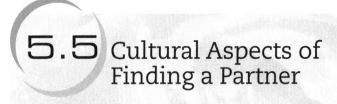

5.5 Cultural Aspects of Finding a Partner

endogamy the cultural expectation to select a marriage partner within one's social group.

endogamous pressures cultural attitudes reflecting approval for selecting a partner within one's social group and disapproval for selecting a partner outside one's social group.

Individuals are not free to marry whomever they please. Indeed, university students routinely assert, "I can marry whomever I want!" Hardly. Rather, their culture and society radically restrict and influence their choice. The best example of mate choice being culturally and socially controlled is the fact that *less than 1 percent* of 60 million marriages in the United States consist of those in which there is one black and one white spouse (*Statistical Abstract of the United States, 2009,* Table 59). Homosexual people are also not free to marry whomever they choose. Indeed, although Maine, Iowa, New Hamsphire, Vermont, Massachusetts, and Connecticut recognize same-sex marriage, federal law does not, reflecting wide societal disapproval.

Independent of the sexual orientation of the partners, endogamy and exogamy are two forms of cultural pressure operative in mate selection.

Endogamy

Endogamy is the cultural expectation to select a marriage partner within one's own social group, such as in the same race, religion, and social class. **Endogamous**

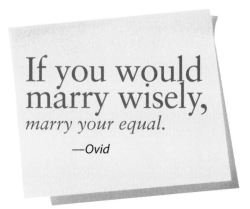

> If you would marry wisely,
> *marry your equal.*
> —*Ovid*

pressures involve social approval and encouragement to select a partner within your own group (for example, someone of your own race and religion) and disapproval for selecting someone outside your own group. The pressure toward an endogamous mate choice is especially strong when race is concerned. Love may be blind, but it knows the color of one's partner. Over 95 percent of individuals of Asian, African, Caucasian, and Native American descent end up selecting someone of the same race to marry.

Exogamy

In addition to the cultural pressure to marry within one's social group, there is also the cultural expectation that one will marry outside the family group. This expectation is known as **exogamy. Exogamous pressures** involve social approval and encouragement to select a partner outside one's own group (for example, someone outside of your own family).

Incest taboos are universal; as well, children are not permitted to marry the parent of the other sex in any society. In the United States, siblings and (in some states) first cousins are also prohibited from marrying each other. The reason for such restrictions is fear of genetic defects in children whose parents are too closely related.

Once cultural factors have determined the general **pool of eligibles** (the population from which a person selects an appropriate mate), individual mate choice becomes more operative. However, even when individuals feel that they are making their own choices, social influences are still operative.

5.6 Sociological Factors Operative in Finding a Partner

Numerous sociological factors are at work in bringing two people together who eventually marry.

Homogamy—Twelve Factors

Whereas endogamy is a concept that refers to cultural pressure, **homogamy** refers to individual initiative toward sameness or "likes attract." The **homogamy theory of mate selection** states that we tend to be attracted to and become involved with those who are similar to ourselves in such characteristics as age, race, religion, and social class. In general, the more couples have in common, the higher the reported relationship satisfaction and the more durable the relationship (Clarkwest 2007; Amato et al. 2007).

exogamy the cultural expectation that one will marry outside the family group.

exogamous pressures cultural attitudes reflecting approval for selecting a partner outside one's family group.

pool of eligibles the population from which a person selects an appropriate mate.

homogamy tendency to select someone with similar characteristics.

homogamy theory of mate selection theory that individuals tend to be attracted to and become involved with those who have similar characteristics.

racial homogamy tendency for individuals to marry someone of the same race.

Race As noted above, **racial homogamy**, the tendency for individuals to marry someone of the same race, operates strongly in selecting a live-in or marital partner (with greater homogamy for marital partners). In a national survey, more than eight in ten American adults (83 percent) agree that "it's all right for blacks and whites to date," up from 48 percent in 1987 (Pew Research Center 2007). Among younger people—those born since 1977—94 percent say it is all right for black people and white people to date. Over a third (36.4 percent) of 1,319 undergraduates reported that they had dated someone of another race. However, 44.2 percent reported that marrying someone of the same race was important for them (Knox and Zusman 2009).

Although homogamy states that similar individuals are more comfortable with each other, racism is also operative. In their book on *Two-Faced Racism*, Picca and Feagin (2007) found that three-fourths of the 9,000 journal entries of college students included reports of racist remarks on campus. The researchers pointed out that, although college students commonly say "I'm not prejudiced" in a public context, racist comments are sometimes made when they are alone (backstage) with their friends. Sometimes racist comments are unintentionally hurtful, as one student reported in this journal entry:

On Thursday . . . my friend Megan (a white female) went on her first date with Steve. As their conversation began they discussed typical first date topics like family, friends, home, etc. Somehow Megan began talking about how squirrels in the Bronx are black and so people call them "squiggers." When Steve did not laugh Megan wondered why he did not think it was funny. As the conversation progressed and Steve began to talk about his family, he revealed to Megan that his dad is black and mom is white. Right then, Megan realized why Steve did not find her joke to be so funny. Megan felt horrible. Without realizing it, Megan hurt someone who appeared to be just as white as she was . . . (p. 195)

Racism, in the form of residential segregation, is also alive. Vesselinov (2008) noted that the "gated" community is yet another form of residential segregation and that "little has changed in segregation levels despite antidiscriminatory legislation and efforts by various groups, social movements, and federal and local institutions" (p. 553).

Although a greater number of white people are available to black people for marriage, black mothers and white fathers have different roles in the respective black and white communities, in terms of setting the norms of interracial relationships. Hence, the black mother who approves of her son's or daughter's interracial relationship may be less likely to be overruled than the white mother (the white husband may be more disapproving than the black husband). Race may also affect one's perceptions. In a study of racial perceptions among college students, DeCuzzi et al. (2006) found that, although both races tended to view women and men of their own and the other race positively, there was a pronounced tendency to view women and men of their own race more positively and members of the other race more negatively.

Hohmann-Marriott and Amato (2008) found that individuals (both men and women) in interethnic (includes interracial relationships such as Hispanic-white, black-Hispanic, and black-white) marriages and cohabitation relationships have lower quality relationships than those in same ethnic relationships. Lower relationship quality was defined in terms of reporting less satisfaction, more problems, higher conflict, and lower commitment to the relationship.

Similarly, Bratter and King (2008) analyzed national data and found higher divorce rates among interracial couples (compared to same-race couples). They also found race and gender variation. Compared to white couples, white female/black male and white female/Asian male marriages were more prone to divorce; meanwhile, those involving nonwhite females and white males and Hispanics and non-Hispanic people had similar or lower risks of divorce.

Age Most individuals select someone who is relatively close in age. Men tend to select women three to five years younger than themselves. The result is the "marriage squeeze," which is the imbalance of the ratio of marriageable-aged men to marriageable-aged women. In effect, women have fewer partners to select from because men choose from not only their same age group but also those younger than themselves. One 40-year-old recently divorced woman said, "What chance do I have with all these guys looking at all these younger women?"

Educational homogamy (selecting a cohabitant or marital partner with similar education) also operates strongly in selecting a live-in and marital partner (with greater homogamy for marital partners) (Kalmijn and Flap 2001). Not only does college provide an opportunity to meet, date, live with, and marry another college student, but it also increases one's chance that only a college-educated partner becomes acceptable as a potential cohabitant or spouse. The very pursuit of education becomes a value to be shared. However, Lewis and Oppenheimer (2000) observed that, when people of similar education are not available, women are par-

EDUCATION IS WHAT YOU HAVE LEFT OVER AFTER YOU HAVE FORGOTTEN EVERYTHING YOU HAVE LEARNED.

—Anonymous

ticularly likely to marry someone with less education. The older the woman, the more likely she is to marry a partner with less education. In effect, the number of educated, eligible males may decrease as she ages.

Open-Mindedness People vary in the degree to which they are **open-minded** (an openness to understanding alternative points of view, values, and behaviors). Homogamous pairings in regard to homogamy are those that reflect partners who are relatively open or closed to new points of view, behaviors, and experiences. For example, an evangelical may not be open to people of alternative religions.

Social Class You have been reared in a particular social class that reflects your parents' occupations, incomes, and educations as well as your residence, language, and values (see Table 5.1). If you were brought up in a home in which both parents were physicians, you probably lived in a large house in a nice residential area—summer vacations and a college education were givens. Alternatively, if your parents dropped out of high school and worked "blue-collar" jobs, your home would be smaller and in a less expensive part of town, and your opportunities would be more limited (for example, education). Social class affects one's comfort in interacting with others—we tend to feel more comfortable with others from our same social class.

open-minded being open to understanding alternative points of view, values, and behaviors.

Table 5.1
112 Million U.S. Family Households by Social Class*

Class Identification	Percentage of Population	Household Income	Education/Occupation	Lifestyle	Example
Upper Class (4%)					
Upper-upper class (old money capitalists)	1%	$500,000	Prestigious schools/ wealth passed down	Large, spacious homes in lush residential areas	Kennedy
Lower-upper class (nouveau riche)	3%	$200,000	Prestigious schools/ investors in or owners of large corporations	(same as above)	Bill Gates, Donald Trump
Middle-Class (45%)					
Upper-middle class	26%	$75,000–$200,000	Postgraduate degrees/ physicians, lawyers, managers of large corporations	Nice homes, nice neighborhoods, send children to state universities	Your physician
Lower-middle class	19%	$50,000– $75,000	College degrees/nurses, elementary or high school teachers	Modest homes, older cars	High school English teacher
Working Class	26%	$25,000–$50,000	High school diploma/ Waitresses, mechanics	Home in lower-income suburb; children get jobs after high school	Employee at fast food restaurant
Working Poor	15%	Below poverty line of $17,604 for family of three	Some high school/service jobs	Live in poorest of housing; barely able to pay rent or buy food	Janitor
Underclass	10%	—	Unemployed/ unemployable; survive via public assistance, begging, hustling, or illegal behavior (e.g., selling drugs)	Transient; contact with mainstream society is via criminal justice system	Homeless

*Appreciation is expressed to Arunas Juska, PhD, for his assistance in the development of this table. Estimates of percentage in each class are in reference to household incomes as published in *Statistical Abstract of the United States, 2009,* Table 684.

mating gradient the tendency for husbands to be more advanced than their wives with regard to age, education, and occupational success.

religion specific fundamental set of beliefs and practices generally agreed upon by a number of people or sects.

religious homogamy tendency for people of similar religious or spiritual philosophies to seek out each other.

The **mating gradient** refers to the tendency for husbands to be more advanced than their wives with regard to age, education, and occupational success. Indeed, husbands are typically older than their wives, have more advanced education, and earn higher incomes (*Statistical Abstract of the United States, 2009*, Table 679).

Physical Appearance Homogamy is operative in regard to physical appearance in that people tend to become involved with those who are similar in degree of physical attractiveness. However, a partner's attractiveness may be a more important consideration for men than for women. In a study of homogamous preferences in mate selection, men and women rated physical appearance an average of 7.7 and 6.8 (out of 10) in importance, respectively (Knox et al. 1997).

Marital Status Never-married people tend to select other never-married people as marriage partners, divorced people tend to select other divorced people, and widowed people tend to select other widowed people. Similar marital status may be more important to women than to men. In the study of homogamous preferences in mate selection, women and men rated similarity of marital status an average of 7.2 and 6.3 (out of 10) in importance, respectively (Knox et al. 1997).

Religion/Spirituality Most adults in the United States tend to be affiliated with a religion. Only 11 percent of two national samples reported no religious affiliation (Amato et al. 2007). **Religion** may be broadly defined as a specific fundamental set of beliefs (in reference to a supreme being, and so on) and practices generally agreed upon by a number of people or sects. Similarly, some individuals view themselves as "not religious" but "spiritual," with spirituality defined as belief in the spirit as the seat of the moral or religious nature that guides one's decisions and behavior. Because religious or spiritual views reflect, in large part, who the person is, they have an enormous impact on one's attraction to a partner, the level of emotional engagement with that partner, and the durability of the relationship and marital happiness (Swenson et al. 2005).

Religious homogamy is operative in that people of similar religion or spiritual philosophy tend to seek out each other. Over 40 percent of 1,319 undergraduates agreed that "It is important that I marry someone of my same religion." Exactly half reported that they had dated someone of another religion (Knox and Zusman 2009).

The phrase "the couple that prays together, stays together" is more than just a cliché. However, it should also be pointed out that religion could serve as a divisive force. For example, when one partner becomes "born again" or "saved," the relationship can be dramatically altered and eventually terminated unless the other partner shares the experience. A former rock-and-roller, hard-drinking, drug-taking wife noted that, when her husband "got saved," it was the end of their marriage. "He gave our money to the church as the 'tithe' when we couldn't even pay the light bill," she said. "And when he told me I could no longer wear pants or lipstick or have a beer, I left." Another example of how religious disharmony has a negative effect on relationships is the marriage of Ted Turner and Jane Fonda. Soon after she became a "Christian" and was "saved," she and Turner split up. In an interview, Fonda noted that she feared telling her husband about her religious

Everyone is a moon and has a dark side which they never show to anyone.

—*Mark Twain, author*

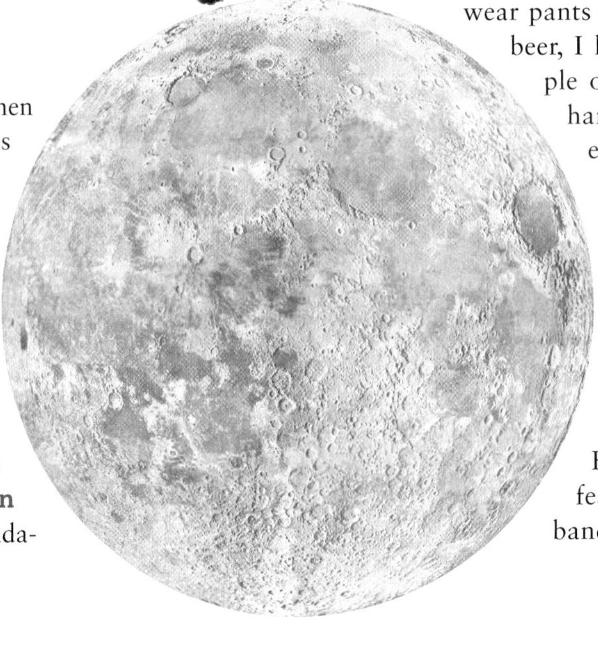

© iStockphoto.com

conversion because she knew it would be the end of their eight-year marriage.

Attachment Individuals who report similar levels of attachment to each other report high levels of relationship satisfaction. This is the conclusion of Luo and Klohnen (2005), who studied 291 newlyweds to assess the degree to which similarity affected marital quality. Indeed, similarity of attachment was *the* variable most predictive of relationship quality.

Personality Helen Fisher (2009) analyzed data from more than 28,000 heterosexual members of chemistry.com. She observed a tendency for certain personality types to select other personality types. For example, "explorers" (curious, creative, adventurous, sexual) were attracted to other "explorers" and "builders" (calm, loyal, traditional) were attracted to other "builders." These pairings are an obvious example of homogamy—like seeking like.

Economic Values, Money Management, and Debt Individuals vary in the degree to which they have money, spend money, and save. Some have very limited resources and are careful about all spending. Others have significant resources and buy whatever they want. Some are deeply in debt, whereas others have no debt. Of 1,319 undergraduates, 9 percent noted that they owed more than a thousand dollars on a credit card (Knox and Zusman 2009). Some carry significant educational debt. The median debt for those with a bachelor's degree was $19,300 (Chu 2007). Money becomes an issue in mate selection in that different economic backgrounds, values, and spending patterns are predictable conflict issues. One undergraduate noted, "There is no way I would get involved with/marry this person as they don't know how to handle money."

5.7 Psychological Factors Operative in Finding a Partner

Psychologists have focused on complementary needs, exchanges, parental characteristics, and personality types with regard to finding a partner.

© Somos/Jupiterimages

Complementary-Needs Theory

"In spite of the Women's Movement and a lot of assertive friends, I am a shy and dependent person," remarked a transfer student. "My need for dependency is met by Warren, who is the dominant, protective type." The tendency for a submissive person to become involved with a dominant person (one who likes to control the behavior of others) is an example of attraction based on complementary needs.

Complementary-needs theory states that we tend to select mates whose needs are opposite and complementary to our own. Partners can also be drawn to each other on the basis of nurturance versus receptivity. These complementary needs suggest that one person likes to give and take care of another, whereas the other likes to be the benefactor of such care. Other examples of complementary needs may involve responsibility versus irresponsibility, peacemaker versus troublemaker, and disorder versus order. In her study of 28,000 participants looking for a partner on chemistry.com, Fisher (2009) also found evidence of complementary needs. In terms of personality, "directors" (analytical, decisive, focused) sought "negotiators" (introspective, verbal, intuitive) and vice versa.

The idea that mate selection is based on complementary needs was suggested by Winch (1955), who noted that needs can be complementary if they are different (for example, dominant and submissive) or if the partners have the same need at different levels of intensity.

As an example of the latter, two individuals may have a complementary relationship if they both want to pursue graduate studies but want

complementary-needs theory tendency to select mates whose needs are opposite and complementary to one's own needs.

exchange theory theory that emphasizes that relations are formed and maintained between individuals offering the greatest rewards and least costs to each other.

to earn different degrees. The partners will complement each other if, for instance, one is comfortable with aspiring to a master's degree and approves of the other's commitment to earning a doctorate.

Winch's theory of complementary needs, commonly referred to as "opposites attract," is based on the observation of twenty-five undergraduate married couples at Northwestern University. Other researchers who have not been able to replicate Winch's study have criticized the findings (Saint 1994). Two researchers said, "It would now appear that Winch's findings may have been an artifact of either his methodology or his sample of married people" (Meyer and Pepper 1977). Singer Carly Simon said of her former husband, James Taylor:

> Our needs are different; it seemed impossible to stay together. James needs a lot more space around him—aloneness, remoteness, more privacy. I need more closeness, more communication. He's more abstract in our relationship. I'm more concrete. He's more of a . . . poet, and I'm more of a . . . reporter. (White 1990, 525)

Three questions can be raised about the theory of complementary needs:

1. **Couldn't personality needs be met just as easily outside the couple's relationship as through mate selection?** For example, couldn't a person who has the need to be dominant find such fulfillment in a job that involved an authoritative role, such as being a supervisor?
2. **What is a complementary need as opposed to a similar value?** For example, is the desire to achieve at different levels a complementary need or a shared value?
3. **Don't people change as they age?** Could dependent people grow and develop self-confidence so that they might no longer need to be involved with a dominant person? Indeed, such a person might no longer enjoy interacting with a dominant person.

Exchange Theory

Exchange theory emphasizes that mate selection is based on assessing who offers the greatest rewards at the lowest cost. The following five concepts help to explain the exchange process in mate selection:

1. **Rewards.** Rewards are the behaviors (your partner looking at you with the eyes of love), words (saying "I love you"), resources (being beautiful or handsome, having a car, condo, and money), and services (cooking for you, typing for you) your partner provides that you value and that influence you to continue the relationship. Increasingly, men are interested in women who offer "financial independence." In a study of Internet ads placed by women, the woman who described herself as "financially independent . . . successful and ambitious" produced 50 percent more responses than the next most popular ad, in which the woman described herself as "lovely . . . very attractive and slim" (Strassberg and Holty 2003).
2. **Costs.** Costs are the unpleasant aspects of a relationship. A woman identified the costs associated with being involved with her partner: "He abuses drugs, doesn't have a job, and lives nine hours away." The costs her partner associated with being involved with this woman included "she nags me," "she doesn't like sex," and "she wants her mother to live with us if we marry." Ingoldsby et al. (2003) assessed the degree to which various characteristics were associated with reducing one's attractiveness on the marriage market. The most undesirable traits were not being heterosexual, having alcohol or drug problems, having a sexually transmitted disease, and being lazy.
3. **Profit.** Profit occurs when the rewards exceed the costs. Unless the couple previously referred to derive a profit from staying together, they are likely to end their relationship and seek someone else with whom there is a higher profit margin.
4. **Loss.** Loss occurs when the costs exceed the rewards.
5. **Alternative.** Is another person currently available who offers a higher profit margin?

Most people have definite ideas about what they are looking for in a mate. For example, Xie et al. (2003) found that men with good incomes were much more likely to marry than men with no or low incomes. The currency used in the marriage market consists of the socially valued characteristics of the people involved, such as age, physical characteristics, and economic status. In our free choice system of mate selection, we typically get as much in return for our social attributes as we have to offer or trade. An unattractive, drug-abusing high school dropout

> Marry yourself.
>
> —*Jack Wright, sociologist*

with no job has little to offer an attractive, drug-free, college student who has just been accepted to graduate school.

Once you identify a person who offers you a good exchange for what you have to offer, other bargains are made about the conditions of your continued relationship. Waller and Hill (1951) observed that the person who has the least interest in continuing the relationship could control the relationship. This **principle of least interest** is illustrated by the woman who said, "He wants to date me more than I want to date him, so we end up going where I want to go and doing what I want to do." In this case, the woman trades her company for the man's acquiescence to her recreational choices. In effect, the person with the least interest controls the relationship.

Parental Characteristics Whereas the complementary-needs and exchange theories of mate selection are relatively recent, Freud suggested that the choice of a love object in adulthood represents a shift in libidinal energy from the first love objects—the parents. **Role theory of partnering** (also known as **modeling theory of partnering**) emphasizes that a son or daughter models after the parent of the same sex by selecting a partner similar to the one the parent selected.

This means that a man looks for a wife who has similar characteristics to those of his mother and that a woman looks for a husband who is very similar to her father.

principle of least interest principle stating that the person who has the least interest in a relationship controls the relationship.

role (modeling) theory of partnering emphasizes that children select partners similar to the one their same sex parent selected.

Desired Personality Characteristics for a Potential Partner

In a study of 700 undergraduates, both men and women reported that the personality characteristics of being warm, kind, and open and having a sense of humor were very important to them in selecting a romantic or sexual partner. Indeed, these intrinsic personality characteristics were rated as more important than physical attractiveness or wealth (extrinsic characteristics) (Sprecher and Regan 2002). Similarly, adolescents wanted intrinsic qualities such as intelligence and humor in a romantic partner but looked for physical appearance and high sex drive in a casual partner (no gender-related differences in responses were found) (Regan and Joshi 2003). Sometimes what individuals say they want and actually select may be dif-

Searching for Homogamy

"Searching for Homogamy" is a way to visualize the courtship process in class as well as trends in the class towards (or away from) homogamy. Here are the rules:

1. Six students need to volunteer; five of one sex, one of the other, or six students of the same sex interested in a same-sex date. Each student must understand and be okay with the high probability of rejection (80%), be uninvolved in a current relationship, and be open to dating people of any racial, ethnic, or religious background.
2. The "dater" then proceeds to ask the candidates questions, one of his choice and one generated by the class. Each date candidate may ask the dater one question. After this question and answer session, the dater chooses which of the candidates he/she would like to accompany him/her on a date.
3. Sometime over the next weekend, the couple goes out to dinner, both paying their own way. They must agree to no alcohol or sex.
4. After the date, both participants report to the class about their dating experience. While one participant speaks, the other should wait outside.

When our classes performed this exercise, we had one man interviewing five women, and no extra credit was awarded for participation. Some of the advantages the participants reported included having fun, seeing guys answer questions asked by girls, and learning how to ask questions. Disadvantages included that the game didn't last long enough, some of the questions could get too personal, and no one liked to be rejected.

Abridged from Knox, D. and K. McGinty. 2009. Searching for homogamy: An In-class exercise. *College Student Journal* 43: 243–247.

ferent. Bogg and Ray (2006) noted that, although women in one study said that they consistently preferred to date and marry men who were egalitarian, it was not unusual for them to select ultramasculine men who were sometimes dominant, mysterious, and rebellious.

Assad et. al. (2007) studied the personality quality of optimism and found that optimistic individuals tended to be involved in satisfying and happy romantic relationships, and a substantial portion of this association was mediated by willingness to be a cooperative problem solver. Hence, selecting a person who is optimistic seems to bode well for the future of the relationship.

In another study, women were significantly more likely than men to identify "having a good job" and "being well-educated" as important attributes in a future mate, whereas significantly more men than women wanted their spouse to be physically attractive (Medora et al. 2002). Toro-Morn and Sprecher (2003) noted that both American and Chinese undergraduate women identified characteristics associated with status (earning potential, wealth) more than physical attractiveness.

The behavior that 60 percent of a national sample of adult single women reported as the most serious fault of a man was his being "too controlling" (Edwards 2000). Women are also attracted to men who have good manners. In a study of 398 undergraduates, women were significantly more likely than men to report that they wanted to "date or be involved with [only] someone who had good manners," that "manners are very important," and that "the more well mannered the person, the more I like the person" (Zusman et al. 2003). Twenge (2006) noted that good manners are being displayed less often.

© Photos.com/Jupiterimages / © Comstock Images/Jupiterimages

Personality Characteristics of Partners to Avoid

Researchers have identified several personality factors predictive of relationships that either do not endure or are unfulfilling (Foster 2008; Wilson and Cousins 2005). Potential partners who are observed to consistently display these characteristics might be avoided.

1. **Narcissism.** Foster (2008) observed that narcissistic individuals view relationships in terms of what they get out of them. When satisfactions wane and alternatives are present, narcissists are the first to go. Because all relationships have difficult times, a narcissist is a high risk for divorce.
2. **Disagreeableness or low positives.** Gattis et al. (2004) studied 132 distressed couples seeking treatment and found that the personality characteristics of "low agreeableness" and "low positive expressions" were associated with their not getting along. Hence, partners who always find something to argue about and who find few opportunities to make positive observations or expressions should be considered with caution.
3. **Poor impulse control.** People who have poor impulse control have little self-restraint and may be prone to aggression and violence (Snyder and Regts 1990). Lack of impulse control is also problematic in relationships because such people are less likely to consider the consequences of their actions. For example, to some people, having an affair might seem harmless but it will have devastating consequences for the partners and their relationship in most cases.
4. **Hypersensitivity.** Hypersensitivity to perceived criticism involves getting hurt easily. Any negative statement or criticism is received with a greater impact than a partner intended. The disadvantage of such hypersensitivity is that a partner may learn not to give feedback for fear of hurting the hypersensitive partner. Such lack of feedback to the hypersensitive partner blocks information about what the person does that upsets the other and what could be done to

make things better. Hence, the hypersensitive one has no way of learning that something is wrong, and the partner has no way of alerting the hypersensitive partner. The result is a relationship in which the partners can't talk about what is wrong, so the potential for change is limited (ibid.).

5. **Inflated ego.** An exaggerated sense of oneself is another way of saying a person has a big ego and always wants things to be his or her way. A person with an inflated sense of self may be less likely to consider the other person's opinion in negotiating a conflict and may prefer to dictate an outcome. Such disrespect for the partner can be damaging to the relationship (Snyder and Regts 1990).

6. **Perfectionism.** Individuals who are perfectionists may require perfection of themselves and others. This attitude is associated with relationship problems (Haring et al. 2003). When the expectation of perfection causes one partner to become critical of the other and to pressure them, the strain can cause irreparable damage to a relationship. It can also create a barrier if the perfectionist's fear of failure makes it difficult to be intimate with their partner.

7. **Insecurity.** Feelings of insecurity also compromise marital happiness. Researchers studied the personality trait of attachment and its effect on marriage in 157 couples at two-time intervals and found that "insecure participants reported more difficulties in their relationships. . . [I]n contrast, secure participants reported greater feelings of intimacy in the relationship at both assessments" (Crowell et al. 2002).

8. **Controlled.** Individuals who are controlled by their parents, grandparents, former partner, child, or whomever compromise the marriage relationship because their allegiance is external to the couple's relationship. Unless the person is able to break free of such control, the ability to make independent decisions will be thwarted, which will both frustrate the spouse and challenge the marriage.

Table 5.2
Personality Types Problematic in a Potential Partner

Type	Characteristics	Impact on Partner
Paranoid	Suspicious, distrustful, thin-skinned, defensive	Partners may be accused of everything.
Schizoid	Cold, aloof, solitary, reclusive	Partners may feel that they can never "connect" and that the person is not capable of returning love.
Borderline	Moody, unstable, volatile, unreliable, suicidal, impulsive	Partners will never know what their Jekyll-and-Hyde partner will be like, which could be dangerous.
Antisocial	Deceptive, untrustworthy, conscienceless, remorseless	Such a partner could cheat on, lie, or steal from a partner and not feel guilty.
Narcissistic	Egotistical, demanding, greedy, selfish	Such a person views partners only in terms of their value. Don't expect such a person to see anything from a partner's point of view; expect such a person to bail in tough times.
Dependent	Helpless, weak, clingy, insecure	Such a person will demand a partner's full time and attention, and other interests will incite jealousy.
Obsessive-compulsive	Rigid, inflexible	Such a person has rigid ideas about how a partner should think and behave and may try to impose them on the partner.

In addition to personality characteristics, Table 5.2 reflects some particularly troublesome personality types and how they may impact you negatively.

5.8 Sociobiological Factors Operative in Finding a Partner

In contrast to cultural, sociological, and psychological aspects of mate selection, which reflect a social learning assumption, the sociobiological perspective suggests that biological or genetic factors may be operative in mate selection.

Definition of Sociobiology

Sociobiology suggests a biological basis for all social behavior—including mate selection. Based on Charles Darwin's theory of natural selection, which states that the strongest of the species survives, sociobiology holds that men and women select each other as mates on the basis

sociobiology suggests a biological basis for all social behavior.

engagement period of time during which committed, monogamous partners focus on wedding preparations and systematically examine their relationship.

of their innate concern for producing offspring who are most capable of surviving.

According to sociobiologists, men look for a young, healthy, attractive, sexually conservative woman who will produce healthy children and invest in taking care of the children. Women, in contrast, look for an ambitious man with good economic capacity who will invest his resources in her children. Earlier in this chapter, we provided data supporting the idea that men seek attractive women and women seek ambitious, financially successful men.

Criticisms of the Sociobiological Perspective

The sociobiological explanation for mate selection is controversial. Critics argue that women may show concern for the earning capacity of men because women have been systematically denied access to similar economic resources, and selecting a mate with these resources is one of their remaining options. In addition, it is argued that both women and men, when selecting a mate, think about their partners more as companions than as future parents of their offspring.

5.9 Engagement

Engagement moves the relationship of a couple from a private love-focused experience to a public, parent-involved experience. Family and friends are invited to enjoy the happiness and commitment of the individuals to a future marriage. Unlike casual dating, **engagement** is a time during which the partners are emotionally committed, are sexually monogamous, and are focused on wedding preparations. The engagement period is the last opportunity before marriage to systematically examine the relationship, ask each other specific questions, find out about the partner's parents and family background, and participate in marriage education or counseling.

Asking Specific Questions

Because partners might hesitate to ask for or reveal information that they feel will be met with disapproval during casual dating, the engagement is a time to be specific about the other partner's thoughts, feelings, values, goals, and expectations. The Involved Couple's Inventory (found on your chapter 5 Self-Assessment Cards) is designed to help individuals in committed relationships learn more about each other by asking specific questions. There are also tools, such as the Right Mate Check-Up from Heartchoice.com, that are designed to help partners assess their relationship.

Visiting a Partner's Parents

Engaged couples should seize the opportunity to discover the family environment in which each partner was reared and consider the implications for their subsequent marriage. When visiting one partner's parents, the other should observe their standard of living and the way they interact and relate (for example, level of affection, verbal and nonverbal behavior, marital roles) with one another. How does their standard of living compare with that of his or her own family? How does the emotional closeness (or distance) of the partner's family compare with that of his or her family? Such comparisons are significant because partners will reflect their respective family or origins. "This is the way we did it in my family" is a phrase that is often repeated. One way to discern how a person will be treated by his or her partner in the future is to observe the way the partner's parent of the same sex treats and interacts with his or her spouse. To know what a partner may be like in the future, one should look at the

© Zefa RF/Alamy

partner's parent of the same sex. There is a tendency for a man to become like his father and a woman to become like her mother. A partner's parent of the same sex and their marital relationship is the model of a spouse and a marriage relationship the person is likely to duplicate in the way he or she relates to the other partner.

Premarital Education Programs

Various **premarital education programs** (also known as premarital prevention programs, premarital counseling, premarital therapy, and marriage preparation), both academic and religious, are formal systematized experiences designed to provide information to individuals and to couples about how to have a good relationship. About 30 percent of couples getting married become involved in some type of premarital education. The greatest predictor of whether a couple will become involved in a premarital education program is the desire and commitment of the female to do so. In effect, she ensures that she and her partner become involved in such a program. A second important predictor is the presence of problems in the relationship of the couple about to marry (Duncan et al. 2007).

Premarital education programs are valuable in that they not only help partners assess the degree to which they are compatible with each other but they may also provide a context to discuss some specific relationship issues. Carroll and Doherty (2003) conducted a review of the various outcome studies of these programs and found that the average participant in a premarital prevention program experienced about a 30 percent increase in measures of outcome success. Specifically, people who attended a premarriage education program were more likely than nonparticipants to experience immediate and short-term gains in interpersonal skills and overall relationship quality. Busby et al. (2007) noted that structured premarital programs such as RELATE were more effective than giving the couple a workbook or participating in a therapist-led program. The greatest value of premarital education programs is that they provide a context for individuals to discuss relationship issues that they may have otherwise avoided.

premarital education programs formal systematized experiences designed to provide information to individuals and couples about how to have a good relationship.

prenuptial agreement formal document specifying how property will be divided if the marriage ends in divorce or by one partner's death.

A HAPPY WIFE SOMETIMES HAS THE BEST HUSBAND, BUT MORE OFTEN MAKES THE BEST OF THE HUSBAND SHE HAS.

Prenuptial Agreement

Former Presidential candidate John McCain has a prenuptial agreement with his wife Cindy. She is heiress to a beer fortune estimated to be worth $100 million. Britney Spears and Kevin Federline had a prenuptial agreement whereby he was awarded only $300,000 of over $100 million in assets. Paul McCartney and Heather Mills did not have a prenuptial agreement. She was awarded almost $50 million. Some couples, particularly those with considerable assets or those in subsequent marriages, might consider discussing and signing a prenuptial agreement. To reduce the chance that the agreement will later be challenged, each partner should hire an attorney (months before the wedding) to develop and/or review the agreement.

The primary purpose of a **prenuptial agreement** (also referred to as a premarital agreement, marriage contract, or antenuptial contract) is to specify how property will be divided if the marriage ends in divorce or when it ends by the death of one partner. In effect, the value of what a partner takes into the marriage is the amount he or she is allowed to take out of the marriage. For example, if a woman brings $250,000 into the marriage and buys the marital home with this amount, her ex-husband is not automatically entitled to half the house at divorce. Some agreements may also contain clauses of no spousal support (alimony) if the marriage ends in divorce (but some states prohibit waiving alimony). Go online to 4ltrpress.cengage.com/mf to see an example of a prenuptial agreement developed by a husband and wife who had both been married before and had assets and children.

Reasons for a prenuptial agreement include the following.

1. **Protecting assets for children from a prior relationship.** People who are in their middle or later years, who have considerable assets, who have been married before, and who have children are often concerned that money and property be kept separate in a second marriage so that the assets at divorce or death go to their own children. Some children encourage their remarrying parent to draw up a prenuptial agreement with the new partner so that their (the offspring's) inheritance, house, or whatever will not automatically go to the new spouse upon the death of their parent.
2. **Protecting business associates.** A spouse's business associate may want a member of a firm or partnership to draw up a prenuptial agreement with a soon-to-be-spouse to protect the firm from intrusion by the spouse if the marriage does not work out.

Prenuptial contracts have a value beyond the legal implications. Their greatest value may be that they facilitate the partners discussing with each other their expectations of the relationship. In the absence of such an agreement, many couples may never discuss the issues they may later face.

There are also disadvantages of signing a prenuptial agreement. They are often legally challenged ("My partner forced me to sign it or call off the wedding"), and not all issues can be foreseen (for example, Who gets the time-share vacation property or the pets?).

Prenuptial agreements also are not very romantic ("I love you, but sign here and see what you get if you don't please me.") and may serve as a self-fulfilling prophecy ("We were already thinking about divorce."). Indeed, 22.8 percent of 1,319 undergraduates agreed "I would not marry someone who required me to sign a prenuptial agreement" and 21.97 percent felt that couples who have a prenuptial agreement are more likely to get divorced (Knox and Zusman 2009). Prenuptial contracts are almost nonexistent in first marriages and are still rare in second marriages. Whether or not signing a prenuptial agreement is a good idea depends on the circumstances. Some individuals who do sign an agreement later regret it. Sherry, then a never-married 22-year-old, signed such an agreement:

> Paul was adamant about my signing the premarriage agreement. He said he loved me but would never consider marrying me unless I signed a prenuptial agreement stating that he would never be responsible for alimony in case of a divorce. I was so much in love, it didn't seem to matter. I didn't realize that basically he was and is a selfish person. Now, five years later after our divorce, I live in a mobile home and he lives in a big house overlooking the lake with his new wife.

The husband viewed it differently. He was glad that she had signed the agreement and that his economic liability to his former wife was limited. He could afford the new house by the lake with his new wife because he was not sending money to Sherry. Billionaire Donald Trump attributed his economic survival of his two divorces to prenuptial agreements with his ex-wives.

Couples who decide to develop a prenuptial agreement need separate attorneys to look out for their respective interests. The laws regulating marriage and divorce vary by state, and only attorneys in those states can help ensure that the document drawn up will be honored. Individuals may not waive child support or dictate child custody. Full disclosure of assets is also important. If one partner hides assets, the prenuptial can be thrown out of court. One husband recommended that the issue of the premarital agreement should be brought up and that it be signed a minimum of six months before the wedding. "This gives the issue time to settle rather than being an explosive emotional issue if it is brought up a few weeks before the wedding." Indeed, as noted previously, if a prenuptial agreement is signed within two weeks of the wedding, that is grounds enough for the agreement to be thrown out of court, because it is assumed that the document was executed under pressure.

© Skip Odonnell/iStockphoto.com

5.10 Marrying for the Wrong Reasons

Some reasons for getting married are more questionable than others. These reasons include the following:

Rebound

A rebound marriage results when people marry someone immediately after another person has ended a relationship with them. It is a frantic attempt on their part to reestablish their desirability in their own eyes and in the eyes of the partner who dropped them. One man said, "After she told me she wouldn't marry me, I became desperate. I called up an old girlfriend to see if I could get the relationship going again. We were married within a month. I know it was foolish, but I was very hurt and couldn't stop myself." To marry on the rebound is questionable because the marriage is made in reference to the previous partner and not to the partner being married. In reality, people who engage in rebound marriage are using the person they intend to marry to establish themselves as the "winner" in the previous relationship.

To avoid the negative consequences of marrying on the rebound, one should wait until the negative memories of a past relationship have been replaced by positive aspects of the current relationship. In other words, one should marry when the satisfactions of being with the current partner outweigh any feelings of revenge. This normally takes between twelve and eighteen months. In addition, care should be taken when getting involved in a relationship with someone who is recently divorced or just emerged from a painful ending of a previous relationship.

Escape

A person might marry to escape an unhappy home situation in which the parents are oppressive, overbearing, conflictual, alcoholic, and/or abusive. One woman said, "I couldn't wait to get away from home. Ever since my parents divorced, my mother has been drinking and watching me like a hawk. 'Be home early, don't drink, and watch out for those horrible men,' she would always say. I admit it. I married the first guy that would have me. Marriage was my ticket out of there."

Marriage for escape is a bad idea. Marriage should be based on mutual love and respect rather than the desire to escape another unhappy situation. Making the decision to marry based on love and respect for one's partner allows for evaluation of the marital relationship in terms of its own potential and not solely as an alternative to an unhappy situation.

Unplanned Pregnancy

Getting married just because a partner becomes pregnant is usually a bad idea. Indeed, the decision of whether to marry should be kept separate from a pregnancy. Adoption, abortion, single parenthood, and unmarried parenthood (the couple can remain together as an unmarried couple and have the baby) are all alternatives to simply deciding to marry if a partner becomes pregnant. Avoiding feelings of being trapped or later feeling that the marriage might not have happened without the pregnancy are a couple reasons for not rushing into marriage because of pregnancy. Couples who marry when the woman becomes pregnant have an increased chance of divorce.

Psychological Blackmail

Some individuals get married because their partner takes the position that "I can't live without you" or "I will commit suicide if you leave me." Because the person fears that the partner may commit suicide, the partner agrees to the wedding. The problem with such a marriage is that one partner has learned to manipulate the relationship to get control. Use of such power often creates resentment in the other partner, who feels trapped in the marriage. Escaping from the marriage becomes even more difficult. One way of coping with a psychological blackmail situation is to encourage couple's therapy to "discuss the relationship." Once inside the therapy room, the pressured partner can tell the counselor that he or she feels pressured to get married because of the suicide threat. Counselors are trained to respond to such a situation.

© Benjamin Goode/iStockphoto.com

Insurance Benefits

In a poll conducted by the Kaiser Family Foundation, a health policy research group, 7 percent of adults said someone in their household had married in the past year to gain access to insurance. "For today's couples, 'in

THE DECISION TO MARRY SHOULD BE BASED ON FACTORS OTHER THAN PITY FOR THE PARTNER.

sickness and in health' may seem less a lover's troth than an actuarial contract. They marry for better or worse, for richer or poorer, for co-pays and deductibles" (Sack 2008). Although selecting a partner who has resources (which may include health insurance) is not unusual, selecting a partner for health benefits is yet another matter. Both parties might be cautious if the alliance is more about "benefits" than the relationship.

Pity

Some partners marry because they feel guilty about terminating a relationship with someone whom they pity. The fiancé of one woman got drunk one Halloween evening and began to light fireworks on the roof of his fraternity house. As he was running away from a Roman candle he had just ignited, he tripped and fell off the roof. He landed on his head and was in a coma for three weeks. A year after the accident, his speech and muscle coordination were still adversely affected. The woman said she did not love him anymore but felt guilty about terminating the relationship now that he had become physically afflicted.

She was ambivalent. She felt marrying her fiancé was her duty, but her feelings were no longer love feelings. Pity may also have a social basis. For example, a partner may fail to achieve a lifetime career goal (for example, the partner may flunk out of medical school). Regardless of the reason, if one partner loses a limb, becomes brain damaged, or fails in the pursuit of a major goal, keeping the issue of pity separate from the advisability of the marriage is important. The decision to marry should be based on factors other than pity for the partner.

Filling a Void

A former student in our classes noted that her father died of cancer. She acknowledged that his death created a vacuum, which she felt driven to fill immediately by getting married so that she would have a man in her life. Because she was focused on filling the void, she had paid little attention to the personality characteristics of her relationship with the man who had asked to marry her.

She reported that she discovered on her wedding night that her new husband had several other girlfriends whom he had no intention of giving up. The marriage was annulled.

In deciding whether to continue or terminate a relationship, couples should listen to what their senses tell them ("Does it feel right?"), listen to their hearts ("Do you love this person or do you question whether you love this person?"), and evaluate their similarities ("Are we similar in terms of core values, goals, view of life?"). Also, they should be realistic. It would be unusual if none of these factors applied to a couple. Indeed, most people have some negative and some positive indicators before they marry.

Consider Calling Off the Wedding If . . .

"No matter how far you have gone on the wrong road, turn back" is a Turkish proverb. If a couple's engagement is characterized by the following factors, they should consider prolonging their engagement and delaying their marriage at least until the most distressing issues have been resolved. Alternatively, they could break the engagement (which happens in 30 percent of formal engagements). Indeed, rather than defend a course of action that does not feel right, they should stop and reverse directions. Breaking an engagement has fewer negative consequences and is less stigmatized than a divorce.

Age 18 or Younger The strongest predictor of getting divorced is getting married during the teen years. Individuals who marry at age 18 or younger have three times the risk of divorce than those who delay marriage into their late twenties or early thirties. Teenagers may be more at risk for marrying to escape an unhappy home and may be more likely to engage in impulsive decision making and behavior. Early marriage is also associated with an end to one's education, social isolation from peer networks, early pregnancy or parenting, and locking one's self into a low income.

Research by Meehan and Negy (2003) on being married while in college revealed higher marital distress among spouses who were also students. In addition, when married college students were compared with single college students, the married students reported more difficulty adjusting to the demands of higher education. The researchers conclude that "these findings suggest that individuals opting to attend college while being married are at risk for compromising their marital happiness and may be jeopardizing their education" (p. 688). Hence, waiting until one is older and through college not only may result in a less stressful marriage but may also be associated with less economic stress.

Partner Known Less Than Two Years Of 1,319 undergraduates, 26 percent agreed, "If I were really in love, I would marry someone I had known for only a short time" (Knox and Zusman 2009). Impulsive marriages in which the partners have known each other for less than a month are associated with a higher-than-average divorce rate. Indeed, partners who date each other for at least twenty-five months before getting married report the highest level of marital satisfaction and are less likely to divorce (Huston et al. 2001). Nevertheless, Dr. Carl Ridley (2009) observed, "I am not sure it is about time but more about attending to self and other needs/desires/preferences and then assessing if one's potential partner can meet these needs or be willing to negotiate so that mutual needs are met. For some, this takes lots of time and for others, not very long."

A short courtship does not allow partners to learn about each other's background, values, and goals and does not permit time to observe and scrutinize each other's behavior in a variety of settings (for example, with one's close friends, parents, or siblings). Indeed, some individuals may be more prone to fall in love at first sight and to want to hurry the partner into a committed love relationship *before* the partner can find out about who they really are. ("Let the buyer beware!") If a person is pressuring his or her partner to marry after dating for a short time and their partner senses that this is too fast, the relationship should be terminated immediately.

To increase the knowledge couples have about each other, they should find out the answers from each other identified in the Couple's Involved Inventory (go online to 4ltrpress.cengage.com/mf to see this inventory), take a five-day "primitive" camping trip, take a fifteen-mile hike together, wallpaper a small room together, or spend several days together when one partner is sick. If the couple plans to have children, they may want to take care of a 6-month-old together for a weekend. Time should also be spent with each other's friends.

Abusive Relationship As we will discuss in Chapter 12, Violence and Abuse in Relationships, partners who emotionally and/or physically abuse their partners while dating and living together continue these behaviors in marriage. Abusive lovers become abusive spouses, with predictable negative outcomes. Though extricating oneself from an abusive relationship is difficult before the wedding, it becomes even more difficult after marriage and even more difficult once the couple has children.

One characteristic of an abusive partner is his or her attempt to systematically detach the intended spouse from all other relationships ("I don't want you spending time with your family and friends—you should be here with me."). This is a serious flag of impending relationship doom, should not be overlooked, and one should seek the exit ramp as soon as possible.

Numerous Significant Differences Relentless conflict often arises from numerous significant differences. Though all spouses are different from each other in some ways, those who have numerous differences in key areas such as race, religion, social class, education, values, and goals are less likely to report being happy and to have durable relationships. Skowron (2000) found that the less couples had in common, the more their marital distress. People who report the greatest degree of satisfaction in durable relationships have a great deal in common (Wilson and Cousins 2005).

On-and-Off Relationship A roller-coaster premarital relationship is predictive of a marital relationship that will follow the same pattern. Partners who break up and get back together several times have developed a pattern in which the dissatisfactions in the relationship become so frustrating that separation becomes the antidote for relief. In courtship, separations are of less social significance than marital separations. "Breaking up" in courtship is called "divorce" in marriage. Couples who routinely break up and get back together should examine the issues that continue to recur in their relationship and attempt to resolve them.

Dramatic Parental Disapproval A parent recalled, "I knew when I met the guy it wouldn't work out. I told my daughter and pleaded that she not marry him. She did, and they divorced." Although parental and in-law dissatisfaction is rare (Amato et al. [2007] found that 13 percent of parents disapprove of the partner their son or daughter plans to marry), such parental predictions (whether positive or negative) often come true. If the predictions are negative, they sometimes contribute to stress and conflict once the couple marries.

Even though parents who reject the commitment choice of their offspring are often regarded as uninformed and unfair, their opinions should not be taken lightly. The parents' own experience in marriage and their intimate knowledge of their offspring combine to put them in a unique position to assess how their child might get along with a particular mate. If the parents of either partner disapprove of the marital choice, the partners should try to evaluate these concerns objectively. The insights might prove valuable. The value of parental approval is illustrated in a study of Chinese marriages. Pimentel (2000) found that higher marital quality was associated with parents' approving of the mate choice of their offspring.

Low Sexual Satisfaction Sexual satisfaction is linked to relationship satisfaction, love, and commitment. Sprecher (2002) found 101 dating couples and sampled them in waves over several years, whether or not they were still together. Sprecher found that low sexual satisfaction (for both women and men) was related to reporting low relationship quality, less love, lower commitment, and breaking up. Hence, couples who are dissatisfied with their sexual relationship might explore ways of improving it (alone or through counseling) or consider the impact of such dissatisfaction on the future of their relationship.

Basically, it is time to end a relationship when the gain or advantages of staying together no longer outweigh the pain and disadvantages of staying—the pain of leaving is less than the pain of staying. Of course, all relationships go through periods of time when the disadvantages outweigh the benefits, so one should not bail out without careful consideration.

A ROLLER-COASTER PREMARITAL RELATIONSHIP IS PREDICTIVE OF A MARITAL RELATIONSHIP THAT WILL FOLLOW THE SAME PATTERN.

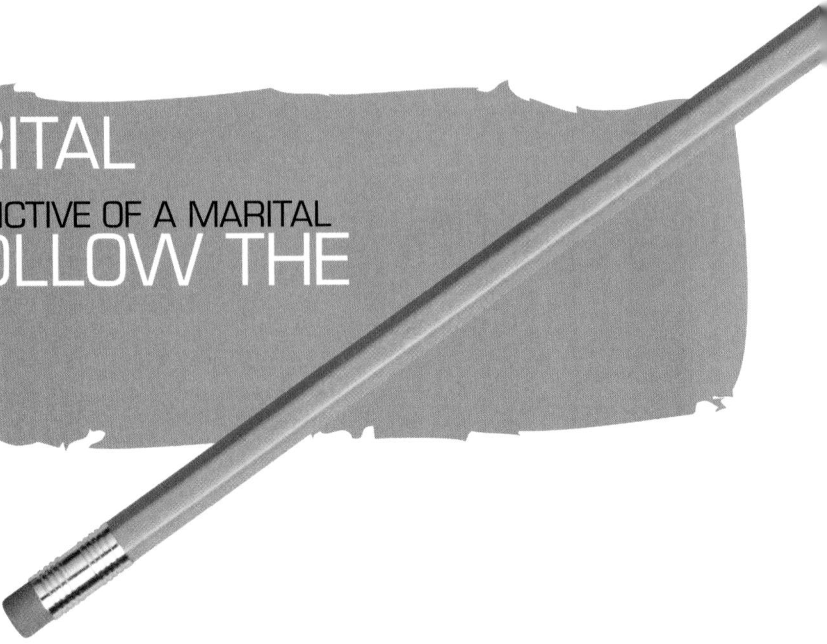

Self-Assessment

Open-Mindedness Scale

The purpose of this scale is to assess the degree to which you are open-minded. Open-mindedness is one's receptiveness to arguments, ideas, suggestions, and opinions. As such, someone who is open-minded typically does not prejudge or have preconceptions of others and is receptive and tolerant of new information. After reading each statement, select the number that best reflects your answer, using the following scale:

1	2	3	4	5	6	7
Strongly Disagree						Strongly Agree

____ 1. I am an open-minded person.

____ 2. I like diversity of thoughts and ideas.

____ 3. I think knowledge of different viewpoints is the only way to find the truth.

____ 4. I have often changed my mind about something after reading more about it.

____ 5. I consider more than one point of view before taking a stand on an issue.

____ 6. It really bothers me if someone makes fun of another person's idea.

____ 7. I am willing to explore a different point of view even if I do not agree with it.

____ 8. I am willing to really listen to any point of view.

____ 9. I evaluate information on the basis of its merit rather than my emotional reaction to it.

____ 10. I try to read about both sides of an issue before I form an opinion.

Scoring

Selecting a 1 reflects the least open-mindedness; selecting a 7 reflects the greatest open-mindedness. Add the numbers you assigned to each item. The lower your total score (10 is the lowest possible score), the more closed-minded you are; the higher your total score (70 is the highest possible score), the greater your open-mindedness. A score of 40 places you at the midpoint between being very closed-minded and very open-minded.

Scores of Other Students Who Completed the Scale

The scale was completed by 44 male and 81 female students at Valdosta State University. They received course credit for their participation. Their ages ranged from 18 to 46 years, with a mean age of 20.50 (standard deviation [SD] = 4.00). The racial and ethnic background of the sample included 64.0 percent white, 29.6 percent black, 1.6 percent Hispanic, and 4.8 percent other. The college classification level of the sample included 61.6 percent freshmen, 27.2 percent sophomores, 8.0 percent juniors, 1.6 percent seniors, and 0.8 percent postbaccalaureate students. Male participants had higher open-mindedness scores (mean [M] = 57.55; SD = 6.85) than did female participants (M = 54.58; SD = 7.75; $p < .05$). Freshman had lower open-mindedness scores (M = 54.01; SD = 7.50) than did upper-classmen (M = 58.04; SD = 6.94; $p < .05$). There were no significant differences in regard to race.

Source

Open-Mindedness Scale. 2006. Mark Whatley, PhD, Department of Psychology, Valdosta State University. Used by permission. Information on the reliability and validity of this scale is available from Dr. Whatley (mwhatley@valdosta.edu).

Marriage Relationships

SECTIONS

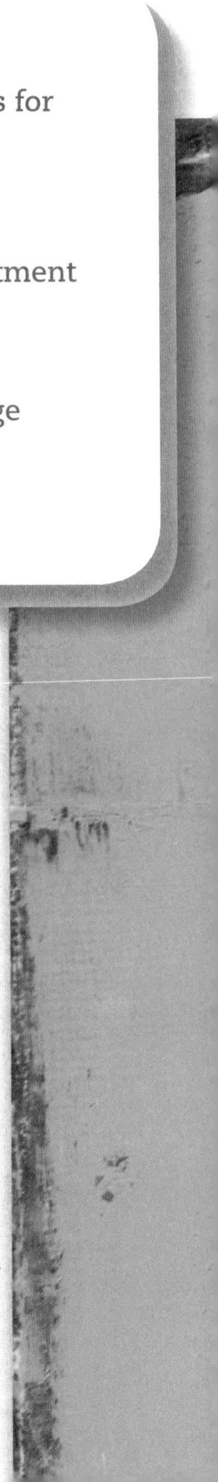

© iStockphoto.com / © Alf Ertsland/iStockphoto.com

Jay Leno's wife, Mavis, was asked the secret of their enduring marriage. She replied:

> One of my beliefs about a happy relationship—whether it's a marriage, a friendship, or a business—is that you let people go their own way. I always want to communicate to Jay that as far as I'm concerned, he can do any damn thing he wants and it's OK with me. Within reasonable limits, you have to mind your own business. If Jay's schedule becomes grueling, I'll just say, "How are you feeling? Don't you think you should cut yourself some slack?" (Burford 2005, 174)

Although the secret of a happy relationship is one thing to Mavis Leno, it is something else to another spouse. "Religion," "monogamy," "children," and so on, are the "secrets" for other couples. The title of this chapter, with plural "relationships," confirms that marriages are different. *Diversity* is the term that best describes relationships, marriages, and families today. No longer is there a one-size-fits-all cultural norm of what a relationship, marriage, or family should be. Rather, individuals, couples, and families select their own paths. In this chapter, we review the diversity of relationships. We begin with looking at some of the different reasons people marry.

6.1 Individual Motivations for Marriage

We have defined marriage as a legal contract between two heterosexual adults that regulates their economic and sexual interaction. However, individuals in the United States tend to think of marriage in more personal than legal terms. The following are some of the reasons people give for getting married.

Love

Many couples view marriage as the ultimate expression of their love for each other—the desire to spend their lives together in a secure, legal, committed relationship. In U.S. society, love is expected to precede marriage—thus, only couples in love consider marriage. Those not in love would be ashamed to admit it.

Personal Fulfillment

We marry because we feel a sense of personal fulfillment in doing so. We were born into a family (family of origin) and want to create a family of our own (family of procreation). We remain optimistic that our marriage will be a good one. Even if our parents divorced or we have friends who have done so, we feel that our relationship will be different.

MEN MARRY BECAUSE THEY ARE TIRED; WOMEN MARRY BECAUSE THEY ARE CURIOUS. BOTH ARE DISAPPOINTED.

— Oscar Wilde

Companionship

Talk show host Oprah Winfrey once said that lots of people want to ride in her limo, but what she wants is someone who will take the bus when the limo breaks down. One of the motivations for marriage is to enter a structured relationship with a genuine companion, a person who will take the bus with you when the limo breaks down.

Although marriage does not ensure it, companionship is the greatest expected benefit of marriage in the United States. Coontz (2000) noted that it has become "the legitimate goal of marriage" (p. 11). Eating meals together is one of the most frequent normative behaviors of spouses. Indeed, although spouses may eat lunch apart, dinner together becomes an expected behavior. **Commensality** is eating with others, and one of the issues spouses negotiate is "who eats with us" (Sobal et al. 2002).

Parenthood

Most people want to have children. In response to the statement, "Someday, I want to have children," 91.7 percent of 1,319 undergraduates at a large southeastern university responded, "yes" (Knox and Zusman 2009). The amount of time parents spend rearing their children has increased. Sayer et al. (2004) documented that, contrary to conventional wisdom, both mothers and fathers report spending greater amounts of time in child-care activities in the late 1990s than in the "family-oriented" 1960s.

Although some people are willing to have children outside marriage (in a cohabitating relationship or in no relationship at all), most Americans prefer to have children in a marital context. Previously, a strong norm existed in our society (particularly for white people) that individuals should be married before they have children. This norm has relaxed, with more individuals willing to have children without being married. An Australian survey revealed that women who elect to remain childfree are viewed more negatively than women who express a desire to have children (Rowland 2006).

commensality eating with others. Most spouses eat together and negotiate who joins them.

Economic Security

Married people report higher household incomes than do unmarried people. Indeed, national data from the Health and Retirement Survey revealed that individuals who were not continuously married had significantly lower wealth than those who remained married throughout the life course. Remarriage offsets the negative effect of marital dissolution (Wilmoth and Koso 2002).

Although individuals may be drawn to marriage for the preceding reasons on a conscious level, unconscious motivations may also be operative. Individuals reared in a happy family of origin may seek to duplicate this perceived state of warmth, affection, and sharing. Alternatively, individuals reared in unhappy, abusive, drug-dependent families may inadvertently seek to recreate a similar family because that is what they are familiar with. In addition, individuals are motivated to marry because of the fear of being alone, to better themselves economically, to avoid birth out of wedlock, and to prove that someone wants them.

Just as most individuals want to marry (regardless of the motivation), most parents want their children to marry. If their children do not marry too young and if they marry someone they approve of, parents feel some relief from the economic responsibility of parenting, anticipate that marriage will have a positive, settling effect on their offspring, and look forward to the possibility of grandchildren.

6.2 Societal Functions of Marriage

As noted in Chapter 1, important societal functions of marriage are to bind a male and female together who will reproduce, provide physical care for their dependent young, and socialize them to be productive members of society who will replace those who die (Murdock 1949). Marriage helps protect children by giving the state legal leverage to force parents to be responsible to their offspring whether or not they stay married. If couples did not have children, the state would have no interest in regulating marriage.

Additional functions include regulating sexual behavior (spouses have less exposure to STIs than singles) and stabilizing adult personalities by providing a companion and "in-house" counselor. In the past, marriage and family have served protective, educational, recreational, economic, and religious functions.

However, as these functions have gradually been taken over by police or legal systems, schools, the entertainment industry, workplace, and church or synagogue, only the companionship-intimacy function has remained virtually unchanged.

The emotional support each spouse derives from the other in the marital relationship remains one of the strongest and most basic functions of marriage (Coontz 2000). In today's social world, which consists mainly of impersonal, secondary relationships, living in a context of mutual emotional support may be particularly important. Indeed, the companionship and intimacy needs of contemporary U.S. marriage have become so strong that many couples consider divorce when they no longer feel "in love" with their partner. Of 1,319 undergraduates, 47.9 percent reported that they would divorce their spouse if they no longer loved the spouse (Knox and Zusman 2009).

The very nature of the marriage relationship has also changed from being very traditional or male-dominated to being very modern or egalitarian. A summary of these differences is presented in Table 6.1. Keep in mind that these are stereotypical marriages and that only a small percentage of today's modern marriages have all the traditional or egalitarian characteristics that are listed.

6.3 Marriage as a Commitment

Marriage represents a multilevel commitment—person-to-person, family-to-family, and couple-to-state.

Table 6.1
Traditional versus Egalitarian Marriages

Traditional Marriage	Egalitarian Marriage
There is limited expectation of husband to meet emotional needs of wife and children.	Husband is expected to meet emotional needs of wife and to be involved with children.
Wife is not expected to earn income.	Wife is expected to earn income.
Emphasis is on ritual and roles.	Emphasis is on companionship.
Couples do not live together before marriage.	Couples may live together before marriage.
Wife takes husband's last name.	Wife may keep her maiden name.
Husband is dominant; wife is submissive.	Neither spouse is dominant.
Roles for husband and wife are rigid.	Roles for spouses are flexible.
Husband initiates sex; wife complies.	Either spouse initiates sex.
Wife takes care of children.	Parents share child rearing.
Education is important for husband, not for wife.	Education is important for both spouses.
Husband's career decides family residence.	Career of either spouse determines family residence.

Person-to-Person Commitment

Commitment is the intent to maintain a relationship. Behavioral indexes of commitment (928 of them) were identified by 248 people who were committed to someone. These behaviors were then coded into ten major categories and included providing affection, providing support, maintaining

commitment an intent to maintain a relationship.

bride wealth also known as bride price or bride payment, the amount of money or payment in goods (e.g., cows) by the groom or his family to the wife's family for giving her up.

rite of passage event that marks the transition from one status to another.

integrity, sharing companionship, making an effort to communicate, showing respect, creating a relational future, creating a positive relational atmosphere, working on relationship problems together, and expressing commitment (Weigel and Ballard-Reisch 2002). Glover et al. (2006) noted that being committed to another was associated with relationship satisfaction and feeling predisposed toward caregiving for the partner.

Family-to-Family Commitment

Whereas love is private, marriage is public. Marriage is the second of three times that one's name can be expected to appear in the local newspaper. When individuals marry, the parents and extended kin also become enmeshed. In many societies (for example, Kenya), the families arrange for the marriage of their offspring, and the groom is expected to pay for his new bride. How much is a bride worth? In some parts of rural Kenya, premarital negotiations include the determination of **bride wealth**—this is the amount of money a prospective groom will pay to the parents of his bride-to-be. Such a payment is not seen as "buying the woman" but compensating the parents for the loss of labor from their daughter. Forms of payment include livestock ("I am worth many cows," said one Kenyan woman), food, and/or money. The man who raises the bride wealth also demonstrates not only that he is ready to care for a wife and children but also that he has the resources to do so (Wilson et al. 2003).

Marriage also involves commitments by each of the marriage partners to the family members of the spouse. Married couples are often expected to divide their holiday visits between both sets of parents.

Couple-to-State Commitment

In addition to making person-to-person and family-to-family commitments, spouses become legally committed to each other according to the laws of the state in which they reside. This means they cannot arbitrarily decide to terminate their own marital agreement.

Just as the state says who can marry (not close relatives, the insane, or the mentally deficient) and when (usually at age 18 or older), legal procedures must be instituted if the spouses want to divorce. The state's interest is that a couple stays married, has children, and takes care of them. Should they divorce, the state will dictate how the parenting is to continue, both physically and economically. Social policies designed to strengthen marriage through divorce law reform reflect the value the state places on stable, committed relationships.

6.4 Marriage as a Rite of Passage

A **rite of passage** is an event that marks the transition from one social status to another. Starting school, getting a driver's license, and graduating from high school or college are events that mark major transitions in status (to student, to driver, and to graduate). The wedding itself is another rite of passage that marks the transition from fiancé to spouse. Preceding the wedding is the traditional bachelor party for the soon-to-be groom. What is new on the cultural landscape is the bachelorette party (sometimes more wild than the bachelor party), which conveys the message of equality and that great changes are ahead (Montemurro 2006).

© Denis Vorob'yev/iStockphoto.com

> *Love is an obsessive delusion that is cured by marriage.*
>
> —*Karl Bowman*

Weddings

The wedding is a rite of passage that is both religious and civil. To the Catholic Church, marriage is a sacrament that implies that the union is both sacred and indissoluble. According to Jewish and most Protestant faiths, marriage is a special bond between the husband and wife sanctified by God, but divorce and remarriage are permitted. Wedding ceremonies still reflect traditional cultural definitions of women as property. For example, the father of the bride usually walks the bride down the aisle and "hands her over to the new husband." In some cultures, the bride is not even present at the time of the actual marriage. For example, in the upper-middle-class Muslim Egyptian wedding, the actual marriage contract signing occurs when the bride is in another room with her mother and sisters. The father of the bride and the new husband sign the actual marriage contract (identifying who is marrying whom, the families they come from, and the names of the two witnesses). The father will then place his hand on the hand of the groom, and the maa'zun, the official presiding, will declare that the marriage has occurred (Sherif-Trask 2003).

Blakely (2008) noted that weddings are increasingly becoming outsourced, commercialized events. The latest wedding expense is to have the wedding webcast so that family and friends afar can actually see and hear the wedding vows in real time. One such website is http://www.webcastmywedding.net/, where the would-be bride and groom can arrange the details. That marriage is a public experience is emphasized by weddings in which the couple invites family and friends of both parties to participate. The wedding is a time for the respective families to learn how to cooperate with each other for the benefit of the couple. Conflicts over number of bridesmaids and ushers, number of guests to invite, and place of the wedding are not uncommon. Though some families harmoniously negotiate all differences, others become so adamant about their preferences that the prospective bride and groom elope to escape or avoid the conflict. However, most families recognize the importance of the event in the life of their daughter or son and try to be helpful and nonconflictual, as seen in the film *Last Chance Harvey*.

To obtain a marriage license, some states require the partners to have blood tests to certify that neither has an STI. The document is then taken to the county courthouse, where the couple applies for a marriage license. Two-thirds of states require a waiting period between the issuance of the license and the wedding. A member of the clergy marries 80 percent of couples; the other 20 percent (primarily in remarriages) go to a justice of the peace, judge, or magistrate.

Brides often wear traditional **artifacts** (concrete symbols that reflect existence of an event): something old, new, borrowed, and blue. The **"old" wedding artifact** is something that represents the durability of the impending marriage (for example, an heirloom gold locket). The **"new" wedding artifact**, perhaps in the form of new, unlaundered undergarments, emphasizes the new life to begin. The **"borrowed" wedding artifact** is something that has already been worn by a currently happy bride (for example, a wedding veil). The **"blue" wedding artifact** represents fidelity (for example, those dressed in blue or in blue ribbons have lovers true). When the bride throws her floral bouquet, it signifies the end of girlhood; the rice thrown by the guests at the newly married couple signifies fertility.

Couples now commonly have weddings that are neither religious nor traditional. In the exchange of

artifact concrete symbol that an event exists; in game theory, evidence that a game exists.

"old" wedding artifact artifact worn by a bride that symbolizes durability of the impending marriage (for example old gold locket).

"new" wedding artifact artifact worn by a bride that symbolizes the new life she is to begin (for example, a new, unlaundered undergarment).

"borrowed" wedding artifact artifact worn by bride that may be a garment owned by a currently happy bride.

"blue" wedding artifact blue artifact worn by bride that is symbolic of fidelity.

honeymoon the time following the wedding whereby the couple become isolated to recover from the wedding and to solidify their new status change from lovers to spouses.

vows, neither partner may promise to obey the other, and the couple's relationship may be spelled out by the partners rather than by tradition. Vows often include the couple's feelings about equality, individualism, humanism, and openness to change. In 2009, the average wedding for a couple getting married for the first time is estimated to be $30,000 (www.theknot.com).

Ways in which couples lower the cost of their wedding include marrying any day but Saturday, or marrying off-season (not June), off-locale (in Mexico or a Caribbean Island where fewer guests will attend), and, as mentioned previously, broadcasting their wedding ceremony live over the Internet (http://www.liveinternetweddings.com/). The latter means that the couple can get married in Hawaii and have their ceremony beamed back to the states where well-wishers can see the wedding without leaving home.

Honeymoons

Traditionally, another rite of passage follows immediately after the wedding—the **honeymoon** (the time following the wedding whereby the couple become isolated to recover from the wedding and to solidify their new status change from lovers to spouses). The functions of the honeymoon are both personal and social. The personal function is to provide a period of recuperation from the usually exhausting demands of preparing for and being involved in a wedding ceremony and reception. The social function is to provide a time for the couple to be alone to solidify the change in their identity from that of an unmarried to a married couple. Now that they are married, their sexual expression and childbearing with each other achieves full social approval and legitimacy.

6.5 Changes after Marriage

After the wedding and honeymoon, the new spouses begin to experience changes in their legal, personal, and marital relationship.

Legal Changes

Unless the partners have signed a prenuptial agreement specifying that their earnings and property will remain separate, after the wedding, each spouse becomes part owner of what the other earns in income and accumulates in property. Although the laws on domestic relations differ from state to state, courts typically award to each spouse half of the assets accumulated during the marriage (even though one of the partners may have contributed a smaller proportion).

For example, if a couple buys a house together, even though one spouse invested more money in the initial purchase, the other will likely be awarded half of the value of the house if they divorce. (Having children complicates the distribution of assets because the house is often awarded to the custodial parent.) In the case of death of the spouse, the remaining spouse is legally entitled to inherit between one-third and one-half of the partner's estate, unless a will specifies otherwise.

Personal Changes

New spouses experience an array of personal changes in their lives. One initial consequence of getting married may be an enhanced self-concept. Imagine you are getting married. Your parents and close friends usually arrange their schedules to participate in your wedding and give gifts to express their approval of you and your marriage. In addition, the strong evidence that your spouse approves of you and is willing to spend a lifetime with you also tells you that you are a desirable person.

Married people also begin adopting new values and behaviors consistent with the married role. Although new spouses often vow that "marriage won't change me," it does. For example, rather than stay out all night at a party, which is not uncommon for single people who may be looking for a partner, spouses (who are already paired off) tend to go home early. Their roles of spouse, employee, and parent result in their adopting more regular, alcohol- and drug-free hours.

Friendship Changes

Marriage also affects relationships with friends of the same and other sex. Less time will be spent with friends because of the new role demands as a spouse. More time will be spent with other married couples who will become powerful influences on the new couple's relationship. Indeed, the degree that married couples have the *same* friends is proportional to the marital quality of the couple. So, the more friends a couple has in common, the higher the quality of the couple's marriage.

What spouses give up in friendships, they gain in developing an intimate relationship with each other. However, abandoning one's friends after marriage may be problematic because one's spouse cannot be expected to satisfy all of one's social needs. Because many marriages end in divorce, friendships that have been maintained throughout the marriage can become a vital source of support for a person adjusting to a divorce.

Marital Changes

A happily married couple of forty-five years spoke to our class and began their presentation with, "Marriage is one of life's biggest disappointments." They spoke of the difference between all the hype and cultural ideal of what marriage is supposed to be . . . and the reality. One effect of getting married is **disenchantment**—the change in relationship from a state of newness and high expectation to a state of mundaneness and boredom in the face of reality. It may not happen in the first few weeks or months of marriage, but it is almost inevitable. Whereas courtship is the anticipation of a life together, marriage is the day-to-day reality that life together does not always fit the dream. "Moonlight and roses become daylight and dishes" is an old adage reflecting the realities of marriage. Disenchantment after marriage is also related to partners shifting their focus away from each other to work or children; each partner usually gives and gets less attention in marriage than in courtship. Most college students do not anticipate a nosedive toward disenchantment; of 1,319 respondents, 17.3 percent agreed that "most couples become disenchanted with marriage within five years." Only 1.4 percent "strongly agreed" (Knox and Zusman 2009).

disenchantment the change in a relationship from a state of newness and high expectation to a state of mundaneness and boredom in the face of reality.

A couple will experience other changes when they marry, such as:

1. **Experiencing loss of freedom.** Single people do as they please. They make up their own rules and answer to no one. Marriage changes that as the expectations of the spouse impact the freedom of the individual. In a study of 1,001 married adults, although 41 percent said that they missed "nothing" about the single life, 26 percent reported that they most missed not being able to live by their own rules (Cadden and Merrill 2007).
2. **Feeling more responsibility.** Single people are responsible for themselves. Spouses are responsible for the needs of each other and sometimes resent it. One wife said she loved when her husband went out of town on business because she then did not feel the responsibility to cook for him. One husband said that he felt burdened by having to help his wife care for her aging parents. In the study

A GREAT MARRIAGE IS NOT WHEN THE "PERFECT COUPLE" COME TOGETHER. IT IS WHEN AN IMPERFECT COUPLE LEARNS TO ENJOY THEIR DIFFERENCES.

—*Dave Meurer*

of 1,001 married adults referred to previously, 25 percent reported that having less responsibility was what they missed most about being single.

3. **Missing alone time.** Aside from the few spouses who live apart, most live together. They wake up together, eat their evening meals together, and go to bed together. Each may feel too much togetherness. "This altogether, togetherness thing is something I don't like," said one spouse. In the study referred to previously, 24 percent reported that "having time alone for myself" was what they missed most about being single (Cadden and Merrill 2007). One wife said, "My best time of the day is at night when everybody else is asleep."

4. **Change in how money is spent.** Entertainment expenses in courtship become allocated to living expenses and setting up a household together. In the same study, 17 percent reported that they missed managing their own money most (Cadden and Merrill 2007).

5. **Discovering that one's mate is different from one's date.** Courtship is a context of deception. Marriage is one of reality. Spouses sometimes say, "He (she) is not the person I married." Jay Leno once quipped that "It doesn't matter who you marry since you will wake up to find that you have married someone else."

© iStockphoto.com / © Maria Toutoudaki/iStockphoto.com

6. **Sexual changes.** The sexual relationship of the couple also undergoes changes with marriage. First, because spouses are more sexually faithful to each other than are dating partners or cohabitants (Treas and Giesen 2000), their number of sexual partners will decline dramatically. Second, the frequency with which they have sex with each other decreases. One wife said the following:

> *"The urgency to have sex disappears after you're married. After a while you discover that your husband isn't going to vanish back to his apartment at midnight. He's going to be with you all night, every night. You don't have to have sex every minute because you know you've got plenty of time. Also, you've got work and children and other responsibilities, so sex takes a lower priority than before you were married.*

Although married couples may have intercourse less frequently than they did before marriage, marital sex is still the most satisfying of all sexual contexts. Of married people in a national sample, 85 percent reported that they experienced extreme physical pleasure and extreme emotional satisfaction with their spouses. In contrast, 54 percent of individuals who were not married or not living with anyone said that they experienced extreme physical pleasure with their partners, and 30 percent said that they were extremely emotionally satisfied (Michael et al. 1994). Fisher and McNulty (2008) studied seventy-two couples just after their weddings and found that high sexual satisfaction was associated with high marital satisfaction one year later.

7. **Power changes.** The distribution of power changes after marriage and across time. The way wives and husbands perceive and interact with each other continues to change throughout the course of the marriage. Two researchers studied 238 spouses who had been married more than thirty years and observed that (over time) men changed from being patriarchal to collaborating with their wives and that women changed from deferring to their husbands' authority to challenging that authority (Huyck and Gutmann 1992). In effect, men tend to lose power and women gain power. However, such power changes may not always occur. In abusive relationships, abusive partners may increase the display of power because they fear the partner will try to escape from being controlled.

Parents and In-Law Changes

Marriage affects relationships with parents. Time spent with parents and extended kin radically increases when a couple has children. Indeed, a major difference between couples with and without children is the amount of time they spend with relatives. Parents and kin rally to help with the newborn and are typically there for birthdays and family celebrations.

Emotional separation from one's parents is an important developmental task in building a successful marriage. When choices must be made between one's parents and one's spouse, more long-term positive consequences for the married couple are associated with choosing the spouse over the parents. However, such choices become more complicated and difficult when one's parents are old, ill, or widowed.

Only a minority of spouses (3 percent to 4 percent) report that they do not get along with their in-laws (Amato et al. 2007). Nuner (2004) interviewed twenty-three daughters-in-law married between five and ten years (with no previous marriages and at least one child from the marriage) and found that most reported positive relationships with their mothers-in-law. The evaluation by the daughters depended on the role of the mothers-in-law within the family (for example, mothers-in-law as grandmothers were perceived more positively). Mother-son relationships were described by the daughters-in-law to be close, "mama's boy," polite, or distant.

The behavior in-laws engage in affects how their children and their spouses like and perceive them. Morr Serewicz and Canary (2008) investigated newlyweds' perceptions of private disclosures received from their in-laws and found that, when these were positive (for example, the in-law talked positively about family members), the spouse perceived the in-law positively. Conversely, when the in-law talked negatively about family members or told negative stories, the spouse viewed the in-law negatively.

Financial Changes

An old joke about money in marriage says that "two can live as cheaply as one as long as one doesn't eat." The reality behind the joke is that marriage involves the need for spouses to discuss and negotiate how they are going to get and spend money in their relationship.

Some spouses bring considerable debt into the marriage. In a sample of 1,319 undergraduates, 8.9 percent reported that they owed "over a thousand dollars on one or more credit cards" (Knox and Zusman 2009). Dew (2008) also observed that debt was associated with marital dissatisfaction. Spouses who were in debt reported spending less leisure time together and arguing more about money. We will discuss more about debt in Chapter 11 on Family and the Economy.

6.6 Diversity in Marriage

The tragedies of September 11, 2001, emphasized the need to understand other cultures and an appreciation for military families who make personal sacrifices for the larger societal good. In this section, we look at examples of diverse marriages: Age-discrepant, interracial, interreligious, cross-national, and military.

Age-Discrepant Relationships and Marriages

Although people in most pairings are of similar age, sometimes the partners are considerably different in age. In marriage, these are referred to as ADMs (age-dissimilar marriages) and are in contrast to ASMs

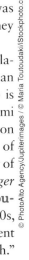

© PhotoAlto Agency/Jupiterimages / © Maria Toutoudaki/iStockphoto.com

(age-similar marriages). ADMs are also known as **May-December marriages**. Typically, the woman is in the spring of her youth (May), whereas the man is in the later years of his life (December). There have been a number of May-December celebrity marriages, including that of Celine Dion, who is twenty-six years younger than René Angelil (in 2011, aged 42 and 68). Larry King is also twenty-six years older than his seventh wife, Shawn (in 2010 he was 75). Michael Douglas is twenty-five years older than his wife, Catherine Zeta Jones, and Ellen DeGeneres is fifteen years older than Portia de Rossi, her partner (they married in 2008).

May-December marriage an age-discrepant marriage, usually in which the younger woman is in the spring of her life (May) and the man is in his later years (December).

cougar a woman, usually in her 30s or 40s, who is financially stable and mentally independent and looking for a younger man with whom to have fun.

One might assume that these marriages are less happy because the spouses were born into such different age contexts. Research shows otherwise. Barnes and Patrick (2004) compared thirty-five ADMs (in which spouses were fourteen or more years apart) and thirty-five ASMs (in which spouses were less than five years apart) and found no difference in reported marital satisfaction between the two groups. As is true in other research, wives reported lower marital satisfaction and more household responsibilities in both groups.

Perhaps the greatest example of a May-December marriage that "worked" is of Oona Chaplin, wife of Charles Chaplin. She married him when she was age 18 (he was age 54). Their May-December alliance was expected to last the requisite six months, but they remained together and raised eight children.

Although less common, some age-discrepant relationships are those in which the woman is older than her partner. Mary Tyler Moore (married 25 years) is 18 years older than her husband, Robert Levine. Demi Moore is sixteen years older than husband Ashton Kutcher (she was 40 and he was 27 at the time of their wedding). Valerie Gibson (2002) is the author of *Cougar: A Guide for Older Women Dating Younger Men*. She noted that the current use of the term **cougars** refers to "women, usually in their 30s and 40s, who are financially stable and mentally independent and looking for a younger man to have fun with." Gibson noted that one-third of women between the ages of 40 and 60 are dating younger men. Financially independent women need not select a man in reference to his breadwinning capabilities. Instead, these "cougars" are looking for men, not to marry but to enjoy. The downside of such relationships comes if the man gets serious and wants to have children, which may spell the end of the relationship.

Interracial Marriages

Interracial marriages may involve many combinations, including American white, American black, Indian, Chinese, Japanese, Korean, Mexican, Malaysian, and Hindu mates. Of a sample of 1,319 undergraduates at a southeastern university, 44.2 percent agreed that "It is important to me that I marry someone of my same race" (Knox and Zusman 2009). However, actual interracial marriages are rare in the United States—fewer than 5 percent of all marriages in the United States are interracial. Of these, fewer than 1 percent are of a black person and a white person (*Statistical Abstract of the United States, 2009*, Table 59). Examples of African American men who are married to Caucasian women are Tiger Woods and Charles Barkley. Segregation in religion (the races worship in separate churches), housing (white and black neighborhoods), and education (white and black colleges), not to speak of parental and peer endogamous pressure to marry within one's own race, are factors that help to explain the low percentage of interracial black and white marriages.

The spouses in black and white couples are more likely to have been married before, to be age-discrepant, to live far away from their families of orientation, to have been reared in racially tolerant homes, and to have educations beyond high school. Some may also belong to religions that encourage interracial unions. The Baha'i religion, which has more than 6 million members worldwide and 84,000 in the United States, teaches that God is particularly pleased with interracial unions. Finally, interracial spouses may tend to seek contexts of diversity. "I have been reared in a military family, been everywhere and met people of different races and nationalities throughout my life. I seek diversity," noted one student. Kennedy (2003) identified three reactions to a black-white couple who cross racial lines to marry: (1) approval (increases racial open-mindedness, decreases social segregation), (2) indifference (interracial marriage is seen as a private choice), and (3) disapproval (reflects racial disloyalty, impedes perpetuation of black culture). As Kennedy notes, "The argument that intermarriage is destructive of racial solidarity has been the principal basis of black opposition" (p. 115). There is also the concern for the biracial identity of offspring of mixed-race parents. Although most mixed-race parents identify their child as having minority race status, there is a trend toward identifying their child as multiracial. This may be an increasing trend because of Barack Obama's racial heritage—he has a black father and a white mother (Brunsma 2005).

Interracial partners sometimes experience negative reactions to their relationship. Black people partnered with white people have their blackness and racial identity challenged by other black people. White people partnered with black people may lose their white status and have their awareness of whiteness heightened more than ever before. At the same time, one partner is not given full status as a member of the other partner's race (Hill and Thomas 2000). Gaines and Leaver (2002) also note that the pairing of a black male and a white female is regarded as "less appropriate" than that of a white male and a black female. In the former, the black male "often is perceived as attaining higher social status (i.e., the white woman is viewed as the black man's 'prize,' stolen from the more deserving white man)" (p. 68). In the latter, when a white male pairs with a black female, "no fundamental change in power within the American social structure is perceived as taking place" (p. 68). Interracial marriages are also more likely to dissolve than same-race marriages (Fu 2006). Disapproval of cross racial relationships begins early. Kreaer (2008) studied adolescents who were dating cross racially—they reported disapproval from peers.

Black-white interracial marriages are likely to increase—slowly. Not only has white prejudice against African Americans in general declined, but also segregation in school, at work, and in housing has decreased, permitting greater contact between the races. The Self-Assessment card for Chapter 6 allows you to assess your openness to involvement in an interracial relationship.

Interreligious Marriages

Although religion may be a central focus of some individuals and their marriage, Americans in general have become more secular, and religion has become less influential as a criterion for selecting a partner as a result. In a survey of 1,319 undergraduates, only 40.2 percent reported that marrying someone of the same religion was important for them (Knox and Zusman 2009).

Are people in interreligious marriages less satisfied with their marriages than those who marry someone of the same faith? The answer depends on a number of factors. First, people in marriages in which one or both spouses profess "no religion" tend to report lower levels of marital satisfaction than those in which at least one spouse has a religious tie. People with no religion are often more liberal and less bound by traditional societal norms and values; they feel less constrained to stay married for reasons of social propriety.

The impact of a mixed religious marriage may also depend more on the devoutness of the partners than on the fact that the partners are of different religions. If both spouses are devout in their religious beliefs, they may expect some problems in the relationship (although not necessarily). Less problematic is the relationship in which one spouse is devout but the partner is not. If neither spouse in an interfaith marriage is devout, problems regarding religious differences may be minimal or nonexistent. In their marriage vows, one interfaith couple who married (he Christian, she Jewish) said that they viewed their different religions as an opportunity to strengthen their connections to their respective faiths and to each other. "Our marriage ceremony seeks to celebrate both the Jewish and Christian traditions, just as we plan to in our life together."

Cross-National Marriages

Of 1,319 undergraduates, 60.4 percent reported that they would be willing to marry someone from another country (Knox and Zusman 2009). The opportunity to meet someone from another country is increasing as more than 600,000 foreign students are studying at American colleges and universities. Because not enough Americans are going into math and engineering, these foreign students are wanted because they may find the cure for cancer or invent a vaccine for HIV (Marklein 2008).

Because American students take classes with foreign students, there is the opportunity for dating and romance between the two groups, which may lead to marriage. Some people from foreign countries marry an American citizen to gain citizenship in the United States, but immigration laws now require the marriage to last two years before citizenship is granted. If the marriage ends before two years, the foreigner must prove good faith (that the marriage was not just to gain entry into the country) or will be asked to leave the country.

When the international student is male, more likely than not his cultural mores will prevail and will clash strongly with his American bride's expectations, especially if the couple should return to his country. One female American student described her experience of marriage to a Pakistani, who violated his parents' wishes by not marrying the bride they had chosen for him in childhood. The marriage produced two children before the four of them returned to Pakistan.

The woman felt that her in-laws did not accept her and were hostile toward her. The in-laws also imposed their religious beliefs on her children and took control of their upbringing. When this situation became intolerable, the woman wanted to return to the United States. Because the children were viewed as being "owned" by their father, she was not allowed to take them with her and was banned from even seeing them. Like many international students, the husband was from a wealthy, high-status family, and the woman was powerless to fight the family. The woman has not seen her children in six years.

Cultural differences do not necessarily cause stress in cross-national marriage; the degree of cultural difference is not necessarily related to degree of stress. Much of the stress is related to society's intolerance of cross-national marriages, as manifested in attitudes of friends and family. Japan and Korea place an extraordinarily high value on racial purity. At the other extreme is the racial tolerance evident in Hawaii, where a high level of out-group marriage is normative.

Military Families

Approximately 1.4 million U.S. troops are active-duty military personnel. Another 1.1 million are in the military reserve and 375,000 in the National Guard (*Statistical Abstract of the United States, 2009*, Tables 493, 499, 501). About 60 percent of military personnel are married and/or have children (NCFR Policy Brief 2004).

There are three main types of military marriages. One, those in which the soldier falls in love with a

high school sweetheart, marries the person, and subsequently joins the military. A second type of military marriage consists of those who meet and marry after one of them has signed up for the military. This is a typical marriage where the partners fall in love on the job and one or both of them happens to be the military. The final and least common military marriage is known as a contract marriage in which a person will marry a civilian to get more money and benefits from the government. For example, a soldier might decide to marry a platonic friend and split the money from the additional housing allowance (which is sometimes a relatively small amount of money and varies depending on geographical location and rank). Other times, the military member keeps the extra money and the civilian takes the benefit of health insurance. Often, in these types of military marriages, the couple does not reside together. There is no emotional connection because the marriage is mercenary. Contract military marriages are not common but they do exist.

Of Americans in a national sample, 60 percent report that one of their family members, close friends, or coworkers is in Iraq (Page 2006). When accounting for spouses, children, and other dependents, there are actually more family members of active duty military personnel than there are military members themselves (Martin and McClure 2000; Military Family Resource Center 2003). Spouses of military members play an important role in the readiness and retention of the active-duty force and, thus, factors that affect the well-being of military spouses are important to study when looking at military families (Easterling 2005). It is also important to note that, although military and civilian families have many similarities, there are also some unique aspects of military families.

Some ways in which military families are unique include:

1. **Traditional sex roles.** Although both men and women are members of the military service, the military has considerably more men than women (Caforio 2003). In the typical military family, the husband is deployed (sent away) and the wife is expected to understand his military obligations and to take care of the family in his absence. Her duties include paying the bills, keeping up the family home, and taking care of the children; a military wife must often play the role of both spouses due to the demands of her husband's military career and obligations. The wife often has to sacrifice her career to follow (or stay behind in the case of deployment) and support her husband in his fulfillment of military duties (Easterling 2005).

   In the case of wives or mothers who are deployed, the rare husband is able to switch roles and become Mr. Mom. One military career wife said of her husband, whom she left behind when she was deployed, "What a joke. He found out what taking care of kids and running a family was really like and he was awful. He fed the kids SpaghettiOs for the entire time I was deployed."

   There are also circumstances in which both parents are military members, and this can blur traditional sex roles because the woman has already deviated from a traditional "woman's job." Military families in which both spouses are military personnel are rare.

2. **Loss of control—deployment.** Military families have little control over their lives as the specter of deployment is ever-present. Where one of the spouses will be next week and for how long are beyond the control of the spouses and parents. Saleska (2004) interviewed wives of Air Force men (enlisted and officers) who emphasized the difficulty of being faced with the constant possibility that their husbands could leave immediately for an indeterminate period of time. Once gone, they may be relegated to a ten-minute phone call every two weeks. The needs of the military come first, and military personnel are expected to be obedient and to do whatever is necessary to comply and get through the ordeal. "You can't believe what it's like to have an empty chair at the dinner table sprung on you and not know where he is or when he'll be back," one respondent said.

   Compounding the loss of control is the fear of being captured, imprisoned, shot, killed by a suicide bomber, or beheaded. Not only may deployed soldiers have such fears, but their spouses, parents, and children may look at the evening news in stark terror and fear that their beloved will be the next to die. Sleeplessness, irritability, and depression may result in those who are left behind to carry on their jobs and parenting. Children may also become anxious and depressed over the absence of their deployed parent, who more often is the father (Cozza et al. 2005).

According to a 2008 survey of military wives, some of the most-reported negative feelings during the deployment of their husbands included loneliness, fear, and sadness. They may go extended periods of time without communicating with their partner and are often in constant worry over the well-being of their deployed spouse. On the positive side, wives of deployed husbands report feelings of independence and strength. They are the sole family member available to take care of the house and children, and they rise to the challenge. The challenges of coping with deployment are enormous.

3. **Infidelity.** Although most spouses are faithful to each other, the context of separation from each other for months (sometimes years) at a time increases the vulnerability of both spouses to infidelity. The double standard may also be operative, whereby "men are expected to have other women when they are away" and "women are expected to remain faithful and be understanding." Separated spouses try to bridge the time they are apart with e-mails, phone calls, and new technology such as Skype (when possible), but sometimes the loneliness becomes more difficult than anticipated. One enlisted husband said that he returned home after a year-and-a-half deployment to be confronted with the fact that his wife had become involved with someone else. "I absolutely couldn't believe it," he noted. "In retrospect, I think the separation was more difficult for her than it was for me."

4. **Frequent moves and separation from extended family or close friends.** Because military couples are often required to move to a new town, parents no longer have doting grandparents available to help them rear their children. As well, although other military families become a community of support for each other, the consistency of such support may be lacking. "We moved seven states away from my parents to a town in North Dakota," said one wife. "It was dreadful."

   Similar to being separated from parents and siblings is the separation from one's lifelong friends. Although new friendships and new supportive relationships develop within the military community to which the family moves, the relationships are sometimes tenuous and temporary as the new families move on. The result is the absence of a stable, predictable social structure of support, which may result in a feeling of alienation and not belonging in either the military or the civilian community. The more frequent the moves, the more difficult the transition and the more likely the alienation of new military spouses. "Enormous stress and frustration among military families is no surprise," notes Donald Wolfe (2006), who is a marriage and family counselor who specializes in military marriages. Among military marriages, white people are as likely to divorce as black people (Lindquist 2004).

5. **Divorce among military families.** There is conflicting research on whether or not military marriages are more likely to end in divorce than are civilian marriages, like most people assume. Preliminary findings in one study indicate that military men are actually less likely to divorce than civilian men. However, military women who marry are more likely to divorce. This indicates that incentives for military men to remain married are greater than are those for military women (Pollard et al. n.d.). Additionally, a study concentrating on military families suggests that military divorce rates are not significantly increasing, even with the wars that have begun since 2001 (Karney and Crown 2007).

6. **Employment of spouses.** Well-documented research indicates that employment is beneficial to one's well-being. Military spouses, however, are at a disadvantage when it comes to finding and maintaining careers or even finding a job they can enjoy. Employers in military communities are often hesitant to hire military spouses because they know the demands that are placed on them in the absence of the deployed military member can be enormous. They are also aware of the frequent moves that military families make and may be reluctant to hire employees for what may be a relatively short amount of time. The result is a disadvantaged wife who has no job and must put her career on hold. Military spouses, when they do find employment, are often underemployed, which can lead to low levels of job satisfaction. They also make less, on average, than their civilian coun-

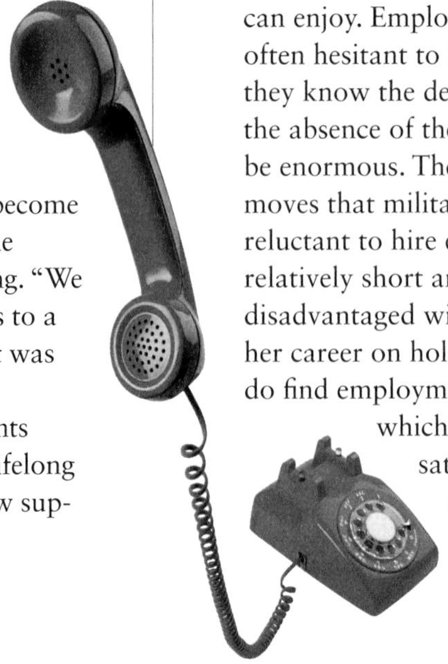

terparts with similar characteristics. All of these factors can contribute to distress among military spouses (Easterling 2005).

7. **Resilient military families.** In spite of these difficulties, there are also enormous benefits to being involved in the military, such as having a stable job (one may get demoted but it is much more difficult to get "fired") and having one's medical bills paid for. In addition, most military families are amazingly resilient. Not only do they anticipate and expect mobilization and deployment as part of their military obligation, they respond with pride. Indeed, some reenlist eagerly and volunteer to return to military life even when retired. One military captain stationed at Fort Bragg, in Fayetteville, North Carolina, noted, "It is part of being an American to defend your country. Somebody's got to do it and I've always been willing to do my part." He and his wife made a presentation in our classes. She said, "I'm proud that he cares for our country and I support his decision to return to Afghanistan to help as needed. And most military wives that I know feel the same way."

    Although military families face great challenges and obstacles, many adopt the philosophy of "whatever doesn't kill us makes us stronger." Facing deployments and frequent moves often forces a military couple to learn to rely on themselves as well as each other. They make it through difficult life events, unique to their lifestyle, which can make day-to-day challenges seem trivial. The strength that is developed within a military marriage through all the challenges they face has the potential to build a strong, resilient marriage.

6.7 Marital Success

Marital success is measured in terms of marital stability and marital happiness. Stability refers to how long the spouses have been married and their view on the permanence of the relationship, whereas marital happiness refers to more subjective aspects of the relationship.

In describing marital success, researchers have used the terms *satisfaction, quality, adjustment, lack of distress,* and *integration.* Marital success is often measured by asking spouses how happy they are, how often they spend their free time together, how often they agree about various issues, how easily they resolve conflict, how sexually satisfied they are, and how often they have considered separation or divorce. The degree to which the spouses enjoy each other's companionship is another variable of marital success. Not all couples, even those recently married, achieve high-quality marriages. In a national sample comparing married people with unmarried people, Princeton Survey Research Associates International found that 43 percent of the spouses reported that they were "very happy," compared with 24 percent of unmarried people (Stuckey and Gonzalez 2006).

marital success relationship in which the partners have spent many years together and define themselves as happy and in love (hence, the factors of time and emotionality).

Corra et al. (Forthcoming) analyzed data collected over a thirty-year period, from the 1972 to 2002 General Social Surveys, to discover the influence of sex (male or female) and race (white or black) on the level of reported marital happiness. Findings indicated greater levels of marital happiness among males and white people than among females and black people. The researchers suggested that males make fewer accommodations in marriage and that white people are not burdened with racism and have less economic stress. Amato et al. 2007 also reported less marital happiness and more marital problems among wives than husbands. Wallerstein and Blakeslee (1995) studied fifty financially secure couples in stable (from ten to forty years), happy marriages with at least one child. These couples defined marital happiness as feeling respected and cherished. They also regarded their marriages as works in progress that needed continued attention to avoid becoming stale. No couple said that they were happy all the time. Rather, a good marriage is a process. Billingsley et al. (1995) interviewed thirty happily married couples who had been wed an average of thirty-two years and had an average of 2.5 children. They found various characteristics associated with couples who stay together and who enjoy each other. These qualities appear to be the same for both husbands and wives (Amato et al. 2007). Based on these studies of couples in stable, happy relationships, the following twelve characteristics emerge:

1. **Personal and emotional commitment to stay married.** Divorce was not considered an option. The spouses were committed to each

© iStockphoto.com / © Maria Toutoudaki/iStockphoto.com

other for personal reasons rather than societal pressure. In addition, the spouses were committed to maintain the marriage out of emotional rather than economic need (DeOllos 2005).

2. **Common interests.** The spouses talked of sharing interests, values, goals, children, and the desire to be together.

3. **Communication.** Gottman and Carrere (2000) studied the communication patterns of couples over an eleven-year period and emphasized that those spouses who stay together are five times more likely to lace their arguments with positives ("I'm sorry I hurt your feelings") and to consciously choose to say things to each other that nurture the relationship rather than harm it. Successful spouses also feel comfortable telling each other what they want and not being defensive at feedback from the partner.

4. **Religiosity.** A strong religious orientation provided the couples with social, spiritual, and emotional support from church members and with moral guidance in working out problems (DeOllos 2005). People with no religious affiliation report more marital problems and are more likely to divorce (Amato et al. 2007).

5. **Trust.** Trust in the partner provided a stable floor of security for the respective partners and their relationship. Neither partner feared that the other partner would leave or become involved in another relationship.

6. **Not materialistic.** Being nonmaterialistic was a characteristic of these happily married couples. Although the couples may have lived in nice houses and had expensive toys (for example, a boat and camper), they were tied to nothing. "You can have my things, but don't take away my people," is a phrase from one husband reflecting his feelings about his family.

7. **Role models.** The couples spoke of positive role models in their parents. Good marriages beget good marriages—good marriages run in families. It is said that the best gift you can give your children is a good marriage.

8. **Sexual desire.** Wilson and Cousins (2005) confirmed that partners' similar rankings of sexual desire as important in predicting long-term relationship success. Earlier, we noted the superiority of marital sex over sex in other relationship contexts in terms of both emotion and physical pleasure.

9. **Equitable relationships.** Amato et al. (2007) observed that the decline in traditional gender attitudes and the increase in egalitarian decision making is related to increased happiness in today's couples.

10. **Absence of negative attributions.** Spouses who do not attribute negative motives to their partner's behavior report higher levels of marital satisfaction than spouses who ruminate about negative motives. Dowd et al. (2005) studied 127 husbands and 132 wives and found that the absence of negative attributions was associated with higher marital quality.

11. **Forgiveness.** At some time in all marriages, each spouse engages in behavior that may hurt the partner. Forgiveness rather than harboring resentment allows spouses to move forward. Spouses who do not "drop the lowest test score" find that they inadvertently create a failing marriage in which they then must live. McNulty (2008) noted the value of forgiveness, particularly

when married to partners who rarely behaved negatively.

12. Economic security. Although money does not buy happiness, having a stable, secure economic floor is associated with marital quality (Amato et al. 2007). North et al. (2008) examined the role of income and social support in predicting concurrent happiness and change in happiness among 274 married adults across a ten-year period. They found that income had a small, positive impact on happiness, which diminished as income increased. In contrast, family social support, as reflected in cohesion, expressiveness, and low conflict showed a substantial, positive association with concurrent happiness, particularly when income was low.

Marital Happiness across the Family Life Cycle

Although a successful marriage is one in which the partners are happy and in love across time, spouses report that some periods are happier than others. Figure 6.1 provides a retrospective of marriage by fifty-two white college-educated husbands and wives over a period of thirty-five years together. The couples reported the most enjoyment with their relationship in the beginning, followed by less enjoyment during the child-bearing stages, and a return to feeling more satisfied after the children left home. Corra et al. (2009) found a similar curvilinear relationship in their analysis of survey data covering a period of thirty years. Whiteman et al. (2007) also observed decreases in marital satisfaction and love, as well as increases in conflict, as the couple's offspring reached puberty and an improvement when their children were 18 to 20—the time children typically leave home for college.

Plagnol and Easterlin (2008) conceptualized happiness as the ratio of aspirations and attainments in the areas of family life and material goods. They studied 47,000 women and men and found that up to age 48, women are happier than men in both domains. After age 48, a shift causes women to become less satisfied with family life (their children are gone) and material goods (some are divorced and have fewer economic resources). In contrast, men become more satisfied with family life (the empty nest is less of a problem) and finances (men typically have more economic resources than women, whether married or divorced). Individuals who are black and/or with lower education report less happiness.

Figure 6.1

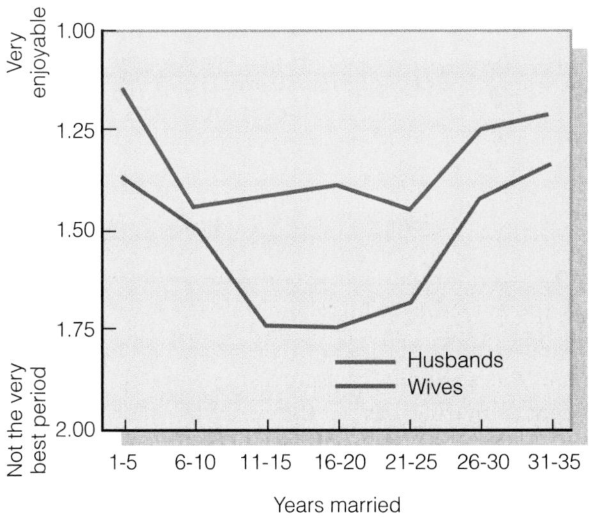

Source: Caroline O. Vaillant and George E. Vaillant. 1993. Is the U-curve of marital satisfaction an illusion? A 40-year study of marriage. *Journal of Marriage and the Family* 55:237 (Figure 6, copyrighted 1993 by the National Council on Family Relations, 3989 Central Avenue N. E., Suite 550, Minneapolis, MN 55421.)

The Value of Marriage Education Programs

A team of researchers (Macomber et al. 2005) evaluated the numerous programs designed to strengthen marriage available in American society. They found an incredible diversity of programs offered through a variety of settings (for example, churches and mental health clinics), focused on a variety of populations (for example, couples and parents), by personnel with a range of academic training (from none to PhD). The lack of systematic content, control groups, and follow-ups makes evaluating the myriad programs impossible, so their value remains an open question. DeMaria (2005) also confirmed the lack of data as to the effectiveness of marriage education programs.

What can we learn from our knowledge about the characteristics of successful marriages? Commitment, common interests, communication skills, and a nonmaterialistic view of life all are factors in maintaining a successful marriage. Couples might strive to include these as part of their relationships.

Same-Sex Couples and Families

SECTIONS

TO DISCRIMINATE AGAINST OUR SISTERS AND BROTHERS WHO ARE LESBIAN OR GAY ON GROUNDS OF THEIR SEXUAL ORIENTATION FOR ME IS AS TOTALLY UNACCEPTABLE AND UNJUST AS APARTHEID EVER WAS.

—Archbishop Desmond Tutu

My wife and I were vacationing in central Mexico and stayed at a bed-and-breakfast in a town west of Mexico City. We befriended a young Mexican male who worked in the office who was responsible for helping guests with various issues such as touring surrounding cities and sites, identifying places in town to eat, and so on. We were also interested to know his perception of the acceptance of homosexual people as a male who was reared in central Mexico. "Their life is over," he said, "the parents will disown them and the mother will be blamed that the child turned out this way." He also gave an example of a hate crime in the area in which a young gay male was found dead . . . his penis had been cut off and stuffed in his mouth with a note on him that said, "This will happen to you if you are gay." The police did nothing.

The police/judicial system looking the other way for the murder of homosexuals has also occurred in the United States. Harvey Milk (portrayed by Sean Penn who won the Best Actor Academy Award for his role in the 2008 film *Milk*) was the first openly gay man elected to any substantial political office anywhere and was assassinated point blank in 1978 by Dan White (a political rival). Although premeditated murder carries the death penalty in California, White was convicted of voluntary manslaughter, which resulted in a lesser sentence.

In the United States, homosexuality remains a subject over which there continues to be wide differences of approval. In 2007, Senator Larry Craig (R-Idaho) lost his job in the wake of a homosexual scandal in the Minneapolis-St. Paul International Airport men's room after he pleaded guilty to misconduct. In regard to marriage, although some states grant marriage licenses to gay couples (for example, Massachusetts and Connecticut), others have overturned a state supreme court decision upholding gay marriage (for example, California's Proposition 8) and defined marriage as the exclusive union between a woman and a man. Hate crimes against homosexuals are not uncommon. Worldwide, approval differences are considerable,

with the Netherlands, Spain, Belgium, Norway, and South Africa granting equal marriage rights to same-sex couples, whereas intense discrimination is the norm in other countries (e.g., Pakistan and Kenya).

In this chapter, we discuss same-sex couples and families—relationships that are, in many ways, similar to heterosexual ones. A major difference, however, is that gay and lesbian couples and families are subjected to **prejudice** and **discrimination**. Although other minority groups also experience prejudice and discrimination, only minorities of sexual orientation are denied federal legal marital status and the benefits and responsibilities that go along with marriage (which we discuss later in this chapter). Also, gay couples are sometimes rejected by their own parents, siblings, and other family members. One father told his son, "I'd rather have a dead son than a gay son."

prejudice negative attitudes toward others based on differences.

discrimination behavior that denies individuals or groups equality of treatment.

sexual orientation a classification of individuals as heterosexual, bisexual, or homosexual, based on their emotional, cognitive, and sexual attractions and self-identity.

heterosexuality the predominance of emotional and sexual attraction to individuals of the opposite sex.

homosexuality predominance of emotional and sexual attraction to individuals of the same sex.

bisexuality emotional and sexual attraction to members of both sexes.

lesbian homosexual woman.

gay homosexual women or men.

lesbigay population collective term referring to lesbians, gays, and bisexuals.

transgendered individuals who express their masculinity and femininity in nontraditional ways consistent with their biological sex.

LGBT (GLBT) refers collectively to lesbians, gays, bisexuals, and transgendered individuals.

Homosexual behavior has existed throughout human history and in most (perhaps all) human societies (Kirkpatrick 2000). In this chapter, we focus on Western views of sexual diversity that define **sexual orientation** as a classification of individuals as heterosexual, bisexual, or homosexual, based on their emotional, cognitive, and sexual attractions and self-identity. **Heterosexuality** refers to the predominance of emotional and sexual attraction to individuals of the opposite sex. **Homosexuality** refers to the predominance of emotional and sexual attraction to individuals of the same sex, and **bisexuality** is emotional and sexual attraction to members of both sexes. The term **lesbian** refers to homosexual women; **gay** can refer to either homosexual women or homosexual men. Lesbians, gays, and bisexuals, sometimes referred to collectively as the **lesbigay population**, are considered part of a larger population referred to as the transgendered community. **Transgendered** individuals are those who express their masculinity and femininity in nontraditional ways consistent with their biological sex. For example, a biological male is not expected to wear a dress. Transgendered individuals include not only homosexuals and bisexuals but also cross-dressers, transvestites, and transsexuals (see Chapter 2 on Gender). Because much of the current literature on the lesbigay population includes other members of the transgendered community, the terms **LGBT** or **GLBT** are often used to refer collectively to lesbians, gays, bisexuals, and transgendered individuals.

© Banana Stock/Jupiterimages

7.1 Prevalence of Homosexuality, Bisexuality, and Same-Sex Couples

Before looking at prevalence data concerning homosexuality and bisexuality in the United States, it is important to understand the ways in which identifying or classifying individuals as heterosexual, homosexual, gay, lesbian, or bisexual can be problematic.

Problems Associated with Identifying and Classifying Sexual Orientation

The classification of individuals into sexual orientation categories (for example heterosexual, homosexual, bisexual) is problematic for a number of reasons (Savin-Williams 2006). First, because of the social stigma associated with nonheterosexual identities, many individuals conceal or falsely portray their sexual-orientation identities to avoid prejudice and discrimination.

Second, not all people who are sexually attracted to or have had sexual relations with individuals of the same sex view themselves as homosexual or bisexual. A final difficulty in labeling a person's sexual orientation is that an individual's sexual attractions, behavior, and identity may change across time. For example, in a longitudinal study of 156 lesbian, gay, and bisexual youth, 57 percent consistently identified as gay or lesbian and

> THE PREJUDICE AGAINST GAYS AND LESBIANS IS MEANER, NASTIER—A VINDICTIVENESS THAT'S ROOTED IN HATRED, NOT IGNORANCE.
>
> *— Betty DeGeneres, mother of comedienne Ellen DeGeneres*

15 percent consistently identified as bisexual over a one-year period, but 18 percent transitioned from bisexual to lesbian or gay (Rosario et al. 2006).

Early research on sexual behavior by Kinsey and his colleagues (1948; 1953) found that, although 37 percent of men and 13 percent of women had had at least one same-sex sexual experience since adolescence, few of the individuals reported exclusive homosexual behavior. These data led Kinsey to conclude that most people are not exclusively heterosexual or homosexual. Rather, Kinsey suggested an individual's sexual orientation may have both heterosexual and homosexual elements. In other words, Kinsey suggested that heterosexuality and homosexuality represent two ends of a sexual-orientation continuum and that most individuals are neither entirely homosexual nor entirely heterosexual, but fall somewhere along this continuum.

The Heterosexual-Homosexual Rating Scale that Kinsey et al. (1953) developed allows individuals to identify their sexual orientation on a continuum. Individuals with ratings of 0 or 1 are entirely or largely heterosexual; 2, 3, or 4 are more bisexual; and 5 or 6 are largely or entirely homosexual (see Figure 7.1). Very few individuals are exclusively a 0 or 6, prompting Kinsey to believe that most individuals are bisexual.

Sexual-orientation classification is also complicated by the fact that sexual behavior, attraction, love, desire, and sexual-orientation identity do not always match. For example, "research conducted across different cultures and historical periods (including present-day Western culture) has found that many individuals develop passionate infatuations with same-gender partners in the absence of same-gender sexual desires . . . whereas others experience same-gender sexual desires that never manifest themselves in romantic passion or attachment" (Diamond 2003, 173).

Consider the findings of a national study of U.S. adults that investigated (1) sexual attraction to individuals of the same sex, (2) sexual behavior with people of the same sex, and (3) homosexual self-identification (Michael et al. 1994). This survey found that 4 percent of women and 6 percent of men said that they are sexually attracted to individuals of the same sex, and 4 percent of women and 5 percent of men reported that they had had sexual relations with a same-sex partner after age 18. What these data tell us is that "those who acknowledge homosexual sexual desires may be far more numerous than those who actually act on those desires" (Black et al. 2000, 140).

Prevalence of Homosexuality, Heterosexuality, and Bisexuality

Despite the difficulties inherent in categorizing individuals' sexual orientation, recent data reveal the prevalence of individuals in the United States who identify as lesbian, gay, or bisexual. In a national survey by Michael et al. (1994), fewer than 2 percent of women and 3 percent of the men identified themselves as homosexual or bisexual. Berg and Lein (2006) estimated that 7 percent of males and 4 percent of females were not heterosexual. A 2004 national poll showed that about 5 percent of U.S. high school students identified themselves as lesbian or gay (Curtis 2004). Tao (2008) analyzed U.S.

Figure 7.1
The Heterosexual-Homosexual Rating Scale

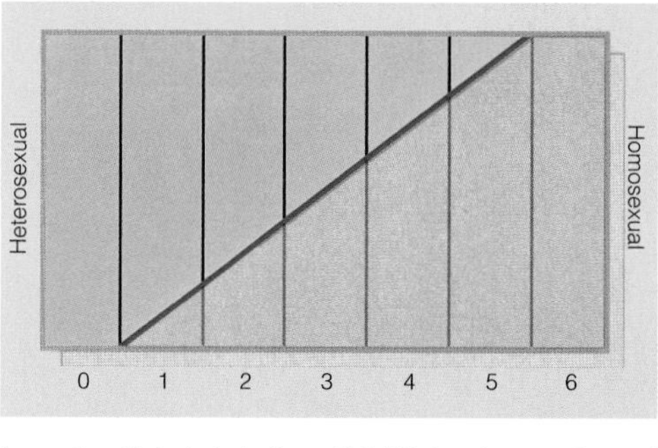

Based on both psychologic reactions and overt experience, individuals rate as follows:

0. Exclusively heterosexual with no homosexual
1. Predominantly heterosexual, only incidentallly homosexual
2. Predominantly heterosexual, but more than incidentally homosexual
3. Equally heterosexual and homosexual
4. Predominantly homosexual, but more than incidentally heterosexual
5. Predominantly homosexual, but incidentally heterosexual
6. Exclusively homosexual

Source: *Sexual Behavior in the Human Male*, W.B. Saunders, 1948. Reprinted by permission of the Kinsey Institute for Research in Sex, Gender, and Reproduction, Inc.

women ages 15 to 44 and found that 1.6 percent and 4 percent, respectively, self-identified as being lesbian and bisexual. Whether women are lesbian or bisexual is relevant to rates of sexually transmitted infections (STIs)—bisexual women are more likely to report having an STI than lesbian women (16 percent to 4.5 percent). In a university sample of 1,319 students, .09 percent, .06 percent, and 1.7 percent reported that they were lesbian, gay male, or bisexual, respectively (Knox and Zusman 2009).

Prevalence of Same-Sex Couple Households

Although U.S. Census surveys do not ask about sexual orientation or gender identity, same-sex cohabiting couples may identify themselves as "unmarried partners." Those couples in which both partners are men or both are women are considered to be same-sex couples or households for purposes of research. In 2006, there were 800,000 unmarried same-sex couple households (*Statistical Abstract of the United States: 2009*, Table 62).

Carpenter and Gates (2008) analyzed data in California and noted that, although 62 percent of heterosexual couples cohabit, about 40 percent of gay males and about 60 percent of lesbians cohabit. Same-sex couples are more likely to live in metropolitan areas than in rural areas. However, the largest proportional increases in the number of same-sex couples self-reporting in 2000 versus 1990 came in rural, sparsely populated states.

Why are data on the numbers of GLBT individuals and couples in the United States relevant? The primary reason is that census numbers on the prevalence of GLBT individuals and couples can influence laws and policies that affect gay individuals and their families. In anticipation of the 2000 census, the National Gay and Lesbian Task Force Policy Institute and the Institute for Gay and Lesbian Strategic Studies conducted a public education campaign urging people to "out" themselves on the 2000 census. The slogan was, "The more we are counted, the more we count" (Bradford et al. 2002, 3). "The fact that the Census documents the actual presence of same-sex couples in nearly every state legislative and U.S. Congressional district means anti-gay legislators can no longer assert that they have no gay and lesbian constituents" (p. 8).

7.2 Origins of Sexual-Orientation Diversity

Much of the biomedical and psychological research on sexual orientation attempts to identify one or more "causes" of sexual-orientation diversity. The driving question behind this research is, "Is sexual orientation inborn or is it learned or acquired from environmental influences?" Although a number of factors have been correlated with sexual orientation, including genetics, gender role behavior in childhood, and fraternal birth order, no single theory can explain diversity in sexual orientation.

Beliefs about What "Causes" Homosexuality

Aside from what "causes" homosexuality, social scientists are interested in what people believe about the "causes" of homosexuality. Most gay people believe that homosexuality is an inherited, inborn trait. In a national study of homosexual men, 90 percent reported that they believed that they were born with their homosexual orientation; only 4 percent believed that environmental factors were the sole cause (Lever 1994).

Individuals who believe that homosexuality is genetically determined tend to be more accepting of homosexuality and are more likely to be in favor of equal rights for lesbians and gays (Tyagart 2002). In contrast, "those who believe homosexuals choose their sexual

THE QUESTION IS NOT WHAT FAMILY FORM OR MARRIAGE ARRANGEMENT WE WOULD PREFER IN THE ABSTRACT BUT HOW CAN WE HELP PEOPLE IN A WIDE VARIETY OF COMMITTED RELATIONSHIPS.

—*Stephanie Coontz, family historian*

orientation are far less tolerant of gays and lesbians and more likely to feel that homosexuality should be illegal than those who think sexual orientation is not a matter of personal choice" (Rosin and Morin 1999, 8).

Although the terms *sexual preference* and *sexual orientation* are often used interchangeably, the term *sexual orientation* avoids the implication that homosexuality, heterosexuality, and bisexuality are determined voluntarily. Hence, those who believe that sexual orientation is inborn more often use the term *sexual orientation*, and those who think that individuals choose their sexual orientation use *sexual preference* more often.

Can Homosexuals Change Their Sexual Orientation?

Individuals who believe that homosexual people choose their sexual orientation tend to think that homosexuals can and should change their sexual orientation. Various forms of **reparative therapy** or **conversion therapy** are dedicated to changing homosexuals' sexual orientation. Some religious organizations sponsor "ex-gay ministries," which claim to "cure" homosexuals and transform them into heterosexuals by encouraging them to ask for "forgiveness for their sinful lifestyle," through prayer and other forms of "therapy." Serovich et al. (2008) reviewed twenty-eight empirically based, peer-reviewed articles and found them methodologically problematic, which threatens the validity of interpreting available data on this topic.

© Banana Stock/Jupiterimages

reparative therapy (conversion therapy) therapy designed to change a person's homosexual orientation to a heterosexual orientation.

Parelli (2007) attended private as well as group therapy sessions to change his sexual orientation from homosexuality to heterosexuality. It did not work. In retrospect, he identified seven reasons for the failure of reparative therapy. In effect, he noted that these therapies focus on outward behavioral change and do not acknowledge the inner yearnings:

> By my mid-forties, I was experiencing a chronic need for appropriately affectionate male touch. It was so acute I could think of nothing else. Every cell of my body seemed relationally isolated and emotionally starved. Life was so completely and fatally ebbing out of my being that my internal life-saving system kicked in and put out a high-alert call for help. I desperately needed to be held by loving, human, male arms. (p. 32)

Critics of reparative therapy and ex-gay ministries take a different approach: "It is not gay men and lesbians who need to change . . . but negative attitudes and discrimination against gay people that need to be abolished" (Besen 2000, 7). The National Association for the Research and Therapy of Homosexuality (NARTH) has been influential in moving public opinion from "gays are sick" to "society is judgmental." The American Psychiatric Association, the American Psychological Association, the American Academy of Pediatrics, the American Counseling Association, the National Association of School Psychologists, the National Association of Social Workers, and the American Medical Association agree that homosexuality is not a mental disorder and needs no cure—that efforts to change sexual orientation do not work and may, in fact, be harmful (Human Rights Campaign 2000; Potok 2005). An extensive review of the ex-gay movement concludes, "There is a growing body of evidence that conversion therapy not only does not work, but also can be extremely harmful, resulting in depression, social isolation from family and friends, low self-esteem, internalized homophobia, and even attempted suicide" (Cianciotto and Cahill 2006, 77). According to the American Psychiatric Association, "clinical experience suggests that any person who seeks conversion therapy may be doing so because of social bias that has resulted in internalized homophobia, and that gay men and lesbians who have accepted their

heterosexism the denigration and stigmatization of any behavior, person, or relationship that is not heterosexual.

homophobia used to refer to negative attitudes toward homosexuality.

homonegativity a construct that refers to antigay responses such as negative feelings (fear, disgust, anger), thoughts ("homosexuals are HIV carriers"), and behavior ("homosexuals deserve a beating").

antigay bias any behavior or statement which reflects a negative attitude toward homosexuals.

sexual orientation are better adjusted than those who have not done so" (quoted by Holthouse 2005, 14).

Close scrutiny of reports of "successful" reparative therapy reveal that (1) many claims come from organizations with an ideological perspective on sexual orientation rather than from unbiased researchers, (2) the treatments and their outcomes are poorly documented, and (3) the length of time that clients are followed after treatment is too short for definitive claims to be made about treatment success (Human Rights Campaign 2000). Indeed, at least thirteen ministries of Exodus International—the largest ex-gay ministry network—have closed because their directors reverted to homosexuality (Fone 2000). Michael Bussy, who helped start Exodus International in 1976, said, "After dealing with hundreds of people, I have not met one who went from gay to straight. Even if you manage to alter someone's sexual behavior, you cannot change their true sexual orientation" (quoted by Holthouse 2005, 14). Michael Bussy worked to help "convert" gay people for three years, until he and another male Exodus employee fell in love and left the organization.

> You commie, *homo-loving sons of guns*
>
> —*Sean Penn, accepting the Best Actor Oscar for his role as gay rights pioneer Harvey Milk*

7.3 Heterosexism, Homonegativity, Homophobia, and Biphobia

The United States, along with many other countries throughout the world, is predominantly heterosexist. **Heterosexism** refers to "the institutional and societal reinforcement of heterosexuality as the privileged and powerful norm." Heterosexism is based on the belief that heterosexuality is superior to homosexuality. Of 1,319 undergraduates, 41.5 percent agreed, "It is better to be heterosexual than homosexual" (Knox and Zusman 2009). Heterosexism results in prejudice and discrimination against homosexual and bisexual people. Prejudice refers to negative attitudes, whereas discrimination refers to behavior that denies equality of treatment for individuals or groups. Before reading further, you may wish to complete the Self-Assessment Card at the end of the book, which assesses behaviors toward individuals perceived to be homosexual.

Homonegativity and Homophobia

The term **homophobia** is commonly used to refer to negative attitudes and emotions toward homosexuality and those who engage in it. Homophobia is not necessarily a clinical phobia (that is, one involving a compelling desire to avoid the feared object despite recognizing that the fear is unreasonable). Other terms that refer to negative attitudes and emotions toward homosexuality include **homonegativity** and **antigay bias**.

The Sex Information and Education Council of the United States states that "individuals have the right to accept, acknowledge, and live in accordance with their

© Words and Meanings/Mark Sykes/Alamy

sexual orientation, be they bisexual, heterosexual, gay or lesbian. The legal system should guarantee the civil rights and protection of all people, regardless of sexual orientation" (SIECUS 2009, retrieved March 23). Nevertheless, negative attitudes toward homosexuality are reflected in the high percentage of the U.S. population who disapprove of homosexuality. According to national surveys by the Gallup Organization, 51 percent of Americans say that homosexuality should be considered an acceptable alternative lifestyle (Saad 2005). Although attitudes toward homosexuality have become more positive, there continues to be more support for same-sex civil unions or domestic partnerships than for same-sex marriage (Avery et al. 2007). In general, individuals who are more likely to have negative attitudes toward homosexuality and to oppose gay rights are those who (1) are men, older, and less educated; (2) attend religious services; (3) live in the South or Midwest; (4) reside in small rural towns; and (5) have had limited contact with someone who is gay or lesbian (Herek 2002; Curtis 2003; Loftus 2001; Page 2003; Mohipp and Morry 2004). Jenkins et al. (2009) found no significant difference between black and white students in reported levels of homophobia in a sample of 551 Midwestern college students.

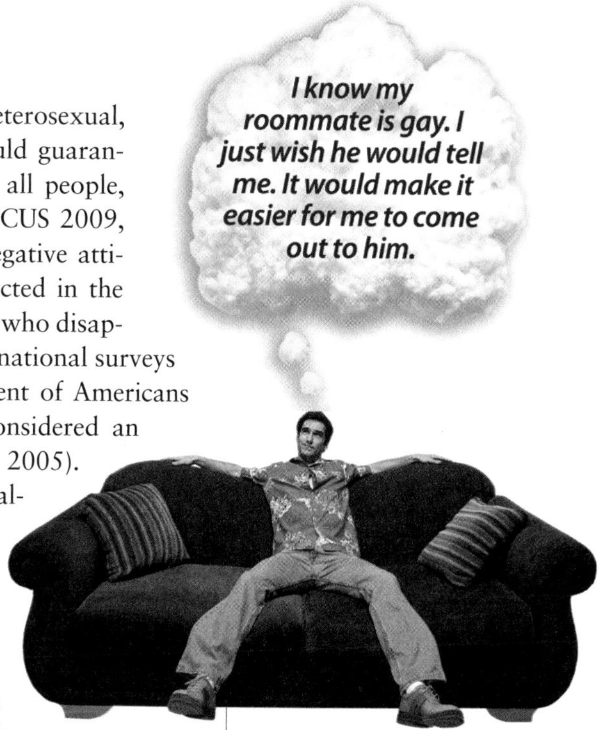

I know my roommate is gay. I just wish he would tell me. It would make it easier for me to come out to him.

© Workbook Stock/Jupiterimages

internalized homophobia a sense of personal failure and self-hatred among lesbians and gay men resulting from social rejection and stigmatization; has been linked to increased risk for depression, substance abuse and addiction, anxiety, and suicidal thoughts.

biphobia (binegativity) refers to a parallel set of negative attitudes toward bisexuality and those identified as bisexual.

Negative social meanings associated with homosexuality can affect the self-concepts of LGBT individuals. **Internalized homophobia**—a sense of personal failure and self-hatred among lesbians and gay men resulting from social rejection and stigmatization—has been linked to increased risk for depression, substance abuse and addiction, anxiety, and suicidal thoughts (Bobbe 2002; Gilman et al. 2001).

Biphobia

Just as the term *homophobia* is used to refer to negative attitudes toward homosexuality, gay men, and lesbians, **biphobia** (also referred to as **binegativity**) refers to a parallel set of negative attitudes toward bisexuality and those identified as bisexual. Although heterosexuals often reject both homosexual- and bisexual-identified individuals, bisexual-identified women and men also face rejection from many homosexual individuals. Thus, bisexuals experience "double discrimination."

Some negative attitudes toward bisexual individuals "are based on the belief that bisexual individuals are really lesbian or gay individuals who are in transition or in denial about their true sexual orientation" (Israel and Mohr 2004, 121). According to this view, bisexual people lack the courage to come out as lesbian or gay, or they are trying to maintain heterosexual privilege. Negative attitudes toward bisexuality are also based on the negative stereotype of bisexuals as incapable of or unwilling to be monogamous. In a review of research on bisexuality, Israel and Mohr (2004) state that, "although bisexual individuals are more likely to value nonmonogamy as an ideal compared to lesbian, gay, and heterosexual individuals, research clearly indicates that some bisexual-identified individuals prefer monogamous relationships" (p. 122). Given the cultural bias against being gay, gay individuals must be careful about "coming out."

7.4 Gay, Lesbian, Bisexual, and Mixed-Orientation Relationships

Research suggests that gay and lesbian couples tend to be more similar than different from heterosexual couples (Kurdek 2005; 2008). However, there are some unique aspects of intimate relationships involving gay, lesbian, and bisexual individuals. In this section, we note the similarities as well as differences between heterosexual, gay male, and lesbian relationships in regard to relationship satisfaction, conflict and conflict resolution, and monogamy and sexuality. We also look at relationship issues involving bisexual individuals and mixed-orientation couples.

Relationship Satisfaction

For both heterosexual and LGBT partners, relationship satisfaction tends to be high in the beginning of the relationship and decreases over time. In a review of literature on lesbian and gay couples, Kurdek (1994) concluded, "The most striking finding regarding the factors linked to relationship satisfaction is that they seem to be the same for lesbian couples, gay couples, and heterosexual couples" (p. 251). These factors include having equal power and control, being emotionally expressive, perceiving many attractions and few alternatives to the relationship, placing a high value on attachment, and sharing decision making. Kurdek (2008) compared relationship quality of cohabitants over a ten-year period of both partners from 95 lesbian, 92 gay male, and 226 heterosexual couples living without children, and both partners from 312 heterosexual couples living with children. Lesbian couples showed the highest levels of relationship quality averaged over all assessments.

Researchers who studied relationship quality among same-sex couples noted that, "in trying to create satisfying and long-lasting intimate relationships, LGBT individuals face all of the same challenges faced by heterosexual couples, as well as a number of distinctive concerns" (Otis et al. 2006, 86). These concerns include if, when, and how to disclose their relationships to others and how to develop healthy intimate relationships in the absence of same-sex relationship models.

In one review of research on gay and lesbian relationships, we concluded that the main difference between heterosexual and nonheterosexual relationships is that, "Whereas heterosexuals enjoy many social and institutional supports for their relationships, gay and lesbian couples are the object of prejudice and discrimination" (Peplau et al. 1996, 268). Both gay male and lesbian couples must cope with the stress created by antigay prejudice and discrimination and by "internalized homophobia" or negative self-image and low self-esteem due to being a member of a stigmatized group. Not surprisingly, higher levels of such stress are associated with lower reported levels of relationship quality among LGBT couples (Otis et al. 2006).

Despite the stresses and lack of social and institutional support LGBT individuals experience, gay men and lesbians experience relationship satisfaction at a level that is at least equal to that reported by married heterosexual spouses (Kurdek 2005). Partners of the same sex enjoy the comfort of having a shared gender perspective, which is often accompanied by a sense of equality in the relationship. For example, contrary to stereotypical beliefs, same-sex couples (male or female) typically do not assign "husband" and "wife" roles in the division of household labor; as well, they are more likely than heterosexual couples to achieve a fair distribution of household labor and at the same time accommodate the different interests, abilities, and work schedules of each partner (Kurdek 2005). In contrast, division of household labor among heterosexual couples tends to be unequal, with wives doing the majority of such tasks.

Same-sex relationships are not without abuse. Bartholomew et al. 2008 studied violence in a random sample of 284 gay and bisexual men and found that almost all reported psychological abuse, more than a third reported physical abuse, and 10 percent reported being forced to have sex. Abuse in gay relationships is less likely to be reported to the police because some gays do not want to be "outed."

Conflict and Conflict Resolution

All couples experience conflict in their relationships, and gay and lesbian couples tend to disagree about the same issues that heterosexual couples argue about. In one study, partners from same-sex and heterosexual couples identified the same sources of most conflict in their relationships: finances, affection, sex, being overly critical, driving style, and household tasks (Kurdek 2004).

However, same-sex couples and heterosexual couples tend to differ in how they resolve conflict. In a study in which researchers videotaped gay, lesbian, and heterosexual couples discussing problems in their relationships, gay and lesbian partners began their discussions more positively and maintained a more positive tone throughout the discussion than did partners in heterosexual marriages (Gottman et al. 2003). Other research has found that, compared with heterosexual married spouses, same-sex partners resolve conflict more positively, argue more

© George Doyle/Stockbyte/Getty Images

effectively, and are more likely to suggest possible solutions and compromises (Kurdek 2004). One explanation for the more positive conflict resolution among same-sex couples is that they value equality more and are more likely to have equal power and status in the relationship than are heterosexual couples (Gottman et al. 2003).

Monogamy and Sexuality

Like many heterosexual women, most gay women value stable, monogamous relationships that are emotionally as well as sexually satisfying. Gay and heterosexual women in U.S. society are taught that sexual expression should occur in the context of emotional or romantic involvement. A common stereotype of gay men is that they prefer casual sexual relationships with multiple partners versus monogamous long-term relationships. However, although most gay men report having more casual sex than heterosexual men (Mathy 2007), most gay men prefer long-term relationships, and sex outside of the primary relationship is usually infrequent and not emotionally involved (Green et al. 1996).

The degree to which gay males engage in casual sexual relationships is better explained by the fact that they are male than by the fact that they are gay. In this regard, gay and straight men have a lot in common: they both tend to have fewer barriers to engaging in casual sex than do women (heterosexual or lesbian). One way that gay men meet partners is through the Internet. Ogilvie et al. (2008) surveyed men who have sex with men (MSM) who found partners using the Internet; he noted that they were more likely to have had ten sexual partners in the last year and to agree with the statement, "I think most guys in relationships have condom-free sex."

Such nonuse of condoms results in the high rate of human immunodeficiency virus (HIV) infection and acquired immunodeficiency syndrome (AIDS). Although most worldwide HIV infections occur through heterosexual transmission, male-to-male sexual contact is the most common mode of HIV transmission in the United States (Centers for Disease Control and Prevention 2005). Women who have sex exclusively with other women have a much lower rate of HIV infection than do men (both gay and straight) and women who have sex with men. Many gay men have lost a love partner to HIV infection or AIDS; some have experienced multiple losses. Those still in relationships with partners who are HIV-positive experience profound changes, such as developing a sense of urgency to "speed up" their relationship because they may not have much time left together (Palmer and Bor 2001).

Relationships of Bisexuals

Individuals who identify as bisexual have the ability to form intimate relationships with both sexes. However, research has found that the majority of bisexual women and men tend toward primary relationships with the other sex (McLean 2004). Contrary to the common myth that bisexuals are, by definition, nonmonogamous, some bisexuals prefer monogamous relationships (especially in light of the widespread concern about HIV). In another study of sixty bisexual women and men, 25 percent of the men and 35 percent of the women were in exclusive relationships; 60 percent of the men and 53 percent of the women were in "open" relationships in which both partners agreed to allow each other to have sexual and or emotional relationships with others, often under specific conditions or rules about how this would occur (McLean 2004). In these "open" relationships, nonmonogamy was not the same as infidelity, and the former did not imply dishonesty. The researcher concluded:

> Despite the stereotypes that claim that bisexuals are deceitful, unfaithful, and untrustworthy in relationships, most of the bisexual men and women I interviewed demonstrated a significant commitment to the principles of trust, honesty, and communication in their intimate relationships and made considerable effort to ensure both theirs and their partner's needs and desires were catered for within the relationship. (McLean 2004, 96)

Monogamous bisexual women and men find that their erotic attractions can be satisfied through fantasy and their affectional needs through nonsexual friendships (Paul 1996). Even in a monogamous relationship, "the partner of a bisexual person may feel that a bisexual person's decision to continue to identify as bisexual . . . is somehow a withholding of full commitment to the relationship. The bisexual person may be perceived as holding onto the possibility of other relationships by maintaining a bisexual identity and, therefore, not fully

ONE'S SEXUAL IDENTITY IS SEPARATE FROM ONE'S CHOICES ABOUT RELATIONSHIP INVOLVEMENT OR MONOGAMY.

sodomy oral and anal sexual acts.

civil union a pair-bonded relationship given legal significance in terms of rights and privileges (more than a domestic relationship and less than a marriage). Vermont recognizes civil unions of same-sex individuals.

domestic partnership a relationship in which individuals who live together are emotionally and financially interdependent and are given some kind of official recognition by a city or corporation so as to receive partner benefits (for example, health insurance).

committed to the relationship" (Ochs 1996, 234). However, this perception overlooks the fact that one's identity is separate from one's choices about relationship involvement or monogamy. Ochs notes that "a heterosexual's ability to establish and maintain a committed relationship with one person is not assumed to falter, even though the person retains a sexual identity as 'heterosexual' and may even admit to feeling attractions to other people despite her or his committed status" (p. 234).

Mixed-Orientation Relationships

Mixed-orientation couples are those in which one partner is heterosexual and the other partner is gay, lesbian, or bisexual. Up to 2 million gay, lesbian, or bisexual people in the United States have been in heterosexual marriages at some point (Buxton 2004). Some lesbigay individuals do not develop same-sex attractions and feelings until after they have been married. Others deny, hide, or repress their same-sex desires.

In a study of twenty gay or bisexual men who had disclosed their sexual orientation to their wives, most of the men did not intentionally mislead or deceive their future wives with regard to their sexuality. Rather, they did not fully grasp their feelings toward men, although they had a vague sense of their same-sex attraction (Pearcey 2004). The majority of the men in this study (14 of 20) attempted to stay married after disclosure of their sexual orientation to their wives, and nearly half (9 of 20) stayed married for at least three years.

Although gay and lesbian spouses in heterosexual marriages are not sexually attracted to their spouses, they may nevertheless love them. However, that is little consolation to spouses, who upon learning that their husband or wife is gay, lesbian, or bisexual, often react with shock, disbelief, and anger. The Straight Spouse Network (http://www.ssnetwk.org) provides support to heterosexual spouses or partners, current or former, of GLBT mates.

7.5 Legal Recognition and Support of Same-Sex Couples and Families

As current divorce rates of heterosexuals suggest (4 in 10 marriages end in divorce), maintaining long-term relationships is challenging. However, the challenge is even greater for same-sex couples who lack the many social supports and legal benefits of marriage. A leading researcher and scholar on LGBT issues noted, "perhaps what is most impressive about gay and lesbian couples is . . . that they manage to endure without the benefits of institutionalized supports" (Kurdek 2005, 253). In this section, we discuss laws and policies designed to provide institutionalized support for same-sex couples and families.

Decriminalization of Sodomy

In the United States, a 2003 Supreme Court decision in *Lawrence v. Texas* invalidated state laws that criminalized **sodomy**—oral and anal sexual acts. The ruling, which found that sodomy laws were discriminatory and unconstitutional, removed the stigma and criminal branding that sodomy laws have long placed on GLBT individuals. Prior to this historic ruling, sodomy was illegal in thirteen states. Sodomy laws, which carried penalties ranging from a $200 fine to twenty years of imprisonment, were usually not used against heterosexuals but were used primarily against gay men and lesbians. Same-sex sexual behavior is still considered a criminal act in many countries throughout the world.

Registered Partnerships, Civil Unions, and Domestic Partnerships

Aside from same-sex marriage (which we discuss later), other forms of legal recognition of same-sex couples exist in a number of countries throughout the world at the national, state, and/or local level. In addition, some workplaces recognize same-sex couples for the purposes of employee benefits. Legal recognition of same-sex couples, also referred to as registered partnerships, **civil unions**, or **domestic partnerships**, conveys most but not all the rights and responsibilities of marriage. Carpenter and Gates (2008) analyzed data in California and noted that half of partnered lesbians are officially registered.

© Michael Krinke/iStockphoto.com / © Valua Vitaly/iStockphoto.com

State and Local Legal Recognition of Same-Sex Couples There is no federal recognition of same-sex couples in the United States. However, a number of U.S. states allow same-sex couples legal status that entitles them to many of the same rights and responsibilities as married opposite-sex couples (see Table 7.1). For example, in New Jersey, same-sex couples can apply for a civil union license, which entitles them to all the rights and responsibilities available under state law to married couples. Unlike marriage for heterosexual couples, the rights of partners in same-sex civil unions are not recognized by U.S. federal law, so they do not have the federal protections that go along with civil marriage, and their legal status is not recognized in other states.

The rights and responsibilities granted to domestic partners vary from place to place but may include coverage under a partner's health and pension plan, rights of inheritance and community property, tax benefits, access to housing for married students, child custody and child and spousal support obligations, and mutual responsibility for debts. The California law (the Domestic Partnership Rights and Responsibilities Act of 2003) provides the broadest array of protections, including eligibility for family leave, other employment and health benefits, the right to sue for wrongful death of partner or inherit from partner as next of kin, and access to the stepparent adoption process (National Gay and Lesbian Task Force 2005–2006). Ten states and the District of Columbia, as well as several dozen U.S. municipalities, offer domestic partner benefits to the same-sex partners of public employees.

One resource for current information on rights for same-sex couples is the interactive gay marriage map produced by the* L.A. Times *and posted at http://www.latimes.com/news/local/la-gmtimeline-fl,0,5345296.htmlstory

Table 7.1

States that Recognize Same-Sex Relationships

State	Same-Sex Relationship Recognition
California	Has a domestic partner registry that confers almost all the state-level spousal rights and responsibilities to registered domestic partners.
Connecticut	Grants same-sex marriage licenses to residents.
Hawaii	Offers "reciprocal beneficiary" status to same-sex registered couples.
Iowa	Grants same-sex marriage licenses to residents.
Maine	Grants same-sex marriage licenses to residents.
Massachusetts	Grants same-sex marriage licenses only to residents. The license is not valid if the couple moves to another state.
New Hampshire	Grants same-sex marriage licenses to residents.
New Jersey	Offers same-sex civil unions.
New York	Recognizes same-sex marriages from couples legally married outside the United States.
Vermont	Grants same-sex marriage licenses to residents.

Source: Human Rights Campaign (2006), updated in 2009; Luther (2006).

Recognition of Same-Sex Couples in the Workplace In 1991, the Lotus Development Corporation became the first major American firm to extend domestic partner recognition to gay and lesbian employees. By the end of 2004, the Human Rights Campaign (2005) identified 8,250 employers that provided domestic partner health insurance benefits to their employees—an increase of 13 percent from the previous year. The percentage of *Fortune* 500 companies that offered health benefits to employees' domestic partners nearly doubled from 25 percent in 2000 to 49 percent in March 2006 (Luther 2006). However, even when companies offer domestic partner benefits to same-sex partners of employees, these benefits are usually taxed as income by the federal government, whereas spousal benefits are not.

> *"Anyone who wishes to examine the 20 years of peer-reviewed studies on the emotional, cognitive and behavioral outcomes of children of gay and lesbian parents will find not one shred of evidence that children are harmed by their parents' sexual orientation.*
>
> *Carol Trust, executive director, National Association of Social Workers*

Same-Sex Marriage

In 2009, Iowa, Vermont, Maine, and New Hampshire approved gay marriage (following Connecticut in 2008). In 2004, Massachusetts became the first U.S. state to offer civil marriage licenses to same-sex couples.

In the first year after the court order went into effect, more than 5,000 same-sex couples were married in Massachusetts (Johnston 2005). Porche and Purvin (2008) interviewed four lesbian and five gay male same-sex couples in Massachusetts who had been together twenty years or more. Seven of the nine couples married soon after same-sex marriage was enacted in Massachusetts. The two who did not marry reaffirmed and maintained their commitment. These data emphasize the value these couples placed on having their relationship sanctioned by the state and culture. (However, unlike marriages between a man and a woman, the same-sex marriages in Massachusetts are not recognized in other states, nor does the federal government recognize them.)

> *"No government has a right to tell its citizens when or whom to love. The only queer people are those who don't love anybody.*
>
> *—Rita Mae Brown*

Attorney Robert Zaleski (2007) coined the term **garriage**. "The 'g' is borrowed from the word gay and connotes the same-sex status of the committed couple. And let the new verb be 'garry,' which would be conjugated in identical fashion with the verb 'marry,' thereby enabling these words to be used interchangeably in conversation." Zaleski also emphasized that the word *garriage* allows heterosexual couples to maintain their uniqueness as it does to gay couples.

Anti-Gay Marriage Legislation In a national survey, 32 percent of U.S. adults support gay marriage, whereas 59 percent are opposed (Pew Research Center 2008). In a national study of first year freshmen in colleges and universities throughout the United States, 72.4% were in favor of same-sex marriage (Pryor et al. 2008). Where disapproval exists, the primary reason is morality. Gay marriage is viewed as "immoral, a sin, against the Bible." However, support for gay marriage varies by age; about half of young adults (18 to 29) in the same survey support gay marriage. A higher percentage of U.S. adults (41 percent) are in favor of equal legal rights for gay relationships (Pew Research Center 2008).

garriage term for relationship of gay individuals who are married or committed.

Defense of Marriage Act (DOMA) legislation passed by Congress denying federal recognition of homosexual marriage and allowing states to ignore same-sex marriages licensed elsewhere.

In 1996, Congress passed and former President Clinton signed the **Defense of Marriage Act (DOMA)**, which states that marriage is a "legal union between one man and one woman" and denies federal recognition of same-sex marriage. In effect, this law allows states to either recognize or not recognize same-sex marriages performed in other states. As of March 2006, thirty-six states have banned gay marriage either through statute or a state constitutional amendment, and seventeen states have passed broader antigay family measures that ban other forms of partner recognition in addition to marriage, such as domestic partnerships and civil unions. These broader measures, known as "Super DOMAs," potentially endanger employer-provided domestic partner benefits, joint and second-parent adoptions, health care decision-making proxies, or any policy or document that recognizes the existence of a same-sex partnership (Cahill and Slater 2004). Some of these "Super DOMAs" ban partner recognition for unmarried heterosexual couples as well.

At the federal level, there are efforts to amend the U.S. Constitution to define marriage as being between a man and a woman. The Federal Marriage Amendment did not pass in 2004 in the Senate or the House, but supporters vowed to continue the fight. The Federal Marriage Amendment was reintroduced and voted on in 2006, and although it garnered more support than in 2004, it failed to reach the two-thirds majority vote necessary for proposal as an amendment. If it had passed, the constitutional amendment would have denied marriage and likely civil union and domestic partnership rights to same-sex couples (LAWbriefs 2005). Such an amendment would also hurt the children in same-sex couple families. Dr. Kathleen Moltz, an assistant professor at Wayne State University School of Medicine, testified against the passage of an antigay constitutional amendment before a U.S. Senate Judiciary Committee, expressing her fears about how such an amendment would affect her family:

Arguments in Favor of Same-Sex Marriage Advocates of same-sex marriage argue that banning or refusing to recognize same-sex marriages granted in other states is a violation of civil rights that denies same-sex couples the many legal and financial benefits that are granted to heterosexual married couples. Rights and benefits that married spouses have include the following:

- The right to inherit from a spouse who dies without a will;
- No inheritance taxes between spouses;

DR. KATHLEEN MOLTZ:
I don't know what harm ... a constitutional amendment might cause. I fear that families like mine, with young children, will lose health benefits; will be denied common decencies like hospital visitation when tragedy strikes; will lack the ability to provide support for one another in old age. I fear that my loving, innocent children will face hatred and insults implicitly sanctioned by a law that brands their family as unequal. I know that these sweet children have already been shunned and excluded by people claiming to represent values of decency and compassion. I also know what such an amendment will not do. It will not help couples who are struggling to stay married. It will not assist any impoverished families struggling to make ends meet or to obtain health care for sick children. It will not keep children with their parents when their parents see divorce as their only option. It will not help any single American citizen to live life with more decency, compassion or morality. (Moltz 2005)

© Michal Rozanski/iStockphoto.com

- The right to make crucial medical decisions for a partner and to take care of a seriously ill partner or parent of a partner under current provisions in the federal Family and Medical Leave Act;
- Social Security survivor benefits; and
- Health insurance coverage under a spouse's insurance plan.

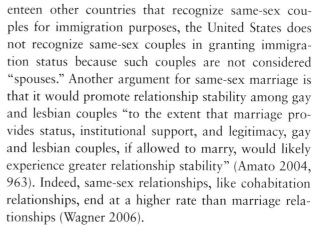

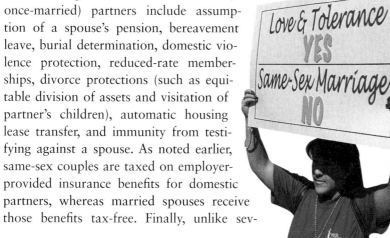

Other rights bestowed on married (or once-married) partners include assumption of a spouse's pension, bereavement leave, burial determination, domestic violence protection, reduced-rate memberships, divorce protections (such as equitable division of assets and visitation of partner's children), automatic housing lease transfer, and immunity from testifying against a spouse. As noted earlier, same-sex couples are taxed on employer-provided insurance benefits for domestic partners, whereas married spouses receive those benefits tax-free. Finally, unlike seventeen other countries that recognize same-sex couples for immigration purposes, the United States does not recognize same-sex couples in granting immigration status because such couples are not considered "spouses." Another argument for same-sex marriage is that it would promote relationship stability among gay and lesbian couples "to the extent that marriage provides status, institutional support, and legitimacy, gay and lesbian couples, if allowed to marry, would likely experience greater relationship stability" (Amato 2004, 963). Indeed, same-sex relationships, like cohabitation relationships, end at a higher rate than marriage relationships (Wagner 2006).

Love & Tolerance
YES

© Chuck Nacke/Alamy

Recognized marriage, argues Amato (2004), would be beneficial to the children of same-sex parents. Without legal recognition of same-sex families, children living in gay- and lesbian-headed households are denied a range of securities that protect children of heterosexual married couples. These include the right to get health insurance coverage and Social Security survivor benefits from a nonbiological parent. In some cases, children in same-sex households lack the automatic right to continue living with their nonbiological parent should their biological mother or father die (Tobias and Cahill 2003). It is ironic that the same pro-marriage groups that stress that children are better off in heterosexual married-couple families disregard the benefits of same-sex marriage to children.

Opponents of gay marriage sometimes suggest that gay marriage leads to declining marriage rates, increased divorce rates, and increased nonmarital births. However, data in Scandinavia reflects that these trends were in place ten years before Scandinavian adopted registered partnership laws, liberalized alternatives to marriage (such as cohabitation), and expanded exit options (such as no-fault divorce) (Pinello 2008).

Finally, there are religious-based arguments in support of same-sex marriage. Although many religious leaders teach that homosexuality is sinful and prohibited by God, some religious groups, such as the Quakers and the United Church of Christ (UCC), accept homosexuality, and other groups have made reforms toward increased acceptance of lesbians and gays. In 2005, the UCC became the largest Christian denomination to endorse same-sex marriages. In a sermon titled, "The Christian Case for Gay Marriage," Jack McKinney (2004) interprets Luke 4: "Jesus is saying that one of the most fundamental religious tasks is to stand with those who have been excluded and marginalized.... [Jesus] is determined to stand with them, to name them beloved of God, and to dedicate his life to seeing them empowered." McKinney goes on to ask, "Since when has it been immoral for two people to commit themselves to a relationship of mutual love and caring? No, the true immorality around gay marriage rests with the heterosexual majority that denies gays and lesbians more than 1,000 federal rights that come with marriage."

Arguments Against Same-Sex Marriage

Whereas advocates of same-sex marriage argue that they will not be regarded as legitimate families by the larger society so long as same-sex couples cannot be legally married, opponents do not want to legitimize same-sex couples and families. Opponents of same-sex marriage who view homosexuality as unnatural, sick, and/or immoral do not want their children to view homosexuality as socially acceptable.

Opponents of same-sex marriage commonly argue that such marriages would subvert the stability and integrity of the heterosexual family. However, Sullivan (1997) suggests that homosexuals are already part of heterosexual families:

> [Homosexuals] are sons and daughters, brothers and sisters, even mothers and fathers, of heterosexuals. The distinction between "families" and "homosexuals" is, to begin with, empirically false; and the stability of existing families is closely linked to how homosexuals are treated within them. (p. 147)

Many opponents of same-sex marriage base their opposition on their religious views. In a Pew Research Center national poll, the majority of Catholics and Protestants opposed legalizing same-sex marriage, whereas the majority of secular respondents favored it (Green 2004). However, churches have the right to deny marriage for gay people in their congregations. Legal marriage is a contract between the spouses and the state; marriage is a civil option that does not require religious sanctioning.

In previous years, opponents of gay marriage have pointed to public opinion polls that suggested that the majority of Americans are against same-sex marriage. However, public opposition to same-sex marriage is decreasing. We previously noted that a 2008 Pew Research Center national poll found that 59 percent of U.S. adults oppose legalizing gay marriage, down from 63 percent in 2004 (Pew Research Center 2006). We also noted that support for gay marriage is higher among young adults.

7.6 GLBT Parenting Issues

Of the more than 600,000 same-sex-couple households identified in the 2000 census, 162,000 had one or more children living in the household. This is a low estimate of children who have gay or lesbian parents, as it does not count children in same-sex households who did not identify their relationship in the census, those headed by gay or lesbian single parents, or those whose gay parent does not have physical custody but is still actively involved in the child's life.

Many gay and lesbian individuals and couples have children from prior heterosexual relationships or marriages. Up to 2 million gay, lesbian, or bisexual people in the United States have been in heterosexual marriages at some point (Buxton 2005). Some of these individuals married as a "cover" for their homosexuality; others discovered their interest in same-sex relationships after they married. Children with mixed-orientation parents may be raised by a gay or lesbian parent, a gay or lesbian stepparent, a heterosexual parent, and a heterosexual stepparent.

A gay or lesbian individual or couple may have children through the use of assisted reproductive technology, including donor insemination, in vitro fertilization, and surrogate mothers. Others adopt or become foster parents.

Less commonly, some gay fathers are part of an emergent family form known as the hetero-gay family. In a hetero-gay family, a heterosexual mother and gay father conceive and raise a child together but reside separately.

Antigay views concerning gay parenting include the belief that homosexual individuals are unfit to be parents and that children of lesbians and gays will not develop normally and/or that they will become homosexual. As the following section suggests, research findings paint a more positive picture of the development and well-being of children with gay or lesbian parents.

Development and Well-Being of Children with Gay or Lesbian Parents

A growing body of research on gay and lesbian parenting supports the conclusion that children of gay and lesbian parents are just as likely to flourish as are children of heterosexual parents. Crowl et al. (2008) reviewed nineteen studies on the developmental outcomes and

> DENYING LEGAL RIGHTS TO SAME-SEX COUPLES INJURES THEIR CHILDREN.
>
> —*J. Asch-Goodkin, scholar*

quality of parent-child relationships among children raised by gay and lesbian parents. The results confirmed previous studies that children raised by same-sex parents fare equally well to children raised by heterosexual parents. For example, one study compared a national sample of forty-four adolescents parented by same-sex couples with forty-four adolescents parented by opposite-sex couples (Wainwright et al. 2004). On an array of assessments, the study showed that the personal, family, and school adjustment of adolescents living with same-sex parents did not differ from that of adolescents living with opposite-sex parents. Self-esteem, depressive symptoms and anxiety, academic achievement, trouble in school, quality of family relationships, and romantic relationships were similar in the two groups of adolescents. Regardless of family type, adolescents were more likely to show positive adjustment when they perceived more caring from adults and when parents described having close relationships with them. Thus, the qualities of adolescent-parent relationships rather than the sexual orientation of the parents were significantly associated with adolescent adjustment.

In another study, researchers examined the quality of parent-child relationships and the socioemotional and gender development of a sample of 7-year-old children with lesbian parents, compared with 7-year-olds from two-parent heterosexual families and with single heterosexual mothers (Golombok et al. 2003). No significant differences between lesbian mothers and heterosexual mothers were found for most of the parenting variables assessed, although lesbian mothers reported spanking their children less and playing more frequently with their children than did heterosexual mothers. No significant differences were found in psychiatric disorders or gender development of the children in lesbian families versus heterosexual families. The findings also suggest that having two parents is associated with more positive outcomes for children's psychological well-being, but the gender of the parents is not relevant.

In addition, the American Psychological Association (2004) noted that "results of research suggest that lesbian and gay parents are as likely as heterosexual parents to provide supportive and healthy environments for their children" and that "the adjustment, development, and psychological well-being of children [are] unrelated to parental sexual orientation and that the children of lesbian and gay parents are as likely as those of heterosexual parents to flourish." Indeed, Pro-Family Pediatricians cheered when a proposal to ban gay marriage was defeated. "Our duty as pediatricians is to see that all children have the same security and protection regardless of the sexual orientation of their parents. Denying legal rights to same-sex couples injures their children," noted Asch-Goodkin (2006).

Discrimination in Child Custody, Visitation, Adoption, and Foster Care

A former student reported that, after she divorced her husband, she became involved in a lesbian relationship. She explained that she would like to be open about her relationship to her family and friends, but she was afraid that if her ex-husband found out that she was in a lesbian relationship, he might take her to court and try to get custody of their children. Although several respected national organizations—including the American Academy of Pediatrics, the Child Welfare League of America, the American Bar Association, the American Medical Association, the American Psychological Association, the American Psychiatric Association, and the National Association of Social Workers—have gone on record in support of treating gays and lesbians without prejudice in parenting and adoption decisions (Howard 2006; Landis 1999), lesbian and gay parents are often discriminated against in child custody, visitation, adoption, and foster care.

second-parent adoption (also called co-parent adoption) a legal procedure that allows individuals to adopt their partner's biological or adoptive child without terminating the first parent's legal status as parent.

Some court judges are biased against lesbian and gay parents in custody and visitation disputes. For example, in 1999, the Mississippi Supreme Court denied custody of a teenage boy to his gay father and instead awarded custody to his heterosexual mother who remarried into a home "wracked with domestic violence and excessive drinking" (Custody and Visitation 2000, 1).

Gay and lesbian individuals and couples who want to adopt children can do so through adoption agencies or through the foster care system in at least twenty-two states and the District of Columbia. However, Florida and Mississippi forbid adoption by gay and lesbian people, Utah forbids adoption by any unmarried couple (which includes all same-sex couples), and Arkansas prohibits lesbians and gay men from serving as foster parents (National Gay and Lesbian Task Force 2004).

Most adoptions by gay people are **second-parent adoptions**. A second-parent adoption (also called co-parent adoption) is a legal procedure that allows individuals to adopt their partner's biological or adoptive child without terminating the first parent's legal status as parent. Second-parent adoption gives children in same-sex families the security of having two legal parents. Second-parent adoption potentially benefits a child by:

- Placing legal responsibility on the parent to support the child;
- Allowing the child to live with the legal parent in the event that the biological (or original adoptive) parent dies or becomes incapacitated;
- Enabling the child to inherit and receive Social Security benefits from the legal parent;
- Enabling the child to receive health insurance benefits from the parent's employer; and
- Giving the legal parent standing to petition for custody or visitation in the event that the parents break up. (Clunis and Dorsey Green 2003)

However, in four states (Colorado, Nebraska, Ohio, and Wisconsin), court rulings have decided that the state adoption law does not allow for second-parent adoption by members of same-sex couples, and it is unclear whether the state adoption laws in twenty-two other states allows second-parent adoption (National Gay and Lesbian Task Force 2005). Second-parent adoption is not possible when a parent in a same-sex relationship has a child from a previous heterosexual marriage or relationship, unless the former spouse or partner is willing to give up parental rights. Although "third-parent" adoptions have been granted in a small number of jurisdictions, this option is not widely available (National Center for Lesbian Rights 2003).

T-shirts for Tolerance

As a junior and senior at Homewood-Flossmoor High School in the suburbs of Chicago, Myka Held played a key role in leading a campaign to promote tolerance of gay and lesbian students. The campaign involved selling T-shirts to students and teachers for them to wear to school on the designated day. The T-shirts say, "gay? fine by me." "I think it's really important for gay people out there to know that there are straight people who support them," Ms. Held said (quoted in Puccinelli 2005). "I have always supported equal rights for every person and have been disgusted by discrimination and prejudice. As a young Jewish woman, I believe it is my duty to stand up and support minority groups. . . . In my mind, fighting for gay rights is a proxy for fighting for every person's rights" (Held 2005). Myka Held's T-shirt campaign illustrates that fighting prejudice and discrimination against sexual-orientation minorities is an issue not just for lesbians, gays, and bisexuals but also for all those who value fairness and respect for human beings in all their diversity.

7.7 Effects of Antigay Bias and Discrimination on Heterosexuals

The antigay and heterosexist social climate of our society is often viewed in terms of how it victimizes the gay population. However, heterosexuals are also victimized by heterosexism and antigay prejudice and discrimination. Some of these effects follow:

1. **Heterosexual victims of hate crimes.** As discussed earlier in this chapter, extreme homophobia contributes to instances of violence against homosexuals—acts known as hate crimes. Hate crimes are crimes of perception, meaning that victims of antigay hate crimes may not be homosexual; they may just be perceived as being homosexual. The National Coalition of Anti-Violence Programs (2005) reported that, in 2004, 192 heterosexual individuals in the United States were victims of antigay hate crimes, representing 9 percent of all antigay hate crime victims.

2. **Concern, fear, and grief over well-being of gay or lesbian family members and friends.** Many heterosexual family members and friends of homosexual people experience concern, fear, and grief over the mistreatment of their gay or lesbian friends and/or family members. For example, heterosexual parents who have a gay or lesbian teenager often worry about how the harassment, ridicule, rejection, and violence experienced at school might affect their gay or lesbian child. Will their child drop out of school, as one-fourth of gay youth do (Chase 2000), to escape the harassment, violence, and alienation they endure there? Will the gay or lesbian child respond to the antigay victimization by turning to drugs or alcohol or by committing suicide? Such fears are not unfounded: lesbian, gay, and bisexual youth who report high levels of victimization at school also have higher levels of substance use and suicidal thoughts than heterosexual peers who report high levels of at-school victimization (Bontempo and D'Augelli 2002). A survey of youths' risk behavior conducted by the Massachusetts Department of Education in 1999, revealed that 30 percent of gay teens had attempted suicide in the previous year, compared with 7 percent of their straight peers (Platt 2001).

   Meyer et al. (2008) studied the lifetime prevalence of mental disorders and suicide attempts of a diverse group of lesbian, gay, and bisexual individuals and found higher rates of substance abuse among bisexual people than lesbians and gay men. Also, Latino respondents attempted suicide more often than white respondents.

   Heterosexual individuals also worry about the ways in which their gay, lesbian, and bisexual family members and friends could be discriminated against in the workplace.

   To heterosexuals who have lesbian and gay family members and friends, lack of family protections such as health insurance and rights of survivorship for same-sex couples can also be cause for concern. Finally, heterosexuals live with the painful awareness that their gay or lesbian family member or friend is a potential victim of antigay hate crime. Imagine the lifelong grief experienced by heterosexual family members and friends of hate crime murder victims, such as Matthew Shepard, a 21-year-old college student who was brutally beaten to death in 1998, for no apparent reason other than he was gay.

3. **Restriction of intimacy and self-expression.** Because of the antigay social climate, heterosexual individuals, especially males, are hindered in their own self-expression and intimacy in same-sex relationships. "The threat of victimization (i.e., antigay violence) . . . causes many heterosexuals to conform to gender roles and to restrict their expressions of (nonsexual) physical affection for members of their own sex" (Garnets et al. 1990, 380). Homophobic epithets frighten youth who do not conform to gender role expectations, leading some youth to avoid activities—such as arts for boys, athletics for girls—that they might otherwise enjoy and benefit from (Gay, Lesbian, and Straight Education Network 2000). A male student in our class revealed that he always wanted to work with young children and had majored in early childhood education. His peers teased him relentlessly about his choice of majors, questioning both his masculinity and his heterosexuality. Eventually, this student changed his major to psychology, which his peers viewed as an acceptable major for a heterosexual male.

4. **Dysfunctional sexual behavior.** Some cases of rape and sexual assault are related to homophobia and compulsory heterosexuality. For example, college men who participate in gang rape, also known as "pulling train," entice each other into the act "by implying that those who do not participate are unmanly or homosexual" (Sanday 1995, 399). Homonegativity also encourages early sexual activity among adolescent men. Adolescent male virgins are often teased by their male peers, who say things like "You mean you don't do it with girls yet? What are you, a fag or something?" Not wanting to be labeled and stigmatized as a "fag," some adolescent boys "prove" their heterosexuality by having sex with girls.

5. **School shootings.** Antigay harassment has also been a factor in many of the school shootings in recent years. In March 2001, 15-year-old Charles Andrew Williams fired more than thirty rounds in a San Diego suburban high school, killing two and injuring thirteen others. A woman who knew Williams reported that the students had teased him and called him gay (Dozetos 2001). According to the Gay, Lesbian, and Straight Education Network (GLSEN), Williams's story is not unusual. Referring to a study of harassment of U.S. students that was commissioned by the American Association of University Women, a GLSEN report concluded, "For boys, no other type of harassment provoked as strong a reaction on average; boys in this study would be less upset about physical abuse than they would be if someone called them gay" (Dozetos 2001).

6. **Loss of rights for individuals in unmarried relationships.** The passage of state constitutional amendments that prohibit same-sex marriage can also result in denial of rights and protections to opposite-sex unmarried couples. For example, in 2005, Judge Stuart Friedman of Cuyahoga County (Ohio) agreed that a man who was charged with assaulting his girlfriend could not be charged with a domestic violence felony because the Ohio state constitutional amendment granted no such protections to unmarried couples (Human Rights Campaign 2005). As discussed earlier, some antigay marriage measures also threaten the provision of domestic partnership benefits to unmarried heterosexual couples.

MY OWN BELIEF IS THAT THERE IS HARDLY ANYONE WHOSE SEXUAL LIFE, IF IT WERE BROADCAST, WOULD NOT FILL THE WORLD AT LARGE WITH SURPRISE AND HORROR.

— *W. Somerset Maugham*

number of people in a vee relationship > **3**

20 years < maximum jail sentence for sodomy prior to the intervention of the U.S. Supreme Court in 2003

percent of gay teens who attempt suicide > **30%**

1978 < year Harvey Milk was killed

same-sex couples married in Massachusetts the year after it became legal in the state > **>5,000**

number of employers that provided domestic partner health benefits in 2004 > **8,250**

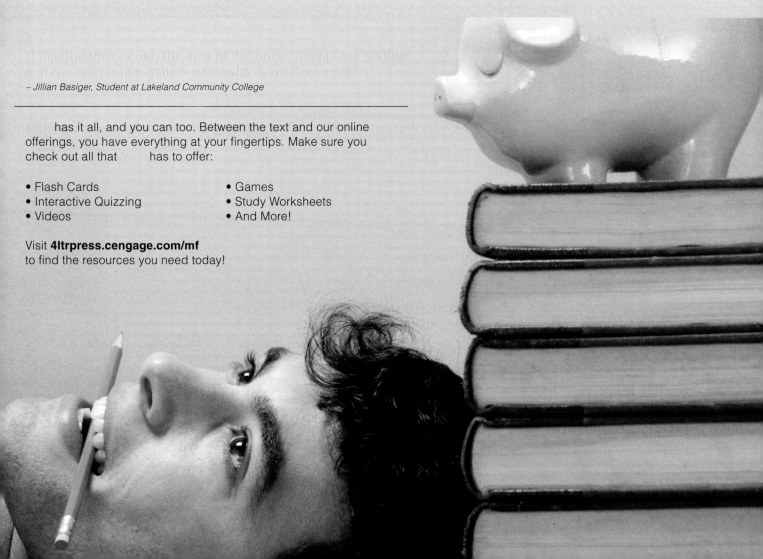

Sexuality in Relationships

SECTIONS

The lyrics to the classic song, "Hold Me, Thrill Me, Kiss Me" (made famous by Mel Carter) begin with, "They told me be sensible with your new love. Don't be fooled thinking this is the last you'll find. But they never stood in the dark alone with you love. . . when you take me in your arms and drive me slowly out of my mind." Being driven out of one's mind in the dark by one's beloved is one of the delights of being involved in a relationship. Indeed, sexuality is a major relationship aspect and the subject of this chapter. We begin by discussing the sexual values lovers bring to the encounter.

8.1 Sexual Values

The following are some examples of choices (reflecting sexual values) with which individuals in a new relationship are confronted:

- How much/how soon in a relationship is sex appropriate?
- Do you require a condom for vaginal or anal intercourse?
- Do you require a condom and/or dental dam (vaginal barrier) for oral sex?
- Before having sex with someone do you require that they have been recently tested for sexually transmitted infections (STIs) and the human immunodeficiency virus (HIV)?
- Do you tell your partner the *actual* number of previous sexual partners?
- Do you tell your partner your sexual fantasies?
- Do you reveal to your partner previous or current same-sex behavior or interests?

sexual values moral guidelines for sexual behavior.

Sexual values are moral guidelines for sexual behavior in nonmarital, marital, heterosexual, and homosexual relationships. Attitudes and values sometimes predict sexual behavior. One's sexual values may be identical to one's sexual choices. For example, a person who values abstinence until marriage may choose to remain a virgin until marriage. One's behavior does not always correspond with one's values. Some who express a value of waiting until marriage have intercourse before marriage. One explanation for the discrepancy between values and behavior is that a person may engage in a sexual behavior, then decide the behavior was wrong, and adopt a sexual value against it.

LEAD US NOT INTO TEMPTATION.
JUST TELL US WHERE IT IS, AND WE'LL FIND IT.

— Sam Levenson, writer

8.2 Alternative Sexual Values

There are at least three sexual value perspectives that guide choices in sexual behavior: absolutism, relativism, and hedonism. People sometimes have different sexual values at different stages of the family life cycle. For example, elderly individuals are more likely to be absolutist, whereas those in the middle years are more likely to be relativistic. Young unmarried adults are more likely than the elderly to be hedonistic.

Absolutism

Absolutism refers to a belief system based on unconditional allegiance to the authority of science, law, tradition, or religion. A religious absolutist makes sexual choices on the basis of moral considerations. To make the correct moral choice is to comply with God's will, and to not comply is a sin. A legalistic absolutist makes sexual decisions on the basis of a set of laws. People who are guided by absolutism in their sexual choices have a clear notion of what is right and wrong.

absolutism belief system based on unconditional allegiance to the authority of science, law, tradition, or religion.

The official creeds of fundamentalist Christian and Islamic religions encourage absolutist sexual values. Intercourse is solely for procreation, and any sexual acts that do not lead to procreation (masturbation, oral sex, homosexuality) are immoral and regarded as sins against God, Allah, self, and community. Waiting until marriage to have intercourse is also an absolutist sexual value. This value is often promoted in the public schools.

"True Love Waits" is an international campaign designed to challenge teenagers and college students to remain sexually abstinent until marriage. Under this program, created and sponsored by the Baptist Sunday School Board, young people are asked to agree to the absolutist position and sign a commitment to the following: "Believing that true love waits, I make a commitment to God, myself, my family, my friends, my future mate, and my future children to be sexually abstinent from this day until the day I enter a biblical marriage relationship" (True Love Waits 2008, http://www.lifeway.com/tlw/students/join.asp).

How effective are these "True Love Waits" and "virginity pledge" programs in delaying sexual behavior until marriage? Data from the National Longitudinal Study of Adolescent Health revealed that although youth who took the pledge were more likely than other youth to experience a later "sexual debut," had fewer partners, and married earlier, most eventually engaged in premarital sex, were less likely to use a condom when they first had intercourse, and were more likely to substitute oral and/or anal sex in the place of vaginal sex. There was no significant difference in the occurrence of STIs between "pledgers" and "nonpledgers" (Brucker and Bearman 2005). The researchers speculated that the emphasis on virginity may have encouraged the pledgers to engage in noncoital (nonintercourse) sexual activities (for example, oral sex), which still exposed them to STIs, and to be less likely to seek testing and treatment for STIs. Similarly, Hollander (2006) collected national data on two waves of adolescents. Half of those who had taken the virginity pledge reported no such commitment a year later. Males and black individuals were particularly likely to retract their pledge.

A similar cultural ritual to encourage and celebrate virginity until one's wedding day is the "Father-Daughter Purity Ball." More than 4,000 such events occur annually that involve fathers and daughters as young as 4 years old taking mutual pledges to be pure. Daughters promise to be pure, which implies the value of absolutism (Gibbs 2008). Fathers promise to protect their daughters and to be faithful and shun pornography themselves.

Attitudes and values sometimes predict sexual behavior.

© SuperStock/Jupiterimages

© Rick Wilking/Reuters/Corbis

More than 4,000 Father-Daughter Purity Balls occur annually and involve fathers and daughters as young as 4 years old taking mutual pledges to be pure.

Some individuals do value virginity. Among a sample of undergraduates, 13 percent of 783 undergraduates at a southeastern university selected absolutism as their primary sexual value (Richey et al. 2009). Black individuals were significantly ($p < .009$) more likely than white individuals to select absolutism as their sexual value (20.6 percent versus 12.1 percent). The fact that significantly more black than white students reported, "I am very religious" (24.2 percent versus 10 percent) helps to account for the fact that black individuals adhered to a more absolutist position. However, Davidson et al. (2008) noted that the conservative sexual values among black individuals may not translate into conservative sexual behavior (see section on Racial Differences in Sex Attitudes and Behaviors of Undergraduates later in this chapter).

> *Some individuals still define themselves as virgins even though they have engaged in oral sex.*

Some individuals still define themselves as virgins even though they have engaged in oral sex. Of 1,319 university students surveyed, 74.2 percent agreed that, "If you have oral sex, you are still a virgin." Hence, according to these undergraduates, having oral sex with someone is not really having sex (Knox and Zusman 2009).

In her paper "Like a Virgin . . . Again?" Carpenter (2003) discussed the concept of **secondary virginity**—the conscious decision of a sexually active person to refrain from intimate encounters for a specified period of time. Secondary virginity closely resembles a pattern scholars have called "regretful" nonvirginity, the chief difference being the adoption of the label *virgin* by the nonvirgin in question. Secondary virginity may be a result of physically painful, emotionally distressing, or romantically disappointing sexual encounters. Of sixty-one young adults Carpenter interviewed, more than half (women more than men) believed that a person could, under some circumstances, be a virgin more than once. Fifteen people contended that people could resume their virginity in an emotional, psychological, or spiritual sense. Terence Duluca, a 27-year-old, heterosexual, white, Roman Catholic, explained:

> There is a different feeling when you love somebody and when you just care about somebody. So I would have to say if you feel that way, then I guess you could be a virgin again. Christians get born all the time again, so. . . . When there's true love involved, yes, I believe that.

A subcategory of absolutism is **asceticism.** Ascetics believe that giving in to carnal lust is unnecessary and attempt to rise above the pursuit of sensual pleasure into a life of self-discipline and self-denial. Accordingly, spiritual life is viewed as the highest good, and self-denial helps one to achieve it. Catholic priests, monks, nuns, and some other celibate people have adopted the sexual value of asceticism.

secondary virginity the conscious decision of a sexually active person to refrain from intimate encounters for a specified period of time.

asceticism the belief that giving in to carnal lusts is wrong and that one must rise above the pursuit of sensual pleasure to a life of self-discipline and self-denial.

relativism sexual value system whereby decisions are made in the context of the situation and the relationship.

What's your attitude *about premarital sex? Take the assessment on* *page 121* *and find out!*

Relativism

Relativism is a value system emphasizing that sexual decisions should be made in the context of the situation and the relationship. Whereas absolutists might feel that having intercourse is wrong for unmarried

© Jennifer Trenchard/iStockphoto.com

friends with benefits (FWB) a relationship between nonromantic friends who also have a sexual relationship.

people, relativists might feel that the moral correctness of sex outside marriage depends on the particular situation. For example, a relativist might feel that in some situations, sex between casual dating partners is wrong (such as when one individual pressures the other into having sex or lies to persuade the other to have sex). However, in other cases—when there is no deception or coercion and the dating partners are practicing "safer sex"—intercourse between casual dating partners may be viewed as acceptable.

Of 783 undergraduates, 62 percent selected "relativism" as their prevailing sexual value. Most of these undergraduates felt that sexual intercourse was justified if they were in a secure, mutual love relationship (Richey et al. 2009). As expected, women were more likely to be relativists than men (72 percent versus 52 percent).

Sexual values and choices that are based on relativism often consider the degree of love, commitment, and relationship involvement as important factors. In a study designed to assess "turn-ons" and "turn-offs" in sexual arousal, women spoke of "feeling desired versus feeling used" by the partner. "Many women talked about how their arousal was increased with partners who seemed particularly interested in them as individual women, rather than someone that they just wanted to have sex with" (Graham et al. 2004).

A disadvantage of relativism as a sexual value is the difficulty of making sexual decisions on a relativistic case-by-case basis. The statement "I don't know what's right anymore" reflects the uncertainty of a relativistic view. Once a person decides that mutual love is the context justifying intercourse, how often and how soon is it appropriate for the person to fall in love? Can love develop after some alcohol and two hours of conversation? How does one know that love feelings are genuine? The freedom that relativism brings to sexual decision making requires responsibility, maturity, and judgment. In some cases, individuals may convince themselves that they are in love so that they will not feel guilty about having intercourse. Though one may feel "in love," "secure," and "committed" at the time first intercourse occurs, only 17 percent of all first intercourse experiences that women reported are with the person they eventually marry (Raley, 2000).

Absolutists and relativists have different views on whether or not two unmarried people should have intercourse. Whereas an absolutist would say that having intercourse is wrong for unmarried people and right for married people, a relativist would say, "It depends on the situation." Suppose, for example, that a married couple do not love each other and intercourse is an abusive, exploitative act. Suppose also that an unmarried couple love each other and their intercourse experience is an expression of mutual affection and respect. A relativist might conclude that, in this particular situation, having intercourse is "more right" for the unmarried couple than the married couple. Students who become involved in a "friends with benefits" relationship reflect a specific expression of relativism.

Friends with Benefits

Friends with benefits is becoming part of the relational sexual landscape of youth. **Friends with benefits (FWB)** is a relationship between nonromantic friends who also have a sexual relationship. Of 1,319 undergraduates, 47.1 percent reported that they have been in an FWB rela-

I THINK MEN TALK TO WOMEN SO THEY CAN SLEEP WITH THEM AND WOMEN SLEEP WITH MEN SO THEY CAN TALK TO THEM.

—Jay McInerney, writer

tionship (Knox and Zusman 2009). Puentes et al. (2008) analyzed data from 1,013 undergraduates (over half of which reported experience with a "friends with benefits" relationship), and compared the background characteristics of participants with nonparticipants in such a relationship. Findings revealed that participants were significantly more likely to be males, casual daters, hedonists, nonromantics, jealous, black, juniors or seniors, those who have had sex without love, and those who regard "financial security" as their top value. The friends with benefits relationship is primarily sexual and engaged in by nonromantic hedonists who have a pragmatic view of relationships.

In a smaller sample of 170 undergraduates at the same university, 57.3 percent of these undergraduates reported that they were or had been involved in an FWB relationship. There were no significant differences between the percentages of women and men reporting involvement in an FWB relationship. This is one of the few studies finding no difference in sexual behavior between women and men (for example, one would expect men to have more FWB relationships than women). However, the percentages of women and men were very similar in their reported rates of FWB involvement—57.1 percent and 57.9 percent, respectively (McGinty et al. 2007). Is a new sexual equality operative in FWB relationships? Analysis of the data revealed other significant differences between female and male college students in regard to various aspects of the FWB relationship.

1. **Women were more emotionally involved.** Women were significantly more likely than men (62.5 percent versus 38.1 percent) to view their current FWB relationship as an emotional relationship. In addition, women were significantly more likely than men to be perceived as being more emotionally involved in the FWB relationship. Of the men, 43.5 percent, compared with 13.6 percent of the women, reported "my partner is more emotionally involved than I am."
2. **Men were more sexually focused.** As might be expected from the first finding, men were significantly more likely than women to agree with the statement, "I wish we had sex more often than we do" (43.5 percent versus 13.6 percent).
3. **Men were more polyamorous.** With polyamory defined as desiring to be involved in more than one emotional or sexual relationship at the same time, men were significantly more likely than women to agree that "I would like to have more than one FWB relationship going on at the same time" (34.8 percent versus 4.5 percent). Serial FWB relationships may already be occurring. Of the men, 52.2 percent, compared with 24.6 percent of women, reported that they had been involved in more than one FWB relationship. Hughes et al. (2005) studied 143 undergraduates in FWB relationships and noted that a ludic, playful, noncommittal love characterized them.

Research data show men in FWB relationships to be more sexually focused.

hedonism belief that the ultimate value and motivation for human actions lie in the pursuit of pleasure and the avoidance of pain.

Hedonism

Hedonism is the belief that the ultimate value and motivation for human actions lie in the pursuit of pleasure and the avoidance of pain. The hedonistic value is

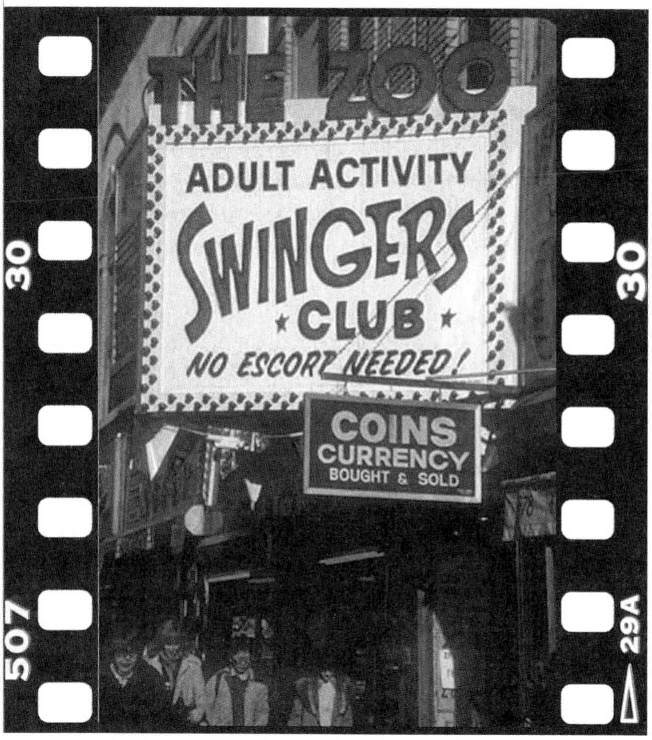

sexual double standard the view that encourages and accepts sexual expression of men more than women.

reflected in the statement, "If it feels good, do it." Hedonism assumes that sexual desire, like hunger and thirst, is an appropriate appetite and its expression is legitimate. Of 1,319 undergraduates surveyed, 29 percent reported that they had hooked up (had oral or sexual intercourse) the first time they met someone (Knox and Zusman 2009).

Of 783 undergraduates, 24.6 percent selected "hedonism" as their primary sexual value. Men were significantly ($p <.001$) more likely than women to select hedonism as their sexual value (36.7 percent versus 12.5 percent). In addition, hedonists were more likely to report having "hooked up," to have been in an FWB relationship, and to be open to living together (Richey et al. 2009).

8.3 Sexual Double Standard

The **sexual double standard**—the view that encourages and accepts sexual expression of men more than women—is reflected in Table 8.1. As noted previously, men were about three times more hedonistic than women (Richey et al. 2009). Acceptance of the double standard is evident in that hedonistic men are thought of as "studs" but hedonistic women as "sluts."

The sexual double standard is also evident in that there is lower disapproval of men having higher numbers of sexual partners but high disapproval of women for having the same number of sexual partners as men. In one study, England and Thomas (2006) noted that the double standard was operative in hooking up. Women who hooked up too often with too many men and had sex too easily were vulnerable to getting bad reputations. Men who did the same thing got a bad reputation among women, but with fewer stigmas. In addition, men gained status among other men for their exploits; women were quieter. Similarly, women who looked at pornography (in contrast to men who viewed pornography) were viewed as "loose" (19.4 percent versus 4.5 percent) (O'Reilly et al. 2007).

Kim et al. (2007) found evidence for the double standard in their review of twenty-five prime-time television programs. In addition, Greene and Faulkner (2005) studied 689 heterosexual couples and found that women were disadvantaged in negotiating sexual issues with their partners, particularly when their traditional gender roles were operative.

Table 8.1
Sexual Value by Sex of Respondent

Respondents	Absolutism	Relativism	Hedonism
Male students	11.6%	51.8%	36.7%
Female students	15.1%	72.4%	12.5%

Source: E. Richey, D. Knox, and M. E. Zusman. 2009. Sexual values of 783 undergraduates. *College Student Journal* 43: 175–80.

8.4 Sources of Sexual Values

The sources of one's sexual values are numerous and include one's school, religion, and family, as well as technology, television, social movements, and the Internet. Halstead (2005) emphasized that schools play a powerful role in shaping a child's sexual values. Previously we noted that public schools in the United States promote absolutist sexual values through abstinence education and that the effectiveness of these programs has been questioned.

Religion is also an important influence. More than 45 percent of 657 undergraduates at a large southeastern university (assessed via random digit dialing) reported that religion had been influential on their sexual choices: "very influential" for 26.9 percent and "somewhat influential" for 18.4 percent (Bristol and Farmer 2005). Bersamin et al. (2008) studied the effects of parental attitudes, practices, and television viewing behavior on adolescent sexual behaviors in a sample of 887 adolescents. They found that adolescents reporting greater parental disapproval and limits on their television viewing were less likely to initiate both oral sex and sexual intercourse. The researchers emphasized that parental attitudes and watching television together can delay potentially risky adolescent sexual behaviors. Similarly, more than 40 percent of 657 undergradu-

© Frédéric Soltan/Sygma/Corbis

Khajuraho has the largest group of medieval Hindu and Jain temples in India, famous for their erotic sculpture.

ates at a large southeastern university (assessed via random digit dialing) reported that their parents had been influential in their sexual choices: "very influential" for 17.5 percent and "somewhat influential" for 24.7 percent (Bristol and Farmer 2005). Similarly, among 918 university students, "mom" was identified as the most influential source of sexual information.

Siblings are also influential. Kornreich et al. (2003) found that girls who had older brothers held more conservative sexual values. "Those with older brothers in the home may be socialized more strongly to adhere to these traditional standards in line with power dynamics believed to shape and reinforce more submissive gender roles for girls and women" (p. 107).

Reproductive technologies such as birth control pills, the morning-after pill, and condoms influence sexual values by affecting the consequences of behavior. Being able to reduce the risk of pregnancy and HIV infection with the pill and condoms allows one to consider a different value system than if these methods of protection did not exist.

The media is also a source of sexual values. A television advertisement shows an affectionate couple with minimal clothes on in a context where sex could occur. "Be ready for the moment" is the phrase of the announcer, and Levitra, the new quick-start Viagra, is the product for sale. The advertiser uses sex to get the attention of the viewer and punches in the product.

However, as a source of sexual values and responsible treatments of contraception, condom usage, abstinence, and consequences of sexual behavior, television is woefully inadequate. Indeed, viewers learn that sex is romantic and exciting but learn nothing about discussing the need for contraception or HIV and STI protection. With few exceptions, viewers are inundated with role models who engage in casual sex without protection.

Social movements such as feminism affect sexual values by empowering women with an egalitarian view of sexuality. This translates into encouraging women to be more assertive about their own sexual needs and giving them the option to experience sex in a variety of contexts (for example, without love or commitment) without self-deprecation. The net effect is a potential increase in the frequency of recreational, hedonistic sex. The gay liberation movement has also been influential in encouraging values that are accepting of sexual diversity.

Another influence on sexual values is the Internet; its sexual content is extensive. The Internet features erotic photos, videos, "live" sex acts and stripping by webcam sex artists. Individuals can exchange nude photos, have explicit sex dialogue, arrange to have "phone sex" or meet in person, or find a prostitute.

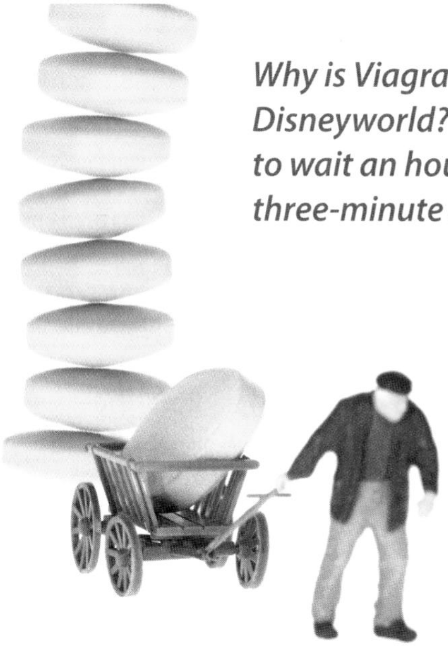

© Christian Kargi/Getty Images

Why is Viagra like Disneyworld? You have to wait an hour for a three-minute ride.

Anonymous

When a man *goes out on a date he wonders if he is going to get lucky. A woman already knows.*

—Frederike Ryder, sex author

8.5 Gender Differences in Sexuality

Gender differences exist in what men and women believe about sex as well as what they do. In this section, we examine these differences as well as how pheromones impact sexual behavior.

Gender Differences in Sexual Beliefs

Knox et al. (2008) analyzed data from 326 undergraduates to discover gender differences in beliefs about sex. Men were more likely to think that oral sex is not sex, that cybersex is not cheating, that men can't tell if a woman is faking orgasm, and that sex frequency drops in marriage. Meanwhile, women tended to believe that oral sex is sex, that cybersex is cheating, that faking orgasm does occur, and that sex frequency stays high in marriage. Little wonder there is frustration and disappointment between men and women as they include sexuality into their relationship.

Gender Differences in Sexual Behavior

In national data based on interviews with 3,432 adults, women reported having fewer sexual partners than men (2 percent versus 5 percent reported having had five or more sexual partners in the previous year), and having orgasm during intercourse less often (29 percent versus 75 percent) (Michael et al. 1994, 102, 128, 156). In a more recent analysis of the responses of 5,385 males and 1,038 females who completed a questionnaire online, males were significantly more likely than females to report frequenting strip clubs, paying for sex, having anonymous sex with strangers, and having casual sexual relations (Mathy 2007).

Pornography use is also higher among males. In a study of 305 undergraduates, 31.7 percent of the men (in contrast to 3.8 percent of the women) reported viewing pornography three to five times a week (O'Reilly et al. 2007).

social script the identification of the roles in a social situation, the nature of the relationship between the roles, and the expected behaviors of those roles.

Men and women also differ in their motivations for sexual intercourse, with men viewing sex more casually (Lenton and Bryan 2005). Earlier, we noted that undergraduate men (in comparison with undergraduate women) were almost three times more likely to report being hedonistic in their sexual values.

Sociobiologists explain males' more casual attitude toward sex, engaging in sex with multiple partners, and being hedonistic as biologically based (that is, due to higher testosterone levels). Social learning theorists, on the other hand, emphasize that the media and peers socialize men to think about and to seek sexual experiences. Men are also accorded social approval and called "studs" for their sexual exploits. Women, on the other hand, are more often punished and labeled "sluts" if they have many sexual partners. Because **social scripts** guide sexual behavior, what individuals think, do, and experience is a reflection of what they have learned (Simon and Gagnon 1998). These scripts operate at the cultural (for example, societal norms for broad sexual conduct—women should not be promiscuous), interpersonal (for example, sexual desires translated into strategies—man should be aggressive in sexual encounters), and intrapsychic (for example, sexual dialogues with self that elicit and sustain arousal—"this will be erotic") levels (DeLamater and Hasday 2007). There are also differences in the perceptions of foreplay and intercourse. Miller and Byers (2004) compared the reported duration of actual foreplay (men = 13 minutes; women = 11 minutes) and desired foreplay (men = 18 minutes; women = 19 minutes), and the reported duration of actual intercourse (men = 8 minutes; women = 7 minutes) and desired intercourse (men = 18 minutes; women = 14 minutes) of heterosexual men and women in long-term relationships with each other. These findings, shown in Table 8.2, suggested that both men and women underestimated their partner's desires for the duration of both foreplay and intercourse; also, they had similar preferences for duration of foreplay, but men wanted longer intercourse than women.

Sexual satisfaction reported by men and women seems to be equal, at least among the French. In a representative sample of 1,002 French respondents (483 men and 519 women) aged 35 years, 83 percent reported relative or full satisfaction with their sex life (Colson et al. 2006).

Mood states typically affect both men and women equally (Lykins et al. 2006). In a study of 663 female college students and 399 college men, the researchers found that individuals who were depressed were less likely to be interested in engaging in sexual behavior.

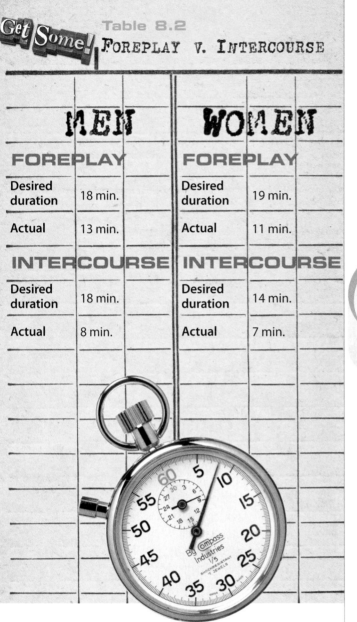

Get Some! Table 8.2
FOREPLAY V. INTERCOURSE

	MEN	WOMEN
FOREPLAY		
Desired duration	18 min.	19 min.
Actual	13 min.	11 min.
INTERCOURSE		
Desired duration	18 min.	14 min.
Actual	8 min.	7 min.

However, this was not always the case, as about 10 percent of women and a higher percentage of men reported that they were interested in engaging in sexual behavior in spite of a negative mood state.

Finally, gender differences may also be influenced by ethnic background. Eisenman and Dantzker (2006) surveyed 128 men and 199 women at a Texas-Mexico border university and found that the genders differed significantly on twenty-six of thirty-eight items. For example, women were less permissive and had more negative attitudes than men in regard to oral sex, premarital intercourse, and masturbation.

8.6 Pheromones and Sexual Behavior

The term *pheromone* comes from the Greek words *pherein*, meaning "to carry," and *hormon*, meaning to "excite." Pheromones are "chemical messengers that are emitted into the environment from the body, where they can then activate specific physiological or behavioral responses in other individuals of the same species" (Grammer et al. 2005, 136). Pheromones are produced primarily by the apocrine glands located in the armpits and pubic region. The functions of pheromones include opposite-sex attractants, same-sex repellents, and mother-infant bonding.

Pheromones typically operate without the person's awareness; researchers disagree about whether pheromones do in fact influence human sociosexual behaviors. Although Levin (2004) reviewed the literature on chemical messengers in attraction, the strongest evidence for the effect of hormones on sexual behavior is that thirty-eight male volunteers who applied a male hormone to their aftershave lotion reported significant increases in sexual intercourse and sleeping next to a partner when compared with men who had a placebo in their aftershave lotion (Cutler et al. 1998).

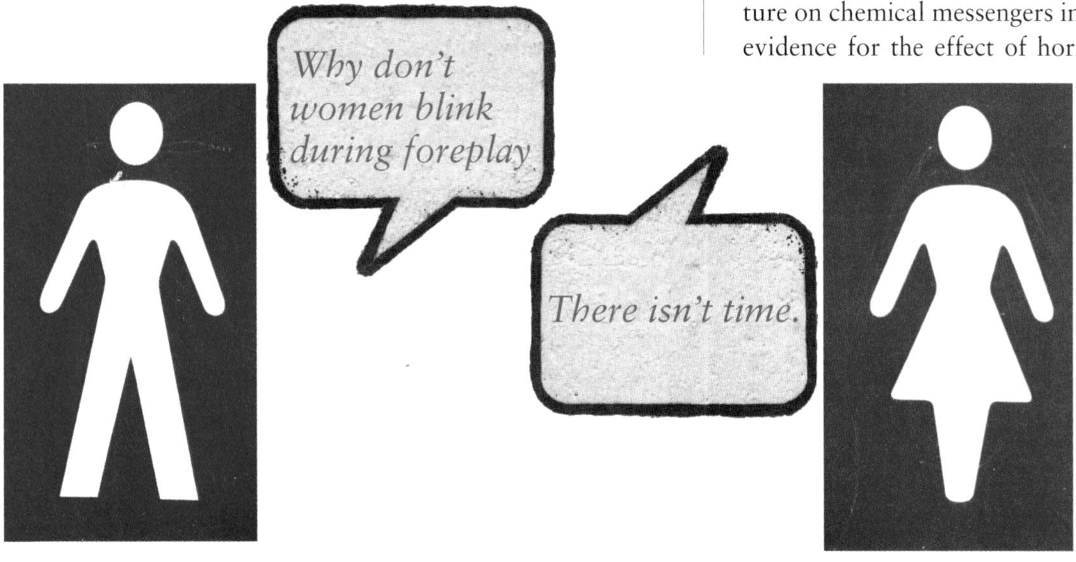
Why don't women blink during foreplay
There isn't time.

Racial Differences in Sex Attitudes and Behaviors of Undergraduates

Davidson et al. (2008) surveyed 1,915 undergraduate women and 1,111 undergraduate men at four universities and found that race was the most influential factor differentiating the sexual attitudes and behavior of the sample. When black people were compared with white people, the former had more permissive attitudes and were more likely to approve of sexual intercourse with casual, occasional, and regular dating partners. Black people also experienced sexual intercourse earlier and with more partners over their lifetimes. Uecker (2008) confirmed that sexual behavior of black people is inconsistent with their close affiliation with religion. One explanation is that the churches attended by black people are often reluctant to address sexual issues.

© Blend Images/Jupiterimages

8.7 Sexuality in Relationships

satiation the state in which a stimulus loses its value with repeated exposure.

Sexuality occurs in a social context that influences its frequency and perceived quality.

Sexual Relationships among Never-Married Individuals

Never-married individuals and those not living together report more sexual partners than those who are married or living together. In one study, 9 percent of never-married individuals and those not living together reported having had five or more sexual partners in the previous twelve months; 1 percent of married people and 5 percent of cohabitants reported the same. However, unmarried individuals, when compared with married individuals and cohabitants, reported the lowest level of sexual satisfaction. One-third of a national sample of people who were not married and not living with anyone reported that they were emotionally satisfied with their sexual relationships. In contrast, 85 percent of the married and pair-bonded individuals reported emotional satisfaction in their sexual relationships. Hence, although never-married individuals have more sexual partners, they are less emotionally satisfied (Michael et al. 1994).

Sexual Relationships among Married Individuals

Marital sex is distinctive for its social legitimacy, declining frequency, and satisfaction (both physical and emotional).

1. **Social legitimacy.** In our society, marital intercourse is the most legitimate form of sexual behavior. Homosexual, premarital, and extramarital intercourse do not have as high a level of social approval as does marital sex. It is not only okay to have intercourse when married, it is expected. People assume that married couples make love and that something is wrong if they do not.

2. **Declining frequency.** Sexual intercourse between spouses occurs about six times a month, which declines in frequency as spouses age. Pregnancy also decreases the frequency of sexual intercourse. In addition to biological changes due to aging and pregnancy, satiation also contributes to the declining frequency of intercourse between spouses and partners in long-term relationships. Psychologists use the term **satiation** to define the state in which a stimulus loses its value with repeated exposure. For example, the first time you listen to a new CD, you derive considerable enjoyment and satisfaction from it. You may play it over and over during the first few days. After a week or so, listening to the same music is no

longer new and does not give you the same level of enjoyment that it first did. So it is with intercourse. The thousandth time that a person has intercourse with the same partner is not as new and exciting as the first few times.

3. **Satisfaction (emotional and physical).** Despite declining frequency and less satisfaction over time (Liu 2003), marital sex remains a richly satisfying experience. Contrary to the popular belief that unattached singles have the best sex, the married and pair-bonded adults enjoy the most satisfying sexual relationships. In a national sample, 88 percent of married people said they received great physical pleasure from their sexual lives, and almost 85 percent said they received great emotional satisfaction (Michael et al. 1994). Individuals least likely to report being physically and emotionally pleased in their sexual relationships are those who are not married, not living with anyone, or not in a stable relationship with one person (ibid.).

Sexual Relationships among Divorced Individuals

Of the almost 2 million people getting divorced, most will have intercourse within one year of being separated from their spouses. The meanings of intercourse for separated or divorced individuals vary. For many, intercourse is a way to reestablish—indeed, repair—their crippled self-esteem. Questions like, "What did I do wrong?" "Am I a failure?" and "Is there anybody out there who will love me again?" loom in the minds of divorced people. One way to feel loved, at least temporarily, is through sex. Being held by another and being told that it feels good give people some evidence that they are desirable. Because divorced people may be particularly vulnerable, they may reach for sexual encounters as if for a lifeboat. "I felt that, as long as someone was having sex with me, I wasn't dead and I did matter," said one recently divorced person.

Because divorced individuals are usually in their thirties or older, they may not be as sensitized to the danger of contracting HIV as people in their twenties. Divorced individuals should always use a condom to lessen the risk of STI, including HIV infection, and AIDS.

8.8 Safe Sex: Avoiding Sexually Transmitted Infections

The Student Sexual Risks Scale (SSRS) on the Self-Assessment Card allows you to assess the degree to which you are at risk for contracting an STI, including HIV infection.

One of the negative consequences of sexual behavior is the risk of contracting a sexually transmitted infection (STI). Also known as sexually transmitted disease, or STD, **STI** refers to the general category of sexually transmitted infections such as chlamydia, genital herpes, gonorrhea, and syphilis. The most lethal of all STIs is that due to human immunodeficiency virus (**HIV**), which attacks the immune system and can lead to acquired immunodeficiency syndrome (**AIDS**). AIDS is the last stage of HIV infection, in which the immune system of a person's body is so weakened that it becomes vulnerable to disease and infection. Because the consequences of contracting HIV are the most severe, we focus on HIV here. For more information on STIs, go to 4ltrpress.cengage.com/mf.

STI sexually transmitted infection.

HIV human immunodeficiency virus, which attacks the immune system and can lead to AIDS.

AIDS acquired immunodeficiency syndrome; the last stage of HIV infection, in which the immune system of a person's body is so weakened that it becomes vulnerable to disease and infection.

MY MESSAGE TO THE BUSINESSMEN OF THIS COUNTRY WHEN THEY GO ABROAD ON BUSINESS IS THAT THERE IS ONE THING ABOVE ALL THEY CAN TAKE WITH THEM TO STOP THEM CATCHING AIDS, AND THAT IS THE WIFE.

—*Edwina Currie, former member of British Parliament*

Transmission of HIV and High-Risk Behaviors

HIV can be transmitted in several ways.

1. **Sexual contact.** HIV is found in several body fluids of infected individuals, including blood, semen, and vaginal secretions. During sexual contact with an infected individual, the virus enters a person's bloodstream through the rectum, vagina, penis (an uncircumcised penis is at greater risk because of the greater retention of the partner's fluids), and possibly the mouth during oral sex. Saliva, sweat, and tears are not body fluids through which HIV is transmitted.
2. **Intravenous drug use.** Drug users who are infected with HIV can transmit the virus to other drug users with whom they share needles, syringes, and other drug-related implements.
3. **Blood transfusions.** HIV can be acquired by receiving HIV-infected blood or blood products. Currently, all blood donors are screened, and blood is not accepted from high-risk individuals. Blood that is accepted from donors is tested for the presence of HIV. However, prior to 1985, donor blood was not tested for HIV. Individuals who received blood or blood products prior to 1985 may have been infected with HIV.
4. **Mother-child transmission.** A pregnant woman infected with HIV has a 40 percent chance of transmitting the virus through the placenta to her unborn child. These babies will initially test positive for HIV as a consequence of having the antibodies from their mother's bloodstream. However, azidothymidine (AZT, alternatively called zidovudine or ZVD) taken by the mother twelve weeks before birth seems to reduce the chance of transmission of HIV to her baby by two-thirds. HIV may also be transmitted, although rarely, from mother to infant through breast-feeding.
5. **Organ or tissue transplants and donor semen.** Receiving transplant organs and tissues, as well as receiving semen for artificial insemination, could involve risk of contracting HIV if the donors have not been tested for HIV. Such testing is essential, and recipients should insist on knowing the HIV status of the organ, tissue, or semen donor.
6. **Other methods of transmission.** For health care professionals, HIV can also be transmitted through contact with amniotic fluid surrounding a fetus, synovial fluid surrounding bone joints, and cerebrospinal fluid surrounding the brain and spinal cord.

STI Transmission—The Illusion of Safety in a "Monogamous" Relationship

Most individuals in a serious "monogamous" relationship assume that they are at zero risk for contracting an STI from their partner. Vail-Smith et al. (Forthcoming) analyzed data from 1,341 undergraduates at a large southeastern university and found that 27.2 percent of the males and 19.8 percent (almost one in five) females reported having oral, vaginal, or anal sex outside of a relationship that their partner considered monogamous. People most likely to cheat were men over the age of 20, those who were binge drinkers, members of a fraternity, male NCAA athletes, or nonreligious people. These data suggest the need for educational efforts to encourage undergraduates in committed relationships to reconsider their STI risk and to protect themselves via condom usage.

Prevention of HIV and STI Transmission

The safest relationship context for avoiding an STI (including HIV) is marriage, with cohabitation a close second (Hattori and Dodoo 2007). Of course, the best way to avoid getting an STI is to avoid sexual contact or to have contact only with partners who are not infected. This means restricting your sexual contacts to those who limit their relationships to one person. The person most likely to get an STI has sexual relations with a number of partners or with a partner who has a variety of partners. Even if you are in a mutually monogamous relationship, you may be at risk for acquiring or transmitting an STI. This is because health officials suggest that when you have sex with someone, you are having sex (in a sense) with everyone that person has had sexual contact with in the past ten years.

Condoms should be used for vaginal, anal, and oral sex and should never be reused. However, in a sample of 1,319 undergraduates, only 24.8 percent reported that they always used a condom when having intercourse (Knox and Zusman 2009). Feeling that the partner is disease-free, having had too much alcohol, and believing that "getting an STI won't happen to me" are reasons

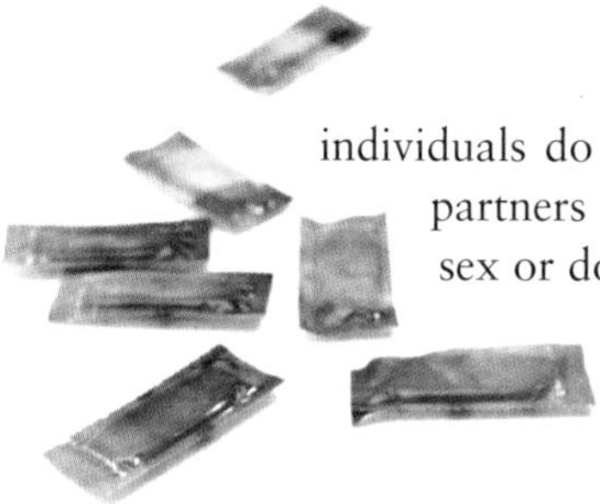

© Zoran Djekic/iStockphoto.com

individuals do not use a condom. Some partners are also forced to have sex or do not feel free to negotiate the use of a condom in their sexual relationship (Heintz and Melendez 2006).

Using the condom properly is also important. Putting on a latex or polyurethane condom before the penis touches the partner's body makes passing STIs from one person to another difficult (natural membrane condoms do not block the transmission of STIs). Care should also be taken to withdraw the penis while it is erect to prevent fluid from leaking from the base of the condom into the partner's genital area. If a woman is receiving oral sex, she should wear a dental dam, which will prevent direct contact between the genital area and her partner's mouth.

Sexuality in an age of HIV and STIs demands talking about safer sex issues with a new potential sexual partner. Bringing up the issue of condom use should be perceived as caring for oneself, the partner, and the relationship rather than as a sign of distrust. Some individuals routinely have a condom available, and it is a "given" in any sexual encounter. Figure 8.1 illustrates that one is more likely to contract an STI through high alcohol use, low condom use, and having sex with multiple partners.

Figure 8.1

Risk of Contracting an STI, as Related to Alcohol, Condom Use, and Number of Partners

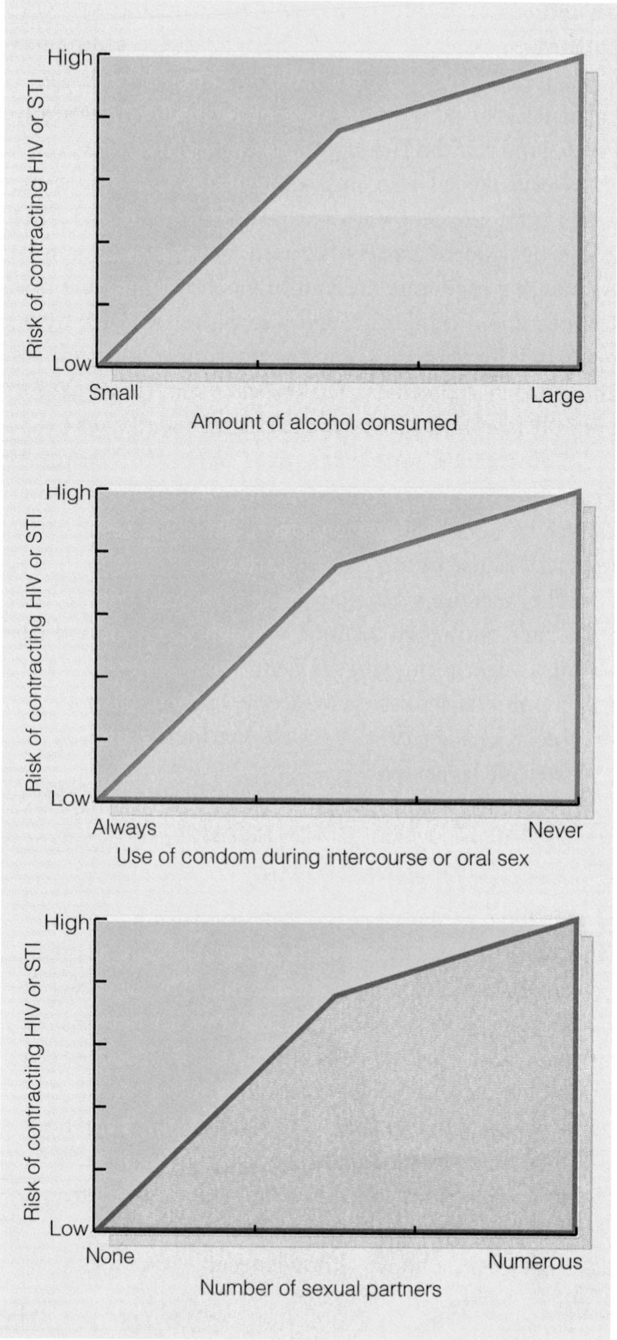

8.9 Sexual Fulfillment: Some Prerequisites

There are several prerequisites for having a good sexual relationship.

Self-Knowledge, Self-Esteem, and Health

Sexual fulfillment involves knowledge about yourself and your body. Such information not only makes it easier for you to experience pleasure but also allows you to give accurate information to a partner about pleasing you. It is not possible to teach a partner what you don't know about yourself.

Sexual fulfillment also implies having a positive self-concept. To the degree that you have positive feelings about yourself and your body, you will regard yourself as a person someone else would enjoy touching, being close to, and making love with. If you do not like yourself or your body, you might wonder why anyone else would.

Effective sexual functioning also requires good physical and mental health. This means regular exercise, good nutrition, lack of disease, and lack of fatigue. Performance in all areas of life does not have to diminish with age—particularly if people take care of themselves physically (see Chapter 15 on Relationships in the Later Years).

Good health also implies being aware that some drugs may interfere with sexual performance. Alcohol is the drug most frequently used by American adults. Although a moderate amount of alcohol can help a person become aroused through a lowering of inhibitions, too much alcohol can slow the physiological processes and deaden the senses. Shakespeare may have said it best: "It [alcohol] provokes the desire, but it takes away the performance" (*Macbeth*, act 2, scene 3). The result of an excessive intake of alcohol for women is a reduced chance of orgasm; for men, overindulgence results in a reduced chance of attaining an erection.

The reactions to marijuana are less predictable than the reactions to alcohol. Though some individuals report a short-term enhancement effect, others say that marijuana just makes them sleepy. In men, chronic use may decrease sex drive because marijuana may lower testosterone levels.

A Good Relationship, Positive Motives

A guideline among therapists who work with couples who have sexual problems is to treat the relationship before focusing on the sexual issue. The sexual relationship is part of the larger relationship between the partners, and what happens outside the bedroom in day-to-day interaction has a tremendous influence on what happens inside the bedroom. The statement, "I can't fight with you all day and want to have sex with you at night" illustrates the social context of the sexual experience.

Women most valued a partner who was open to discussing sex, who was knowledgeable about sex, who clearly communicated his desires, who was physically attractive, and who paid her compliments during sex. Being easily sexually aroused and being uninhibited were also important.

Sexual interaction communicates how the partners are feeling and acts as a barometer for the relationship. Each partner brings to a sexual encounter, sometimes unconsciously, a motive (pleasure, reconciliation, procreation, duty), a psychological state (love, hostility, boredom, excitement), and a physical state (tense, exhausted, relaxed, turned on). The combination of these factors will change from one encounter to another. Tonight a wife may feel aroused and loving and seek pleasure, but her husband may feel exhausted and hostile and have sex only out of a sense of duty. Tomorrow night, both partners may feel relaxed and have sex as a means of expressing their love for each other.

One's motives for a sexual encounter are related to the outcome. Impett et al. (2005) found that, when individuals have intercourse out of the desire to enhance personal and interpersonal or relationship pleasure, the personal and interpersonal effect on well-being is very positive. However, when sexual motives were to avoid conflict, the personal and interpersonal effects did not result in similar positive outcomes. In a study of 1,002 French adults, sexuality was more synonymous with pleasure (44.0 percent) and love (42.1 percent) than with procreation, children, or motherhood (7.8 percent) (Colson et al. 2006).

An Equal Relationship

Laumann et al. (2006) surveyed 27,500 individuals in twenty-nine countries and found that reported sexual satisfaction was higher where men and women were considered equal. Austria topped the list, with 71 percent reporting sexual satisfaction; only 25.7 percent of those surveyed in Japan reported sexual satisfaction. The United States was among those countries in which a high percentage of the respondents reported sexual satisfaction.

SEXUALLY FULFILLED PARTNERS TAKE THE GUESSWORK OUT OF THEIR RELATIONSHIP BY COMMUNICATING PREFERENCES AND GIVING FEEDBACK.

Open Sexual Communication and Feedback

Sexually fulfilled partners are comfortable expressing what they enjoy and do not enjoy in the sexual experience. Unless both partners communicate their needs, preferences, and expectations to each other, neither is ever sure what the other wants. In essence, the Golden Rule ("Do unto others as you would have them do unto you") is *not* helpful, because what you like may not be the same as what your partner wants. A classic example of the uncertain lover is the man who picks up a copy of *The Erotic Lover* in a bookstore and leafs through the pages until the topic on how to please a woman catches his eye. He reads that women enjoy having their breasts stimulated by their partner's tongue and teeth. Later that night in bed, he rolls over and begins to nibble on his partner's breasts. Meanwhile, she wonders what has possessed him and is unsure what to make of this new (possibly unpleasant) behavior.

Sexually fulfilled partners take the guesswork out of their relationship by communicating preferences and giving feedback. This means using what some therapists call the touch-and-ask rule. Each touch and caress may include the question, "How does that feel?" It is then the partner's responsibility to give feedback. If the caress does not feel good, the partner should say what does feel good.

Guiding and moving the partner's hand or body are also ways of giving feedback. What women and men want each other to know about sexuality is presented in Table 8.3.

Having Realistic Expectations

To achieve sexual fulfillment, expectations must be realistic. A couple's sexual needs, preferences, and expectations may not coincide. It is unrealistic to assume that your partner will want to have sex with the same frequency and in the same way that you do on all occasions. It may also be unrealistic to expect the level of sexual interest and frequency of sexual interaction in long-term relationships to remain consistently high.

spectatoring mentally observing one's own and one's partner's sexual performance.

Sexual fulfillment means not asking things of the sexual relationship that it cannot deliver. Failure to develop realistic expectations will result in frustration and resentment. One's health, feelings about the partner, age, and previous sexual experiences (including child sexual abuse, rape, and so on) will have an effect on one's sexuality and one's sexual relationship.

Avoiding Spectatoring

One of the obstacles to sexual functioning is **spectatoring**, which involves mentally observing your sexual performance and that of your partner. When the researchers in one extensive study observed how individuals actually behave during sexual intercourse, they reported a tendency for sexually dysfunctional partners to act as spectators by mentally observing their own and their partners' sexual performance. For example, the man would focus on whether he was having an erection, how complete it was, and whether it would last. He might also watch to see whether his partner was having an orgasm (Masters and Johnson 1970).

Spectatoring, as Masters and Johnson

Table 8.3

What Women and Men Want Each Other to Know about Sexuality

What Women Want Men to Know about Sex
Women tend to like a loving, gentle, patient, tender, and understanding partner. Rough sexual play can hurt and be a turnoff.
It does not impress women to hear about other women in the man's past.
If men knew what it is like to be pregnant, they would not be so apathetic about birth control.
Most women want more caressing, gentleness, kissing, and talking before and after intercourse.
Some women are sexually attracted to other women, not to men.
Sometimes the woman wants sex even if the man does not. Sometimes she wants to be aggressive without being made to feel that she shouldn't be.
Intercourse can be enjoyable without orgasm.
Many women do not have an orgasm from penetration only; they need direct stimulation of their clitoris by their partner's tongue or finger.
Men should be interested in fulfilling their partner's sexual needs.
Most women prefer to have sex in a monogamous love relationship.
When a woman says "no," she means it.
Women do not want men to expect sex every time they are alone with their partner.
Many women enjoy sex in the morning, not just at night.
Sex is *not* everything.
Women need to be lubricated before penetration.
Men should know more about menstruation.
Many women are no more inhibited about sex than men are.
Women do not like men to roll over, go to sleep, or leave right after orgasm.
Intercourse is more of a love relationship than a sex act for some women.
The woman should not always be expected to supply a method of contraception. It is also the man's responsibility.
Men should always have a condom with them and initiate putting it on.

What Men Want Women to Know about Sex
Men do not always want to be the dominant partner; women should be aggressive.
Men want women to enjoy sex totally and not be inhibited.
Men enjoy tender and passionate kissing.
Men really enjoy fellatio and want women to initiate it.
Women need to know a man's erogenous zones.
Men enjoy giving oral sex; it is not bad or unpleasant.
Many men enjoy a lot of romantic foreplay and slow, aggressive sex.
Men cannot keep up intercourse forever. Most men tire more easily than women.
Looks are not everything.
Women should know how to enjoy sex in different ways and different positions.
Women should not expect a man to get a second erection right away.
Many men enjoy sex in the morning.
Pulling the hair on a man's body can hurt.
Many men enjoy sex in a caring, loving, exclusive relationship.
It is frustrating to stop sex play once it has started.
Women should know that not all men are out to have intercourse with them. Some men like to talk and become friends.

conceived it, interferes with each partner's sexual enjoyment because it creates anxiety about performance, and anxiety blocks performance. A man who worries about getting an erection reduces his chance of doing so. A woman who is anxious about achieving an orgasm probably will not. The desirable alternative to spectatoring is to relax, focus on and enjoy your own pleasure, and permit yourself to be sexually responsive.

Spectatoring is not limited to sexually dysfunctional couples and is not necessarily associated with psychopathology. It is a reaction to the concern that the performance of one's sexual partner is consistent with expectations. We all probably have engaged in spectatoring to some degree. When such spectatoring is continuous, performance is impaired.

Debunking Sexual Myths

Sexual fulfillment also means not being victim to sexual myths. Some of the more common myths include that sex equals intercourse and orgasm, that women who love sex don't have values, and that the double standard is dead. Another common sexual myth is that the elderly have no interest in sex. Beckman et al. (2006) examined the sexual interests and needs of 563 70-year-olds and found that 95 percent reported the continuation of such interests and needs as they aged. Almost 70 percent (69 percent) of the married men and 57 percent of the married women reported continued sexual behavior. Nobre et al. (2006) noted that belief in sexual myths makes one vulnerable to sexual dysfunctions. Table 8.4 presents some other sexual myths.

Table 8.4
Common Sexual Myths

Masturbation is sick.
Women who love sex are sluts.
Sex education makes children promiscuous.
Sexual behavior usually ends after age 60.
People who enjoy pornography end up committing sexual crimes.
Most "normal" women have orgasms from penile thrusting alone.
Extramarital sex always destroys a marriage.
Extramarital sex will strengthen a marriage.
Simultaneous orgasm with one's partner is the ultimate sexual experience.
My partner should enjoy the same things that I do sexually.
A man cannot have an orgasm unless he has an erection.
Most people know a lot of accurate information about sex.
Using a condom ensures that you won't get HIV.
Most women prefer a partner with a large penis.
Few women masturbate.
Women secretly want to be raped.
An erection is necessary for good sex.
An orgasm is necessary for good sex.

Self-Assessment: Attitudes toward Premarital Sex Scale

Premarital sex is defined as engaging in sexual intercourse prior to marriage. The purpose of this survey is to assess your thoughts and feelings about intercourse before marriage. Read each item carefully and consider how you feel about each statement. There are no right or wrong answers to any of these statements, so please give your honest reactions and opinions. Please respond by using the following scale:

1	2	3	4	5	6	7
Strongly Disagree						Strongly Agree

____ 1. I believe that premarital sex is healthy.
____ 2. There is nothing wrong with premarital sex.
____ 3. People who have premarital sex develop happier marriages.
____ 4. Premarital sex is acceptable in a long-term relationship.
____ 5. Having sexual partners before marriage is natural.
____ 6. Premarital sex can serve as a stress reliever.
____ 7. Premarital sex has nothing to do with morals.
____ 8. Premarital sex is acceptable if you are engaged to the person.
____ 9. Premarital sex is a problem among young adults.
____ 10. Premarital sex puts unnecessary stress on relationships.

Scoring

Selecting a 1 reflects the most negative attitude toward premarital sex; selecting a 7 reflects the most positive attitude toward premarital sex. Before adding the numbers you assigned to each item, change the scores for items #9 and #10 as follows: replace a score of 1 with a 7; 2 with a 6; 3 with a 5; 4 with a 4; 5 with a 3; 6 with a 2; and 7 with a 1. After changing these numbers, add your ten scores. The lower your total score (10 is the lowest possible score), the less accepting you are of premarital sex; the higher your total score (70 is the highest possible score), the greater your acceptance of premarital sex. A score of 40 places you at the midpoint between being very disapproving of premarital sex and very accepting of premarital sex.

Source: M. Whatley. Attitudes toward Premarital Sex Scale. 2006. Valdosta, Georgia: Department of Psychology, Valdosta State University. Used by permission.

Planning to Have Children

SECTIONS

The cultural message on having children is mixed. Although young married individuals are encouraged to "have fun and travel before they begin their family" and to "strap on their seat belts when their children become teenagers," spouses are also told that "marriage has no real meaning without children" and "aren't they precious?" In spite of these dichotomous messages, having children continues to be a major goal of college students. In a nonrandom sample of 1,319 undergraduates at a large southeastern university, 91 percent agreed, "Someday I want to have children" (Knox and Zusman 2009).

Planning for children, or failing to do so, is a major societal issue. Planning when to become pregnant has benefits for both the mother and the child. Having several children at short intervals increases the chances of premature birth, infectious disease, and death of the mother or the baby. Would-be parents can minimize such risks by planning fewer children with longer intervals in between. Women who plan their pregnancies can also modify their behaviors and seek preconception care from a health care practitioner to maximize their chances of having healthy pregnancies and babies. For example, women planning pregnancies can make sure they eat properly and avoid alcohol and other substances (such as cigarettes) that could harm developing fetuses. Partners who plan their children also benefit from family planning by pacing the financial demands of their offspring. Having children four years apart helps to avoid having more than one child in college at the same time. Conscientious family planning will also help to reduce the number of unwanted pregnancies. Schwarz et al. (2008) asked 192 women how they would feel if they learned they were pregnant. Of them, 9 percent reported that they would feel like they were dying, and 28 percent said they would trade time from the end of their life.

Your choices in regard to children and contraception have important effects on your happiness, lifestyle, and resources. These choices, in large part, are influenced by social and cultural factors that may operate without your awareness. We now discuss these influences.

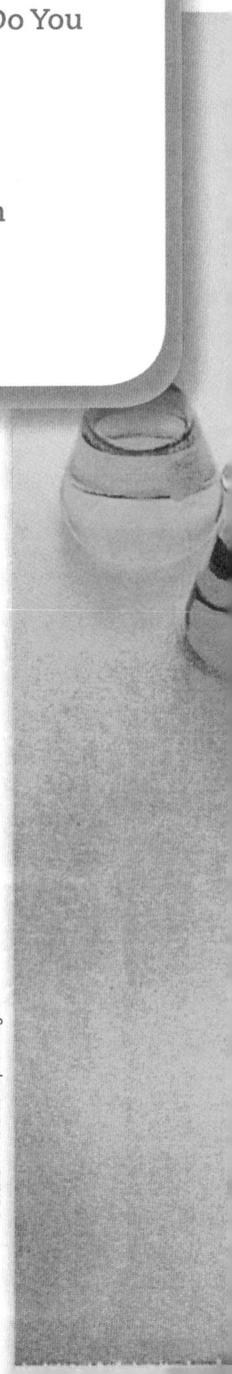

© Brand X Pictures/Jupiterimages

9.1 Do You Want to Have Children?

Beyond a biological drive to reproduce, societies socialize their members to have children. This section examines the social influences that motivate individuals to have children, the lifestyle changes that result from such a choice, and the costs of rearing children.

CHAPTER 9

Social Influences Motivating Individuals to Have Children

Our society tends to encourage childbearing, an attitude known as **pronatalism**. Our family, friends, religion, and government help to develop positive attitudes toward parenthood. Cultural observances also function to reinforce these attitudes.

pronatalism view that encourages having children.

HE CHANGED EVERYTHING, BUT IN THE MOST WONDERFUL WAY. EVERYTHING THAT SHOULD MATTER, MATTERS. HE'S ABSOLUTELY THE CENTER OF MY LIFE.

— Angelina Jolie, on her adopted son, Maddox

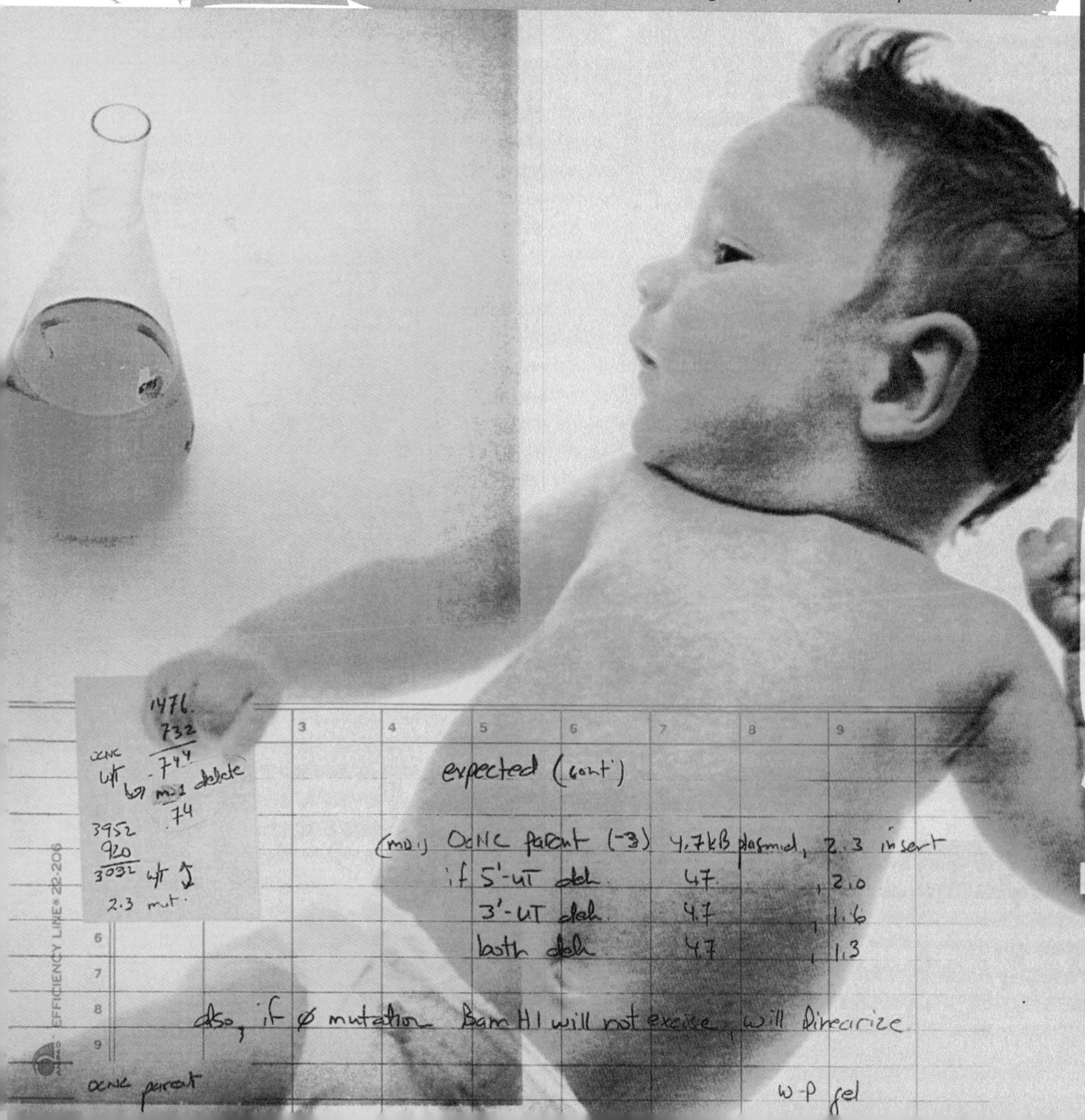

Family Our experience of being reared in families encourages us to have families of our own. Our parents are our models. They married; we marry. They had children; we have children. Some parents exert a much more active influence. "I'm 73 and don't have much time. Will I ever see a grandchild?" asked the mother of an only child.

Friends Our friends who have children influence us to do likewise. After sharing an enjoyable weekend with friends who had a little girl, one husband wrote to the host and hostess, "Lucy and I are always affected by Karen—she is such a good child to have around. We haven't made up our minds yet, but our desire to have a child of our own always increases after we leave your home." This couple became parents sixteen months later.

Religion Religion is a powerful influence on the decision to have children. Catholics are taught that having children is the basic purpose of marriage and gives meaning to the union. Mormonism and Judaism also have a strong family orientation.

Race Although Hispanics have the highest fertility rate, their numbers are not sufficient to account for the fact that fertility rates for the United States are increasing (Martin 2008).

Government The tax structures that our federal and state governments impose support parenthood. Married couples without children pay higher taxes than couples with children, although the reduction in taxes is not sufficient to offset the cost of rearing a child and is not large enough to be a primary inducement to have children.

Economy Times of affluence are associated with a high birth rate. Postwar expansion of the 1950s resulted in the oft-noted "baby boom" generation. Similarly, couples are less likely to decide to have a child during economically depressed times. In addition, the necessity of two wage earners in our postindustrial economy is associated with a reduction in the number of children.

Cultural Observances Our society reaffirms its approval of parents every year by identifying special days for Mom and Dad. Each year on Mother's Day and Father's Day (and now Grandparents' Day), parenthood is celebrated across the nation with cards, gifts, and embraces. People choosing not to have children have no cultural counterpart (for example, Childfree Day). In addition to influencing individuals to have children, society and culture also influence feelings about the age parents should be when they have children. Recently, couples have been having children at later ages.

© Daniel Loiselle/iStockphoto.com

Individual Motivations for Having Children

Individual motivations, as well as social influences, play an important role in the decision to have children. Some of these are conscious, as in the desire to love and to be loved by one's own child, companionship, and the desire to be personally fulfilled as an adult by having a child. Some also want to recapture their own childhood and youth by having a child. Unconscious motivations for parenthood may also be operative. Examples include wanting a child to avoid career tracking and to gain the acceptance and approval of one's parents and peers. Teenagers sometimes want to have a child to have someone to love them. Later in the chapter we detail teenage motherhood as a major social issue.

Lifestyle Changes and Economic Costs of Parenthood

Although becoming a parent has numerous potential positive outcomes, parenting also has drawbacks. Every parent knows that parenthood involves difficulties as well as joys. Some of the difficulties associated with parenthood are discussed next.

THERE NEVER WAS A CHILD SO LOVELY, BUT HIS MOTHER WAS GLAD TO GET HIM ASLEEP.

— Ralph Waldo Emerson

Lifestyle Changes Becoming a parent often involves changes in lifestyle. Daily living routines become focused around the needs of the children. Living arrangements change to provide space for another person in the household. Some parents change their work schedule to allow them to be home more. Food shopping and menus change to accommodate the appetites of children. A major lifestyle change is the loss of freedom of activity and flexibility in one's personal schedule. Lifestyle changes are particularly dramatic for women. The time and effort required to be pregnant and rear children often compete with the time and energy needed to finish one's education. Building a career is also negatively impacted by the birth of children. Parents learn quickly that being both involved, on-the-spot parents and climbing the career ladder are difficult. The careers of women may suffer most.

Financial Costs Meeting the financial obligations of parenthood is difficult for many parents. The costs begin with prenatal care and continue at childbirth. For an uncomplicated vaginal delivery, with a two-day hospital stay, the cost may total $10,000, whereas a cesarean section birth may cost $14,000. The annual cost of a child less than 2 years old for middle-income parents ($45,800 to $77,100)—which includes housing ($4,010), food ($1,280), transportation ($1,390), clothing ($410), health care ($780), child care ($2,000), and miscellaneous ($1,090)—is $10,960. For a 15- to 17-year-old, the cost is $12,030 (*Statistical Abstract of the United States, 2009*, Table 667). These costs do not include the wages lost when a parent drops out of the workforce to provide child care. But in spite of the costs children incur, most people look forward to having children.

Most parents anticipate their children attending college. The price varies depending on whether a child attends to a public or private college, and websites such as www.collegeboard.com identify the cost of a specific college. On average, the annual cost for a child attending a four-year public college in the state of residence is around $14,203 (including tuition, board, dorm); for a private college the cost is around $40,000 annually (*Statistical Abstract of the United States, 2009*, Table 282).

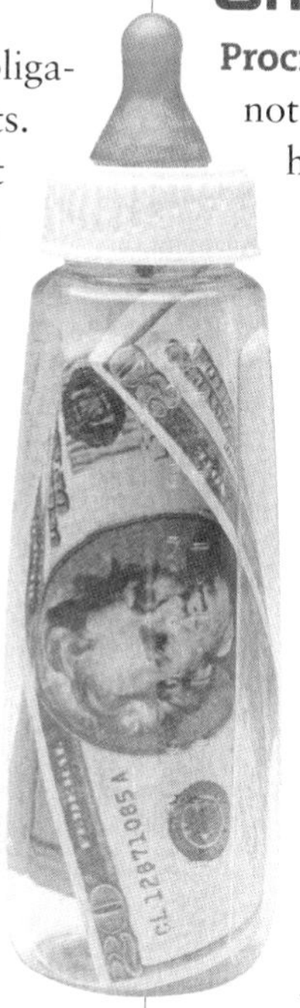

© Dynamic Graphics/Creatas Images/Jupiterimages

9.2 How Many Children Do You Want?

Childfree Marriage

Procreative liberty is the freedom to decide whether or not to have children. More women are deciding not to have children or to have fewer children.

Koropeckyj-Cox and Pendell (2007a) examined attitudes about childlessness in the United States. They used a national sample and found that college educated, white females had the most favorable attitudes toward childlessness. Those who were most negative were not college educated, black, male, and held conservative religious beliefs. In general, there seems to be an acceptance of childlessness, not an endorsement of the lifestyle. The data reflecting adults in a national study revealed that 55.9 percent of females compared to 48.1% of males disagreed that "People who have never had children have empty lives." Hence, a stigma is still associated with not having children, and men buy into this more than women (Koropeckyj-Cox and Pendell 2007b).

When a couple does not have children, is it by choice or because of infertility? Aside from infertility, typical reasons couples give for not having children include the freedom to spend their time and money as they choose, to enjoy their partner without interference, to continue in school or pursue their career, to avoid health problems associated with pregnancy, and to avoid passing on genetic disorders to a new generation.

Some people simply do not like children. Aspects of our society reflect **antinatalism** (opposition to having children). Indeed, there is a

> For most women, *including women who want to have children, contraception is not an option; it is a basic health care necessity.*
>
> —*Louise Slaughter, Congresswoman from New York*

procreative liberty the freedom to decide whether or not to have children.

antinatalism opposition to having children.

continuous fight for corporations to implement or enforce any family policies (from family leaves to flex time to on-site day care). Profit and money—not children—are priorities. In addition, although people are generally tolerant of their own children, they often exhibit antinatalistic behavior in reference to the children of others. Notice the unwillingness of some individuals to sit next to a child on an airplane.

One Child

Some couples have an only child because they want the experience of parenthood without children markedly interfering with their lifestyle and careers. Still others have an only child because of the difficulty in pregnancy or birthing the child. "I threw up every day for nine months including on the delivery table." Another said, "I was torn up giving birth to my child." "It took two years for my body to recover. Once is enough for me." There are also those who have only one child because they can't get pregnant a second time. Couples in China typically have one child due to China's One Child Policy—there are penalties for having more than one.

Two Children

The most preferred family size in the United States (for non-Hispanic white women) is the two-child family (1.9 to be exact!). Reasons for this preference include feeling that a family is "not complete" without two children, having a companion for the first child, having a child of each sex, and repeating the positive experience of parenthood enjoyed with their first child. Some couples may not want to "put all their eggs in one basket." They may fear that, if they have only one child and that child dies or turns out to be disappointing, they will not have another opportunity to enjoy parenting.

competitive birthing pattern in which a woman will want to have the same number of children as her peers.

Three Children

Religion is a strong influence in the number of children a couple have. Twenty percent of Mormons and 15 percent of Muslims have at least three children (Pew Research 2008). In addition to religious influences, couples are more likely to have a third child, and to do so quickly, if they already have two girls rather than two boys. They are least likely to bear a third child if they already have a boy and a girl. Some individuals may want three children because they enjoy children and feel that "three is better than two." In some instances, a couple that has two children may simply want another child because they enjoy parenting and have the resources to do so.

Having a third child creates a "middle child." This child is sometimes neglected because parents of three children may focus more on the "baby" and the firstborn than on the child in between. However, an advantage to being a middle child is the chance to experience both a younger and an older sibling. Each additional child also has a negative effect on the existing children by reducing the amount of parental time available to existing children. The economic resources for each child are also affected for each subsequent child.

Hispanics are more likely to want larger families than are white or African American people. Larger families have complex interactional patterns and different values. The addition of each subsequent child dramatically increases the possible relationships in the family. For example, in a one-child family, four interpersonal relationships are possible: mother-father, mother-child, father-child, and father-mother-child. In a family of four, eleven relationships are possible; in a family of five, 26; and in a family of six, 57.

Four Children—New Standard for the Affluent?

Smith (2007) noted that, among affluent couples, four children may be the new norm. Fueled by competitive career moms who have opted out of the workforce and who find themselves in suburbia surrounded by other moms with resources and time on their hands, having a large family is being reconsidered. A pattern has begun called **competitive birthing**, where "keeping up with

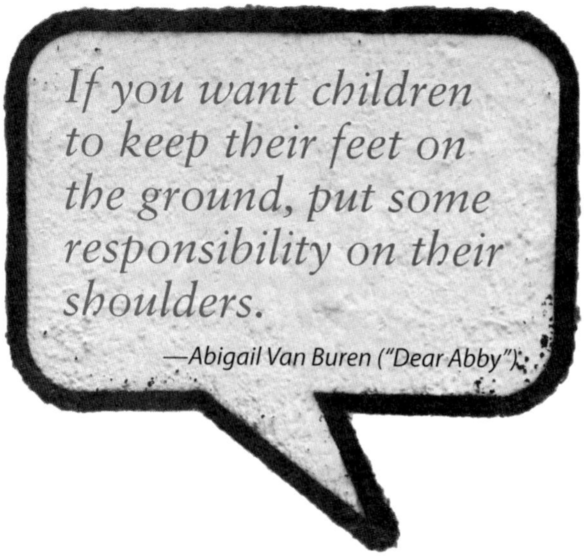
If you want children to keep their feet on the ground, put some responsibility on their shoulders.
—Abigail Van Buren ("Dear Abby")

© iStockphoto.com

the Joneses" now means having the same number of children. Subsequent research will need to confirm that the pattern is widespread.

9.3 Teenage Motherhood

Reasons for teenagers having a child include not being socialized as to the importance of contraception, having limited parental supervision, and perceiving few alternatives to parenthood. Indeed, motherhood may be one of the only remaining meaningful roles available to them. In addition, some teenagers feel lonely and unloved and have a baby to create a sense of being needed and wanted. In contrast, in Sweden, eligibility requirements for welfare payments make it almost necessary to complete an education and get a job before becoming a parent.

Problems Associated with Teenage Motherhood

Teenage parenthood is associated with various negative consequences, including the following:

1. **Stigmatization and marginalization.** Wilson and Huntington (2006) noted that, because teen mothers resist the typical life trajectory of their middle-class peers, they are stigmatized and marginalized. In effect, they are a threat to societal goals of economic growth through higher education and increased female workforce participation. In spite of such stigmatization and marginalization, McDermott and Graham (2005) noted the resilient behaviors of teen mothers: they invest in the "good" mother identity, maintain kin relations, and prioritize the mother-child dyad in their life. Rolfe (2008) interviewed thirty-three young women who were mothers before the age of 21 and discovered three themes of their experience of teenage motherhood—as "hardship and reward," "growing up and responsibility," and "doing things differently." The researcher noted that the respondents were "active in negotiating and constructing their own identities as mothers, careers and women" (p. 299).
2. **Poverty among single teen mothers and their children.** Many teen mothers are unwed. Livermore and Powers (2006) studied a sample of 336 unwed mothers and found them plagued with financial stress; almost 20 percent had difficulty providing food for themselves and their children (18.5 percent), had their electricity cut off for nonpayment (19.7 percent), and had no medical care for their children (18.2 percent). Almost half (47 percent) reported experiencing "one or more financial stressors" (p. 6).
3. **Poor health habits.** Teenage unmarried mothers are less likely to seek prenatal care and more likely than older and married women to smoke, drink alcohol, and take other drugs. These factors have an adverse effect on the health of the baby. Indeed, babies born to unmarried teenage mothers are more likely to have low birth weights (less than five pounds, five ounces) and to be born prematurely. Children of teenage unmarried mothers are also more likely to be developmentally delayed. These outcomes are largely a result of the association between teenage unmarried childbearing and persistent poverty.
4. **Lower academic achievement.** Poor academic achievement is both a contributing factor and a potential

© iStockphoto.com

infertility the inability to achieve a pregnancy after at least one year of regular sexual relations without birth control, or the inability to carry a pregnancy to a live birth.

fertilization (conception) the fusion of the egg and sperm.

pregnancy a condition that begins five to seven days after conception, when the fertilized egg is implanted (typically in the uterine wall).

Fertell an at-home fertility kit that allows women to measure the level of their follicle-stimulating hormone on the third day of their menstrual cycle and men to measure the concentration of motile sperm.

outcome of teenage parenthood. Some studies note that between 30 percent and 70 percent of teen mothers drop out of high school before graduation (the schools may push them out and/or they may no longer feel motivated). Mollborn (2007) confirmed that teen parenthood diminished the chance that the mother would complete high school.

Zachry (2005) interviewed nineteen mothers and noted that, although all dropped out of school, each evidenced a new appreciation for education as a way of providing a better future for their child. Wendy, one of the mothers, said, "I want to better my education for my kids, and myself . . . because I'm their role model. And they're only gonna learn from what they see from me" (p. 2566).

Although it is assumed that children of teen mothers do poorly in school, Levine et al. 2007 provided longitudinal data and noted no causal relationship between having a teenager as a mother and doing poorly on academic tests. Hence, dire predictions for youngsters of teen mothers might be revised.

9.4 Infertility

Infertility is defined as the inability to achieve a pregnancy after at least one year of regular sexual relations without birth control, or the inability to carry a pregnancy to a live birth. Different types of infertility include the following:

There is also an epidemic of infertility in this country. There are more women who have put off childbearing in favor of their professional lives.

—*Iris Chang, historian*

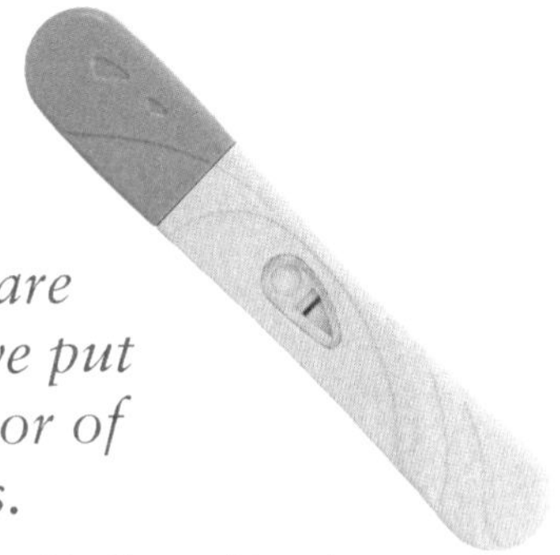

© Libby Chapman/iStockphoto.com

1. **Primary infertility.** The woman has never conceived even though she wants to and has had regular sexual relations for the past twelve months.
2. **Secondary infertility.** The woman has previously conceived but is currently unable to do so even though she wants to and has had regular sexual relations for the past twelve months.
3. **Pregnancy wastage.** The woman has been able to conceive but has been unable to produce a live birth.

Causes of Infertility

Although popular usage does not differentiate between the terms *fertilization* and the *beginning of pregnancy*, **fertilization** or **conception** refers to the fusion of the egg and sperm, whereas **pregnancy** is not considered to begin until five to seven days later, when the fertilized egg is implanted (typically in the uterine wall). Hence, not all fertilizations result in a pregnancy. An estimated 30 percent to 40 percent of conceptions are lost prior to or during implantation.

Forty percent of infertility problems are attributed to the woman, 40 percent to the man, and 20 percent to both of them. Some of the more common causes of infertility in men include low sperm production, poor semen motility, effects of STIs (such as chlamydia, gonorrhea, and syphilis), and interference with passage of sperm through the genital ducts due to an enlarged prostate. The causes of infertility in women include blocked fallopian tubes, endocrine imbalance that prevents ovulation, dysfunctional ovaries, chemically hostile cervical mucus that may kill sperm, and effects of STIs.

An at-home fertility kit, **Fertell**, allows women to measure the level of their follicle-stimulating hormone on the third day of their menstrual cycles. An abnormally high level means that egg quality is low. The test takes thirty minutes and involves a urine stick. The same kit allows men to measure the concentration of motile sperm. Men provide a sample of sperm (for example, via masturbation) that swim through a solution similar to cervical mucus. This procedure takes about eighty minutes. Fertell has been approved by the Food and Drug Administration (FDA), no prescription is necessary, and costs around $100.

Being infertile (for the woman) may have a negative lifetime effect. Wirtberg et al. (2007) interviewed fourteen Swedish women twenty years after their infertility treatment and found that childlessness had had a major impact on all the women's lives and remained a major life theme. The effects were both personal (sad) and interpersonal (half were separated and all reported negative effects on their sex lives). The effects of childlessness were especially increased at the time the study was conducted, as the women's peer group was entering the "grandparent phase." The researchers noted that infertility has lifetime consequences for the individual woman and her relationships.

Assisted Reproductive Technology

A number of technological innovations are available to assist women and couples in becoming pregnant. These include hormonal therapy, artificial insemination, ovum transfer, in vitro fertilization, gamete intrafallopian transfer, and zygote intrafallopian transfer.

Hormone Therapy Drug therapies are often used to treat hormonal imbalances, induce ovulation, and correct problems in the luteal phase of the menstrual cycle. Frequently used drugs include Clomid, Pergonal, and human chorionic gonadotropin (HCG), a hormone extracted from human placenta. These drugs stimulate the ovary to ripen and release an egg. Although they are fairly effective in stimulating ovulation, hyperstimulation can occur, which may result in permanent damage to the ovaries.

Hormone therapy also increases the likelihood that multiple eggs will be released, resulting in multiple births. The increase in triplets and higher order multiple births over the past decade in the United States is largely attributed to the increased use of ovulation-inducing drugs for treating infertility. Infants of higher order multiple births are at greater risk of having low birth weight and their mortality rates are higher. Mortality rates have improved for these babies, but these low birth-weight survivors may need extensive neonatal medical and social services.

Artificial Insemination When the sperm of the male partner are low in count or motility, sperm from several ejaculations may be pooled and placed directly into the cervix. This procedure is known as *artificial insemination by husband* (AIH). When sperm from someone other than the woman's partner are used to fertilize a woman, the technique is referred to as *artificial insemination by donor* (AID).

Lesbians who want to become pregnant may use sperm from a friend or from a sperm bank (some sperm banks cater exclusively to lesbians). Regardless of the source of the sperm, it should be screened for genetic abnormalities and STIs, quarantined for 180 days, and retested for human immunodeficiency virus (HIV); also, the donor should be younger than 50 to diminish hazards related to aging. These precautions are not routinely taken—let the buyer beware.

How do children from donor sperm feel about their fathers? A team of researchers (Scheib et al. 2005) studied twenty-nine individuals (41 percent from lesbian couples, 38 percent from single women, and 21 percent from heterosexual couples) and found that most (75 percent) always knew about their origin and were comfortable with it. All but one reported a neutral to positive impact with the birth mother. Most (80 percent) indicated a moderate interest in learning more about the donor. No youths reported wanting money, and only 7 percent reported wanting a father-child relationship. Berger and Paul (2008) studied the effects of disclosing or not disclosing to the child that he or she is from a donor sperm. The results were inconclusive but favored disclosure.

Artificial Insemination of a Surrogate Mother In some instances, artificial insemination does not help a woman get pregnant. (Her fallopian tubes may be blocked, or her cervical mucus may be hostile to sperm.) The couple that still wants a child and has decided against adoption may consider parenthood through a surrogate mother. There are two types of surrogate mothers. One is the contracted surrogate mother who supplies the egg, is impregnated with the male partner's sperm, carries the child to term, and gives the baby to the man and his partner. A second type is the surrogate mother who carries to term a baby to whom she is not genetically related (a fertilized egg from the "infertile couple" who can't carry a baby to term is implanted in her uterus). As with AID, the motivation of the prospective parents is to have a child that is genetically related to at least one of them. For the surrogate mother, the primary motivation is to help childless couples achieve their aspirations of parenthood and to make money. Although some American women are

cryopreservation fertilized eggs are frozen and implanted at a later time.

willing to "rent their wombs," women in India have also begun to provide this service. For $5,000, an Indian wife who already has a child will carry a baby to term for an infertile couple (for a fraction of the cost of an American surrogate).

California is one of twelve states in which entering into an arrangement with a surrogate mother is legal. The fee to the surrogate mother is $20,000 to $25,000. Other fees (travel, hospital, lawyers, and so on) can run the figure to as high as $125,000. Surrogate mothers typically have their own children, making giving up a child that they carried easier. For information about the legality of surrogacy in your state, see http://www.surrogacy.com/legals/map.html.

In Vitro Fertilization About 2 million couples cannot have a baby because the woman's fallopian tubes are blocked or damaged, preventing the passage of eggs to the uterus. In some cases, blocked tubes can be opened via laser surgery or by inflating a tiny balloon within the clogged passage. When these procedures are not successful (or when the woman decides to avoid invasive tests and exploratory surgery), *in vitro* (meaning "in glass") *fertilization* (IVF), also known as test-tube fertilization, is an alternative.

Using a laparoscope (a narrow, telescope-like instrument inserted through an incision just below the woman's naval to view tubes and ovaries), the physician is able to see a mature egg as it is released from the woman's ovary. The time of release can be predicted accurately within two hours. When the egg emerges, the physician uses an aspirator to remove the egg, placing it in a small tube containing stabilizing fluid. The egg is taken to the laboratory, put in a culture petri dish, kept at a certain temperature-acidity level, and surrounded by sperm from the woman's partner (or donor). After one of these sperm fertilizes the egg, the egg divides and is implanted by the physician in the wall of the woman's uterus. Usually, several eggs are implanted in the hope one will survive. This was the case of Nadya Suleman, who ended up giving birth to 8 babies. Eight embryos were transferred in 2008 at Duke University's in vitro fertilization program into her body with the thought that some would not survive . . . all did (Rochman 2009).

Some couples want to ensure the sex of their baby. In a procedure called "family balancing," which is used by couples who already have several children of one sex, the eggs of a woman are fertilized and the sex of the embryos three and eight days old is identified. Only those of the desired sex are then implanted in the woman's uterus.

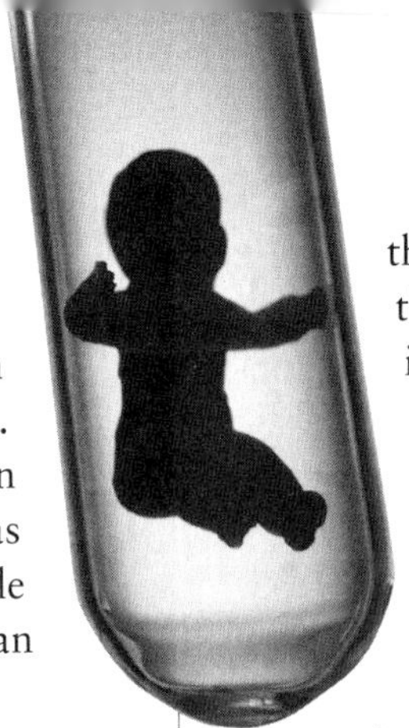

© BananaStock/Jupiterimages

Alternatively, the Y chromosome of the male sperm can be identified and implanted. The procedure is accurate 75 percent of the time for producing a boy baby and 90 percent of the time for a girl baby. The Genetics and IVF Institute, in Fairfax, Virginia, specializes in the sperm-sorting technique.

Occasionally, some fertilized eggs are frozen and implanted at a later time, if necessary. This procedure is known as **cryopreservation**. Separated or divorced couples may disagree over who owns the frozen embryos, and the legal system is still wrestling with the fate of their unused embryos, sperm, or ova after a divorce or death.

Ovum Transfer In conjunction with in vitro fertilization is ovum transfer, also referred to as embryo transfer. In this procedure, an egg is donated, fertilized in vitro with the husband's sperm, and then transferred to his wife. Alternatively, a physician places the sperm of the male partner in a surrogate woman. After about five days, her uterus is flushed out (endometrial lavage), and the contents are analyzed under a microscope to identify the presence of a fertilized ovum.

The fertilized ovum is then inserted into the uterus of the otherwise infertile partner. Although the embryo can also be frozen and implanted at another time, fresh embryos are more likely to result in successful implantation. Infertile couples that opt for ovum transfer do so because the baby will be biologically related to at least one of them (the father) and the partner will have the experience of pregnancy and childbirth. As noted earlier, the surrogate woman participates out of her desire to help an infertile couple or to make money.

Other Reproductive Technologies A major problem with in vitro fertilization is that only about 15 percent to 20 percent of the fertilized eggs will implant on the uterine wall. To improve this implant percentage (to between 40 percent and 50 percent), physicians place the egg and the sperm directly into the fallopian tube, where they meet and fertilize. Then the fertilized egg travels down into the uterus and implants.

Because the term for sperm and egg together is *gamete*, this procedure is called *gamete intrafallopian transfer*, or GIFT. This procedure, as well as in vitro fertilization, is not without psychological costs to the couple.

Gestational surrogacy, another technique, involves fertilization in vitro of a woman's ovum and transfer to a surrogate. Trigametic IVF also involves the use of sperm in which the genetic material of another person has been inserted. This technique allows lesbian couples to have a child genetically related to both women. Infertile couples hoping to get pregnant through one of the more than 400 in vitro fertilization clinics should make informed choices by asking questions such as, "What is the center's pregnancy rate for women with a similar diagnosis?"

What percentage of these women has a live birth? According to the Centers for Disease Control and Prevention, the typical success rate (live birth) for infertile couples who seek help in one of the 400 fertility clinics is 28 percent (Lee 2006). Beginning assisted-reproductive technology as early after infertility is suspected is important. Wang et al. (2008) analyzed data on 36,412 patients to assess success of actual births for infertile women using assisted-reproductive technology and found that, for women age 30 and above, each additional year in age was associated with an 11 percent reduction in the chance of achieving pregnancy and a 13 percent reduction in the chance of a live delivery. If women aged 35 years or older would have had their first treatment one year earlier, 15 percent more live deliveries would be expected.

> The typical success rate (live birth) for infertile couples who seek help in one of the 400 fertility clinics is 28 percent (Lee 2006).

Finally, Hammarberg et al. (2008) studied 166 women who had conceived through assisted-reproductive technology to identify any differences in birthing. They did find that ART participants were more likely to have a cesarean birth (51 percent versus 25 percent) and to report disappointment with the birth event when compared with those who had a vaginal birth.

9.5 Planning for Adoption

Angelina Jolie and Brad Pitt are celebrities who have given national visibility to adopting children. They are not alone in their desire to adopt children. The various routes to adoption are public (children from the child welfare system), private agency (children placed with nonrelatives through agencies), independent adoption (children placed directly by birth parents or through an intermediary such as a physician or attorney), kinship (children placed in a family member's home), and stepparent (children adopted by a spouse). Motives for adopting a child include wanting a child because of an inability to have a biological child (infertility), a desire to give an otherwise unwanted child a permanent loving home, or a desire to avoid contributing to overpopulation by having more biological children. Some couples may seek adoption for all of these motives. Adoption is actually quite rare, with less than 5 percent of couples adopting; 15 percent of these adoptions will be children from other countries.

Demographic Characteristics of People Seeking to Adopt a Child

Whereas demographic characteristics of those who typically adopt are white, educated, and high-income, adoptees are being increasingly placed in nontraditional families including with older, gay, and single individuals. Sixteen states have taken steps to ban adoption by gay couples on the grounds that, because "marriage" is "heterosexual marriage," children do not belong in homosexual relationships (Stone 2006). Leung et al. (2005) compared children adopted or reared by gay or lesbian and heterosexual parents. They found no negative effects when the adoptive parents were gay or lesbian. Approval for adoption by same-sex couples is evident in the population, as half of the respondents in surveys in the United States report approval of same-sex adoptions (Maill and March 2005).

Characteristics of Children Available for Adoption

Adoptees in the highest demand are healthy, white infants. Those who are older, of a racial or ethnic group different from that of the adoptive parents, of a sibling group, or with physical or developmental disabilities have been difficult to place. Flower Kim (2003) noted that, because the waiting period for a healthy white infant is from five to ten years, couples are increasingly open to cross-racial adoptions. Of the 1.6 million adopted children younger than 18 living in U.S. households, the percentages adopted from other countries are as follows:

transracial adoption the practice of parents adopting children of another race.

24 percent from Korea; 11 percent from China; 10 percent from Russia; and 9 percent from Mexico. International or cross-racial adoptions may complicate the adoptive child's identity. Children adopted after infancy may also experience developmental delays, attachment disturbances, and post-traumatic stress disorder (Nickman et al. 2005). Baden and Wiley (2007) reviewed the literature on adoptees as adults and found that the mental health of most was on par with those who were not adopted. However, a small subset of the population showed concerns that may warrant therapeutic intervention.

Costs of Adoption

Adopting from the U.S. foster care system is generally the least expensive type of adoption, usually involving little or no cost, and states often provide subsidies to adoptive parents. However, a couple can become foster care parents to a child and become emotionally bonded with the child, and the birth parents can reappear and request their child back.

Stepparent and kinship adoptions are also inexpensive and have less risk of the child being withdrawn. Agency and private adoptions can range from $5,000 to $40,000 or more, depending on travel expenses, birth mother expenses, and requirements in the state. International adoptions can range from $7,000 to $30,000 (see http://costs.adoption.com/).

Transracial Adoption

Transracial adoption is defined as the practice of parents adopting children of a different race—for example, a white couple adopting a Korean or African American child. In a study on transracial adoption attitudes of college students, the scores of the 188 respondents reflected overwhelmingly positive attitudes toward transracial adoption. Overall, women, people willing to adopt a child at all, interracially experienced daters, and those open to interracial dating were more willing to adopt transracially than were men, people rejecting adoption as an optional route to parenthood, people with no previous interracial dating experience, and people closed to interracial dating (Ross et al. 2003).

Ethiopia has become a unique country from which to adopt a child. Not only is the adoption time shorter (four months) and less expensive, but the children there are also psychologically very healthy. "You don't hear crying babies [in the orphanages] . . . they are picked up immediately" (Gross and Conners 2007, A16). In addition, "adoption families are encouraged to meet birth families and visit the villages where the children are raised . . ." (ibid.). Ethiopian adoptions have received considerable visibility in the United States due to the involvement of celebrity Angelina Jolie, who adopted there.

Transracial adoptions are controversial. Wolters et al. (Forthcoming) analyzed data from 1,027 undergraduates at a large undergraduate university and found that females were 17.5 percent more supportive of transracial adoption than males ($p > .001$). Black females were the most favorable toward transracial adoption, whereas white males were the least favorable. Kennedy (2003) noted, "Whites who seek to adopt black children are widely regarded with suspicion. Are they ideologues, more interested in making a political point than in actually being parents?" (p. 447). Another controversy is whether it is beneficial for children to be adopted by parents of the same racial background. In regard to the adoption of African American children by same-race parents, the National Association of Black Social Workers (NABSW) passed a resolution against transracial adoptions, citing that such adoptions prevented black children from developing a positive sense of themselves

© Nathan Gleave/iStockphoto.com / © Comstock Images/Jupiterimages

"that would be necessary to cope with racism and prejudice that would eventually occur" (Hollingsworth 1997, 44).

The counterargument is that healthy self-concepts, an appreciation for one's racial heritage, and coping with racism or prejudice can be learned in a variety of contexts. Legal restrictions on transracial adoptions have disappeared, and social approval for transracial adoptions is increasing. However, a substantial number of studies conclude that "same-race placements are preferable and that special measures should be taken to facilitate such placements, even if it means delaying some adoptions" (Kennedy 2003, 469).

One 26-year-old black female was asked how she felt about being reared by white parents and replied, "Again, they are my family and I love them, but I am black. I have to deal with my reality as a black woman" (Simon and Roorda 2000, 41). A black man reared in a white home advised white parents considering a transracial adoption, "Make sure they have the influence of blacks in their lives; even if they have to go out and make friends with black families—it's a must" (p. 25). Indeed, Huh and Reid (2000) found that positive adjustment by adoptees was associated with participation in the cultural activities of the race of the parents who adopted them. Thomas and Tessler (2007) found that American parents intent on keeping the Chinese cultural heritage of their adopted child alive take specific steps (for example, establish friendships with Chinese adults and families).

Open versus Closed Adoptions

Another controversy is whether adopted children should be allowed to obtain information about their biological parents. Surveys in both Canada and the United States reveal that about three-fourths of the respondents approve of some form of open adoption and of giving adult adoptees unlimited access to confidential information about their birth parents (Maill and March 2005). In general, there are considerable benefits for having an open adoption—the biological parent has the opportunity to stay involved in the child's life. Adoptees learn early that they are adopted and who their biological parents are. Birth parents are more likely to avoid regret and to be able to stay in contact with their child. Adoptive parents have information about the genetic background of their adopted child. Ge et al. (2008) studied birth mothers and adoptive parents and found that increased openness between the two sets of parents was positively associated with greater satisfaction for both birth mothers and adoptive parents.

Another controversy is whether adopted children should be allowed to obtain information about their biological parents

9.6 Foster Parenting

Some individuals seek the role of parent via foster parenting. A **foster parent**, also known as a **family caregiver**, is neither a biological nor an adoptive parent but is a person who takes care of and fosters a child taken into custody. A foster parent has made a contract with the state for the service, has judicial status, and is reimbursed by the state. Foster parents are screened for previous arrest

foster parent (family caregiver) a person who either alone or with a spouse, takes care of and fosters a child taken into custody.

induced abortion the deliberate termination of a pregnancy through chemical or surgical means.

spontaneous abortion (miscarriage) an unintended termination of a pregnancy.

abortion rate the number of abortions per 1,000 women aged 15 to 44.

abortion ratio the number of abortions per 1,000 live births.

parental consent woman needs permission from parent to get an abortion if under a certain age, usually 18.

parental notification woman required to tell parents she is getting an abortion if she is under a certain age, usually 18; but she does not need parental permission.

records and child abuse and neglect. Foster parents are licensed by the state; some states require a "foster parent orientation" program. Rhode Island, for example, provides a twenty-seven-hour course. Brown (2008) asked sixty-three foster parents what they needed to allow them to have a successful foster parenting experience. They reported that they needed the right personality (for example, patience and nurturance), information about the foster child, a good relationship with the fostering agency, linkages to other foster families, and supportive immediate and extended families. Other research has found the need for formal foster parent organizations.

Children placed in foster care have typically been removed from parents who are abusive, who are substance abusers, and/or who are mentally incompetent. Although foster parents are paid for taking care of children in their home, they are also motivated by love of children. The goal of placing children in foster care is to remove them from a negative family context, improve that context, and return them, or find a more permanent home than foster care. Some couples become foster parents in hopes of being able to adopt a child who is placed in their custody.

Due to tighter restrictions on foreign adoptions (for example, it typically takes three years to complete a foreign adoption; China excludes people seeking to adopt who are unmarried, obese, and over age 50, and due to the limited number of domestic infants, more couples are considering adoption of a foster child. Tax credits are available for up to $11,650 for adopting a child with special needs (Block 2008).

9.7 Abortion

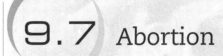

An abortion may be either an **induced abortion**, which is the deliberate termination of a pregnancy through chemical or surgical means, or a **spontaneous abortion (miscarriage)**, which is the unintended termination of a pregnancy. In this text, the term *abortion* refers to an induced abortion. In general, abortion is legal in the United States, though varying support from different administrations can affect levels of federal funding to groups offering abortions or abortion advice.

Incidence of Abortion

Table 9.1 reflects the abortion rates and the number of abortions in the United States for selected years. **Abortion rate** refers to the number of abortions per 1,000 women aged 15 to 44; **abortion ratio** refers to the number of abortions per 1,000 live births. Whether looking at the rate, ratio, or actual number, abortions are decreasing. Reasons for the decrease in abortions include increased access to contraceptives, a reduced rate of unintended pregnancies, more supportive attitudes toward women becoming single parents, and an increase in restrictive abortion policies, such as those requiring parental consent and parental notification. **Parental consent** means that a woman needs permission from a parent to get an abortion if under a certain age, usually 18. **Parental notification** means that a woman has to tell a parent she is getting an abortion if she is under a certain age, usually 18, but she doesn't need parental permission. Laws vary by states. Table 9.2 lists each state's consent laws, but you can also call the National Abortion Federation Hotline at 1-800-772-9100 for laws in your state.

Table 9.1
Abortion Rate and Number for Selected Years

Year	Number of abortions per 1,000 women aged 15 to 44	Number of abortions per year
2000	21.3	1,312,990
2004	19.7	1,222,100
2005	19.4	1,206,200

Source: *Statistical Abstract of the United States, 2009.* 128th ed. Washington, DC. U.S. Bureau of the Census. Table 99.

Table 9.2
Parental Consent by State

Key
PC = Parental Consent
PN = Parental Notification
NONE = No PC or PN are required now
2 = Both parents required

State	Status	Comments
Alabama	PC	
Alaska	NONE	PN law stopped by court
Arizona	PC	
Arkansas	PC	
California	NONE	PC law stopped by court
Colorado	PN	
Connecticut	NONE	
Delaware	PN	Applies only to girls under 16 years old; notice may also be to grandparent or counselor; doctor can bypass
Washington	NONE	
Florida	PN	
Georgia	PN	
Hawaii	NONE	
Idaho	PC	PC law is back as of March 2007
Illinois	NONE	PN law stopped by court
Indiana	PC	
Iowa	PN	Also allows consent of grandparent instead
Kansas	PN	
Kentucky	PC	
Louisiana	PC	
Maine	NONE	
Maryland	PN	Doctor can bypass
Massachusetts	PC	
Michigan	PC	
Minnesota	PN2	
Mississippi	PC2	
Missouri	PC	
Montana	NONE	PN stopped by court
Nebraska	PN	
Nevada	NONE	PN stopped by court
New Hampshire	NONE	
New Jersey	NONE	PN law stopped by court
New Mexico	NONE	PC law stopped by court
New York	NONE	
North Carolina	PC	Also allows consent of grandparent instead
North Dakota	PC2	
Ohio	PC	
Oklahoma	PN and PC	Law requiring both started November 2006
Oregon	NONE	
Pennsylvania	PC	
Rhode Island	PC	
South Carolina	PC	Women under 17 years old; also allows for consent of grandparent
South Dakota	PN	
Tennessee	PC	
Texas	PC	Law enacted by Gov. George W. Bush in October 1999
Utah	PN and PC	
Virginia	PC	Also allows consent of grandparent instead
Vermont	NONE	
Washington	NONE	
West Virginia	PN	Doctor can bypass
Wisconsin	PC	Also allows other family members over 25 to consent; doctor can bypass
Wyoming	PC	

Source: Coalition for Positive Sexuality. http://www.positive.org/Resources/consent.html (retrieved August 18, 2008).

therapeutic abortion an abortion performed to protect the life or health of a woman.

Reasons for an Abortion

A team of researchers (Finer et al. 2005) surveyed 1,209 women who reported having had an abortion. The most frequently cited reasons were that having a child would interfere with a woman's education, work, or ability to care for dependents (74 percent); that she could not afford a baby now (73 percent); and that she did not want to be a single mother or was having relationship problems (48 percent). Nearly four in ten women said they had completed their childbearing, and almost one-third of the women were not ready to have a child. Fewer than 1 percent said their parents' or partner's desire for them to have an abortion was the most important reason.

Abortions performed to protect the life or health of the woman are called **therapeutic abortions**. However, there is disagreement over this definition. Garrett et al. (2001) noted, "Some physicians argue that an abortion is therapeutic if it prevents or alleviates a serious physical or mental illness, or even if it alleviates temporary emotional upsets. In short, the health of the pregnant woman is given such a broad definition that a very large number of abortions can be classified as therapeutic" (p. 218).

Some women with multifetal pregnancies (a common outcome of the use of fertility drugs) may have a procedure called *transabdominal first-trimester selective termination*. In this procedure, the lives of some fetuses are terminated to increase the chance of survival for the others or to minimize the health risks associated with multifetal pregnancy for the woman. For example, a woman carrying five fetuses may elect to abort three of them to minimize the health risks of the other two.

Pro-Life and Pro-Choice Abortion Positions

A dichotomy of attitudes toward abortion is reflected in two opposing groups of abortion activists. Individuals and groups who oppose abortion are commonly referred to as "pro-life" or "antiabortion."

Pro-Life Of 657 undergraduates in a random sample at a large southeastern university, 40 percent reported that they were pro-life in regard to their feelings about abortion (Bristol and Farmer 2005). Pro-life groups favor anti-abortion policies or a complete ban on abortion. They essentially believe the following:

© Larry W. Smith/epa/Corbis

- The unborn fetus has a right to live and that right should be protected.
- Abortion is a violent and immoral solution to unintended pregnancy.
- The life of an unborn fetus is sacred and should be protected, even at the cost of individual difficulties for the pregnant woman.

Individuals who are over the age of 44, female, mothers of three or more children, married to white-collar workers, affiliated with a religion, and Catholic are most likely to be pro-life (Begue 2001). Pro-life individuals emphasize the sanctity of human life and the moral obligation to protect it. The unborn fetus cannot protect itself so is literally dependent on others for life. Naomi Judd noted that if she had had an abortion she would have deprived the world of one of its greatest singers—Wynonna Judd.

Pro-Choice In a large, nonrandom sample, 60.1 percent of 1,319 undergraduates at a southeastern university reported that, "abortion is acceptable under certain conditions" (Knox and Zusman 2009). Pro-choice advocates support the legal availability of abortion for all women. They essentially believe the following:

- Freedom of choice is a central value—the woman has a right to determine what happens to her own body.
- Those who must personally bear the burden of their moral choices ought to have the right to make these choices.
- Procreation choices must be free of governmental control.

People most likely to be pro-choice are female, are mothers of one or two children, have some college education, are employed, and have annual income of more than $50,000. Although many self-proclaimed feminists and women's organizations, such as the National Organization for Women (NOW), have been active in promoting abortion rights, not all feminists are pro-choice.

Physical Effects of Abortion

Part of the debate over abortion is related to the presumed effects of abortion. In regard to the physical effects, legal abortions, performed under safe medical conditions, are safer than continuing the pregnancy. The earlier in the pregnancy the abortion is performed, the safer it is. Vacuum aspiration, a frequently used method in early pregnancy, does not increase the risks to future childbearing. However, late-term abortions do increase the risks of subsequent miscarriages, premature deliveries, and babies of low birth weight.

Postabortion complications include the possibility of incomplete abortion, which occurs when the initial procedure misses the fetus and the procedure must be repeated. Other possible complications include uterine infection; excessive bleeding; perforation or laceration of the uterus, bowel, or adjacent organs; and an adverse reaction to a medication or anesthetic. After having an abortion, women are advised to expect bleeding (usually not heavy) for up to two weeks and to return to their health care provider thirty days after the abortion to check that all is well.

Psychological Effects of Abortion

Of equal concern are the psychological effects of abortion. The American Psychological Association reviewed all outcome studies on the mental health effects of abortion and concluded, "Based on our comprehensive review and evaluation of the empirical literature published in peer-reviewed journals since 1989, this Task Force on Mental Health and Abortion concludes that the most methodologically sound research indicates that among women who have a single, legal, first-trimester abortion of an unplanned pregnancy for nontherapeutic reasons, the relative risks of mental health problems are no greater than the risks among women who deliver an unplanned pregnancy" (Major et al. 2008, 71). Steinberg and Russo (2008) also looked at national data and did not find a significant relationship between first pregnancy abortion and subsequent rates of generalized anxiety disorder, social anxiety, or post-traumatic stress disorder.

Postabortion Attitudes of Men

Researchers Kero and Lalos (2004) conducted interviews with men four and twelve months after their partners had had an abortion. Overwhelmingly, the men (at both time periods) were happy with the decision of their partners to have an abortion. More than half accompanied their partner to the abortion clinic (which they found less than welcoming); about a third were not using contraception a year later.

Parenting

SECTIONS

parenting the provision by an adult or adults of physical care, emotional support, instruction, and protection from harm in an ongoing structural (home) and emotional context to one or more dependent children.

There is a story about Picasso's mother, who had high ambitions for him. She told him if he were to become a soldier that he would be a general and if he were to be a monk he would end up as the pope. But alas, he became a painter and ended up Picasso. Such is the enormous influence of parents (Fadiman 1985, 451). Although there are guidelines for effective parenting (and we will review them in this chapter), a number of influential factors (e.g. genetics, peers, health, economics) are beyond our control. Nevertheless, our focus in this chapter is on wise parenting choices, with the goal of facilitating happy, economically independent, socially contributing members to our society. We begin by looking at the various roles of parenting.

10.1 Roles Involved in Parenting

Although finding one definition of **parenting** is difficult, there is general agreement about the various roles parents play in the lives of their children. New parents assume at least seven roles:

A happy family is but an earlier heaven.
—*George Bernard Shaw*

1. **Caregiver.** A major responsibility of parents is the physical care of their children. From the moment of birth, when infants draw their first breath, parents stand ready to provide nourishment (milk), cleanliness (diapers), and temperature control (warm blanket). The need for such sustained care continues and becomes an accepted and anticipated role of parents. Parents who

excuse themselves early from a party because they "need to check on the baby" are alerting the hostess of their commitment to the role of caregiver.

2. **Emotional resource.** Beyond providing physical care, parents are sensitive to the emotional needs of children in terms of their need to belong, to be loved, and to develop positive self-concepts. In hugging, holding, and kissing an infant, parents not only express their love for the infant but also reflect an awareness that such display of emotion is good for the child's sense of self-worth. Kouros et al. (2008) found that children exhibit increased emotional insecurity when their parents are in conflict or depression. The family context is the emotional context for children. Strife or depression in this context does occur without a negative effect on the children.
3. **Teacher.** All parents think they have a philosophy of life or set of principles their children will benefit from. Parents later discover that their children may not be interested in their religion or philosophy—indeed, they may rebel against it. This possibility does not

deter them from their role as teacher. Children are forever learning from their parents, most often by observing their behavior.

Parents also feel that their role is made more difficult by an increasingly liberal society. A sample of 2,020 Americans noted that one of the biggest problems confronting parents today is the societal influence on their children. These include drugs and alcohol; peer pressure; television, Internet, and movies; and crime and gangs (Pew Research Center 2007).

What the world needs is not romantic lovers but husbands and wives who willingly give their time and attention to their children.

—Margaret Mead, anthropologist

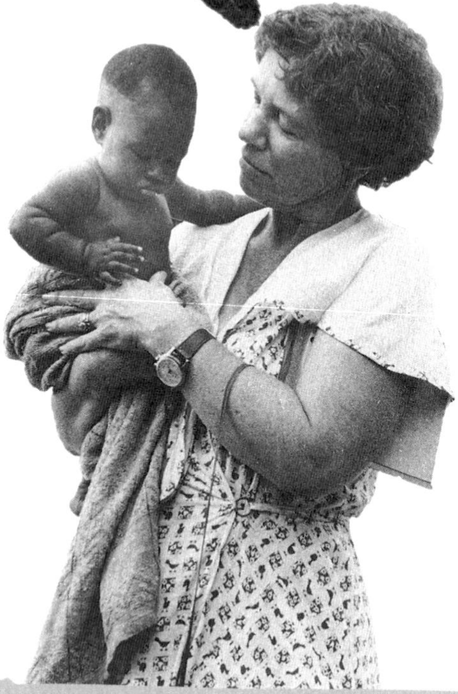

© iStockphoto.com / © AP Photo

Parents may also teach without awareness. Cui et al. (2008) noted that parents (whether together or divorced) who reported high conflict in their marriage tended to have young adult offspring who also reported high conflict and low quality in their own romantic relationships.

4. **Economic resource.** New parents are also acutely aware of the costs for medical care, food, and clothes for infants, and seek ways to ensure that such resources are available to their children. Working longer hours, taking second jobs, and cutting back on leisure expenditures are some of the common attempts parents make to ensure that money is available to meet the needs of the child. Sometimes the pursuit of money for the family has a negative consequence for children. Rapoport and Le Bourdais (2008) investigated the effects of parents' working schedules on the time they devoted to their children and confirmed that the more parents worked, the less time they spent with their children. In view of extensive work schedules, parents are under pressure to spend "quality time" with their children, and it is implied that putting children in day care robs children of this time. However, Booth et al. (2002) compared children in day care with those in home care in terms of time the mother and child spent together per week. Although the mothers of children in day care spent less time with their children than the mothers who cared for their children at home, the researchers concluded that the "groups did not differ in the quality of mother-infant interaction" and that the difference in the "quality of the mother-infant interaction may be smaller than anticipated" (p. 16).

   Parents provide an economic resource for their children by providing free room and board for them. Some young adults continue to live with their parents well into adulthood and may return at other times, such as following a divorce, job loss, and so on.

5. **Protector.** Parents also feel the need to protect their children from harm. This role may begin in

WHILE WE TRY TO TEACH OUR CHILDREN ALL ABOUT LIFE, OUR CHILDREN TEACH US WHAT LIFE IS ALL ABOUT.

— Angela Schwindt, author

MAMA SAYS

Things your mother taught you (Internet humor):

1. My mother taught me to *appreciate a job well done*: "If you're going to kill each other, do it outside. I just finished cleaning."
2. My mother taught me *religion*: "You better pray that will come out of the carpet."
3. My mother taught me about *time travel*: "If you don't straighten up, I'm going to knock you into the middle of next week!"
4. My mother taught me *logic*: "Because I said so, that's why."
5. My mother taught me *more logic*: "If you fall out of that swing and break your neck, you're not going to the store with me."
6. My mother taught me *foresight*: "Make sure you wear clean underwear, in case you're in an accident."
7. My mother taught me *irony*: "Keep crying, and I'll give you something to cry about."
8. My mother taught me about the science of *osmosis*: "Shut your mouth and eat your supper."
9. My mother taught me about *contortionism*: "Will you look at that dirt on the back of your neck!"
10. My mother taught me about *stamina*: "You'll sit there until all that spinach is gone."
11. My mother taught me about *weather*: "This room of yours looks as if a tornado went through it."
12. My mother taught me about *hypocrisy*: "If I told you once, I've told you a million times. Don't exaggerate!"
13. My mother taught me the *circle of life*: "I brought you into this world, and I can take you out."
14. My mother taught me about *behavior modification*: "Stop acting like your father!"
15. My mother taught me about *envy*: "There are millions of less fortunate children in this world who don't have wonderful parents like you do."
16. My mother taught me about *anticipation*: "Just wait until we get home."
17. My mother taught me about *receiving*: "You are going to get it when you get home!"
18. My mother taught me *medical science*: "If you don't stop crossing your eyes, they are going to get stuck that way."
19. My mother taught me about *ESP*: "Put your sweater on; don't you think I know when you are cold?"
20. My mother taught me about my *roots*: "Shut that door behind you. Do you think you were born in a barn?"

oppositional defiant disorder a condition in which children do not comply with requests of authority figures.

pregnancy. Castrucci et al. (2006) interviewed 1,451 women about their smoking behavior during pregnancy. Although 89 percent reduced their smoking during pregnancy, 24.9 percent stopped smoking during pregnancy. Other expressions of the protective role include insisting that children wear seat belts, protecting them from violence or nudity in the media, and protecting them from strangers. Diamond et al. (2006) studied forty middle-class mothers of young children and identified fifteen strategies they used to protect their children. Their three principal strategies were to educate, control, and remove risk. The strategy used depended on the age and temperament of the child. For example, some parents feel protecting their children from certain television content is important. This ranges from families that do not allow a television in their home, to setting the V-chip on their television or allowing only G-rated movies. Research confirms that parents do find media ratings helpful (Bushman and Cantor 2003).

Some parents feel that protecting their children from harm implies appropriate discipline for inappropriate behavior. Galambos et al. (2003) noted that parents who intervened when they saw a negative context developing were able to help their children avoid negative peer influences. Kolko et al. (2008) compared clinically referred boys and girls (ages 6 to 11) diagnosed with **oppositional defiant disorder** (children do not comply with requests of authority figures) to a matched sample of healthy control children and found that the former had greater exposure to delinquent peers. Hence, parents who monitor the peer relationships with their children and minimize the exposure to delinquent models are making a wise time investment.

Increasingly, parents are joining the technological age and learning how to text message. In their role as protector, this allows parents to text-message their child to tell them to come home, phone home, or to work out a logistical problem—"meet me at Gap in the mall." Children can also use text-messaging to their parents to let them know that they arrived safely at a destination, when they need to be picked up, or when they will be home.

6. **Health promotion.** The family is a major agent for health promotion. Children learn from the family context about healthy food. Indeed, one-third of U.S. children are overweight. A major cause of overweight children is parents who do not teach healthy food choices, let their children watch television all day, and are bad models (eat junk food and don't exercise). Health promotion also involves sunburn protection, responsible use of alcohol, and safe driving skills.

7. **Ritual bearer.** To build a sense of family cohesiveness, parents often foster rituals to bind members together in emotion and in memory. Prayer at meals and before bedtime, birthday celebrations, and vacationing at the same place (beach, mountains, and so on) provide predictable times of togetherness and sharing.

10.2 The Choices Perspective of Parenting

Although both genetic and environmental factors are at work, the choices parents make have a dramatic impact on their children. In this section, we review the nature of parental choices and some of the basic choices parents make.

Nature of Parenting Choices

Parents might keep the following points in mind when they make choices about how to rear their children:

1. **Not to make a parental decision is to make a decision.** Parents are constantly making choices even when they think they are not doing so. When a child is impolite and the parent does not provide feedback and encourage polite behavior, the parent has chosen to teach the child that being impolite is acceptable. When a child makes a promise ("I'll call you when I get to my friend's house") and does not do as promised, the parent has chosen to allow the child to not take commitments seriously. Hence, parents cannot choose not to make choices in their parenting, because their inactivity is a choice that has as much impact as a deliberate decision to reinforce politeness and responsibility.

2. **All parental choices involve trade-offs.** Parents are also continually making trade-offs in the parenting choices they make. The decision to take on a second job or to work overtime to afford the larger house will come at the price of having less time to spend with one's children and being more exhausted when such time is available. The choice to enroll one's child in the highest-quality day care (which may also be the most expensive) will mean less money for family vacations. The choice to have an additional child will provide siblings for the existing children but will mean less time and fewer resources for those children. Parents should increase their awareness that no choice is without a tradeoff and should evaluate the costs and benefits in making such decisions.

3. **Reframe "regretful" parental decisions.** All parents regret a previous parental decision (e.g., whether or not they should have held their child back a year in school; they should have intervened in a bad peer relationship; they should have handled their child's drug use differently). Whatever the issue, parents chide themselves for their mistakes. Rather than berate themselves as parents, they might emphasize the positive outcome of their choices: not holding the child back made the child the "first" to experience some things among his or her peers; they made the best decision they could at the time; and so on. Children might also be encouraged to view their own decisions positively.

Five Basic Parenting Choices

The five basic choices parents make include deciding (1) whether to have a child, (2) the number of children, (3) the interval between children, (4) one's method of discipline and guidance, and (5) the degree to which one will be invested in the role of parent. Though all of these decisions are important, the relative importance one places on parenting as opposed to one's career will have implications for the parents, their children, and their children's children. Parents continually make choices in reference to their children, including whether their children will sleep with them in the "family bed."

CAREFUL THE THINGS YOU SAY,
CHILDREN WILL LISTEN.
CAREFUL THE THINGS YOU DO.
CHILDREN WILL SEE. AND LEARN.

—*Stephen Sondheim,* Into the Woods

10.3 Transition to Parenthood

The **transition to parenthood** refers to that period from the beginning of pregnancy through the first few months after the birth of a baby. The mother, father, and couple all undergo changes and adaptations during this period.

Transition to Motherhood

transition to parenthood the period of time from the beginning of pregnancy through the first few months after the birth of a baby.

oxytocin a hormone released from the pituitary gland during the expulsive stage of labor that has been associated with the onset of maternal behavior in lower animals.

baby blues transitory symptoms of depression in mothers twenty-four to forty-eight hours after a baby is born.

Of her transition to motherhood, Madonna noted a dramatic effect, "The whole idea of giving birth and being responsible for another life put me in a different place, a place I'd never been before. I feel like I'm starting life over in many ways. My daughter's birth was like a rebirth for me" (Morton 2003, 213). In effect, the role of mother represents one of the most profound role changes a person ever experiences.

Although childbirth is sometimes thought of as a painful ordeal, some women describe the experience as fantastic, joyful, and unsurpassed. A strong emotional bond between the mother and her baby usually develops early, and both the mother and infant resist separation.

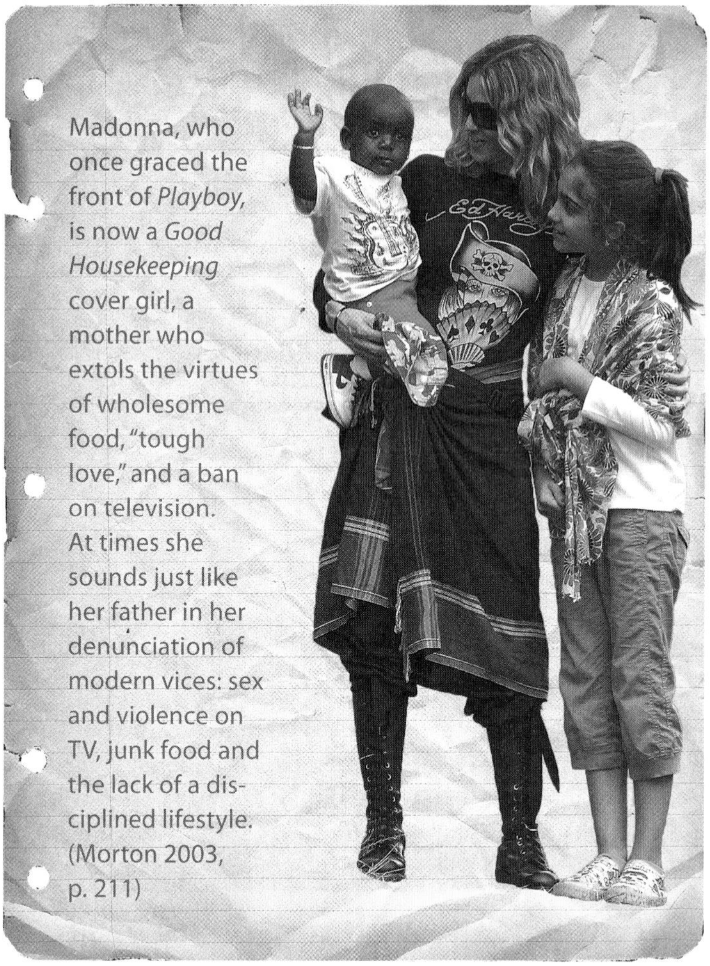

Madonna, who once graced the front of *Playboy,* is now a *Good Housekeeping* cover girl, a mother who extols the virtues of wholesome food, "tough love," and a ban on television. At times she sounds just like her father in her denunciation of modern vices: sex and violence on TV, junk food and the lack of a disciplined lifestyle. (Morton 2003, p. 211)

Sociobiologists suggest that the attachment between a mother and her offspring has a biological basis (one of survival). The mother alone carries the fetus in her body for nine months, lactates to provide milk, and produces **oxytocin**—a hormone from the pituitary gland—during the expulsive stage of labor that has been associated with the onset of maternal behavior in lower animals.

Not all mothers feel joyous after childbirth. Emotional bonding may be temporarily impeded by feeling overworked, exhaustion, mild depression, irritability, crying, loss of appetite, and difficulty in sleeping. Many new mothers experience **baby blues**—transitory symp-

THERE ARE TIMES WHEN PARENTHOOD SEEMS NOTHING MORE THAN FEEDING THE HAND THAT BITES YOU.

— *Peter De Vries, American editor and novelist*

toms of depression twenty-four to forty-eight hours after the baby is born. A few, about 10 percent, experience postpartum depression—a more severe reaction than baby blues.

Postpartum depression is believed to be a result of the numerous physiological and psychological changes occurring during pregnancy, labor, and delivery. Although the woman may become depressed during pregnancy or in the hospital, she more often experiences these feelings within the first month after returning home with her baby (sometimes the woman does not experience postpartum depression until a couple of years later). Most women recover within a short time; some (between 5 percent and 10 percent) become suicidal (Pinheiro et al. 2008).

In *Down Came the Rain*, actress Brooke Shields (2005) recounts her experience with postpartum depression, including this excerpt from the back cover:

> I started to experience a sick sensation in my stomach; it was as if a vise were tightening around my chest. Instead of nervous anxiety that often accompanies panic; a feeling of devastation overcame me. I hardly moved. Sitting on my bed, I let out a deep, slow, guttural wail. I wasn't simply emotional or weepy . . . this was something quite different. This was sadness of a shockingly different magnitude. It felt as if it would never go away.

To minimize baby blues and postpartum depression, antidepressants such as Paxil have been used. Brooke Shields benefited from Paxil, and eventually gave visibility to the issue of postpartum depression, its physiological basis, and the value of medication.

Postpartum psychosis, a reaction in which a new mother wants to harm her baby, is experienced by only one or two women per 1,000 births (British Columbia Reproductive Mental Health Program 2005). One must recognize that having misgivings about a new infant is normal. In addition, a woman who has negative feelings about her new role as mother should elicit help with the baby from her family or other support network so that she can continue to keep up her social contacts with friends and to spend time by herself and with her partner. Regardless of the cause, a team of researchers noted that maternal depression is associated with subsequent antisocial behavior in the child (Kim-Cohen et al. 2005).

Hammarberg et al. (2008) assessed the different parenting experiences of those who had difficulty getting pregnant or who used assisted reproductive technology (ART) compared to those who did not use ART. The researchers concluded that, although the evidence is inconclusive, those couples who become pregnant via ART may possibly idealize parenthood and this might then hinder adjustment and the development of a confident parental identity. Is transition to motherhood similar for lesbian and heterosexual mothers? Not according to Cornelius-Cozzi (2002), who interviewed lesbian mothers and found that the egalitarian norm of the lesbian relationship had been altered; for example, the biological mother became the primary caregiver, and the co-parent, who often heard the biological mother refer to the child as "her child," suffered a lack of validation.

postpartum depression a reaction more severe than the "baby blues" to the birth of one's baby, characterized by crying, irritability, loss of appetite, and difficulty in sleeping.

postpartum psychosis a reaction in which a new mother wants to harm her baby.

gatekeeper role refers to a mother's encouragement or criticism of the father, which influences the degree to which he is involved with his children.

Transition to Fatherhood

Schoppe-Sullivan et al. (2008) emphasized that mothers are the "gatekeepers" of the father's involvement with his children. A father may be involved or not involved with his children to the degree that a mother encourages or discourages a father's involvement. The **gatekeeper role** is particularly pronounced after divorce if the mother ends up with custody of the children (the role of father may be severely limited). When a mother is not present, the role of the father may be enormous. Such was the case of Joe Biden's involvement with his three children. His first wife and young child were killed in an automobile accident that resulted in him becoming the sole parent until his remarriage. One of Biden's sons noted that, "Our dad had his job in Washington but his family in his heart."

The importance of the father in the lives of his children is enormous and goes beyond his economic contribution (Bronte-Tinkew et al. 2008; Flouri and Buchanan 2003; Knox and Legget 2000). Children from homes in which their fathers maintain an active involvement in their lives tend to:

- Make good grades
- Be less involved in crime
- Have good health/ self-concept
- Have a strong work ethic
- Have durable marriages
- Have a strong moral conscience
- Have higher life satisfaction
- Have higher education levels
- Have higher incomes as adults
- Have higher cognitive functioning
- Exhibit fewer anorectic symptoms
- Have fewer premarital births
- Have lower incidences of child sex abuse
- Have stable jobs

Gavin et al. (2002) noted that parental involvement was predicted most strongly by the quality of the parents' romantic relationship. If the father was emotionally and physically involved with the mother, he was more likely to take an active role in the child's life. Fathers whose wives worked more hours than the fathers worked also reported more involvement with their children (McBride et al. 2002).

Thomas et al. (2008) noted the inadequacy of the stereotype of the "uninvolved" African American father due to data that they often do not live in the household with the mother. The researchers sought to redefine father presence in the context of feelings of closeness to the father as well as frequency of father visitation. Their findings confirmed that a considerable portion of African American nonresident fathers visit their children on a daily or weekly basis. In addition, African American adult children with nonresident fathers often feel significantly closer to their fathers than do their white peers. Finally, African American adult children were more likely than their white peers to believe that their mothers supported their relationship with their father and to have positive perceptions of their parents' relationship.

Transition from a Couple to a Family

Recent research suggests that parenthood decreases marital happiness. Bost et al. (2002) interviewed 137 couples before the birth of their first child and then at 3-, 12-, and 24-month periods. The spouses consistently reported depression and adjustment through 24 months postpartum. Twenge et al. (2003) reviewed 148 samples representing 47,692 individuals in regard to the effect children have on marital satisfaction. They found that (1) parents (both women and men) reported lower marital satisfaction than nonparents; (2) mothers of infants reported the most significant drop in marital satisfaction; (3) the higher the number of children, the lower the marital satisfaction; and (4) the factors in depressed marital satisfaction were conflict and loss of freedom. Claxton and Perry-Jenkins (2008) confirmed a decrease in marital leisure time after the birth of a baby but an increase when the wife went back to work. In addition, they noted that wives who reported high levels of prenatal joint leisure reported greater marital love and less conflict the first year after the baby's birth.

For parents who experience a pattern of decreased happiness, it bottoms out during the teen years. Facer and Day (2004) found that adolescent problem behavior, particularly that of a daughter, is associated with increases in marital conflict. Of even greater impact was their perception of the child's emotional state. Parents who viewed their children as "happy" were less maritally affected by their adolescent's negative behavior.

Regardless of how children affect the feelings that spouses have about their marriage, spouses report more commitment to their relationship once they have children (Stanley and Markman 1992). Figure 10.1 illustrates that the more children a couple has, the more likely the couple will stay married. A primary reason for this increased commitment is the desire on the part of both parents to provide a stable family context for their children. In addition, parents of dependent children may keep their marriage together to maintain continued access to and a higher standard of living for their children. Finally, people (especially mothers) with small

Figure 10.1
Percentage of Couples Getting Divorced by Number of Children

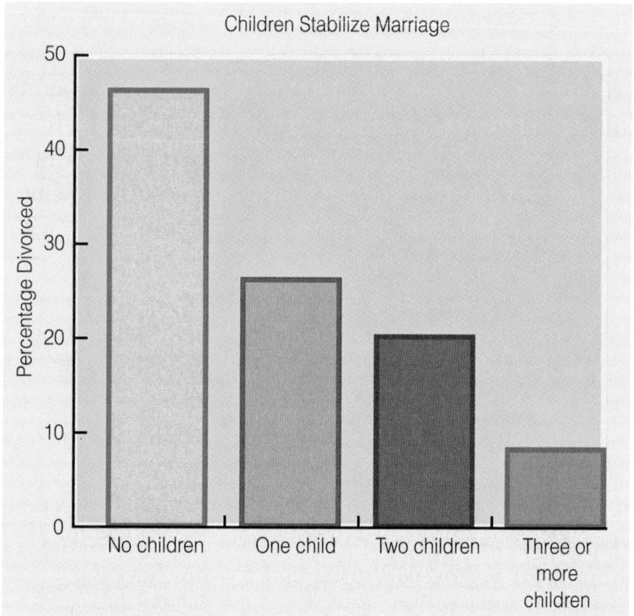

children feel more pressure to stay married (if the partner provides sufficient economic resources) regardless of how unhappy they may be. Hence, though children may decrease happiness, they increase stability because pressure exists to stay together.

> *The family is one of nature's masterpieces.*
>
> —*George Santayana,* The Life of Reason

10.4 Parenthood: Some Facts

Parenting is only one stage in an individual's or couple's life (children typically live with an individual 30 percent of that person's life and with a couple 40 percent of their marriage). Parenting involves responding to the varying needs of children as they grow up, and parents require help from family and friends in rearing their children.

Some additional facts of parenthood follow.

Views of Children Differ Historically

Whereas children of today are thought of as dependent, playful, and adventurous, they have been historically viewed quite differently (Mayall 2002). Indeed, the concept of childhood, like gender, has been socially constructed rather than a fixed life stage. From the thirteenth through the sixteenth centuries, children were viewed as innocent, sweet, and a source of amusement for adults. From the sixteenth through the eighteenth centuries, they were viewed as in need of discipline and moral training. In the nineteenth century, whippings were routine as a means of breaking children's spirits and bringing them to submission. Although remnants of both the innocent and moralistic views of children exist today, the lives of children are greatly improved. Child labor laws protect children from early forced labor, education laws ensure a basic education, and modern medicine has been able to increase the life span of children.

Each Child Is Unique

Children differ in their genetic makeup, physiological wiring, intelligence, tolerance for stress, capacity to learn, comfort in social situations, and interests. Parents soon become aware of the uniqueness of each child—of the child's difference from every other child they know. Parents of two or more children are often amazed at how children who have the same parents can be so different.

Children also differ in their mental and physical health. Mental and physical disabilities of children present emotional and financial challenges to their parents. Green (2003) discussed the potential stigma associated with disability and how parents cope with their children who are stigmatized.

Although parents often contend, "we treat our children equally," Tucker et al. (2003) found that parents treat children differently, with firstborns usually receiving more privileges than children born later. Suitor and Pillemer (2007) found that elderly mothers reported that they tended to establish a closer relationship with the firstborn child, whom they are more likely to call on in later life when there is a crisis.

> *Having a child is like throwing a hand grenade into a marriage.*
>
> —*Nora Ephron, writer, producer, and director*

Birth Order Effects on Personality

Psychologist Frank Sulloway (1996) examined how a child's birth order influenced the development of various personality characteristics. His thesis is that children with siblings develop different strategies to maximize parental investment in them. For example, children can promote parental favor directly by "helping and obeying parents" (p. 67). Sulloway identified the following personality characteristics that have their basis in a child's position in the family:

1. **Conforming or traditional.** Firstborns are the first on the scene with parents and always have the "inside track." They want to stay that way so they are traditional and conforming to their parent's expectations.
2. **Experimental or adventurous.** Children born later learn quickly that they enter an existing family constellation where everyone is bigger and stronger. They cannot depend on having established territory so must excel in ways different from the firstborn. They are open to experience, adventurousness, and trying new things because their status is not already assured.
3. **Neurotic or emotionally unstable.** Because firstborns are "dethroned" by younger children to whom parents had to divert their attention, they tend to be more jealous, anxious, and fearful. Children born later are never number one to begin with so do not experience this trauma.

As support for his ideas, Sulloway (1996) cited 196 controlled birth order studies and contended that, although there are exceptions, one's position in the family is a factor influencing personality outcomes. Nevertheless, he acknowledged that researchers disagree on the effects of birth order on the personalities of children and that birth order research is incomplete in that it does not consider the position of each child in families that vary in size, gender, and number. Indeed, Sulloway (2007) has continued to conduct research, the findings of which do not always support his prediction. For example, he examined intelligence and birth order among 241,310 Norwegian 18- and 19-year-olds and found no relationship. He recommended that researchers look to how a child was reared rather than birth order for understanding a child's IQ.

Parents Are Only One Influence in a Child's Development

Although parents often take the credit—and the blame—for the way their children turn out, they are only one among many influences on child development. Although parents are the first significant influence, peer influence becomes increasingly important during adolescence. Pinquart and Silbereisen (2002) studied seventy-six dyads of mothers and their 11- to 16-year-old adolescents and observed a decrease in connectedness between the children and their mothers and a movement toward their adolescent friends.

Siblings also have an important and sometimes lasting effect on each other's development. Siblings are social mirrors and models (depending on the age) for each other. They may also be sources of competition and can be jealous of each other.

Teachers are also significant influences in the development of a child's values. Some parents send their children to religious schools to ensure that they will have teachers with conservative religious values. This may continue into the child's college and university education.

Media in the form of television—replete with MTV and "parental discretion advised" movies—are a major source of language, values, and lifestyles for children that may be different from those of the parents. Parents are also concerned about the violence to which television and movies expose their children.

Another influence of concern to parents is the Internet. Though parents may encourage their children to conduct research and write term papers using the Internet, they may fear their children are accessing por-

THE POPULAR CULTURE IS INCREASINGLY ORIENTED TO FULFILLING THE X-RATED FANTASIES AND DESIRES OF ADULTS. CHILD-REARING VALUES—SACRIFICE, STABILITY, DEPENDABILITY, MATURITY—SEEM STALE AND MUSTY BY COMPARISON.

—Barbara Defoe Whitehead, National Marriage Project

nography and related sex sites. Parental supervision of teens on the Internet and the right of the teen for privacy remain potential conflict issues.

Parenting Styles Differ

Diana Baumrind (1966) developed a typology of parenting styles that has become classic in the study of parenting. She noted that parenting behavior has two dimensions: responsiveness and demandingness. **Responsiveness** refers to the extent to which parents respond to and meet the needs of their children. In other words, how supportive are the parents? Warmth, reciprocity, person-centered communication, and attachment are all aspects of responsiveness. **Demandingness**, on the other hand, is the degree to which parents place expectations on children and use discipline to enforce the demands. How much control do they exert over their children? Monitoring and confrontation are also aspects of demandingness. Categorizing parents in terms of their responsiveness and their demandingness creates four categories of parenting styles: permissive (also known as indulgent), authoritarian, authoritative, and uninvolved.

1. Permissive parents are high on responsiveness and low on demandingness. They are very lenient and allow their children to largely regulate their own behavior.
2. Authoritarian parents are high on demandingness and low in responsiveness. They feel that children should obey their parents no matter what, and they provide a great deal of structure in the child's world.
3. Authoritative parents are both demanding and responsive. They impose appropriate limits on their children's behavior but emphasize reasoning and communication. This style offers a balance of warmth and control.
4. Uninvolved parents are low in responsiveness and demandingness. These parents are not invested in their children's lives.

responsiveness the extent to which parents respond to and meet the needs of their children. Refers to such qualities as warmth, reciprocity, person-centered communication, and attachment.

demandingness the degree to which parents place expectations on children and use discipline to enforce the demands.

McKinney and Renk (2008) identified the differences between maternal and paternal parenting styles, with mothers tending to be authoritative and fathers tending to be authoritarian. Mothers and fathers also use different parenting styles for their sons and daughters, with fathers being more permissive with their sons than daughters. Overall, this study emphasizes the importance of examining the different parenting styles of parents and adolescent outcome and suggests that having one authoritative parent may be a protective factor for late adolescents. Recall that the authoritative parenting style is the combination of warmth, guidelines, and discipline—"I love you but you need to be in by midnight or lose privileges of having a cell phone and a car."

In summary, parenting does not come as naturally to humans as we might hope. After many decades of exploring how to provide an optimal environment for the development of children, social scientists are narrowing in on the essential elements, with more emphasis now on the specific ways in which elements of effective parenting are related to specific aspects of development for children and adolescents.

10.5 Principles of Effective Parenting

Numerous principles are involved in being effective parents. We begin with the most important of these, which involves giving time, love, praise, and encouragement to children.

Give Time, Love, Praise, and Encouragement

Children most need to feel that they are worth spending time with and that someone loves them. Because children depend first on their parents for the development of their sense of emotional security, it is critical that parents provide a warm emotional context in which the children can develop. Feeling loved as an infant also affects one's capacity to become involved in adult love relationships.

As children mature, positive reinforcement for prosocial behavior also helps to encourage desirable behavior and a positive self-concept. Instead of focusing only on correcting or reprimanding bad behavior, parents should frequently comment on and reinforce good behavior. Comments like, "I like the way you shared your toys," "You asked so politely," and "You did such a good job cleaning your room" help to reinforce positive social behavior and may enhance a child's self-concept. However, parents need to be careful not to overpraise their children, as too much praise may lead to the child's striving to please others rather than trying to please himself.

Praise focuses on other people's judgments of a child's actions, whereas encouragement focuses more on the child's efforts. For example, telling a child who brings you his painting, "I love your picture; it is the best one that I have ever seen" is not as effective in building the child's confidence as saying, "You worked really hard on your painting. I notice that you used lots of different colors." Some parents feel that rewarding positive behavior is not a good idea.

overindulgence giving children too much, too soon, too long; a form of child neglect in which children are not allowed to develop their own competences.

Be Realistic Parents should rid themselves of the illusions of childhood honesty. Talwar and Lee (2008) confirmed that children lie. In an experiment, 138 children (3 to 8 years) were told not to peek at a toy—82 percent peeked in the experimenter's absence, and 64 percent lied about their transgression.

Avoid Overindulgence **Overindulgence** is defined as giving children too much, too soon, too long—it is a form of child neglect in which children are not allowed to develop their own competences (see www.overindulgence.info). Overindulgence has serious ramifications for those children as they grow up. Adults other than parents are less likely to cater to an overindulged child as he or she grows into adulthood. This clash of perspectives is perhaps most obvious in the workplace. Corporate employers have discovered that, compared to their Baby Boomer parents, Generation Y youth (the 79.8 million people born between 1977 and 1995) are less anxious to work and expect to get a job. They are assertive in asking about benefits and less likely to put in sixty-hour weeks than earlier generations. Furthermore, "when it comes to loyalty, the companies they work for are last on their list—behind their families, their friends, their communities, their co-workers and, of course, themselves" (Hira 2007).

Parents typically overindulge because they feel guilty or because they did not have certain material goods in their own youth. In a study designed to identify who overindulged, a researcher found mothers were four times more likely to overindulge than fathers (Clarke 2004). Overindulgence can result in children growing up without consequences, making it difficult for them to cooperate with others in a group-oriented work environment where compromise is necessary. It can also lead to a misplaced sense of entitlement, where the child (or adult) expects to receive praise (or a raise), when they may not have earned one.

Monitor Child's Activities or Drug Use

Abundant research suggests that parents who monitor their children and teens—know where their children are, who they are with, what they are doing, and so on—are less likely to report that their adolescents receive low grades or are engaged in early sexual activity, delinquent behavior, and drug use. Regarding drug use, Brook et al. (2008) studied the association of marijuana use during the transition from late adolescence to early adulthood with reported relationship quality with significant others. The community-based sample consisted of 534 young adults (mean age = 27) from upstate New York who were interviewed at four points in time at the mean ages 14, 16, 22, and 27 years. Marijuana use during the transition from late adolescence to early adulthood was associated with less relationship cohesion and harmony, and with more relationship conflict. The researchers concluded that marijuana use during emerging adulthood predicts diminished relationship quality with a partner in the mid- to late twenties. Hence, parents might well monitor the whereabouts of their teens.

Parents who used marijuana or other drugs themselves wonder how to go about encouraging their own children to be drug-free. Drugfree.org has some recommendations for parents, including being honest about previous drug use, making clear that you do not want your children to use drugs, and explaining that although not all drug use leads to negative consequences, staying clear of such possibilities is the best course of action.

Set Limits and Discipline Children for Inappropriate Behavior

The goal of guidance is self-control. Parents want their children to be able to control their own behavior and to make good decisions without their parents.

Guidance may involve reinforcing desired behavior or providing limits to children's behavior. This sometimes involves disciplining children for negative behavior. Unless parents provide negative consequences for lying, stealing, and hitting, children can grow up to be dishonest, to steal, and to be inappropriately aggressive.

Time-out (a noncorporal form of punishment that involves removing the child from a context of reinforcement to a place of isolation recommended for one minute for each year of the child's age) has been shown to be an effective consequence for inappropriate behavior. Withdrawal of privileges (watching television, playing with friends), pointing out the logical consequences of the misbehavior ("you were late; we won't go"), and positive language ("I know you meant well but . . .") are also effective methods of guiding children's behavior.

Physical punishment is less effective in reducing negative behavior; it teaches the child to be aggressive and encourages negative emotional feelings toward the parents. When using time-out or the withdrawal of privileges, parents should make clear to the child that they disapprove of the child's behavior, not the child. Some evidence suggests that consistent discipline has positive outcomes for children. Lengua et al. (2000) studied 231 mothers of 9- to 12-year-olds and found that inconsistent discipline was related to adjustment problems, particularly for children high in impulsivity.

Obstinacy in children is like a kite; it is kept up just as long as we pull against it.
—*Marlene Cox, coach/mentor*

© Jodi Matthews/iStockphoto.com

Provide Security

Predictable responses from parents, a familiar bedroom or playroom, and an established routine help to encourage a feeling of security in children. Security provides children with the needed self-assurance to venture beyond the family. If the outside world becomes too frightening or difficult, a child can return to the safety of the family for support. Knowing it is always possible to return to an accepting environment enables a child to become more involved with the world beyond the family.

As children become teenagers and young adults, they may fall victim to an abusive partner whom they protect (out of "love") so that the parents will not find out. In the case of a college student, parents might be alert to their child being isolated or cut off from them (the child calls or comes home less often), a change in their behavior (grades drop), or frequent breakups (and getting back together). Helping a child to extricate from such a relationship is difficult. Criticizing their son or daughter's partner is usually ineffective because the child may become defensive. Rather, the key may be focusing on what a good relationship is (love, mutual support) and is not (fear, guilt, anger).

Encourage Responsibility

Giving children increased responsibility encourages the autonomy and independence they need to be assertive and independent. Giving children more responsibility as they

time-out a discipline procedure whereby a child is removed from an enjoyable context to an isolated one and left there.

Today everybody talks about rights and privileges. Twenty-five years ago everybody talked about their obligations and responsibilities.

—*Lou Holtz, television commentator*

© iStockphoto.com

grow older can take the form of encouraging them to choose healthy snacks and letting them decide what to wear and when to return from playing with a friend (of course, the parents should praise appropriate choices).

Children who are not given any control and responsibility for their own lives remain dependent on others. Successful parents can be defined in terms of their ability to rear children who can function as independent adults. A dependent child is a vulnerable child.

Some American children remain in their parents' house into their twenties and thirties, primarily because it is cheaper to do so. Ward and Spitze (2007) analyzed data from the National Survey of Families and Households in regard to children aged 18 and older who lived with their parents. Findings revealed that, although disagreements between parents and children increased, the quality of parent-child or husband-wife relations did not change.

Provide Sex Education

Usher-Seriki et al. (2008) studied mother-daughter communication about sex and sexual intercourse in a sample of 274 middle- to upper-income African American adolescent girls. They found that daughters were more likely to delay sexual intercourse when they had close relationships with their mother, when they perceived that their mother disapproved of premarital sex for moral reasons, and when their mothers emphasized the negative consequences of premarital sex.

How often do parents communicate about sexual issues? Wyckoff et al. (2008) assessed the degree to which 135 African American parents communicated with their 9- to 12-year-olds about sexual issues such as the risks of sexual activity, sexual risk prevention, and so on, and found that most had communicated on such topics. Mothers and fathers were equally likely to communicate with sons whereas mothers were more likely to communicate with daughters than were fathers.

menarche the time of first menstruation.

In regard to adolescent females and **menarche** (the time of their first period), Lee (2008) reported that, of 155 young women, most said that their mothers were supportive and emotionally engaged with them regarding menarche. These emotionally connected mothers were, for the most part, able to mitigate feelings of shame and humiliation associated with the discourses of menstruation in contemporary culture.

Express Confidence

"One of the greatest mistakes a parent can make," confided one mother, "is to be anxious all the time about your child, because the child interprets this as your lack of confidence in his or her ability to function independently." Rather, this mother noted that it is best for parents to convey to a child that they know that he will be all right and that they are not going to worry about the child because they have confidence in him. "The effect on the child," said this mother, "is a heightened sense of self-confidence" (Rhea). Another way to conceptualize this parental principle is to think of the self-fulfilling prophecy as a mechanism that facilitates self-confidence. If parents show a child that they have confidence in him or her, the child begins to accept these social definitions as real and becomes more self-confident.

Shellenbarger (2006) noted that some parents have become "helicopter parents" in that they are constantly hovering at school and in the workplace to ensure their child's "success." The workplace has become the new field where parents negotiate their child's benefits and salaries with employers they feel are out to take advan-

© Flirt/Jupiterimages / © David Falk/iStockphoto.com

S
E
X

© iStockphoto.com

tage of inexperienced workers. However, employers may not appreciate the tampering, and the parents risk hampering their child's "ability to develop self-confidence" (p. D1). Other employers are engaging parents and know that, unless the parents are convinced, the offspring won't sign on (Hira 2007).

Keep in Touch with Nature

Louv (2006) used the term **nature-deficit disorder** to denote that children today are encouraged to detach themselves from direct contact with nature—playing in the woods, wading through a stream, catching tadpoles—since they are overscheduled with planned "activities" such as swimming, tennis, gymnastic, soccer, and piano lessons, which leaves little time for anything else. He states that children need such contact with nature just as they need good nutrition and adequate sleep.

Respond to the Teen Years Creatively

Parenting teenage children presents challenges that differ from those in parenting infants and young children. The teenage years have been characterized as a time when adolescents defy authority, act rebellious, and search for their own identity. Teenagers today are no longer viewed as innocent, naive children.

Conflicts between parents and teenagers often revolve around money and independence. The desires for a cell phone, DVD player, and high-definition television can outstrip the budget of many parents. Teens also increasingly want more freedom. However, neither of these issues needs to result in conflicts. When they do, the effect on the parent-child relationship may be inconsequential. One parent tells his children, "I'm just being the parent, and you're just being who you are; it is OK for us to disagree—but you can't go."

Sound travels slowly. *Sometimes the things you say when your kids are teenagers don't reach them til they're in their forties.*

—Michael Hodgkin

Sometimes teenagers present challenges with which the parents feel unable to cope. Aside from careful monitoring of the child's behavior, family therapy may be helpful. A major focus of such therapy is to increase the emotional bond between the parents and the teenagers and to encourage positive consequences for desirable behavior (for example, concert tickets for good grades) and negative consequences for undesirable behavior (for example, loss of car privileges for getting a speeding ticket).

10.6 Single-Parenting Issues

At least half of all children will spend one-fourth of their lives in a female-headed household (Webb 2005). The stereotype of the single parent is the unmarried black mother. In reality, 40 percent of single mothers are white and only 33 percent are black (Sugarman 2003).

Distinguishing between a single-parent "family" and a single-parent "household" is important. A single-parent family is one in which there is only one parent—the other parent is completely out of the child's life through death, sperm donation, or complete abandonment, and no contact is ever made with the other parent. In contrast, a single-parent household is one in which one parent typically has primary custody of the child or children but the parent living out of the house is still a part

nature-deficit disorder result when children are encouraged to detach themselves from direct contact with nature.

Put Out the Fire

Keeping conflicts with teens to a low level

The following suggestions can help to keep conflicts with teenagers at a low level:

1. **Catch them doing what you like rather than criticizing them for what you don't like. Adolescents are like everyone else—they don't like to be criticized but they do like to be noticed for what they do that is good.**
2. **Be direct when necessary. Though parents may want to ignore some behaviors of their children, addressing some issues directly may also be effective. Regarding the avoidance of STI or HIV infections and pregnancy, Dr. Louise Sammons (2008) tells her teenagers, "It is utterly imperative to require that any potential sex partner produce a certificate indicating no STIs or HIV infection and to require that a condom or dental dam be used before intercourse or oral sex."**
3. **Provide information rather than answers. When teens are confronted with a problem, try to avoid making a decision for them. Rather, providing information on which they may base a decision is helpful. What courses to take in high school and what college to apply for are decisions that might be made primarily by the adolescent. The role of the parent might best be that of providing information or helping the teenager to obtain information.**
4. **Be tolerant of high activity levels. Some teenagers are constantly listening to loud music, going to each other's homes, and talking on cell phones for long periods of time. Parents often want to sit in their easy chairs and be quiet. Recognizing that it is not realistic to expect teenagers to be quiet and sedentary may be helpful in tolerating their disruptions.**
5. **Engage in some activity with your teenagers. Whether renting a video, eating a pizza, or taking a camping trip, structuring some activities with your teenagers is important. Such activities permit a context in which to communicate with them.**

of the child's family. For example, a child who remains connected to her father but lives with her mother is part of a single-parent household. If that child's father dies, she would become part of a single-parent family.

Single Mothers by Choice

Single parents enter their role though divorce or separation, widowhood, adoption, or deliberate choice to rear a child or children alone. Jodie Foster, Academy Award–winning actress, has elected to have children without a husband. She now has two children and smiles when asked, "Who's the father?" The implication is that she has a right to her private life and that choosing to have a single-parent family is a viable option. An organization for women who want children and who may or may not marry is Single Mothers by Choice.

Bock (2000) noted that single mothers by choice are, for the most part, in the middle to upper class, mature, well-employed, politically aware, and dedicated to motherhood. Interviews with twenty-six single mothers by choice revealed their struggle to avoid stigmatization and to seek legitimization for their choice. Most felt that their age (older), sense of responsibility, maturity, and fiscal capability justified their choice. Their self-concepts were those of competent, ethical, mainstream mothers.

Challenges Faced by Single Parents

The single-parent lifestyle involves numerous challenges, including some of the following issues:

1. **Responding to the demands of parenting with limited help.** Perhaps the greatest challenge for single parents is taking care of the physical, emotional, and disciplinary needs of their children—alone. Many single parents resolve this problem by getting help from their parents or extended family.
2. **Adult emotional needs.** Single parents have emotional needs of their own that children are often incapable of satisfying. The unmet need to share an emotional relationship with an adult can weigh heavily on a single parent. One single mother said, "I'm working two jobs, taking care of my kids, and trying to go to school. Plus my mother has cancer. Who am I going to talk to about my life?" Many single women solve the dilemma with a network of friends.
3. **Adult sexual needs.** Some single parents regard their parental role as interfering with their sexual relationships. They may be concerned that their children will find out if they have a sexual encounter at home or be frustrated if they have to go away from home to enjoy a sexual relationship. Some choices with which they are confronted include, "Do I wait until my children are asleep and then ask my lover to leave before morning?" or "Do I openly acknowledge my lover's presence in my life to my children and ask them not to tell anybody?" and "Suppose my kids get attached to my lover, who may not be a permanent part of our lives?"
4. **Lack of money.** Single-parent families, particularly those headed by women, report that money is always lacking.
5. **Guardianship.** If the other parent is completely out of the child's life, the single parent needs to appoint a guardian to take care of the child in the event of the parent's death or disability.
6. **Prenatal care.** Single women who decide to have a child have poorer pregnancy outcomes than married women. The reason for such an association may be the lack of economic funds (no male partner with economic resources available) as well as the lack of social support for the pregnancy or the working conditions of the mothers, all of which result in less prenatal care for their babies.
7. **Absence of a father.** Another consequence for children of single-parent mothers is that they often do not have the opportunity to develop an emotionally supportive relationship with their father. Barack Obama noted in one of his campaign speeches, "I know what it is like to grow up without a father." The late Rodney Dangerfield said in his entire life he spent an average of two hours a year with his father and that he missed such love/nurturing. In contrast, Ansel Adams, the late photographer, attributed his personal and life success to his father, who was steadfastly involved in his life and guided his development.
8. **Negative life outcomes for the child in a single-parent family.** Researcher Sara McLanahan, herself a single mother, set out to prove that children reared by single parents were just as well off as those reared by two parents. McLanahan's data on 35,000 children of single parents led her to a different conclusion—children of only one parent were twice as likely as those reared by two married parents to drop out of high school, get pregnant before marriage, have drinking problems, and experience a host of other difficulties, including getting divorced themselves (McLanahan and Booth 1989; McLanahan 1991). Lack of supervision, fewer economic resources, and less extended family support were among the culprits. Other research suggests that negative outcomes are reduced or eliminated when income levels remain stable (Pong and Dong 2000).

Though the risk of negative outcomes is higher for children in single-parent homes, most are happy and well-adjusted. Benefits to single parents themselves include a sense of pride and self-esteem that results from being independent.

Family and the Economy

SECTIONS

11.1 Effects of Employment on Spouses

A couple's marriage is organized around the work of each. Where the couple lives is determined by where the spouses can get jobs. Jobs influence what time spouses eat, which family members eat with whom, when they go to bed, and when, where, and for how long they vacation. In this section, we examine some of the various influences of work on a couple's relationship. We begin by looking at how the income that results from work impacts the power distribution in the relationship.

Money as Power in a Couple's Relationship

Money is a central issue in relationships because of its association with power, control, and dominance. Generally, the more money a partner makes, the more power that person has in the relationship. Males make considerably more money than females and generally have more power in relationships. However, Morin and Cohn (2008) reported on a national sample and found that, although two-thirds of all husbands in dual-income families say they make more money than their wives, women are still more likely to make the decisions in more areas (42 percent versus 30 percent).

When a wife earns an income, her power increases in the relationship. We know of a married couple in which the wife recently began to earn an income. Before doing so, her husband's fishing boat was in the protected carport. With

Work keeps us from three great evils—boredom, vice and need.
—Voltaire

© Hulton Archive/Getty Images / © iStockphoto.com / © Maria Toutoudaki/iStockphoto.com

© Big Cheese Photo/Jupiterimages / © Abel Leão/iStockphoto.com

her new job and increased power in the relationship, she began to park her car in the carport and her husband put the fishing boat underneath the pine trees to the side of the house. Money also provides an employed woman the power to be independent and to leave an unhappy marriage. Indeed, the higher a wife's income, the more likely she is to leave an unhappy relationship (Schoen et al. 2002). Similarly, because adults are generally the only source of money in a family, they have considerable power over children, who have no money.

To some individuals, money also means love. While admiring the engagement ring of her friend, a woman said, "What a big diamond! He must really love you." A cultural assumption is that a big diamond equals an expensive diamond and a lot of sacrifice and love. Similar assumptions are often made when gifts are given or received. People tend to spend more money on presents for the people they love, believing that the value of the gift symbolizes the depth of their love. People receiving gifts may make the same assumption. "She must love me more than I thought," mused one man. "I gave her a Blu-Ray movie for Christmas, but she gave me a Blu-Ray player. I felt embarrassed."

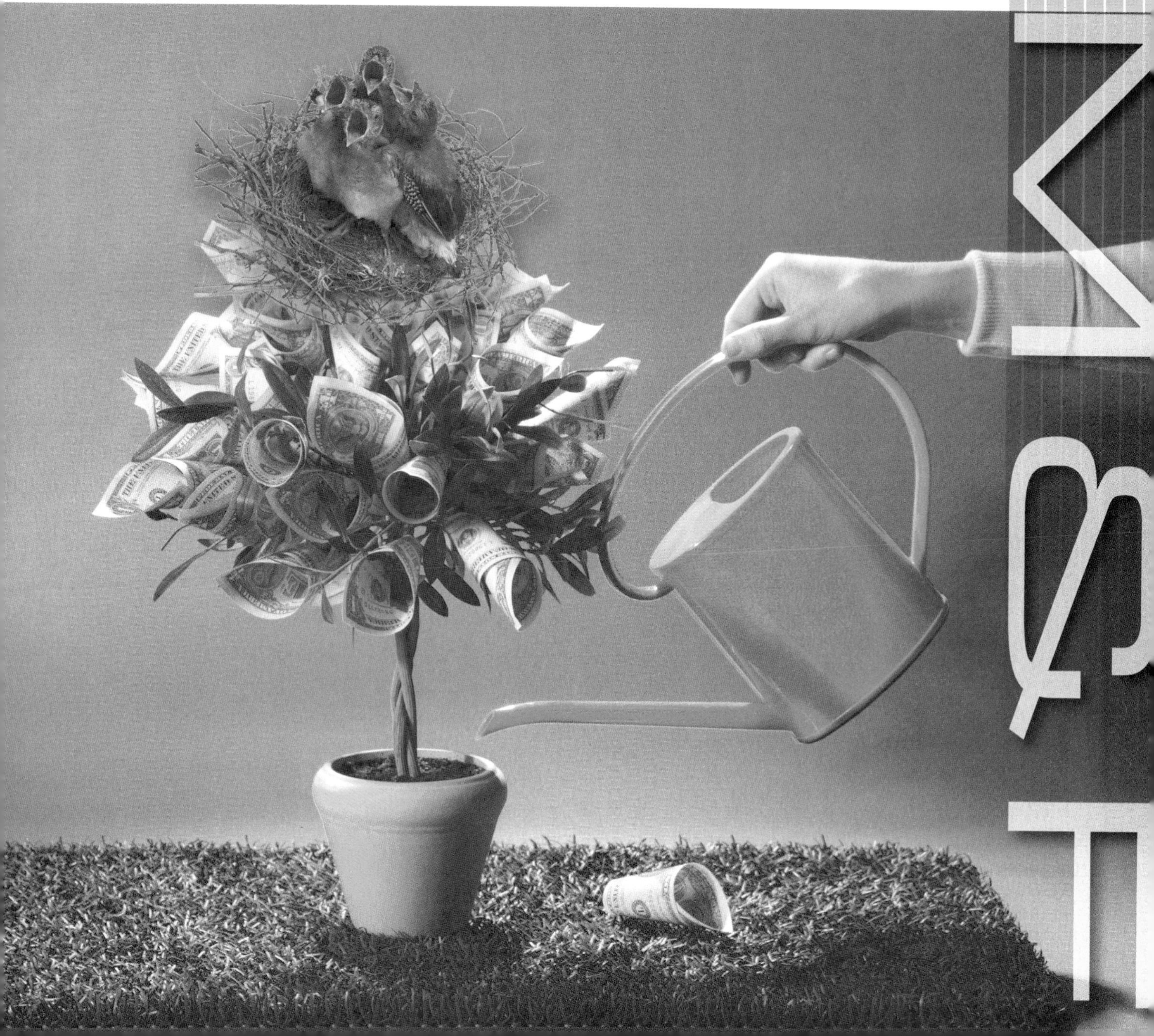

Working Wives

Driven primarily by the need to provide income for the family, 69 percent of all U.S. wives are in the labor force. Most have children. The time wives are most likely to be in the labor force is when their children are teenagers (between the ages of 14 and 17), the time when food and clothing expenses are the highest (*Statistical Abstract of the United States, 2009*, Tables 578 and 579). The stereotypical family consisting of a husband who earns the income and a wife who stays at home with two or more children is no longer the norm. Only 13 percent are "traditional" in the sense of a breadwinning husband, a stay-at-home wife, and their children (Stone 2007). In contrast, most marriages may be characterized by a dual-earner couple (Amato et al. 2007).

> Mother, food, love, and career are the four major guilt groups.
>
> —*Cathy Guisewite*

Because women still take on more child care and household responsibilities than men, women in **dual-earner marriages** (marriages in which both spouses work outside the home to provide economic support for the family) are more likely than men to want to be employed part-time rather than full-time. If this is not possible, many women prefer to work only a portion of the year (the teaching profession allows employees to work about ten months and to have two months in the summer free). Although many low-wage earners need two incomes to afford basic housing and a minimal standard of living, others have two incomes to afford expensive homes, cars, vacations, and educational opportunities for their children. Whether it makes economic sense is another issue.

dual-earner marriage both husband and wife work outside the home to provide economic support for the family.

mommy track stopping paid employment to spend time with young children.

Some parents wonder if the money a wife earns by working outside the home is worth the sacrifices to earn it. Not only is the mother away from their children but she must also pay for strangers to care for their children. Sefton (1998) calculated that the value of a stay-at-home mother is $36,000 per year in terms of what a dual-income family spends to pay for all services that she provides (domestic cleaning, laundry, meal planning and preparation, shopping, providing transportation to activities, taking the children to the doctor, and running errands). Adjusting for changes in the consumer price index, this figure was $47,982 in 2009. The value of a house husband would be the same. However, because males have higher incomes than females, the loss of income would be greater than for a female.

This estimated figure suggests that working outside the home may not be as economically advantageous as one might think—that women may work outside the home for psychological (enhanced self-concept) and social (enlarged social network) benefits. Some may also enjoy a lifestyle that is made possible by earning a significant income by outside employment. One wife noted that the only way she could afford the home she wanted was to help earn the money to pay for it.

The **mommy track** (stopping paid employment to spend time with young children) is another potential cost to those women who want to build a career. Taking time out to rear children in their formative years can derail a career. Noonan and Corcoran (2004) found that lawyers who took time out for child responsibilities were less likely to make partner and more likely to earn less money if they did make partner. Aware that executive women have found it difficult to reenter the workforce after being on the mommy track, the Harvard Business School created an executive training program for mothers to improve their technical skills and to help them return to the workforce (Rosen 2006).

Wives Who "Opt Out"

Pamela Stone (2007) published *Opting Out*, which revealed content from fifty-four interviews with women

> OH, YOU HATE YOUR JOB? WHY DIDN'T YOU SAY SO? THERE'S A SUPPORT GROUP FOR THAT. IT'S CALLED EVERYBODY, AND THEY MEET AT THE BAR.
>
> —*Drew Carey, actor and comedian*

representing a broad spectrum of professions—doctors, lawyers, scientists, bankers, management consultants, editors, and teachers. **Opting out** involved women leaving their careers and returning home to take care of their children for a variety of reasons. However, two reasons stand out: (1) husbands who were unavailable or unable to "shoulder significant portions of caregiving and family responsibilities" (p. 68); and (2) employers who had a lot of policies on the books to encourage and support women parental leave "but not much in the way of making it possible for them to return or stay once they had babies" (p. 119). Part-time work didn't work out for these women because the work was not really "part-time"—the employer kept wanting more.

> *There is mounting evidence to suggest that younger generations—women and men—are challenging the work-centric-mindset of their elders, wanting to find more time for their family and personal lives.*
>
> —*Pamela Stone, author of* Opting Out

Keller (2008) revealed a similar story, that career women who left work for the home found a lower salary and a demotion, and were sidelined on their return. They were, indeed, punished at work for prioritizing family. Nevertheless, women who are flexible, creative, and determined can find work that meets their needs and can be a good fit for their employer.

Types of Dual-Career Marriages

A **dual-career marriage** is defined as one in which both spouses pursue careers and maintain a life together that may or may not include dependents. A career is different from a job in that the former usually involves advanced education or training, full-time commitment, working nights and weekends "off the clock," and a willingness to relocate. Dual-career couples operate without a person who stays home to manage the home and care for dependents.

Nevertheless, four types of dual-career marriages are those in which the husband's career takes precedence (HIS/her), the wife's career takes precedence (HER/his), both careers are regarded equally (HIS/HER), or both spouses share a career or work together (THEIR career).

When couples hold traditional gender role attitudes, the husband's career is likely to take precedence (**HIS/her career**). This situation translates into the wife being willing to relocate and to disrupt her career for the advancement of her husband's career. In this arrangement, the wife may also have children early, which has an effect on the development of her career. Gordon and Whelan-Berry (2005) interviewed thirty-six professional women and found that in 22 percent of the marriages, the husband's career took precedence. The primary reasons for this arrangement were that the husband earned a higher salary, going where the husband could earn the highest income was easier because the wife could more easily find a job wherever he went (than vice versa), and ego needs (the husband needed to have the dominant career). This arrangement is sometimes at the expense of the wife's career. Mason and Goulden (2004) studied women in academia and found that those who had a child within five years of earning their PhDs were less likely to achieve tenure.

For couples who do not have traditional gender role attitudes, the wife's career may take precedence (**HER/his career**). Of the marriages in the Gordon and Whelan-Berry study (2005) mentioned previously, 19 percent could be categorized as giving precedence to a wife's career. In such marriages, the husband is willing to relocate and to disrupt his career for his wife's. Such a pattern is also likely to occur when a wife earns considerably more money than her husband. In some cases, the husband who is downsized or who prefers the role of full-time parent becomes "Mr.

© Jason Stitt/iStockphoto.com

opting out when professional women leave their careers and return home to care for their children.

dual-career marriage a marriage in which both spouses pursue careers and maintain a life together that may or may not include dependents.

HIS/her career a husband's career is given precedence over a wife's career.

HER/his career a wife's career is given precedence over a husband's career.

HIS/HER career a husband's and wife's careers are given equal precedence.

THEIR career a career shared by a couple who travel and work together (for example, journalists).

Mom." More than 100,000 husbands (and parents of at least one child under the age of 6) are married to wives who work full-time in the labor force (Tucker 2005). The incidence of Mr. Moms has increased almost 30 percent in the past ten years (Society for the Advancement of Education, 2005); almost 80 percent of men in Europe report that they would be happy to stay at home with the kids (Pepper 2006). A major advantage of men assuming this role is a more lasting emotional bond with their children (Tucker 2005).

When the careers of both the wife and husband are given equal precedence in the relationship (**HIS/HER career**), they may have a commuter marriage in which they follow their respective careers wherever they lead. Alternatively, Deutsch et al. (2007) surveyed 236 undergraduate senior women to assess their views on husbands, managing children, and work. Most envisioned two egalitarian scenarios in which both spouses would cut back on their careers and/or both would arrange their schedules to devote time to child care. Still other couples may hire domestic child care help so that neither spouse functions in the role of housekeeper. In reality, equal status to HIS and HER careers is not a dominant pattern (Stone 2007).

Finally, some couples have the same career and may travel and work together (**THEIR career**). Some news organizations hire both spouses to travel abroad to cover the same story. These careers are rare.

In the following sections, we look at the effects on women, men, their marriage, and their children when a wife is employed outside the home.

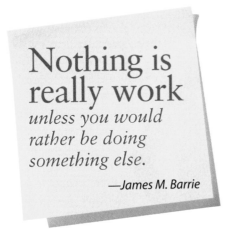

Nothing is really work *unless you would rather be doing something else.*

—James M. Barrie

11.2 Effects of Employment on Children

Independent of the effect on the wife, husband, and marriage, what is the effect of the wife earning an income outside of the home on the children? Individuals disagree on the effects of maternal employment on children.

Quality Time

Dual-income parents struggle with having "quality time" with their children. The term *quality time* has become synonymous with good parenting. Snyder (2007) studied 220 parents from 110 dual-parent families and found that "quality time" is defined in different ways. Some parents (structured-planning parents) saw quality time as planning and executing family activities. Mormons set aside "Monday home evenings" as a time to bond, pray, and sing together (thus, quality time). Other parents (child-centered parents) noted that "quality time" occurred when they were having heart-to-heart talks with their children. Still other parents believed that all the time they were with their children was quality time. Whether they were having dinner together or riding to the post office, quality time was occurring if they were together. As might be expected, mothers assumed greater responsibility for "quality time."

Day-Care Considerations

Parents going into or returning to the workforce are intent on finding high-quality day care. Rose and Elicker (2008) surveyed the various characteristics of day care that are important to 355 employed mothers of children under 6 years of age and found that warmth of caregivers, a play-based curriculum, and educational level of caregivers emerge as the first-, second-, and third-most important factors in selecting a day-care center.

Quality of Day Care Most mothers prefer relatives for the day-care arrangement for their children. Researchers Gordon and Högnäs (2006) analyzed data from the National Institute of Child Health and Human Development Study of Early Child Care and found that nearly two-thirds of the mothers reported such a preference for relatives, including 15 percent who preferred their spouse or partner, 28 percent who preferred another relative in their home, and 22 percent who preferred another relative in another home. An additional 16 percent preferred care by a nonrelative in their own home and 11 percent preferred care by a nonrelative in another private home. Just 9 percent expressed a preference for center-based care. However, more than half of U.S. children are in center-based child-care programs.

Employed parents are concerned that their children get good-quality care. Their concern is warranted. Warash et al. (2005) reviewed the literature on day-care centers and found that the average quality of such centers is mediocre—"unsafe, unsanitary, non-educational, and inadequate in regard to the teacher-child ratio for a classroom." Care for infants was particularly lacking. Of 225 infant or toddler rooms observed, 40 percent were rated "less than minimal" with regard to hygiene and safety. Because of the low pay and stress of the occupation, the rate of turnover for family child-care providers is very high (estimated at between 33 and 50 percent) (Walker 2000). However, De Schipper et al. (2008) noted that day-care workers who engage in high-frequency positive behavior engender secure attachments with the children they work with. Hence, children in the care of such workers don't feel they are on an assembly line but bond with their caretakers. Parents concerned about the quality of day care their children receive might inquire about the availability of webcams. Some day-care centers offer full-time webcam access so that parents or grandparents can log onto their computers and see the interaction of the day-care worker with their children.

Ahnert and Lamb (2003) emphasized that attentive, sensitive, loving parents can mitigate any potential negative outcomes in day care. "Home remains the center of children's lives even when children spend considerable amounts of time in child care. . . . [A]lthough it might be desirable to limit the amount of time children spend in child care, it is much more important for children to spend as much time as possible with supportive parents" (pp. 1047– 48).

Aside from quality time issues, Gordon et al. (2007) studied the effects of maternal paid work and nonmaternal child care on injuries and infectious disease for children aged 12 to 36 months. They found no statistically significant adverse effects on the incidence of infectious disease and injury. However, greater time spent by children in center-based care was associated with increased rates of respiratory problems for children aged 12 to 36 months and increased rates of ear infections for children aged 12 to 24 months.

Cost of Day Care Day-care costs are a factor in whether a low-income mother seeks employment, because the cost can absorb her paycheck. Even for dual-earner families, cost is a factor in choosing a day-care center. Day-care costs vary widely—from nothing, where friends trade off taking care of the children, to very expensive institutionalized day care in large cities. We e-mailed a dual-earner metropolitan couple (who use day care for their two children and are planning for them to enter school) to inquire about the institutional costs in 2009 of day care in the Baltimore area.

The father's eye-opening reponse is on page 212.

11.3 Balancing Work and Family

Work is definitely stressful on relationships. Lavee and Ben-Ari (2007) noted that one of the effects of work stress is that spouses have greater emotional distance between them. We suggested that such stress may signal a deterioration of a relationship or a way that couples minimize interaction so as to protect the relationship.

One of the major concerns of employed parents and spouses is how to juggle the demands of work and family simultaneously and achieve a sense of accomplishment and satisfaction in each area. Cinamon (2006) noted that women experience work interfering with family at higher levels than men. Women are more likely to resolve the conflict by giving precedence to family. Kiecolt (2003) examined national data and concluded that employed women with young children are "more likely to find home a haven, rather than finding work a haven" (p. 33). Recall the research by Stone (2007) in regard to women opting out of high-income or high-status work in favor of taking care of their children.

Nevertheless, the conflict between work and family is substantial, and various strategies are employed to cope with the stress of role overload and role conflict, including (1) the superperson strategy, (2) cognitive restructuring, (3) delegation of responsibility, (4) planning and time management, and (5) role compartmentalization (Stanfield 1998).

DAY CARE COSTS ARE ONLY THE BEGINNING

There is a sliding scale of costs based on quality of provider, age of child, full-time or part-time. Full-time infant care can be hard to find in this area. Many people put their names on a waiting list as soon as they know they are pregnant. We had our first child on a waiting list for about nine months before we got him into our first choice of providers.

Infant care runs $1,250 per month for high-quality day care. That works out to $15,000 per year—and you thought college was expensive.

The cost goes down when the child turns 2—to around $900–$1,000 per month. In Maryland, this is because the required ratio of teachers to children increases at age 2. This cost break doesn't last long. Preschool programs (more academic in structure than day care) begin at age 3 or 4. When our second child turned 4, he started an academic preschool in September. The school year lasts nine and a half months, and it costs about $14,000. Summer camps or summer day-care costs must be added to this amount to get the true annual cost for the child. My wife and I have budgeted about $30,000 total for our two children to attend private school and summer camps this year.

The news only gets worse when the children get older. A good private school for grades K–5 runs $15,000 per school year. Junior high (grades 6–8) is about $20,000, and private high school here goes for $20,000 to $35,000 per nine-month school year.

Religious private schools run about half the costs above. However, they require that you be a member in good standing in their congregation and, of course, your child undergoes religious indoctrination.

There is always the argument of attending a good public school and thus not having to pay for private school. However, we have found that home prices in the "good school neighborhoods" were out of our price range. Some of the better-performing public schools also now have waiting lists—even if you move into that school's district.

In most urban and suburban areas, cost is secondary to admissions. Getting into any good private school is difficult. In order to get into the good high schools, it's best to be in one of the private elementary or middle schools that serves as a "feeder" school. Of course, to get into the "right" elementary school, you must be in a good feeder kindergarten. And of course to get into the right kindergarten, you have to get into the right feeder preschool. Parents have A LOT of anxiety about getting into the right preschool— because this can put your child on the path to one of the better private high schools.

Getting into a good private preschool is not just about paying your money and filling out applications. Yes, there are entrance exams. Both of our children underwent the following process to get into preschool: First you must fill out an application. Second, your child's day-care records/transcripts are forwarded for review. If your child gets through this screening, you and your child are called in for a visit. This visit with the child lasts a few hours and your child goes through evaluation for physical, emotional, and academic development. Then you wait several agonizing months to see if your child has been accepted.

Though quality day care is expensive, parents delight in the satisfaction that they are doing what they feel is best for their child.

Superperson Strategy

The superperson strategy involves working as hard and as efficiently as possible to meet the demands of work and family. The person who uses the superperson strategy often skips lunch and cuts back on sleep and leisure to have more time available for work. Women are particularly vulnerable because they feel that if they give too much attention to child-care concerns, they will be sidelined into lower-paying jobs with no opportunities.

Hochschild (1989) noted that the terms **superwoman** or **supermom** are cultural labels that allow a mother who is experiencing role overload to regard herself as very efficient, bright, and confident. However, Hochschild noted that this is a "cultural cover-up" for an overworked and frustrated woman. Not only does the woman have a job in the workplace (first shift), she comes home to another set of work demands in the form of house care and child care (second shift). Finally, she has a "third shift" (Hochschild 1997).

The **third shift** is the expense of emotional energy by a spouse or parent in dealing with various issues in family living. An example is the emotional energy needed for children who feel neglected by the absence of quality time. Although young children need time and attention, responding to conflicts and problems with teenagers also involves a great deal of emotional energy—the third shift.

© Image 99/Image100/Jupiterimages

Cognitive Restructuring

Another strategy used by some women and men experiencing role overload and **role conflict** is cognitive restructuring, which involves viewing a situation in positive terms. Exhausted dual-career earners often justify their time away from their children by focusing on the benefits of their labor—their children live in a nice house in a safe neighborhood and attend the best schools. Whether these outcomes offset the lack of "quality time" may be irrelevant—the beliefs serve simply to justify the two-earner lifestyle.

supermom (superwoman) a cultural label that allows a mother who is experiencing role overload to regard herself as particularly efficient, energetic, and confident.

third shift the emotional energy expended by a spouse or parent in dealing with various family issues.

role conflict being confronted with incompatible role obligations.

Delegation of Responsibility and Limiting Commitments

A third way couples manage the demands of work and family is to delegate responsibility to others for performing certain tasks. Because women tend to bear most of the responsibility for child care and housework, they may choose to ask their partner to contribute more or to take responsibility for these tasks. Although some husbands are involved, cooperative, and contributing, Stone (2007) revealed that fifty-three respondents in her study opted out of their careers to return home because they had a lack of available help either from husbands who were not at home or who did not do much when they were home.

Another form of delegating responsibility involves the decision to reduce one's current responsibilities and not take on additional ones. For example, women and men may give up or limit agreeing to volunteer responsibilities or commitments. One woman noted that her life was being consumed by the responsibilities of her church; she had to change churches because the demands were relentless. In the realm of paid work, women and men can choose not to become involved in professional activities beyond those that are required.

Time Management

The use of time management is another strategy for minimizing the conflicting demands of work and family.

shift work having one parent work during the day and the other parent work at night so that one parent can always be with the children.

role compartmentalization separating the roles of work and home so that an individual does not dwell on the problems of one role while physically being at the place of the other role.

This involves prioritizing and making lists of what needs to be done each day. Time planning also involves trying to anticipate stressful periods, planning ahead for them, and dividing responsibilities with the spouse. Such division of labor allows each spouse to focus on an activity that needs to be done (grocery shopping, picking up children at day care) and results in a smoothly functioning unit.

Having flexible jobs and/or careers is particularly beneficial for two-earner couples. Being self-employed, telecommuting, or working in academia permits flexibility of schedule so that individuals can cooperate on what needs to be done. Alternatively, some dual-earner couples attempt to solve the problem of child care by **shift work**, or having one parent work during the day and the other parent work at night so that one parent can always be with the children. Shift workers often experience sleep deprivation and fatigue, which may make fulfilling domestic roles as a parent or spouse difficult for them. Similarly, shift work may have a negative effect on a couple's relationship because of their limited time together.

Presser (2000) studied the work schedules of 3,476 married couples and found that recent husbands (married less than five years) who had children and who worked at night were six times more likely to divorce than those who worked days.

Role Compartmentalization

Some spouses use **role compartmentalization**—separating the roles of work and home so that they do not think about or dwell on the problems of one when they are at the physical place of the other. Spouses unable to compartmentalize their work and home feel role strain, role conflict, and role overload, with the result that their efficiency drops in both spheres. Some families look to the government and their employers for help in balancing the demands of family and work.

© Mark Evans/iStockphoto.com

11.4 Debt

Soaring gas prices, home foreclosures, job loss, inflation, and health care costs result in families unable to pay their bills and accumulating more debt (Zibel 2008). Dew (2008) analyzed data on 1,078 couples and found that couples in debt spend less time together and argue more over money, both of which are associated with decreases in marital satisfaction. Boushey and Weller (2008) confirmed that debt and distress are associated. Similarly, Bryant et al. (2008) studied 962 African Americans and 560 black Caribbean individuals and found that debt in both groups depressed levels of marital satisfaction (African Americans more than black Caribbean individuals).

Income Distribution and Poverty

During the presidential debates, both candidates were asked, "What is your definition of rich?" Obama responded, "I would argue that if you are making more than $250,000, then you are in the top 3, 4 percent of this country. You are doing well." McCain responded, "I think if you're just talking about income, how about $5 million?" What are the facts? Families vary in the amount of income they have. Table 11.1 shows the percentage of families in the United States at various income levels.

> Money is a reality, a needed currency for every person every day of an adult life, but it is also a metaphorical currency for power, control, acknowledgment, self-worth, competence, caring, security, commitment, and feeling loved and accepted.
>
> —*Margaret Shapiro, family therapist*

Table 11.1
Distribution of Income Level in U.S. Families in 2006

Income Level of Family	Percentage at This Level
Less than $15,000	8.4
$15,000–$24,999	9.2
$25,000–$34,999	10.5
$35,000–$49,999	14.5
$50,000–$74,999	19.8
$75,000–$99,999	13.5
$100,000 or more	24.2

Median family income = $58,407.

Source: *Statistical Abstract of the United States, 2009.* 128th ed. Washington, DC: U.S. Bureau of the Census, Table 673.

What is the definition of poverty in terms of actual dollars in the United States? Table 11.2 reflects the Department of Health and Human Services' various poverty level guidelines by size of family and where the family lives. A significant proportion of families in the United States continue to be characterized by unemployment and low wages.

Poverty has traditionally been defined as the lack of resources necessary for material well-being—most importantly food and water, but also housing, land, and health care. This lack of resources that leads to hunger and physical deprivation is known as **absolute poverty**. In contrast, **relative poverty** refers to a deficiency in material and economic resources compared with some other population. Although many lower-income Americans, for example, have resources and a level of material well-being that millions of people living in absolute poverty can only dream of (for example, those in third-world countries), they are relatively poor compared with the American middle and upper classes.

One of the factors driving Americans into poverty is the cost of medical care. Indeed, the primary cause of bankruptcy is the inability to pay hospital bills. Compounding the problem is the inability to afford health care insurance.

How much it costs to live varies by the country one lives in. In rural Mexico, the costs of food and housing are substantially below the costs in rural America. For example, lunch in the former may cost less than a dollar whereas lunch in the latter would more likely be three or four times as much. Similarly, due to the low value of the dollar in Europe, a hamburger in America may be $1.50 but could be $8.00 in Paris.

Effects of Poverty on Marriages and Families

> Poverty is no shame, but being ashamed of it is.
>
> —*Benjamin Franklin, statesman*

Poverty is devastating to couples and families. Those living in poverty have poorer physical and mental health, report lower personal and marital satisfaction, and die sooner. The anxiety over lack of money may result in relationship conflict. Money is the most common problem that couples report. Stanley and Einhorn (2007) suggested that the reason money is such a profound issue in marriage is its symbolic significance (for example, power and control) as well as the fact that individuals do something daily in reference to money—spend, save, or worry about it. The potential for conflict is endless. One couple in marriage therapy reported that they argued over whether to buy a new air conditioner for their car. The husband thought it necessary; the wife thought they could roll down the windows.

poverty the lack of resources necessary for material well-being.

absolute poverty the lack of resources that leads to hunger and physical deprivation.

relative poverty a deficiency in material and economic resources compared with some other population.

Table 11.2
2009 HHS Poverty Guidelines

People in Family or Household	48 Contiguous States and D.C.	Alaska	Hawaii
1	$10,830	$13,530	$12,460
2	14,570	18,210	16,760
3	18,310	22,890	21,060
4	22,050	27,570	25,360
5	25,790	32,250	29,660
6	29,530	36,930	33,960
7	33,270	41,610	38,260
8	37,010	46,290	42,560
For each additional person, add	3,740	4,680	4,300

Source: *Federal Register*, vol. 74, no. 14, pp. 4199–4201. January 23, 2009. http://aspe.hhs.gov/poverty/09poverty.shtml.

The stresses associated with low income also contribute to substance abuse, domestic violence, child abuse and neglect, divorce, and questionable parenting practices. For example, economic stress is associated with greater marital discord, and couples with incomes less than $25,000 are 30 percent more likely to divorce than couples with incomes greater than $50,000 (Whitehead and Popenoe 2004). Child neglect is more likely to be found with poor parents who are unable to afford child care or medical expenses and leave children at home without adult supervision or fail to provide needed medical care. Poor parents are more likely than other parents to use harsh physical disciplinary techniques, and they are less likely to be nurturing and supportive of their children.

Another family problem associated with poverty is teenage pregnancy. Poor adolescent girls are more likely to have babies as teenagers or to become young single mothers. Early childbearing is associated with numerous problems, such as increased risk of having premature babies or babies of low birth weight, dropping out of school, and earning less money as a result of lack of academic achievement.

Global Inequality

Global economic inequality has reached unprecedented levels. The most comprehensive study on the world distribution of household wealth offers the following facts on wealth inequality worldwide:

- The richest 1 percent of adults in the world own 40 percent of global household wealth; the richest 2 percent of adults own more than half of global wealth; and the richest 10 percent of adults own 85 percent of total global wealth.
- The poorest half of the world's adult population owns barely 1 percent of global wealth.
- Households with assets of $2,200 per adult are in the top half of the world wealth distribution; assets of $61,000 per adult places a household in the top 10 percent, and assets of more than $500,000 per adult places a household in the richest 1 percent worldwide.
- Although North America has only 6 percent of the world's adult population, it accounts for one-third (34 percent) of all household wealth worldwide. More than one-third (37 percent) of the richest 1 percent of individuals in the world resides in the United States.
- The degree of wealth inequality in the world is as if one person in a group of ten takes 99 percent of the total pie and the other nine people in the group share the remaining 1 percent (Davies et al. 2006).

Being Wise about Credit

One way to keep from slipping deeper into debt or poverty is to use credit wisely. The "free" credit cards that college students receive in the mail are Trojan horses and can plunge them into massive debt from which recovering will take years.

You use credit when you take an item from the store home today and pay for it later. The amount you pay later will depend on the arrangement you make with the seller. Suppose you want to buy a flat panel, 42-inch high definition television which sells for $2,000. Unless you pay cash, the seller will set up one of three types of credit accounts with you: installment, revolving charge, or open charge.

Sheriff's Deputy Rick Ferguson supervises as an eviction team removes household items from a house foreclosed upon in Lafayette, Colorado. The owners had stopped paying their mortgage payments and the bank foreclosed on the house with a court order. As the national unemployment rate continues to rise nationwide, many Americans are falling further behind in their rent and mortgage payments, resulting in more evictions and foreclosures.

Under the **installment plan**, you sign a contract to pay for the item with regular installments over an agreed upon period of time. You and the seller negotiate the period of time over which the payments will be spread and the amount you will pay each month. The seller adds a finance charge to the cash price of the television set and remains the legal owner of the set until you have made your last payment. Most department stores, appliance and furniture stores, and automobile dealers offer installment credit. The total cost of buying the $2,000 high-definition television will actually be $2,175 as shown in Table 11.3.

Instead of buying the flat panel $2,000 high-definition television set on the installment plan, you might want to buy it on the **revolving charge** plan. Most credit cards, such as Visa and MasterCard, represent revolving charge accounts that permit you to buy on credit up to a stated amount during each month. At the end of the month, you may pay the total amount you owe, any amount over the stated minimum payment due, or the minimum payment. If you choose to pay less than the full amount, the cost of the credit on the unpaid amount is approximately 1.5 percent per month, or 18 percent per year. Avoid these finance charges by paying off the credit card monthly.

You can also purchase items on an **open charge** account. Under this system, you agree to pay in full within the agreed amount of time. Because this type of account has no direct service charge or interest, the television set would cost only the purchase price. For example, Sears and JCPenney offer open charge accounts. If you do not pay the full amount in thirty days, a finance charge is placed on the remaining balance.

The use of both revolving charge and open charge accounts is wise if you pay off the bill before finance charges begin. In deciding which type of credit account to use, remember that credit usually costs money; the longer you take to pay for an item, the more the item will cost you.

installment plan repayment plan whereby one signs a contract to pay for an item with regular installments over an agreed upon period of time.

revolving charge the repayment plan whereby you may pay the total amount you owe, any amount over the stated minimum payment due, or the minimum payment.

open charge repayment plan whereby one agrees to pay the amount owed in full within the agreed amount of time.

Table 11.3

Calculating the Cost of Installment Credit

Amount to be financed	
Cash price	$2,000
Amount to be paid	
Monthly payments	$75.
X number of payments	x 29
Total amount to be repaid	$2,175.

Credit Rating and Identity Theft

When you apply for a loan, the lender will seek a credit report from credit bureaus such as Equifax, TransUnion, and Experian. In effect, you will have a "credit score"—also referred to as a NextGen score or a FICO (Fair Isaac Credit Company) score—calculated as follows:

35 percent based on late payments, bankruptcies, judgments
30 percent based on current debts
15 percent based on how long accounts have been opened and established
10 percent based on type of credit (credit cards, loan for house or car)
10 percent based on applications for new credit or inquiries

As these percentages indicate, the way to improve your credit is to make payments on time and reduce your current debt. Your score will range from 620 to 850 (a score of excellent is 750 to 850; a score of 660 to 749 is good; a score of 620 to 659 is fair; a score of 400 to 619 is poor). The higher your score, the lower your rate of interest. For example, on a $150,000 thirty-year, fixed-rate mortgage, a score above 760 would result in a 5.5 percent interest rate with a monthly payment of $852. In contrast, a score of 639 and below would result in an interest rate of 7.09 percent, with a monthly payment of $1,007 (see http://www.myfico.com/ or type in "credit report" on www.google.com). Taking all credit cards to the limit can lead to financial trouble and eventual bankruptcy.

When someone poses as you and uses your credit history to buy goods and services, that person has stolen your identity. More than 10 million Americans are

identity theft one person using the Social Security number, address, and bank account numbers of another, posing as that person to make purchases.

victims of **identity theft**, which can destroy your credit, plunge you into debt, and keep you awake at night with lawsuits from creditors. Identity theft happens when someone gets access to personal information such as your Social Security number, credit card number, or bank account number and goes online to pose as you to buy items or services. Identity theft is the number one fraud complaint in the United States.

Safeguards include (1) never giving such information over the phone or online unless you initiate the contact, and (2) shredding bank and credit card statements and preapproved credit card offers. (A shredder can cost as little as $20.) Also, avoid paying your bills by putting an envelope in your mailbox with the flag up—use a locked box or the post office. Finally, check your credit reports, scrutinize your bank statements, and guard your personal identification number (PIN) at automatic teller machines (ATMs). If you use the Internet, protect your safety by installing firewall software.

Discussing Debt/Money: Do It Now

Shapiro (2007) noted that sex, religion, and money are sensitive issues in polite society and that couples may shy away from discussing such details. Indeed, once partners define themselves in a serious relationship, Shapiro (a marriage and family therapist) recommended that they talk about how they feel about running up debt on their credit card, late charges, or how much they are bothered by debt. The issue becomes relevant because the debt of one partner may become the debt of the other if the relationship continues. If John thinks nothing of charging a high-definition television on his third credit card (because the other two have been maxed out), Mary might legitimately be concerned. The following are other issues that couples might discuss:

- At the time of engagement, what do the partners feel about a ring? Should one be bought? How much should be spent on it? And what is the symbolic meaning of the "size of the ring"—does a smaller, less expensive ring mean less love and less commitment?
- When the couple has their first child, will the wife stop working? What is the implication in terms of her access to money? Does she have to ask for money, or is "his" money deposited in an account so that both have access?
- As children get older, what are the respective feelings of the spouses in regard to sending their children to college versus letting them take out student loans? What about sending money to one's parents who may need financial help with health care bills (including a nursing home for one or both sets of parents)?
- As retirement comes, does the couple save their money or travel around the world?

At each stage of the family life cycle, financial decisions can cause very deep-seated feelings about money to surface. Couples might also discuss the importance of education and its association with higher income. The "more you learn, the more you earn" slogan is true. Table 11.4 shows the increased income associated with increased education.

Table 11.4
Women's and Men's Median Income with Similar Education

	Bachelor's	Master's	Doctoral Degree
Men	$54,403	$67,425	$90,511
Women	$35,094	$46,250	$61,091

Source: *Statistical Abstract of the United States, 2009,* 128th ed. Washington, DC: U.S. Bureau of the Census, Table 680.

Money can be a central issue in a relationship, so couples should talk about how they feel about money and debt.

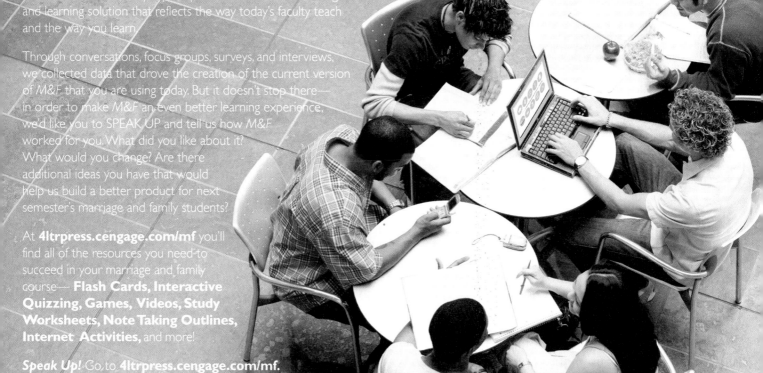

Violence and Abuse in Relationships

SECTIONS

12.1 Nature of Relationship Abuse

On February 8, 2009, Chris Brown allegedly beat girlfriend Rihanna, prompting her to call 911. He later turned himself in to the police and was released on $50,000 bail. Other examples of partner abuse are more subtle than a beating. Angie, a first-year university student, became alarmed by the way her boyfriend began to treat her. After a year of dating, she noticed that he began to get upset when she would spend time with her friends. He would also say that she had put on weight and accuse her of "looking like a slut" when she went out. His emotionally abusive behavior resulted in her feeling hopeless to reverse his treatment of her, feeling trapped by his threats, and feeling depressed. She dropped out of school. Angie is not alone. Unfortunately, abuse is not uncommon, and there are several ways abuse manifests itself in relationships.

Violence

physical abuse intentional infliction of physical harm by either partner on the other.

intimate-partner violence an all-inclusive term that refers to crimes committed against current or former spouses, boyfriends, or girlfriends.

Also referred to as **physical abuse**, violence may be defined as the intentional infliction of physical harm by either partner on the other. Examples of physical violence include pushing, throwing something at the partner, slapping, hitting, and forcing sex on the partner. **Intimate-partner violence** (IPV) is an all-inclusive term that refers to crimes committed against current or former spouses, boyfriends, or girlfriends. John Gottman (2007) identified two types of violence. One type is where conflict escalates over an issue and one or both partners lose control. The trigger seems to be feeling disrespected and a loss of one's dignity. The person feels threatened and seeks to defend his or her dignity by posturing and threatening the partner while in a very agitated state. Control is lost and the partner strikes out. Because both partners may lose control at the same time, this type of violence is symmetrical. Couples can

© Courtney Weittenhiller/iStockphoto.com

prevent this type of violence by recognizing the sequence of interaction that leads to the violent behavior (Gottman 2007).

A second type of violence is based on one partner controlling the other, creating a clear perpetrator and victim. The best way to avoid the second type of violence is for the victim to leave the relationship.

A syndrome related to violence is **battered-woman syndrome**, which refers to the general pattern of battering that a woman is subjected to and is defined in terms of the frequency, severity, and injury she experiences. Battering is severe if the person's injuries require medical treatment or the perpetrator could be prosecuted.

battered-woman syndrome general pattern of battering that a woman is subjected to, defined in terms of the frequency, severity, and injury.

uxoricide the murder of a woman by her romantic partner.

emotional abuse (verbal abuse; symbolic aggression; psychological abuse) the denigration of an individual with the purpose of reducing the victim's status and increasing the victim's vulnerability so that he or she can be more easily controlled by the abuser.

Battering may lead to murder. **Uxoricide** is the murder of a woman by her romantic partner. The murder of Laci Peterson by her convicted husband Scott Peterson is an example of uxoricide. As a prelude to murder, the man may hold his wife hostage (holding one or more people against their will with the actual or implied use of force). A typical example is an estranged spouse who reenters the house of a former spouse and holds his wife and children hostage. These situations are potentially dangerous because the perpetrator often has a weapon and may use it on the victims, himself, or both. A negotiator is usually called in to resolve the situation.

Emotional Abuse

In addition to being physically violent, partners may also engage in **emotional abuse** (also known as **verbal abuse, symbolic aggression**, or **psychological abuse**). Whereas more than 10 percent of 1,319 undergraduates (10.7 percent) reported being involved in a physically abusive relationship, 31.6 percent reported that they "had been involved in an *emotionally* abusive relationship with a partner" (Knox and Zusman 2009). Although emotional abuse does not involve physical harm, it is designed to make the partner feel bad—to denigrate the partner, reduce the partner's status, and make the partner vulnerable to being controlled by the abuser. Although there is debate about what constitutes psychological abuse, examples of various categories of emotional abuse include the following:

Criticism and ridicule—being called obese, stupid, crazy, ugly, pitiful, and repulsive

Isolation—being prohibited by the partner from spending time with friends, siblings, and parents

Control—being told how to dress and/or accused of dressing like a slut; having one's money controlled

Silent treatment—having a partner who refuses to talk to or to touch the victim

Yelling—being yelled and screamed at by one's partner

Accusation—being told (unjustly) that one is unfaithful

Threats—being threatened by the partner with abandonment or threats of harm to one's self, family, or pets

Demeaning behavior—being insulted by the partner in front of others

Restricting behavior—having one's mobility restricted (for example, use of the car)

Demanding behavior—being required to do as the partner wishes (for example, have sex)

Although women may be abusers, more often abusers are men. Of 227,941 restraining orders issued to adults in California, most were for domestic violence, and most (72.2 percent) of the abusers were men, African Americans, and 25- to 34-year-olds (Shen and Sorenson 2005). Similarly, according to National Incident-Based Reporting System (NIBRS) data on intimate-partner

Nothing good ever comes of violence.
—Martin Luther King Jr.

© Nonstock/Jupiterimages

CHET BAKER WOULD SING THE TENDEREST LOVE SONG OF ROMANCE, THEN BEAT UP THE WOMAN HE PROFESSED LOVE TO.

—*James Gavin,* Deep in a Dream: The Long Night of Chet Baker

violence, the victims tended to be young, female, and minority members (Vazquez et al. 2005).

Female Abuse of Partner

Women also abuse their partners. Swan et al. (2008) reviewed the literature on women who are violent toward their intimate partners and found that women's physical violence may be just as prevalent as men's violence but is more likely to be motivated by self-defense and fear, whereas men's physical violence is more likely than women's to be driven by control motives. Hence, women who are violent toward their partners tend to be striking back rather than throwing initial blows. An abusive female wrote the following:

> When people think of physical abuse in relationships, they get the picture of a man hitting or pushing the woman. Very few people, including myself, think about the "poor man" who is abused both physically and emotionally by his partner—but this is what I do to my boyfriend. We have been together for over a year now and things have gone from really good to terrible.
>
> The problem is that he lets me get away with taking out my anger on him. No matter what happens during the day, it's almost always his fault. It started out as kind of a joke. He would laugh and say "oh, how did I know that this was going to somehow be my fault." But then it turned into me screaming at him, and not letting him go out and be with his friends as his "punishment."
>
> Other times, I'll be talking to him or trying to understand his feelings about something that we're arguing about and he won't talk. He just shuts off and refuses to say anything but "alright." That makes me furious, so I start verbally and physically attacking him just to get a response.
>
> Or, an argument can start from just the smallest thing, like what we're going to watch on TV, and before you know it, I'm pushing him off the bed and stepping on his stomach as hard as I can and throwing the remote into the toilet. That really gets to him.

Abuse of a partner can escalate rapidly from dismissiveness to violence.

stalking unwanted following or harassment that induces fear in a target person.

cyber-victimization being sent unwanted e-mail, spam, viruses, or being threatened online.

Stalking in Person

Abuse may take the form of stalking. **Stalking** is defined as unwanted following or harassment that induces fear in a target person. Stalking is a crime. A stalker is one who has been rejected by a previous lover or is obsessed with a stranger or acquaintance who fails to return the stalker's romantic overtures. In about 80 percent of cases, the stalker is a heterosexual male who follows his previous lover. Women who stalk are more likely to target a married male. Stalking is a pathological emotional or motivational state (Meloy and Fisher 2005). More than a third (33.7 percent) of 1,319 university students reported that they had been stalked (followed and harassed) (Knox and Zusman 2009). Such stalking is usually designed either to seek revenge or to win a partner back.

Stalking Online—Cybervictimization

Cybervictimization includes being sent threatening e-mail, unsolicited obscene e-mail, computer viruses, or junk mail (spamming). It may also include flaming

obsessive relational intrusion the relentless pursuit of intimacy with someone who does not want it.

corporal punishment the use of physical force with the intention of causing a child to experience pain, but not injury, for the purpose of correction or control of the child's behavior.

(online verbal abuse), and leaving improper messages on message boards. Cybervictimization in reference to Internet dating is "minimal." This is the conclusion of Jerin and Dolinsky (2007) and is based on a study of 134 people who completed profiles on a variety of Internet dating sites. Cyberangels.org is a website that promotes online safety.

Prior to stalking is **obsessive relational intrusion** (ORI), the relentless pursuit of intimacy with someone who does not want it. The person becomes a nuisance but does not have the goal of harm as does the stalker. When ORI and stalking are combined, as many as 25 percent of women and 10 percent of men can expect to be pursued in unwanted ways (Spitzberg and Cupach 2007).

People who cross the line in terms of pursuing an ORI relationship or responding to being rejected (stalking) engage in a continuum of eight forms of behavior. Spitzberg and Cupach (2007) have identified these:

1. **Hyperintimacy**—telling a person that they are beautiful or desirable to the point of making them uncomfortable.
2. **Relentless electronic contacts**—flooding the person with e-mail messages, cell phone calls, text messages, or faxes.
3. **Interactional contacts**—showing up at the person's work or gym. The intrusion may also include joining the same volunteer groups as the pursued.
4. **Surveillance**—monitoring the movements of the pursued, such as by following the person or driving by the person's house.
5. **Invasion**—breaking into the person's house and stealing objects that belong to the person; identity theft; or putting Trojan horses (viruses) in the person's computer. One woman downloaded child pornography on her boyfriend's computer and called the authorities to arrest him. He is now in prison.
6. **Harassment or intimidation**—leaving unwanted notes on one's desk or a dead animal on one's doorstep.
7. **Threat or coercion**—threatening physical violence or harm to the person or one's family or friends.
8. **Aggression or violence**—carrying out a threat by becoming violent (for example, kidnapping or rape).

12.2 Explanations for Violence and Abuse in Relationships

Research suggests that numerous factors contribute to violence and abuse in intimate relationships. These factors include those that occur at the cultural, community, and individual and family levels.

Cultural Factors

In many ways, American culture tolerates and even promotes violence. Violence in the family stems from the acceptance of violence in our society as a legitimate means of enforcing compliance and solving conflicts at interpersonal, familial, national, and international levels. Violence and abuse in the family may be linked to cultural factors, such as violence in the media, acceptance of corporal punishment, gender inequality, and the view of women and children as property. The context of stress is also conducive to violence.

Violence in the Media One need only watch the evening news to see the violence in countries such as Iraq. In addition, feature films and television movies regularly reflect themes of violence. In 2007, the film *No Country for Old Men* won four Academy Awards including Best Picture. The film is ultraviolent, depicting incessant merciless violence and murder. Its selection as Best Picture reflects a society that finds value in violence. However, media violence is only the beginning. Football is a very violent sport. Not only football players, but also males who have football players as friends, are more likely to get into fights (Kreager 2007).

Corporal Punishment of Children **Corporal punishment** is defined as the use of physical force with the intention of causing a child to experience pain, but

not injury, for the purpose of correction or control of the child's behavior. In the United States, it is legal in all fifty states for a parent to spank, hit, belt, paddle, whip, or otherwise inflict punitive pain on a child so long as the corporal punishment does not meet the individual state's definition of child abuse. Violence has become a part of our cultural heritage through the corporal punishment of children. In a review of the literature, 94 percent of parents of toddlers reported using corporal punishment (Straus 2000). Spankers are more likely to be young parents and mothers who have been hit when they were children (Walsh 2002). Children who are victims of corporal punishment display more antisocial behavior, are more violent, and have an increased incidence of depression as adults. Straus (2000) recommended an end to corporal punishment to reduce the risk of physical abuse and other harm to children.

Gender Inequality Domestic violence and abuse may also stem from traditional gender roles. Nayak et al. (2003) found that individuals in countries espousing very restrictive roles for women (for example, Kuwait) tend to be more accepting of violence against women. Traditionally, men have also been taught that they are superior to women and that they may use their aggression toward women, believing that women need to be "put in their place." The greater the inequality and dependence of the woman on the man, the more likely the abuse.

Some occupations lend themselves to contexts of gender inequality. In military contexts, men notoriously devalue, denigrate, and sexually harass women. In spite of the rhetoric about gender equality in the military, half of the women in the Army, Navy, and Air Force academies in a 2004 Pentagon survey reported being sexually harassed (Komarow 2005). These male perpetrators may not separate their work roles from their domestic roles. One student in our classes noted that she was the wife of a Navy Seal and that "he knew how to torment someone, and I was his victim."

Being a police officer may also be associated with abuse. Johnson et al. (2005) studied 413 officers and found that burnout, authoritarian style, alcohol use, and department withdrawal were associated with domestic violence. A team of researchers (Erwin et al. 2005) compared 106 police officers who had been charged with intimate partner violence with 105 police officers without such an offense and found that minority status, being on the force for more than seven years, and being assigned to a high-crime district were associated with domestic violence. The combined data from these two studies suggest that the stress of being a police officer seems to make one vulnerable to domestic abuse.

honor crime (honor killing) killing a daughter because she brought shame to the family by having sex while not married. The killing is typically overlooked by the society. Jordan is a country where honor crimes occur.

Marital violence is found to occur at a higher rate among those with less education (Verma 2003), which is another context for inequality. Similarly, women who have higher incomes than their husbands report a higher frequency of beatings by their husbands (ibid.).

In cultures where a man's "honor" is threatened if his wife is unfaithful, the husband's violence toward her is tolerated. In some cases, formal, legal traditions defend a man's right to beat or even to kill his wife in response to her infidelity (Vandello and Cohen 2003). Unmarried women in Jordan who have intercourse are viewed as bringing shame on their parents and siblings and may be killed; this is referred to as an **honor crime** or **honor killing**. The legal consequence is minimal to nonexistent.

View of Women and Children as Property Prior to the late nineteenth century, a married woman was considered the property of her husband. A husband had a legal right and marital obligation to discipline and control his wife through the use of physical force.

> NO WOMAN HAS TO BE A VICTIM OF PHYSICAL ABUSE. WOMEN HAVE TO FEEL LIKE THEY ARE NOT ALONE.
>
> —*Salma Hayek, actress*

Stress Our culture is also a context of stress. The stress associated with getting and holding a job, rearing children, staying out of debt, and paying bills may predispose one to lash out at others. Haskett et al. (2003) found that parenting stress was particularly predictive of child abuse. People who have learned to handle stress by being abusive toward others are vulnerable to becoming abusive partners.

Community Factors

Community factors that contribute to violence and abuse in the family include social isolation, poverty, and inaccessible or unaffordable health care, day care, elder care, and respite care services and facilities.

Social Isolation Living in social isolation from extended family and community members increases the risk of being abused. Spouses whose parents live nearby are least vulnerable.

Poverty Abuse in adult relationships occurs among all socioeconomic groups. However, poverty and low socioeconomic development are associated with crime and higher incidences of violence. This violence may spill over into interpersonal relationships as well as the frustration of living in poverty.

Inaccessible or Unaffordable Community Services Failure to provide medical care to children and elderly family members sometimes results from the lack of accessible or affordable health care services in the community. Failure to provide supervision for children and adults may result from inaccessible day-care and eldercare services. Without eldercare and respite care facilities, families living in social isolation may not have any help with the stresses of caring for elderly family members and children.

Individual Factors

Individual factors associated with domestic violence and abuse include psychopathology, personality characteristics, and alcohol or substance abuse.

Personality Factors A number of personality characteristics have been associated with people who are abusive in their intimate relationships. Some of these characteristics follow:

1. **Dependency.** Therapists who work with batterers have observed that they are extremely dependent on their partners. Because the thought of being left by their partners induces panic and abandonment anxiety, batterers use physical aggression and threats of suicide to keep their partners with them.
2. **Jealousy.** Along with dependence, batterers exhibit jealousy, possessiveness, and suspicion. An abusive husband may express his possessiveness by isolating his wife from others; he may insist she stay at home, not work, and not socialize with others. His extreme, irrational jealousy may lead him to accuse his wife of infidelity and to beat her for her presumed affair. Indeed, O.J. Simpson said, "If I killed my wife it would be because I was jealous and loved her—right?"
3. **Need to control.** Abusive partners have an excessive need to exercise power over their partners and to control them. The abusers do not let their partners make independent decisions, and they want to know where they are, whom they are with, and what they are doing. They like to be in charge of all aspects of family life, including finances and recreation.
4. **Unhappiness and dissatisfaction.** Abusive partners often report being unhappy and dissatisfied with their lives, both at home and at work. Many abusers have low self-esteem and high levels of anxiety, depression, and hostility. They may expect their partner to make them happy.
5. **Anger and aggressiveness.** Abusers tend to have a history of interpersonal aggressive behavior. They have poor impulse control and can become instantly enraged and lash out at the partner. Battered women report that episodes of

violence are often triggered by minor events, such as a late meal or a shirt that has not been ironed.

6. **Quick involvement.** Because of feelings of insecurity, the potential batterer will move his partner quickly into a committed relationship. If the woman tries to break off the relationship, the man will often try to make her feel guilty for not giving him and the relationship a chance.

7. **Blaming others for problems.** Abusers take little responsibility for their problems and blame everyone else. For example, when they make mistakes, they will blame their partner for upsetting them and keeping them from concentrating on their work. A man may become upset because of what his partner said, hit her because she smirked at him, and kick her in the stomach because she poured him too much alcohol.

8. **Jekyll-and-Hyde personality.** Abusers have sudden mood changes so that a partner is continually confused. One minute an abuser is nice, and the next minute angry and accusatory. Explosiveness and moodiness are typical.

9. **Isolation.** An abusive person will try to cut off a partner from all family, friends, and activities. Ties with anyone are prohibited. Isolation may reach the point at which an abuser tries to stop the victim from going to school, church, or work.

10. **Alcohol and other drug use.** Whether alcohol reduces one's inhibitions to display violence, allows one to avoid responsibility for being violent, or increases one's aggression, alcohol and substance abuse is associated with violence and abuse.

11. **Emotional deficit.** Some abusing spouses and parents may have been reared in contexts that did not provide them with the capacity to love, nurture, or be emotionally engaged. Rosenbaum and Leisring (2003) found that men who abuse women report having had limited love from their mothers, more punishment from their mothers, and less attention from their fathers than men who do not abuse their partners.

Family Factors

Family factors associated with domestic violence and abuse include being abused as a child, having parents who abused each other, and not having a father in the home.

> Those children who are beaten *will, in turn, give beatings; those who are intimidated will be intimidating; those who are humiliated will impose humiliation, and those whose souls are murdered will murder.*
>
> —*Alice Miller*

> Violence is the last refuge of the incompetent.
>
> —*Salvor Hardin, Mayor of Terminus*

Child Abuse in Family of Origin Individuals who were abused as children were more likely to be abusive toward their partners as adults.

Family Conflict Schaeffer et al. (2005) found that high conflict between spouses, parents, and children was predictive of child abuse. Fathers who were not affectionate were also more vulnerable to being abusive.

Parents Who Abused Each Other Busby et al. (2008) reconfirmed in a study of 30,600 individuals that the family of orientation is the context in which abusive individuals learn violent behavior. That is, individuals who observe their parents being violent with each other and who perpetuate the violence as children end up being more likely to be violent in their adult relationships. However, a majority of children who witness abuse do not continue the pattern. A family history of violence is only one factor out of many that may be associated with a greater probability of adult violence.

12.3 Sexual Abuse in Undergraduate Relationships

> **Rape** *is the only crime in which the victim becomes the accused.*
>
> —*Freda Adler*

Flack et al. (2008) assessed sexual abuse among undergraduates and found that 44 percent of the women and 7 percent of the men in their sample reported at least one unwanted sexual encounter while at the university. The researchers found no evidence of the **red zone**, the first month of the first year of college when women are most likely to be victims of sexual abuse. In a sample of 1,319 undergraduates at a large southeastern university, 35 percent reported being pressured to have sex by a partner they were dating (Knox and Zusman 2009). Katz et al. (2008) noted that some women experience sexual abuse in addition to a larger pattern of physical abuse and that the combination is associated with less general satisfaction, less sexual satisfaction, more conflict, and more psychological abuse from the partner. In effect, women in these covictimization relationships are miserable.

red zone the first month of the first year of college when women are particularly vulnerable to unwanted sexual advances.

acquaintance rape nonconsensual sex between adults who know each other.

date rape nonconsensual sex between two people who are dating or on a date.

Acquaintance and Date Rape

The word *rape* often evokes images of a stranger jumping out of the bushes or a dark alley to attack an unsuspecting victim. However, most rapes are perpetrated not by strangers but by people who have a relationship with the victim. About 85 percent of rapes are perpetrated by someone the woman knows. This type of rape is known as **acquaintance rape**, which is defined as nonconsensual sex between adults (of same or other sex) who know each other. The behaviors of sexual coercion occur on a continuum from verbal pressure and threats to the use of physical force to obtain sexual acts, such as kissing, petting, or intercourse.

Men are also raped. Chapleau et al. (2008) found that about 13 percent of men reported being raped. Rape myths abound as it is assumed that "men can't be raped," "men who are raped are gay," and that "men always want sex." Not only is there a double standard of perceptions that only women can be raped but a double standard is operative in the perception of the gender of the person engaging in sexual coercion. Men who rape are aggressive; women who rape are promiscuous (Oswald and Russell 2006).

One type of acquaintance rape is **date rape**, which refers to nonconsensual sex between people who are dating or on a date. Women who dress seductively, even wives, are viewed by both undergraduate men and women as partly responsible for being raped by their partners (Whatley 2005).

Women are also vulnerable to repeated sexual force. Daigle et al. (2008) noted that 14 percent to 25 percent of college women experience repeat sexual victimization during the same academic year—they are victims of rape or other unwanted sexual force more than once, often in the same month. The primary reason is that they do not change the context; they may stay in the relationship with the same person and continue to use alcohol or drugs in that context.

Both women and men may pressure a partner to have sex. Buddie and Testa (2005) studied sexual aggressiveness in women (both in and out of college) and found

> WHEN A GIRL TELLS ME TO STOP, I STOP.
>
> —*Holden Caulfield,* The Catcher in the Rye

I Was Raped by My Boyfriend

I was 13, a freshman in high school. He was 16, a junior. We were both in band, that's where we met. We started dating and I fell head over heels in love. He was my first real boyfriend. I thought I was so cool because I was a freshman dating a junior. Everything was great for about 6 maybe 8 months and that's when it all started. He began to verbally and emotionally abuse me. He would tell me things like I was fat and ugly and stupid. It hurt me but I was young and was trying so hard to fit in that I didn't really do anything about it.

When he began to drink and smoke marijuana excessively everything got worse. I would try to talk to him about it and he would only yell at me and tell me I was stupid. He began to get physical with the abuse about this time too. He would push me around, grab me really hard, and bruise me up, things like that.

I tried to break up with him and he would tell me that if I broke up with him I would be alone because I was so ugly and fat that no one else would want me. He would say that I was lucky to have him. He began to pressure me for sex. I wasn't ready for sex. I was only 14 at that time. He told everyone that we were having sex and I never said anything different. I was scared to say anything different. It got to the point that I was trying my best to avoid being around him alone. I was afraid of him; I never knew what he would do to me.

Our school band had a competition in Myrtle Beach, South Carolina. When we were there we stayed overnight in a hotel. At the hotel he asked me to come by his room so we could go to dinner together. When I went by his room he pulled me into the room. He locked the door behind me. He pushed me into the bathroom and shut and locked the door behind us. He had his hand over my mouth and told me not to make a sound or he would hurt me. He said that it was time for him to get what everyone already thought he had.

He pushed me up against the counter where the sink was and pulled my clothes off. He put a condom on and began to rape me. I closed my eyes, tears streamed down my face. I had no idea what to do. When he was finished he cleaned himself up and turned on the shower. He pushed me into the shower and told me to take a shower and get ready for dinner. Before he left the room he told me that if I had given in to him and had sex earlier he wouldn't have had to force me. He also told me not to tell anyone or he'd hurt me plus no one would believe me because everyone already thought we had been having sex.

I sat in the shower and cried for a while. I thought it was my fault. I didn't tell anyone about it for two years. I decided not to press charges; I didn't want to go through it again. I regret that decision still to this day. I was an emotional wreck after I was raped. I didn't know how to act. He acted normal, like nothing had happened. A few months later he broke up with me for another girl. It was a relief to me and it made me angry all at the same time. It angered me that he would rape me and then break up with me.

I didn't know what to do with myself. I started working out all the time, several hours a day, every day. After high school I didn't have the time to work out so I developed an eating disorder. This went on for almost 6 years. I am just now dealing with moving past the rape and rebuilding my self-esteem, ability to trust men, and ability to enjoy sex in a new relationship. I encourage anyone who has been raped to speak up about it. Do not let the person get away with it. I also encourage anyone who has been raped to get immediate professional help, don't try to get past it on your own.

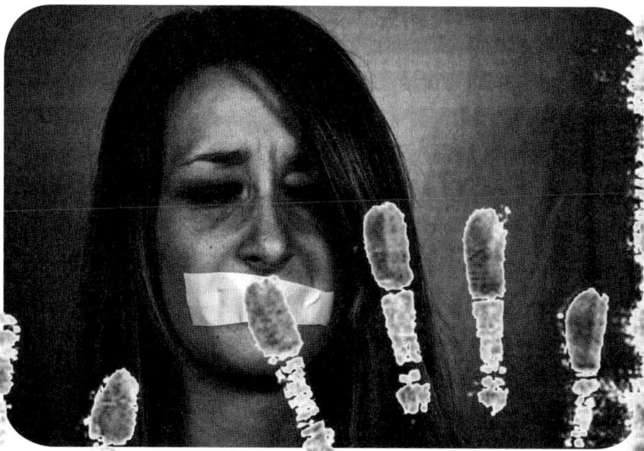

rohypnol date rape drug that causes profound, prolonged sedation and short-term memory loss.

that women who engaged in heavy episodic drinking and who had a high number of sexual partners were most likely to be sexually aggressive and rape or attempt to rape their partners. Sexual arousal is one method a woman uses to be sexually aggressive with a male (see the following quote).

> I locked the room door that we were in. I kissed and touched him. I removed his shirt and unzipped his pants. He asked me to stop. I didn't. Then I sat on top of him. He had had two beers but wasn't drunk. (Struckman-Johnson et al. 2003, 84)

Although acts of sexual aggression of women on men do occur, they are less frequent than women being raped. The feature "I Was Raped by My Boyfriend" on the previous page relays an example provided by one of our students.

Rohypnol—The Date Rape Drug

Rohypnol—also known as the date rape drug, rope, roofies, Mexican Valium, or the "forget (me) pill"—causes profound, prolonged sedation and short-term memory loss. Similar to Valium but ten times as strong, Rohypnol is a prescription drug in Europe and used as a potent sedative. It is sold in the United States for about $5, is dropped in a drink (where it is tasteless and odorless), and causes victims to lose their memory for eight to ten hours. During this time, victims may be raped yet be unaware until they notice signs of it. A former student in our classes reported being drugged ("he put something in my drink") by a "family friend" when she was 16. She noticed blood in her panties the next morning but had no memory of the previous evening. The "friend" is currently being prosecuted.

The Drug-Induced Rape Prevention and Punishment Act of 1996 makes it a crime to give a controlled substance to anyone without their knowledge and with the intent of committing a violent crime (such as rape). Violation of this law is punishable by up to twenty years in prison and a fine of $250,000.

Women are defenseless when drugged. When an assault begins when the woman is not drugged, her responses may vary from pleading to resistance including "turning cold" and "running away." Gidycz et al. (2008) discussed the various responses in terms of background. Women who had been victimized as children were more likely to "freeze and turn cold."

The effect of rape, whether drug-induced or not, is negative to devastating. In addition to the loss of self-esteem, loss of trust, and ability to be sexual, Campbell and Wasco (2005) emphasized that rape also affects the family, friends, and significant others of the victim. In regard to preventing rape, Brecklin and Ullman (2005) analyzed data on 1,623 women and noted that those who had had self-defense or assertiveness training reported that their resistance stopped the offender or made him less aggressive than victims without such training. Women with the training also noted that they were less scared during the attack.

Oprah Winfrey experienced trauma resulting from child abuse. She shares her personal life experience of rape and sexual molestation at a young age by three different family members, spreading information and hope to others who have experienced abuse in their lives.

12.4 Abuse in Marriage Relationships

The chance of abuse in a relationship increases with marriage. Indeed, the longer individuals know each other and the more intimate the relationship, the greater the abuse.

General Abuse in Marriage

Abuse in marriage differs from unmarried abuse in that the husband may feel "ownership" of the wife and feel the need to "control her." The feature "Twenty-Three Years in an Abusive Marriage," on page 232, reveals the experience of horrific marital abuse by a woman who survived it.

Rape in Marriage

Marital rape, now recognized in all states as a crime, is forcible rape by one's spouse (Rousseve 2005). The forced sex may take the form of sexual intercourse, fellatio, or anal intercourse. Sexual violence against women in an intimate relationship is often repeated. Rand (2003) analyzed data from the National Violence Against Women Survey and noted that about half of the women raped by an intimate partner and two-thirds of the women physically assaulted by an intimate partner had been victimized multiple times.

12.5 Effects of Abuse

Abuse affects the physical and psychological well-being of victims. Abuse between parents also affects the children.

Effects of Partner Abuse on Victims

Sarkar (2008) identified the negative impact of intimate partner violence. IPV affected the woman's physical and mental health, and increased the risk for unintended pregnancy and multiple abortions. These women also reported high levels of anxiety and depression that often led to alcohol and drug abuse. Violence on pregnant women significantly increased the risk for infants of low birth weight, preterm delivery, and neonatal death. Katz and Myhr (2008) noted that 21 percent of 193 female undergraduates were experiencing verbal sexual coercion in their current relationships. The effects included feeling psychologically abused, arguing, decreased relationship satisfaction, and decreased sexual functioning.

marital rape forcible rape by one's spouse.

Effects of Partner Abuse on Children

Abuse between adult partners affects children. In the most dramatic effect, some women are abused during their pregnancy, resulting in a high rate of miscarriage and birth defects. Negative effects may also accrue to children who just witness domestic abuse. Kitzmann et al. (2003) analyzed 118 studies to identify outcomes for children who were and were not exposed to violence between their parents. The researchers found more negative outcomes (for example, arguing, withdrawing, avoidance, overt hostility) among children who had witnessed such behavior than those who had not.

Howard et al. (2002) found that adolescents who had witnessed violence reported "intrusive thoughts, distraction, and feeling a lack of belonging" (p. 455). Children who witness high levels of parental violence are also more likely to blame themselves for the violence (Grych et al. 2003), to be violent toward their parents (particularly their mother) (Ulman and Straus 2003), and to engage in aggressive delinquent behavior (Kernic et al. 2003).

It is not unusual for children to observe and to become involved in adult domestic violence. One-fourth of 114 battered mothers noted that their children yelled, called for help, or intervened when an adult partner was physically abusing the mother (Edleson et al. 2003).

12.6 The Cycle of Abuse

The cycle of abuse exemplified in "I Got Flowers Today" (p 233) begins when a person is abused and the perpetrator feels regret, asks for forgiveness, and starts acting nice (e.g., gives flowers). The victim, who perceives

23 YEARS
IN AN ABUSIVE MARRIAGE

My name is Jane and I am a survivor (though it took me forever to get out) of a physically and mentally abusive marriage. I met my husband-to-be in high school when I was 15. He was the most charming person that I had ever met. He would see me in the halls and always made a point of coming over to speak. I remember thinking how cool it was that an older guy would be interested in a little freshman like me. He asked me out but my parents would not let me go out with him until I turned 16 so we met at the skating rink every weekend until I was old enough to go out on a date with him. Needless to say, I was in love and we didn't date other people.

When we married I remember the wedding day very well as I was on top of the world because I was marrying the man of my dreams whom I loved very much. My family (they did not like him) tried for the longest time to talk me out of marrying him. My dad always told me that something was not right with him but I did not listen. I wish to God that I had listened to my dad and walked out of the church before I said "I do."

I remember the first slap, which came six months after we had been married. I had burned the toast for his dinner. I was shocked when he slapped me, cried, and when he saw the blood coming from my lip he started crying and telling me he was sorry and that he would never do that again. But the beatings never stopped for twenty-three years. His most famous way of torturing me was to put a gun to my head while I was sleeping and wake me up and pull the trigger and say next time you might not be so lucky.

I left this man a total of seventeen times but I always went back after I got well from my beatings. At one point during this time every bone in my body had been broken with the exception of my neck and back. Some bones more than once and sometimes more than one at a time. I was pushed down a flight of stairs and still suffer from the effects of it to this day. I was stabbed several times and I had to be hospitalized a total of six times.

After our children were born I thought the beatings would stop but they didn't. When our children got older he started on them so I took many beatings for my children just so he would leave them alone. Three years ago he beat me so badly that I almost died. I remember being in the hospital wanting to die because if I did I would not hurt anymore and I would finally be safe. At that time I was thinking that death would be a blessing since I would not have to go back to live with this monster.

While I was in the hospital a therapist came to see me and told me that I had two choices—go home and be killed or fight for my life and live. Instead of going back to the monster, I moved my children in with my mother. It took me three years to get completely clear of him, and even today he harasses me but I am free. I had to learn how to think for myself and how to love again. I am now remarried to a wonderful man.

If you are in an abusive relationship there is help out there and please don't be afraid to ask for it. I can tell you this much—the first step (deciding to leave) is the hardest to take. If and when you do leave I promise your life will be better—it may take awhile but you will get there.

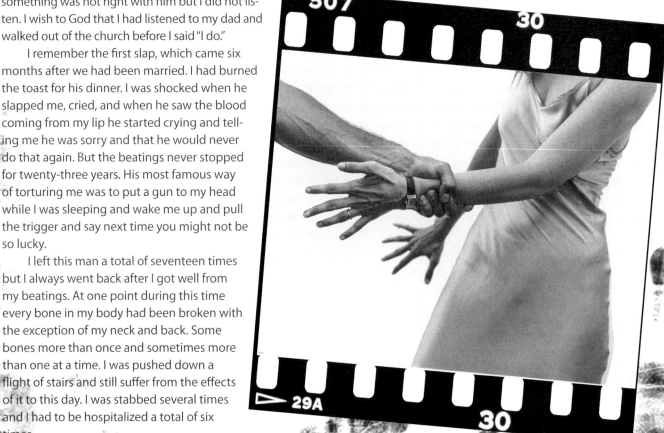

few options and feels guilty terminating the relationship with the partner who asks for forgiveness, feels hope for the relationship at the contriteness of the abuser and does not call the police or file charges. Forgiving the partner and taking him back usually occurs seven times before the partner leaves for good. Shakespeare (in *As You Like It*) said of such forgiveness, "Thou prun'st a rotten tree."

After the forgiveness, couples usually experience a period of making up or honeymooning, during which the victim feels good again about the partner and is hopeful for a nonabusive future. However, stress, anxiety, and tension mount again in the relationship, which is relieved by violence toward the victim. Such violence is followed by the familiar sense of regret and pleadings for forgiveness, accompanied by being nice (a new bouquet of flowers, and so on).

Figure 12.1 illustrates this cycle. In the rest of this section, we discuss reasons people stay in an abusive relationships and how to get out of such relationships.

As the cycle of abuse reveals, some victims do not prosecute their partners who abuse them. To deal with this problem, Los Angeles has adopted a "zero tolerance" policy toward domestic violence. Under the law, an arrested person is required to stand trial and his victim required to testify against the perpetrator. The sentence in Los Angeles County for partner abuse is up to six months in jail and a fine of $1,000.

Why People Stay in Abusive Relationships

One of the most frequently asked questions of people who remain in abusive relationships is, "Why do you stay?" Few and Rosen (2005) interviewed twenty-five women who had been involved in abusive dating relationships from three months to nine years (average = 2.4 years) to find out why they stayed. The researchers

Figure 12.1
The Cycle of Abuse

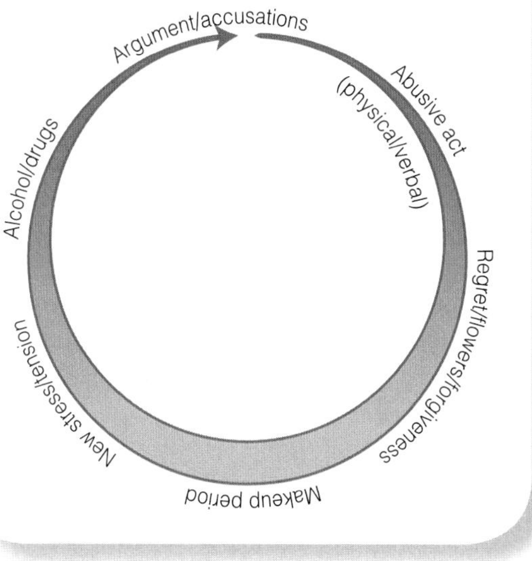

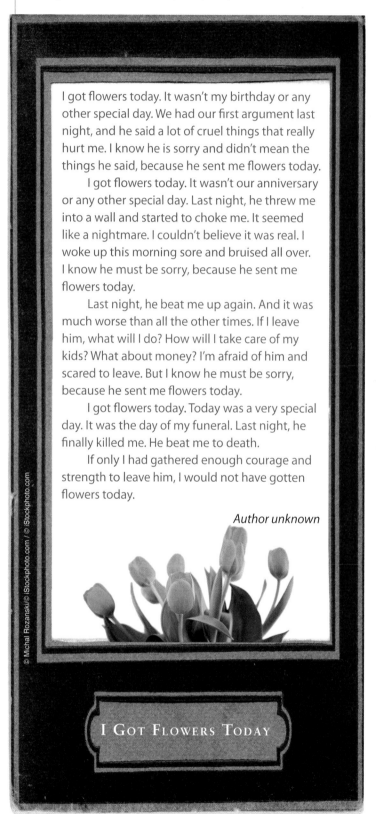

I got flowers today. It wasn't my birthday or any other special day. We had our first argument last night, and he said a lot of cruel things that really hurt me. I know he is sorry and didn't mean the things he said, because he sent me flowers today.

I got flowers today. It wasn't our anniversary or any other special day. Last night, he threw me into a wall and started to choke me. It seemed like a nightmare. I couldn't believe it was real. I woke up this morning sore and bruised all over. I know he must be sorry, because he sent me flowers today.

Last night, he beat me up again. And it was much worse than all the other times. If I leave him, what will I do? How will I take care of my kids? What about money? I'm afraid of him and scared to leave. But I know he must be sorry, because he sent me flowers today.

I got flowers today. Today was a very special day. It was the day of my funeral. Last night, he finally killed me. He beat me to death.

If only I had gathered enough courage and strength to leave him, I would not have gotten flowers today.

Author unknown

I GOT FLOWERS TODAY

© Michal Rozanski/© iStockphoto.com / © iStockphoto.com

entrapped stuck in an abusive relationship and unable to extricate one's self from the abusive partner.

conceptualized the women as **entrapped**—stuck in an abusive relationship and unable to extricate one's self from the abusive partner. Indeed, these women escalated their commitment to stay in hopes that doing so would eventually pay off. Among the factors of their perceived investment were the time they had already spent in the relationship, the sharing of their emotional self with their partner, and the relationships to which they were connected because of the partner. In effect, they had invested time with a partner they were in love with and wanted to turn the relationship around into a safe, nonabusive one. The following are some of the factors explaining how abused women become entrapped:

- Fear of loneliness ("I'd rather be with someone who abuses me than alone")
- Love ("I love him")
- Emotional dependency ("I need him")
- Commitment to the relationship ("I took a vow 'for better or for worse'")
- Hope ("He will stop")
- A view of violence as legitimate ("All relationships include some abuse")
- Guilt ("I can't leave a sick man")
- Fear ("He'll kill me if I leave him")
- Economic dependence ("I have no place to go")
- Isolation ("I don't know anyone who can help me")

Battered women also stay in abusive relationships because they rarely have escape routes related to educational or employment opportunities, their relatives are critical of plans to leave the partner, they do not want to disrupt the lives of their children, and they may be so emotionally devastated by the abuse (anxious, depressed, or suffering from low self-esteem) that they feel incapable of planning and executing their departure.

How One Leaves an Abusive Relationship

Leaving an abusive partner begins with the decision to do so. Such a choice often follows the belief that one will die or one's children will be harmed by staying. A plan comes into being and the person acts (for example, moves in with a sister, mother, friend, or goes to a homeless shelter). If the alternative is better than being in the abusive context, the person will stay away. Otherwise, the person may go back and start the cycle all over. As noted previously, this leaving and returning typically happens seven times.

Sometimes the woman does not just disappear while the abuser is away but calls the police and has the man arrested for violence and abuse. While the abuser is in jail, she may move out and leave town. In either case, disengagement from the abusive relationship takes a great deal of courage. Calling the National Domestic Violence Hotline (800-799-7233 [SAFE]), available twenty-four hours, is a point of beginning.

Some women who withdraw go to abuse shelters. A team of researchers (Ham-Rowbottom et al. 2005) conducted a follow-up of eighty-one abuse victims who had escaped their situations and graduated from abuse shelters. Although all of the women were now living independently, 43 percent and 75 percent reported clinical levels of depression and trauma symptoms, respectively. The researchers noted that earlier child sexual abuse accounted for much of the continued psychological depression and trauma.

Kress et al. (2008) noted that involvement with an intimate partner who is violent may be life-threatening. Particularly if the individual decides to leave the violent partner, the abuser may react with more violence and murder the person who has left. Indeed, a third of mur-

Leaving an abusive partner begins with the decision to do so. Unfortunately, returning is common and typically happens seven times.

ders that occur in domestic violence cases occur shortly after a breakup. Specific signs that could be precursors to someone about to murder an intimate partner are stalking, strangulation, forced sex, physical abuse, perpetrator suicidality, gun ownership, and drug or alcohol use on the part of the violent partner.

Individuals in such relationships should be cautious about how they react and develop a safe plan of withdrawal. Safety plans will vary but include the following:

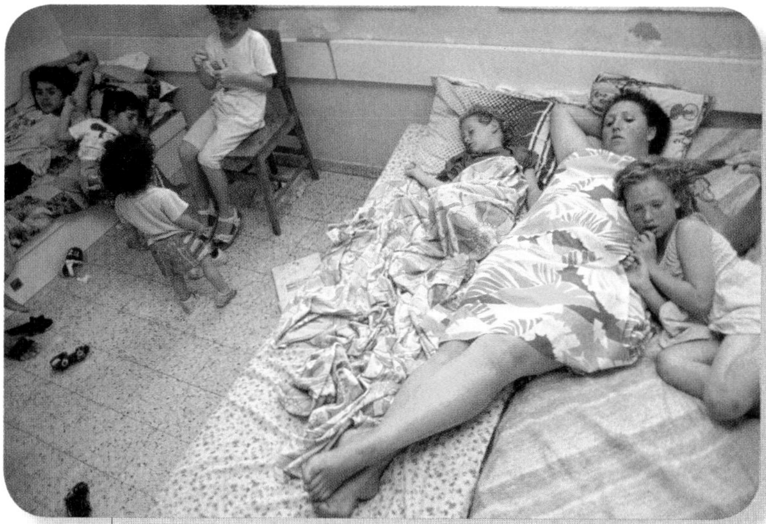

© David H. Wells/Corbis

- Identifying a safe place an individual can go the next time she needs to leave the house. A friend or women's shelter must be set up in advance. The victim needs to stay in a protected context.
- Telling friends or neighbors about the violence and requesting that they call the police if they hear suspicious noises or witness suspicious events.
- Storing an escape kit (for example, keys, money, checks, important phone numbers, medications, Social Security cards, bank documents, birth certificates, change of clothes, and so on) somewhere safe (and usually not in the house) (Kress et al. 2008).

Above all, individuals should trust their instincts and do what they can to de-escalate the situation.

Strategies to Prevent Domestic Abuse

Family violence and abuse prevention strategies are focused at three levels: the general population, specific groups thought to be at high risk for abuse, and families who have already experienced abuse. Public education and media campaigns aimed at the general population convey the criminal nature of domestic assault, suggest ways abusers might learn to prevent abuse (seek therapy for anger, jealousy, or dependency), and identify where abuse victims and perpetrators can get help. Rothman and Silverman (2007) noted that preventative abuse programs for college students were effective in reducing violence for both men and women. However, people with a prior history of assault; who were gay, lesbian, or bisexual; or who were binge drinkers did not benefit.

Preventing or reducing family violence through education necessarily involves altering aspects of American culture that contribute to such violence. For example, violence in the media must be curbed or eliminated (although not easy with nightly video clips of bombing assaults in other countries and reruns of the violent *Sopranos*, it will eventually disappear), and traditional gender roles and views of women and children as property must be replaced with egalitarian gender roles and respect for women and children.

Another important cultural change is to reduce violence-provoking stress by reducing poverty and unemployment and by providing adequate housing, nutrition, medical care, and educational opportunities for everyone. Integrating families into networks of community and kin would also enhance family well-being and provide support for families under stress.

> It is easier to stay out than to get out.
> — *Mark Twain*

Treatment of Partner Abusers

Silvergleid and Mankowski (2006) identified learning a new way of masculinity and making a personal commitment to change as crucial to the rehabilitation of the male batterer. These new behaviors can be learned in individual or group therapy. Because alcohol or drug abuse and violence toward a partner are often related, addressing one's alcohol or substance abuse problem is often a prerequisite for treating partner abuse (Stuart 2005).

In addition, some men stop abusing their partners only when their partners no longer put up with it. One abusive male said that his wife had to leave him before he learned not to be abusive toward women. "I've never touched my second wife," he said.

Stress and Crisis in Relationships

SECTIONS

stress a nonspecific response of the body to demands made on it.

Ann Landers was once asked what she would consider the single most useful bit of advice all people could profit from. She replied, "Expect trouble as an inevitable part of life, and when it comes, hold your head high, look it squarely in the eye and say, 'I will be bigger than you.'" Life indeed brings both triumphs and tragedies. Nearly half of all adults report experiencing at least one traumatic event at some point in their lives (Ozer et al. 2003). These are often turning points that may come at any time (Leonard and Burns 2006). This chapter is about experiencing and coping with these events and the day-to-day stress we feel during the in-between time.

13.1 Personal Stress and Crisis Events

In this section, we review the definitions of crisis and stressful events, the characteristics of resilient families, and a framework for viewing a family's reaction to a crisis event.

The problem is not that there are problems. The problem is expecting otherwise and thinking that having problems is a problem.

—Theodore Rubin, psychiatrist

Definitions of Stress and Crisis Events

Stress is a nonspecific response of the body to demands made on it (physical, environmental, or interpersonal). Stress is particularly acute in the military. Duparcq (2008) wrote that "U.S. soldiers in Iraq can find stress deadlier than the enemy," and cited deployments as the cause of increasing post-traumatic stress syndrome, divorce, and suicide. Stress is often accompanied by irritability, high blood pressure, and depression. Stress is a process rather than a state. For example, a person will experience different levels of stress through-

© Charles Taylor/iStockphoto.com / © Rubberball/Jupiterimages out a divorce—the stress involved in acknowledging that one's marriage is over, telling the children, leaving the family residence, getting the final decree, and seeing one's ex may all result in varying levels of stress.

A **crisis** is a crucial situation resulting in a sharp change for which typical patterns of coping are not adequate and new patterns must be developed. A family crisis is a situation that upsets the normal functioning of the family and requires a new set of responses to the

crisis a sharp change for which typical patterns of coping are not adequate and new patterns must be developed.

YOU WIN SOME, YOU LOSE SOME, AND SOME GET RAINED OUT, BUT YOU GOTTA SUIT UP FOR THEM ALL.

— *J. Askenberg*

> # Misfortune
> *does not always come to injure.*
>
> —*Robert Frost, poet*

stressor. Sources of stress and crises can be external, such as Hurricane Ike that devastated the coast of Texas in 2008 or the earthquake in Italy in 2009. Other examples of an external crisis are economic recession, tornados, downsizing, or military deployment to Afghanistan. Stress and crisis events may also be induced internally (for example, alcoholism, an extramarital affair, or Alzheimer's disease of spouse or parents). Ingram et al. (2008) noted that, of 300,000 crisis calls to a national hotline over a five-year period, 73 percent were about parenting, youth, and mental health concerns. Across the lifespan, calls about loneliness increased and calls about depression decreased.

Stressors or crises may also be categorized as expected or unexpected. Examples of expected family stressors include the need to care for aging parents and the death of one's parents. Unexpected stressors include contracting HIV, a miscarriage, or the suicide of one's teenager.

Both stress and crises are normal parts of family life and sometimes reflect a developmental sequence. Pregnancy, childbirth, job changes or loss, children leaving home, retirement, and widowhood are all stressful and predictable for most couples and families. Crisis events may have a cumulative effect: the greater the number in rapid succession, the greater the stress. Stress that spouses experience spills over into and negatively affects the marital relationship, with husbands being more affected by the wife's stress than vice versa (Neff and Karney 2007).

resiliency the ability of a family to respond to a crisis in a positive way.

family resilience the successful coping of family members under adversity that enables them to flourish with warmth, support, and cohesion.

Resilient Families

Just as the types of stress and crisis events vary, individuals and families vary in their abilities to respond successfully to crisis events. **Resiliency** refers to a family's strengths and ability to respond to a crisis in a positive way. Black and Lobo (2008) defined **family resilience** as the successful coping of family members under adversity that enables them to flourish with warmth, support, and cohesion. The key factors that promote family resiliency include positive outlook, spirituality, flexibility, communication, financial management, shared family recreation, routines or rituals, and support networks. A family's ability to bounce back from a crisis (from loss of one's job to the death of a family member) reflects its level of resiliency. Resiliency may also be related to individuals' perceptions of the degree to which they are in control of their destiny.

13.2 Positive Stress-Management Strategies

Researchers Burr and Klein (1994) administered an eighty-item questionnaire to seventy-eight adults to assess how families experiencing various stressors such as bankruptcy, infertility, a disabled child, and a troubled teen used various coping strategies and how useful they evaluated these strategies. In the following, we detail some helpful stress-management strategies.

© iStockphoto.com / © Kenneth C. Zirkel/iStockphoto.com

Changing Basic Values and Perspective

The strategy that the highest percentage of respondents reported as being helpful was changing basic values as a result of the crisis situation. Survivors of Hurricane Kyle, which dev-

> Our remedies oft in ourselves do lie.
>
> —*Shakespeare,* All's Well That Ends Well

WHEN WRITTEN IN CHINESE, THE WORD 'CRISIS' IS COMPOSED OF TWO CHARACTERS. ONE REPRESENTS DANGER AND THE OTHER REPRESENTS OPPORTUNITY.

—*President John Fitzgerald Kennedy*

astated Galveston in 2008, noted that focusing on the sparing of their lives and those close to them rather than the loss of their home or material possessions was an essential view in coping with the crisis. Sharpe and Curran (2006) confirmed that finding positive meaning in a crisis situation is associated with positive adjustment to the situation. Similarly, Waller (2008) studied new parents at two time frames (when the child was 1 and 4), and noted that whether they survived the crisis of having a child was a function of their perception. She found that "parents in stable unions framed tensions as manageable within the context of a relationship they perceived to be moving forward, whereas those in unstable unions viewed tensions as intolerable in relationships they considered volatile."

Some crisis events provide an opportunity for positive growth. In responding to the crisis of bankruptcy, people may reevaluate the importance of money and conclude that relationships are more important. In coping with unemployment, people may decide that the amount of time they spend with family members is more valuable than the amount of time they spend making money. Buddhists have the saying, "Pain is inevitable; suffering is not." This is another way of emphasizing that how one views a situation, not the situation itself, determines its impact on you.

Now, if you are going to win any battle, you have to do one thing. You have to make the mind run the body.

—*General George S. Patton*

Exercise

The Centers for Disease Control and Prevention (CDC) and the American College of Sports Medicine (ACSM) recommend that people aged 6 years and older engage regularly, preferably daily, in light to moderate physical activity for at least thirty minutes at a time. These recommendations are based on research that has shown the physical, emotional, and cognitive benefits of exercise. Tetlie et al. (2008) confirmed that a structured exercise program lasting eight to twelve weeks was associated with individuals reporting improved feelings of well-being. In addition, Taliaferro et al. (2008) noted that vigorous exercise and involvement in sports are associated with lower rates of suicide among adolescents.

Friends and Relatives

A network of relationships is associated with successful coping with various life transitions (Levitt et al. 2005).

WE CANNOT DIRECT THE WIND, BUT WE CAN ADJUST THE SAILS.

—*Bertha Calloway, founder Great Plains Black History Museum*

biofeedback a process in which information that is relayed back to the brain enables people to change their biological functioning.

Women are more likely to feel connected to others and to feel that they can count on others in times of need (Weckwerth and Flynn 2006). News media covering hurricanes, earthquakes, and tsunamis emphasized that, as long as one's family was still together, individuals were less concerned about the loss of material possessions.

Love

A love relationship also helps individuals cope with stress. Being emotionally involved with another and sharing the experience with that person helps to insulate individuals from being devastated by a crisis event. Love is also viewed as helping resolve relationship problems. More than 85 percent (85.9 percent) of undergraduate males and 72.5 percent of undergraduate females agreed that, "If you love someone enough, you will be able to resolve your problems with that person" (Dotson-Blake et al. 2009).

Religion and Spirituality

Religion may be helpful in adjusting to a crisis. Park (2006) found that religion is associated with providing meaning for experiencing a crisis and for adjusting to it. In addition, religion provides a framework for being less punitive. Rather than be in a rage and seek revenge against a perceived aggressor, religious individuals might "turn the other cheek" (Unnever et al. 2005).

Humor

A sense of humor is related to lower anxiety and a happier mood. Indeed, a team of researchers compared the effects of humor, aerobic exercise, and listening to music on the reduction of anxiety and found humor to be the most effective. Just sitting quietly seemed to have no effect on reducing one's anxiety (Szabo et al. 2005).

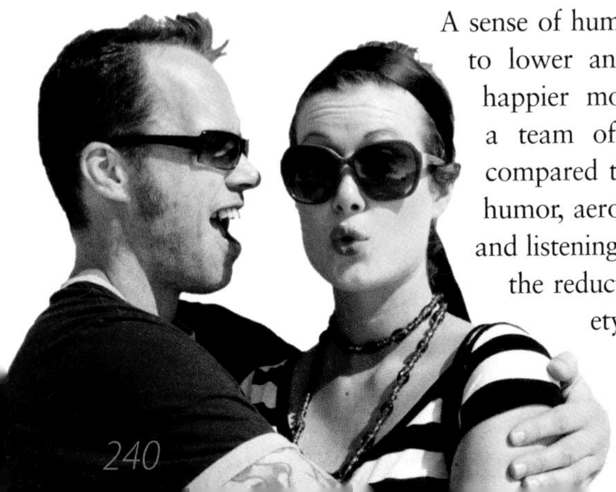

Sleep

Getting an adequate amount of sleep is also associated with lower stress levels (Mostaghimi et al. 2005). Even midday naps are associated with positive functioning, particularly memory performance (Schabus et al. 2005). Indeed, adequate sleep helps one to respond to daily stress and to crisis events.

Biofeedback

Biofeedback is a process in which information that is relayed back to the brain enables people to change their biological functioning. Biofeedback treatment teaches a person to influence biological responses such as heart rate, nervous system arousal, muscle contractions, and even brain wave functioning. Biofeedback is used at about 1,500 clinics and treatment centers worldwide. A typical session lasts about an hour and costs $60 to $150. The following are several types of biofeedback:

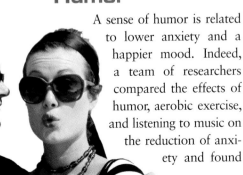

1. **Electromyographic (EMG) biofeedback.** EMG measures electrical activity created by muscle contractions, and is often used for relaxation training and for stress and pain management.
2. **Thermal or temperature biofeedback.** Because stress causes blood vessels in the fingers to constrict, reducing blood flow and leading to cooling, thermal biofeedback uses a temperature sensor to detect changes in temperature of the fingertips or toes. This trains people to quiet the nervous system arousal mechanisms that produce hand and/or foot cooling, and is often used for stress, anxiety, and pain management.
3. **Galvanic skin response (GSR) biofeedback.** GSR utilizes a finger electrode to measure sweat gland activity. This measure is very useful for relaxation and stress management training and is also

used in the treatment of attention-deficit/hyperactivity disorder.

4. **Neurofeedback.** Also called *neurobiofeedback* or *EEG (electroencephalogram) biofeedback*, neurofeedback may be particularly helpful for individuals coping with a crisis. It trains people to enhance their brain-wave functioning and has been found to be effective in treating a wide range of conditions, including anxiety, stress, depression, tension and migraine headaches, addictions, and high blood pressure.

Because neurofeedback is the fastest-growing field in biofeedback, let's take a closer look at this treatment modality. Neurofeedback involves a series of sessions in which a client sits in a comfortable chair facing a specialized game computer. Small sensors are placed on the scalp to detect brain-wave activity and transmit this information to the computer. The neurofeedback therapist (in the same room) also sits in front of a computer that displays the client's brain-wave patterns in the form of an electroencephalogram (EEG). After a clinical assessment of the client's functioning, the therapist determines what kinds of brain-wave patterns are optimal for the client. During neurofeedback sessions, clients learn to produce desirable brain waves by controlling a computerized game or task, similar to playing a video game, but the client's brain waves—instead of a joystick—control the game.

Neurofeedback is like an exercise of the brain, helping it to become more flexible and effective. Unlike body exercise, which will lose its benefits over time when training is stopped; brain-wave training generally does not. Once the brain is trained to function in its optimal state (which may take an average of twenty to twenty-five sessions), it generally remains in this more healthy state. Neurofeedback therapists liken the process to that of learning to ride a bicycle: once you learn to ride a bike, you can do so even if you have not in years. One neurofeedback therapist explains, "clients speak often of their disorders—panic attacks, chronic pain, etc.—as if they were stuck in a certain pattern of response. Consistently in clinical practice, EEG biofeedback helps 'unstick' people from these unhealthy response patterns" (Carlson-Catalano 2003).

Deep Muscle Relaxation

Tensing and relaxing one's muscles have been associated with an improved state of relaxation. Calling it abbreviated progressive muscle relaxation (APMR), Termini (2006) found that the cognitive benefits were particularly evident; people who tensed and relaxed various muscle groups noticed a mental relaxation more than a physical relaxation.

Education

Sometimes becoming informed about a family problem helps to cope with the problem. Friedrich et al. (2008) studied how siblings cope with the fact that a brother or sister is schizophrenic. Education and family support were the primary mechanisms. Becoming informed about schizophrenia helped siblings understand that the "parents were not to blame."

Counseling for Children

Baggerly and Exum (2008) reviewed the value of counseling for children experiencing natural disasters such as hurricanes. Such counseling involves providing a safe context for children and having them remind themselves that they are safe (cognitive behavior therapy).

13.3 Harmful Stress-Management Strategies

Some coping strategies not only are ineffective for resolving family problems but also add to the family's stress by making the problems worse. Respondents in the Burr and Klein (1994) research identified several strategies they regarded as harmful to overall family functioning. These included keeping feelings inside, taking out frustrations on or blaming others, and denying or avoiding the problem.

Burr and Klein's research also suggests that women and men differ in their perceptions of the usefulness of various coping strategies. Women were more likely than men to view as helpful such strategies as sharing concerns with relatives and friends, becoming more involved in religion, and expressing emotions. Men were more likely than women to use potentially harmful strategies such as using alcohol, keeping feelings inside, or keeping others from knowing how bad the situation was.

13.4 Family Crisis Examples

Some of the more common crisis events that spouses and families face include physical illness, mental challenges, an extramarital affair, unemployment, substance abuse, and death.

Physical Illness and Disability

Absence of physical illness is important for individuals to define themselves as being healthy overall. Although approximately 60 percent of those aged 85 and older define themselves as healthy (Ostbye et al. 2006), physical problems are not unusual.

Reacting to Prostate Cancer: One Husband's Experience

For men, prostate cancer is an example of a medical issue that can rock the foundation of their personal and marital well-being. In the following section, a student described his experience with prostate cancer as follows:

> Since my father had prostate cancer, I was warned that it is genetic and to be alert as I reached age 50. At the age of 56, I noticed that I was getting up more frequently at night to urinate. My doc said it was probably just one of my usual prostate infections, and he prescribed the usual antibiotic. When the infection did not subside, a urologist did a transrectal ultrasound (TRUS) needle biopsy. The TRUS gives the urologist an image of the prostate while he takes about ten tissue samples from the prostate with a thin, hollow needle.
>
> Two weeks later I learned the bad news (I had prostate cancer) and the good news (it had not spread). Since the prostate is very close to the spinal column, failure to act quickly can allow time for cancer cells to spread from the prostate to the bones. My urologist outlined a number of treatment options, including traditional surgery, laparoscopic surgery, radiation treatment, implantation of radioactive "seeds," cryotherapy (in which liquid nitrogen is used to freeze and kill prostate cancer cells), and hormone therapy, which blocks production of the male sex hormones that stimulate growth of prostate

IN THIS AGE, WHICH BELIVES THAT THERE IS A SHORT CUT TO EVERYTHING, THE GREATEST LESSON TO BE LEARNED IS THAT THE MOST DIFFICULT WAY IS, IN THE LONG RUN, THE EASIEST.

—Henry Miller, novelist

cancer cells. He recommended traditional surgery ("radical retropubic prostatectomy"), in which an incision is made between the belly button and the pubic bone to remove the prostate gland and nearby lymph nodes in the pelvis. This surgery is generally considered the "gold standard" when the disease is detected early. Within three weeks I had the surgery.

Every patient awakes from radical prostate surgery with urinary incontinence and impotence—which can continue for a year, two years, or forever. Such patients also awake from surgery hoping that they are cancer-free. This is determined by laboratory analysis of tissue samples taken during surgery. The patient waits for a period of about two weeks, hoping to hear the medical term "negative margins" from his doctor. That finding means that the cancer cells were confined to the prostate and did not spread past the margins of the prostate. A finding of negative margins should be accompanied by a PSA [prostate specific antigen] score of zero, confirming that the body no longer detects the presence of cancer cells. I cannot describe the feeling of relief that accompanies such a report, and I am very fortunate to have heard those words used in my case.

The psychological effects have been devastating—more for me than for my partner. I have only been intimate with one woman—my wife—and having intercourse with her was one of the greatest pleasures in my life. For a year, I was left with no erection and an inability to have an orgasm. Afterwards I was able to have an erection (via Caverject [$25], an injection) and an orgasm. Dealing with urinary incontinence (I refer to myself as Mr. Drippy) is a "wish it were otherwise" on my psyche.

When faced with the decision to live or die, the choice for most of us is clear. In my case, I am alive, cancer-free, and enjoying the love of my life (now in our 44th year together).

In addition to prostate cancer, chronic degenerative diseases (autoimmune dysfunction, rheumatoid arthritis, lupus, Crohn's disease, and chronic fatigue syndrome) can challenge a mate's and couple's ability to cope. These illnesses are particularly invasive in that conventional medicine has little to offer besides pain medication. For example, spouses with chronic fatigue syndrome may experience financial consequences ("I could no longer meet the demands of my job so I quit"), gender role loss ("I couldn't cook for my family" or "I was no longer a provider"), and changed perceptions by their children ("They have seen me sick for so long they no longer ask me to do anything"). In addition, osteoporosis is a health threat for 44 million older women and results in a decline in their ability to perform routine activities (Roberto 2005).

In those cases in which the illness is fatal, **palliative care** is helpful. This term describes the health care for the individual who has a life-threatening illness (focusing on relief of pain and suffering) and support for them and their loved ones. Such care may involve the person's physician or a palliative care specialist who works with the physician, nurse, social worker, and chaplain. Pharmacists or rehabilitation specialists may also be involved. The effects of such care are to approach the end of life with planning (how long should life be sustained on machines?) and forethought to relieve pain and provide closure.

palliative care health care focused on the relief of pain and suffering of the individual who has a life-threatening illness and support for them and their loved ones.

Another physical issue with which some parents cope is that of their children being overweight. Body mass index (BMI) is calculated as weight in kilograms divided by height in meters squared; overweight is indicated by a BMI of 25.0 to 29.9, and obese by 30.0 or higher (Pyle et al. 2006). Routh and Rao (2006) studied 225 9- to 10-year-olds in primary schools and found that 20 percent were overweight and 5 percent were obese. National figures reflect that 31.5 percent are at risk for being overweight and that 16.5 percent could be classified as overweight. Minority females and individuals of lower socioeconomic status are more prone to being overweight or obese. Not only does this chronic physical health issue have medical, life-threatening implications (Pyle et al. 2006) in both industrial and developing countries (Flynn et al. 2006), but there is also prejudice and discrimination toward overweight children. To assess your attitude toward overweight children, take the self-assessment online at 4ltrpress.cengage.com/mf.

Mental Illness

Everyone experiences problems in living. In a national survey of 9,282 respondents, the reported psychological problems and their percentages of lifetime incidence include: anxiety (29 percent), impulse control disorder (25 percent), mood disorder (21 percent), and substance abuse (15 percent) (Mahoney 2005). Five percent of adults in the United States between the ages of 18 and 35 report being depressed (Pratt and Brody 2008).

Insel (2008) noted the enormous economic costs of serious mental illness that involve a high rate of emergency room care (for example, suicide attempts), high prevalence of pulmonary disease (people with serious mental illness smoke 44 percent of all cigarettes in the United States), and early mortality (a loss of thirteen to thirty-two years).

The toll of mental illness on a relationship can be immense. A major initial attraction of partners to each other includes intellectual and emotional qualities. Butterworth and Rodgers (2008) surveyed 3,230 couples to assess the degree to which mental illness of a spouse or spouses affects divorce and found that couples in which either men or women reported mental health problems had higher rates of marital disruption than couples in which neither spouse experienced mental health problems. For couples in which both spouses reported mental health problems, rates of marital disruption reflected the additive combination of each spouse's separate risk. As an aside, mental illness is also a disadvantage for those seeking a remarriage. Teitler and Reichman (2008) found that unmarried mothers with mental illness were about two-thirds as likely as mothers without mental illness to marry.

Families may also experience the added stress associated with caring for mentally ill children. Typical mental illnesses in children include autism (three to four times more common in boys), attention-deficit/hyperactivity disorder (4 percent of children, ages 3 to 17), and antisocial behavior (*Statistical Abstract of the United States, 2009*, Table 179). These difficulties can stress spouses to the limit of their coping capacity, which may put an enormous strain on their marriage.

The reverse is also true; children must learn how to cope with the mental illness of their parents. Mordoch and Hall (2008) studied twenty-two children between 6 and 16 years of age who were living part- or full-time with a parent with depression, schizophrenia, or bipolar illness. They found that the children learned to maintain connections with their parents by creating and keeping a safe distance between themselves and their parents so as not to be engulfed by their parents' mental illnesses.

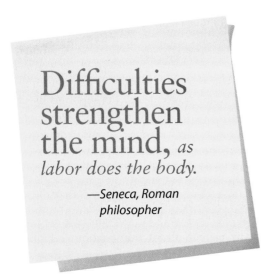

Difficulties strengthen the mind, *as labor does the body.*

—Seneca, Roman philosopher

Middle-Age Crazy (Midlife Crisis)

The stereotypical explanation for 45-year-old people who buy convertible sports cars, have affairs, marry 20-year-olds, or adopt a baby is that they are "having a midlife crisis." The label conveys that they feel old, think that life is passing them by, and seize one last great chance to do what they have always wanted. Indeed, one father (William Feather) noted, "Setting a good example for your children takes all the fun out of middle age."

However, a ten-year study of close to 8,000 U.S. adults aged 25 to 74 by the MacArthur Foundation Research Network on Successful Midlife Development revealed that, for most respondents, the middle years brought no crisis at all but a time of good health, productive activity, and community involvement. Less than a quarter (23 percent) reported a "crisis" in their lives. Those who did experience a crisis were going through a divorce. Two-thirds were accepting of getting older; one-third did feel some personal turmoil related to the fact that they were aging (Goode 1999).

Of those who initiated a divorce in midlife, 70 percent had no regrets and were confident that they did the right thing. This is the result of a study of 1,147 respondents aged 40 to 79 who experienced a divorce in their forties, fifties, or sixties. Indeed, midlife divorcers' levels of happiness or contentment were similar to those of single individuals their own age and those who remarried (Enright 2004).

Some people embrace middle age. The Red Hat Society (http://www.redhatsociety.com/) is a group of women who have decided to "greet middle age with verve, humor, and élan. We believe silliness is the comedy relief of life [and] share a bond of affection, forged by common life experiences and a genuine enthusiasm for wherever life takes us next." The society traces its beginning to Sue Ellen Cooper buying a bright red hat because of a poem written by Jenny Joseph in 1961 titled "The Warning Poem." It says:

> When I am an old woman I shall wear purple
> With a red hat which doesn't go and doesn't suit me.

Cooper gave red hats to friends as they turned 50. The group wore their red hats and purple dresses out to tea, and that's how it got started. Now there are more than 1 million members worldwide.

The Red Hat Society is a group of women who have decided to "greet middle age with verve, humor, and élan."

© Evan Hurd/Corbis

In the rest of this chapter, we examine how spouses cope with the crisis events of an extramarital affair, unemployment, drug abuse, and death. Each of these events can be viewed either as devastating and the end of meaning in one's life or as an opportunity and challenge to rise above.

Extramarital Affair (and Successful Recovery)

Extramarital affair refers to a spouse's sexual involvement with someone outside the marriage. Affairs are of different types, which may include the following:

1. **Brief encounter (situationally determined affairs).** A spouse hooks up with or meets a stranger at a conference. In this case, the spouse is usually out of town, and alcohol is involved.
2. **Periodic sexual encounters.** A spouse is sexually unsatisfied in the marriage and seeks external sex, often with a hooker (for example, former New York governor Eliot Spitzer). A married person of bisexual orientation may also seek a periodic encounter outside the marriage.
3. **Instrumental or utilitarian affair.** This is sex in exchange for a job or promotion, to get back at one's spouse, to evoke jealousy, or to transition out of a marriage.
4. **Coping mechanism.** Sex can be used to enhance one's self-concept or feeling of sexual inadequacy, compensate for failure in business, cope with the death of a family member, test out one's sexual orientation, and so on.
5. **Paraphiliac affairs.** In these, spouses act out sexual fantasies that most people would consider to be bizarre or abnormal sexual practices, such as sexual masochism, sexual sadism, or transvestite fetishism.
6. **New love.** A spouse may be in love with the new partner and may plan marriage after divorce (Bagarozzi 2008).

extramarital affair a spouse's sexual involvement with someone outside the marriage.

In addition, the computer or Internet affair is another type of affair. Although legally an extramarital affair does not exist unless two people (one being married) have intercourse, an online computer affair can be just as disruptive to a marriage or a couple's relationship. Computer friendships may move to feelings of intimacy, involve secrecy (one's partner does not know the level of involvement), include sexual tension (even though there is no overt sex), and take time, attention, energy, and affection away from one's partner. Schneider (2000) studied ninety-one women who experienced serious adverse consequences from their partner's cybersex involvement, including loss of interest in relational sex, and feeling hurt, betrayed, rejected, abandoned, lonely, jealous, and angry over being constantly lied to. These women noted that the cyber affair was as emotionally painful as an off-line affair and that their partners' cybersex addiction was a major reason for their separation or divorce. Cramer et al. (2008) also noted that women become more upset when their man is emotionally unfaithful with another woman (although men become more upset when their partner is sexually unfaithful with another man). Measure your attitude towards infidelity by taking the Self-Assessment on the Chapter 13 Assessment card at the end of the book.

WE ACT AS THOUGH COMFORT AND LUXURY WERE THE CHIEF REQUIREMENTS OF LIFE, WHEN ALL WE NEED TO MAKE US REALLY HAPPY IS SOMETHING TO BE ENTHUSIASTIC ABOUT.

— *Charles Kingsley*

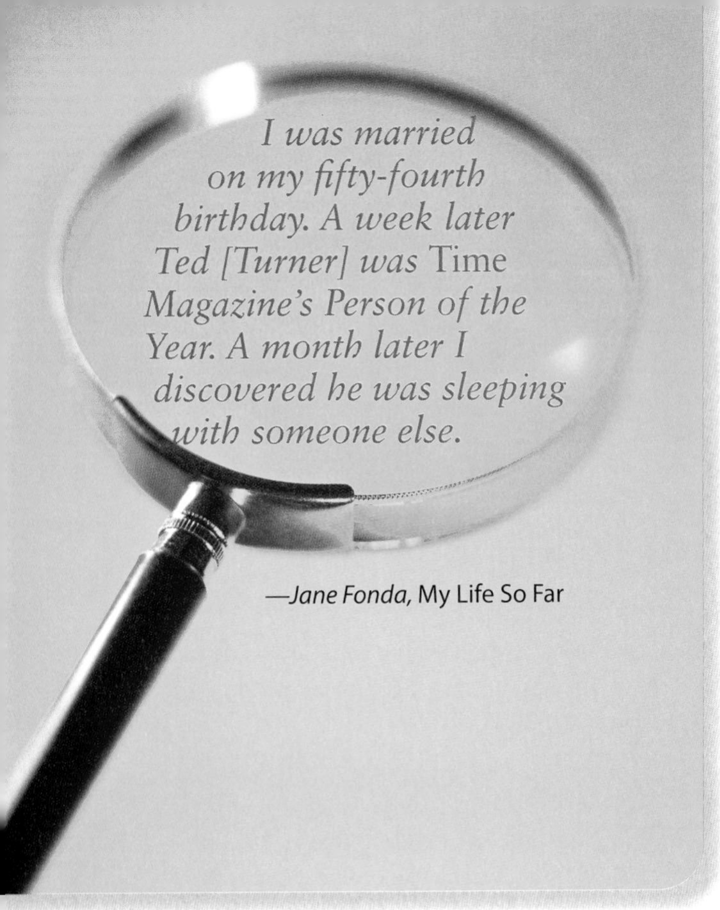
I was married on my fifty-fourth birthday. A week later Ted [Turner] was Time Magazine's Person of the Year. A month later I discovered he was sleeping with someone else.
—Jane Fonda, My Life So Far

© BananaStock/Jupiterimages

extradyadic (extrarelational) involvement emotional or sexual involvement between a member of a pair and someone other than the partner.

Coolidge effect term used to describe the waning of sexual excitement and the effect of novelty and variety on sexual arousal.

Extradyadic (extrarelational) involvement refers to sexual involvement of a pair-bonded individual with someone other than the partner. Extradyadic involvements are not uncommon. Of 1,319 undergraduates, 37.4 percent agreed, "I have cheated on a partner I was involved with." More than half (53 percent) agreed, "A partner I was involved with cheated on me" (Knox and Zusman 2009).

Marriage and family therapists rank an extramarital affair as the second most stressful crisis event for a couple (physical abuse is number one) (Olson et al. 2002). Characteristics associated with spouses who are more likely to have extramarital sex include male gender, a strong interest in sex, permissive sexual values, low subjective satisfaction in the existing relationship, employment outside the home, low church attendance, greater sexual opportunities, higher social status (power and money), and alcohol abuse (Hall et al. 2008; Smith 2005; Olson et al. 2002). Elmslie and Tebaldi (2008) noted that women's infidelity behavior is influenced by religiosity (less religious = more likely), city size (urban = more likely), and happiness (less happy = more likely), whereas men's infidelity behavior is affected by race (white = less likely), happiness status (happy = less likely), city size (same for women), and employment status (employed = less likely). For more information regarding extramarital affairs in other countries, see page 253.

Reasons for Extramarital Affair Spouses report a number of reasons why they become involved in a sexual encounter outside their marriage:

1. **Variety, novelty, and excitement.** Extradyadic sexual involvement may be motivated by the desire for variety, novelty, and excitement. One of the characteristics of sex in long-term committed relationships is the tendency for it to become routine. Early in a relationship, the partners cannot seem to have sex often enough. However, with constant availability, partners may achieve a level of satiation, and the attractiveness and excitement of sex with the primary partner seem to wane. A high-end call girl said the following of the Eliot Spitzer affair:

> Almost all of my clients are married. I would say easily over 90 percent. I'm not trying to justify this business, but these are men looking for companionship. They are generally not men that couldn't have an affair [if they wanted to], but men who want this tryst with no strings attached. They're men who want to keep their lives at home intact. (Kottke 2008)

The **Coolidge effect** is a term used to describe this waning of sexual excitement and the effect of novelty and variety on sexual arousal:

> One day President and Mrs. Coolidge were visiting a government farm. Soon after their arrival, they were taken off on separate tours. When Mrs. Coolidge passed the chicken pens, she paused to ask the man in charge if the rooster copulated more than once each day. "Dozens of times," was the reply. "Please tell that to the President," Mrs. Coolidge requested. When the President passed the pens and was told about the rooster, he asked, "Same hen every time?" "Oh no, Mr. President, a different one each time." The President nodded slowly and then said, "Tell that to Mrs. Coolidge." (Bermant 1976, 76–77)

© Chapel House Photography

Whether or not individuals are biologically wired for monogamy continues to be debated. Monogamy among mammals is rare (from 3 percent to 10 percent), and monogamy tends to be the exception more often than the rule (Morell 1998). Even if such biological wiring for plurality of partners does exist, it is equally debated whether such wiring justifies nonmonogamous behavior—that individuals are responsible for their decisions.

2. **Workplace friendships.** A common place for extramarital involvements to develop is the workplace. Neuman (2008) noted that four in ten of the affairs men reported began with a woman they met at work. Coworkers share the same world eight to ten hours a day and, over a period of time, may develop good feelings for each other that eventually lead to a sexual relationship. Former Democratic presidential candidate John Edwards became involved with a campaign employee when they were on the campaign trail together. Tabloid reports regularly reflect that romances develop between married actors making a movie together (for example, Brad Pitt and Angelina Jolie).

3. **Relationship dissatisfaction.** It is commonly believed that people who have affairs are not happy in their marriage. Spouses who feel misunderstood, unloved, and ignored sometimes turn to another who offers understanding, love, and attention. Djamba et al. (2005) analyzed General Social Survey data and found that unhappiness in one's marriage was the primary reason for becoming involved in an extramarital affair. Neuman (2008) confirmed that being emotionally dissatisfied in one's relationship is the primary culprit leading to an affair.

   One source of relationship dissatisfaction is an unfulfilling sexual relationship. Some spouses engage in extramarital sex because their partner is not interested in sex. Others may go outside the relationship because their partners will not engage in the sexual behaviors they want and enjoy. The unwillingness of the spouse to engage in oral sex, anal intercourse, or a variety of sexual positions sometimes results in the other spouse's looking elsewhere for a more cooperative and willing sexual partner.

4. **Revenge.** Some extramarital sexual involvements are acts of revenge against one's spouse for having an affair. When partners find out that their mate has had or is having an affair, they are often hurt and angry. One response to this hurt and anger is to have an affair to get even with the unfaithful partner.

down low
African-American "heterosexual" man who has sex with men.

5. **Homosexual relationship.** Some individuals marry as a front for their homosexuality. Cole Porter, a composer known for such songs as "I've Got You Under My Skin," "Night and Day," and "Easy to Love," was a homosexual who feared no one would buy or publish his music if his sexual orientation were known. He married Linda Lee Porter (alleged to be a lesbian), and they had an enduring marriage for thirty years.

   Other gay individuals marry as a way of denying their homosexuality. These individuals are likely to feel unfulfilled in their marriage and may seek involvement in an extramarital homosexual relationship. Other individuals may marry and then discover later in life that they desire a homosexual relationship. Such individuals may feel that (1) they have been homosexual or bisexual all along, (2) their sexual orientation has changed from heterosexual to homosexual or bisexual, (3) they are unsure of their sexual orientation and want to explore a homosexual relationship, or (4) they are predominately heterosexual but wish to experience a homosexual relationship for variety. The term **down low** refers to African American heterosexual men who have sex with men (Browder 2005).

6. **Aging.** A frequent motive for intercourse outside marriage is the desire to return to the feeling of youth. Ageism, which is discrimination against the elderly, promotes the idea that being young is good and being old is bad. Sexual attractiveness is equated with youth, and having an affair may confirm to older partners that they are still sexually desirable. Also, people may try to recapture the love, excitement, adventure, and romance associated with youth by having an affair.

7. **Absence from partner.** One factor that may predispose a spouse to an affair is prolonged separation from the partner. Some wives whose husbands are away for military service report that the loneliness can become unbearable. Some husbands who are away say that remaining faithful is difficult. Partners in commuter relationships may also be vulnerable to extradyadic sexual relationships.

© iStockphoto.com

Effects of an Affair Reactions to the knowledge that one's spouse has been unfaithful vary. For most, the revelation is difficult. The following is an example of a wife's reaction to her husband's affairs:

> My husband began to have affairs within six months of our being married. Some of the feelings I experienced were disbelief, doubt, humiliation and outright heart-wrenching pain! When I confronted my husband he denied any such affair and said that I was suspicious, jealous and had no faith in him. In effect, I had the problem. He said that I should not listen to what others said because they did not want to see us happy but only wanted to cause trouble in our marriage. I was deeply in love with my husband and knew in my heart that he was guilty as sin; I lived in denial so I could continue our marriage.
>
> Of course, my husband continued to have affairs. Some of the effects on me included:
>
> 1. I lost the ability to trust my husband and, after my divorce, other men.
> 2. I developed a negative self concept—the reason he was having affairs is that something was wrong with me.
> 3. He robbed me of my innocence and my virginity—clearly he did not value the opportunity to be the only man to have experienced intimacy with me.
> 4. I developed an intense hatred for my husband.
>
> It took years for me to recover from this crisis. I feel that through faith and religion I have emerged "whole" again. Years after the divorce my husband made a point of apologizing and letting me know that there was nothing wrong with me, that he was just young and stupid and not ready to be serious and committed to the marriage.

Affairs may also have negative effects on children, which may result from hearing conflicts between the parents, from an absence of attention (because the father is not at home), and from the breakup of the marriage (Schneider 2003).

Successful Recovery from Infidelity *Sex and the City: The Movie* featured Steve's infidelity to Miranda, his wife. They ended up getting back together, as do most couples once an affair is discovered. Olson et al. (2002) identified three phases of successful recovery from the discovery of a partner's affair (all of these phases were beautifully illustrated in the movie). The "roller-coaster" phase involves agony at the initial discovery, which elicits an array of feelings including rage or anger, self-blame, the desire to give up, and the desire to work on the marital relationship. The second phase, "moratorium," involves less emotionality and a decision to work it out. The partners settle into a focused though tenuous commitment to get beyond the current crisis. The third phase, "trust building," involves taking responsibility for the infidelity, reassurance of commitment, increased communication, and forgiveness. Couples in this phase "reengage," "open up," and focus on problems leading up to the infidelity. Working through a discovered affair takes time, commitment to new behavior in the primary relationship, and forgiveness.

Bagarozzi (2008) defined forgiveness as a conscious decision on the part of the offended spouse to grant a pardon to the offending spouse, to give up feeling angry, and to relinquish the right to retaliate against the offending spouse. In exchange, offending spouses take responsibility for the affair, agree not to repeat the behavior, and grant their partner the right to check up on them to regain trust.

Positive outcomes of having experienced and worked through infidelity include a closer marital relationship, increased assertiveness, placing higher value on each other, and realizing the importance of good marital communication (Olson et al. 2002; Linquist and Negy 2005).

Spouses who remain faithful to their partners have decided to do so. They avoid intimate conversations with members of the other sex and a context (for example, where alcohol drinking and/or being alone are involved) that are conducive to physical involvement. The best antidote is for the spouse to be direct, simply telling the person he or she is not interested.

Prevention of Infidelity Allen et al. (2008) identified the premarital factors predictive of future infidelity. The primary factor for both partners was a negative pattern of interaction. Partners who end up being unfaithful were in relationships where they did not connect, argued, and criticized each other. Hence, spouses least vulnerable are in loving, nurturing, communicative relationships where each affirms the other. Neuman (2008) also noted that avoiding friends who have affairs and establishing close relationships with married couples who value fidelity further insulates one from having an affair.

Unemployment

In the wake of Hurricanes Katrina and Ike, there was massive unemployment in the New Orleans and Galveston areas. The jobs of thousands were literally washed away. Coupled with these job losses, corporate America continues to downsize and outsource jobs to India and Mexico. The result is massive layoffs and insecurity in the lives of American workers. Forced unemployment or the threatened loss of one's job is a major stressor for individuals, couples, and families (Legerski et al. 2006). Also, when spouses or parents lose their jobs as a result of physical illness or disability, a family experiences a double blow—loss of income combined with high medical bills. Unless an unemployed spouse is covered by the partner's medical insurance, unemployment can also result in loss of health insurance for the family. Insurance for both health care and disability is very important to help protect a family from an economic disaster. Prolonged unemployment may result in bankruptcy (Miller 2005).

The effects of unemployment may be more severe for men than for women. Our society expects men to be the primary breadwinners in their families and equates masculine self-worth and identity with job and income. Stress, depression, suicide, alcohol abuse, and lowered self-esteem, as well as increased cigarette smoking (Falba et al. 2005), are all associated with unemployment. Macmillan and Gartner (2000) also observed that men who are unemployed are more likely to be violent toward their working wives.

Women tend to adjust more easily to unemployment than men. Women are not burdened with the cultural expectation of the provider role, and their identity is less tied to their work role. Hence, women may view unemployment as an opportunity to spend more time with their families; many enjoy doing so.

Although unemployment can be stressful, increasing numbers of workers are also experiencing job stress. With the recession, layoffs, and downsizing, workers are given the work of two and told, "we'll hire someone soon." Meanwhile, employers have learned to pay for one and get the work of two. The result is lowered morale, exhaustion, and stress that can escalate into violence.

The effects of unemployment may be more severe for men than for women.

Substance Abuse

Spouses, parents, and children who abuse drugs contribute to the stress and conflict experienced in their respective marriages and families. Although some individuals abuse drugs to escape from unhappy relationships or the stress of family problems, substance abuse inevitably adds to the individual's marital and family problems when it results in health and medical problems, legal problems, loss of employment, financial ruin, school failure, divorce, and even death (due to accidents or poor health). The story titled "Marriage Cancelled" on the next page reflects the experience of the wife of a man addicted to crack.

Although getting married is associated with significant reductions in cigarette smoking, heavy drinking, and marijuana use for both men and women (Merline et al. 2008), family crises involving alcohol and/or drugs are not unusual. Alcohol is also a major problem on campus (more than half of 772 college students who reported drinking alcohol reported having blacked out) (White et al. 2002). The context of the most intense drinking is at a fraternity social. In a study of 306 university students, the average number of drinks at a fraternity social was 5.91 compared to 4.04 at campus parties (Miley and Frank 2006). University officials have attempted (mostly unsuccessfully) to curb excessive drinking. Miley and Frank (2006) suggested that fraternity socials might become a focus because alcohol consumption is the highest.

Some students have also grown up in homes where one or both parents abused alcohol. Haugland (2005) studied alcohol abuse by the father and found that a drunken father simply was not around the children during drinking episodes.

MARRIAGE CANCELLED

Testimonial of a wife married to a man addicted to crack.

Marriage needs a foundation of trust and open communication, but when you are married to an addict, you won't find either. When I first met "Nate," I thought he was everything I wanted in a partner—good looks, a great sense of humor, intelligence, and ambition. It didn't take me long to realize that I was extremely attracted to this man. Since we knew some of the same people, we ended up at several parties together. We both drank and occasionally used cocaine but it wasn't a problem. I loved being with him and whenever I wasn't with him, I was thinking about him.

After only four months of dating, we realized we were falling madly in love with each other and decided to get married. Neither of us had ever been married before but I had a five-year-old son from a previous relationship. Nate really took to my son and it seemed like we had the perfect marriage—that changed drastically and quickly. I didn't know it but my husband had been using crack cocaine the entire time we dated. I wasn't extremely familiar with this new drug and I knew even less about its addictive power. Of course, I soon found out more about it than I cared to know.

Less than three months after we married, Nate went on his first binge. After leaving work one Friday night, he never came home. I was really worried and was calling all of his friends trying to find out where he was and what was wrong. I talked to the girlfriend of one of his best friends and found out everything. She told me that Nate had been smoking crack off and on for a long time and that he was probably out using again. I didn't know what to do or where to turn. I just stayed home all weekend waiting to hear from my husband.

Finally, on Sunday afternoon, Nate came home. He looked like hell and I was mad as hell. I sent my son to play with his friends next door and as soon as he was out of hearing range, I lost it. I began screaming and crying, asking why he never came home over the weekend. He just hung his head in absolute shame. I found out later that he had spent his entire paycheck, pawned his wedding ring, and had written checks off of a closed checking account. He went through over $1,000 of "our" money in less than three days.

I was devastated and in shock. After I calmed down and my anger subsided, we discussed his addiction. He told me he loved me with all his heart and that he was so sorry for what he had done. He also swore he would never do it again. Well, this is when I became an enabler and I continued to enable my husband for three years. He would stay clean for a while and things would be great between us, then he would go on another binge. It was the roller-coaster ride from hell. Nate was on a downward spiral and he was dragging my son and me down along with him.

By the end of our third year together, we were more like roommates. Our once wonderful sex life was virtually nonexistent and what love I still had for him was quickly fading away. I knew I had to leave before my love turned to hate. It was obvious that my son had been pulling away from Nate emotionally so it was a good time to end the nightmare. I packed our belongings and moved in with my sister.

Less than three months after I left, Nate went into a rehabilitation program. I was happy for him but I knew I could never go back; it was too little, too late. That was nine years ago and the last I heard, he was still struggling with his addiction, living a life of misery. I have no regrets about leaving but I am saddened by what crack cocaine had done to my once wonderful husband—it turned him into a thief and a liar and ended our marriage.

Table 13.1

Drug Use by Type of Drug and Age Group

Type of Drug Used	Age 12 to 17	Age 18 to 25	Age 26 to 34
Marijuana and hashish	7.0%	16.0%	9.0%
Cocaine	.4%	2.2%	1.7%
Alcohol	17.0%	62.0%	no data
Cigarettes	10.0%	38.0%	no data

Source: Adapted from *Statistical Abstract of the United States, 2009,* 128th ed. Washington, DC: U.S. Bureau of the Census, Table 199.

As indicated in Table 13.1, drug use is most prevalent among 18- to 25-year-olds. Drug use among teenagers under age 18 is also high. Because teenage drug use is common, it may compound the challenge parents may have with their teenagers.

Drug Abuse Support Groups Although treatments for alcohol abuse are varied, a combination of medications (naltrexone and acamprosate) and behavioral interventions (for example, control for social context) is a favored approach (Mattson and Litten 2005). The support group Alcoholics Anonymous (AA; www.alcoholics-anonymous.org) has also been helpful. There are more than 15,000 AA chapters nationwide; one in your community can be found through the Yellow Pages. The only requirement for membership is the desire to stop drinking.

> And yet they, who passed away long ago, still exist in us.
>
> —*Rainer Maria Rilke,* Letters to a Young Poet

Former abusers of drugs (other than alcohol) also meet regularly in local chapters of Narcotics Anonymous (NA) to help each other continue to be drug-free. Patterned after Alcoholics Anonymous, the premise of NA is that the best person to help someone stop abusing drugs is someone who once abused drugs. NA members of all ages, social classes, and educational levels provide a sense of support for each other to remain drug-free.

Al-Anon an organization that provides support for family members and friends of alcohol abusers.

Al-Anon is an organization that provides support for family members and friends of alcohol abusers. Spouses and parents of substance abusers learn how to live with and react to living with a substance abuser.

Parents who abuse drugs may also benefit from the Strengthening Families Program, which provides specific social skills training for both parents and children. After families attend a five-hour retreat, parents and children are involved in face-to-face skills training over a four-month period. A twelve-month follow-up has revealed that parenting skills remained improved and that reported heroin and cocaine use had declined.

> All I know is *that I am sick of living; I'm through. I'm drowned and contented on the bottom of a bottle.*
>
> —*Larry Slade*

Death of Family Member

Even more devastating than drug abuse are family crises involving death—of one's child, parent, or loved one (we discuss the death of one's spouse in Chapter 15 on Relationships in the Later Years). The crisis is particularly acute when the death is a suicide.

Death of One's Child A parent's worst fear is the death of a child. Most people expect the death of their parents but not the death of their children. The following reflects the anguish of a father and mother who experienced the death of their daughter when she was 17.

Emily: Our Daughter's Death at 17

Our lives changed forever when we received a phone call that our daughter had been in an accident. She had been on a trip to the beach with friends and was on the way home when the van flipped, crossed the divided highway and severely injured her. She then died on the side of the highway from head injuries. We discovered that the driver had put the cruise control on 70 and was playing "switch the driver" with his girlfriend, the front passenger. One of them hit the steering wheel inadvertently and our daughter (the only one of the six to die) was gone forever.

Emily was a month shy of her 17th birthday. She was a lovely child. We, as her parents, haven't been the same since her death. Even after all these years, our first thought upon awakening and last thought before we sleep is of our daughter, and how we miss her. At the time of her death, we were not alone in our grief. When we told Emily's grandfather of her death, he was overwhelmed with grief, and two weeks later died of a heart attack. To this day, we wonder if we had been protective enough. The grief goes on.

Also, our marriage has been strained due to the trauma. Since all couples have different grieving trajectories, we found one of us would be battling depression, while the other was half way level ... or vice versa. Her death will always remain a nightmare from which we cannot awaken. Time, therapy, support groups and the like have helped to manage our loss of our child, but we will forever continue to deal with our personal tragedy within our family ... we hope for peace ... but twelve years after her death we know our loss can only be managed.

Mothers and fathers sometimes respond to the death of their child in different ways. When they do, the respective partners may interpret these differences in negative ways, leading to relationship conflict and unhappiness. For example, after the death of their 17-year-old son, one wife accused her husband of not sharing in her grief. The husband explained that, although he was deeply grieved, he poured his grief into working more as a way of distraction. To deal with these differences, spouses might need to be patient and practice tolerance in allowing each to grieve in his or her own way.

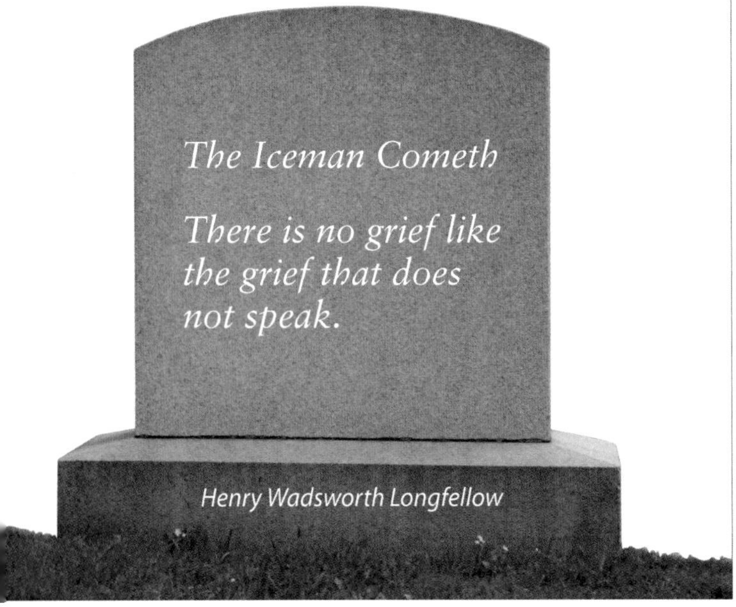
The Iceman Cometh
There is no grief like the grief that does not speak.
Henry Wadsworth Longfellow

Death of One's Parent Terminally ill parents may be taken care of by their children. Such care over a period of years can be emotionally stressful, financially draining, and exhausting. Hence, by the time the parent dies, a crisis has already occurred.

Reactions to the death of one's parents include depression, loss of concentration, and anger (Ellis and Granger 2002). Michael and Snyder (2005) studied college students who had experienced the death of a loved one (most often a parent) and found that their constant ruminations about the deceased correlated with a lower sense of psychological well-being. Whether the death is that of a child or a parent, Burke et al. (1999) noted that grief is not a one-time experience that people adjust to and move on from. Rather, for some, there is "chronic sorrow," where grief-related feelings occur periodically throughout the lives of those left behind. The late Paul Newman was asked how he got over the death of his son who overdosed. He replied, "You never get over it." Johnny Carson, Ed McMahon, and Dean Martin also experienced the death of a child. Grief feelings may be particularly acute on the anniversary of the death or when the bereaved individual thinks of what might have been had the person lived. Burke et al. (1999) noted that 97 percent of the individuals in one study who had experienced the death of a loved one two to twenty years earlier met the criteria for chronic sorrow. Field et al. (2003) also observed bereavement-related distress five years after the death of a spouse.

Suicide of Family Member

Suicide is a devastating crisis event for families, and not that unusual. We know personally of five suicides, some of which have occurred within the same family. Annually there are 31,000 suicides (750,000 attempts), and each suicide immediately affects at least six other people in that person's life. These effects include depression (for example, grief), physical disorders (for example, shingles due to stress), and social stigma (for example, the person is viewed as weak and the family as a failure in not being able to help with the precipitating emotional problems) (De Castro and Guterman 2008). Schum (2007) identified individuals who experience the suicide of a family member as having "the worst day of their lives."

People between 15 and 19, homosexual, or male, and those with a family history of suicide or mood disorder, substance abuse, or past history of child abuse and parental sex abuse are more vulnerable to suicide than others (Melhem et al. 2007). As noted earlier, Taliaferro et al. (2008) found that vigorous exercise and involvement in sports are associated with lower rates of suicide among adolescents.

Suicide is viewed as a "rational act" in that the person feels that suicide is the best option available at the time. Therapists view suicide as a "permanent solution to a temporary problem" and routinely call 911 to have people hospitalized or restrained who threaten suicide or who have been involved in an attempt.

Adjustment to the suicide of a family member takes time. The son of physician T. Schum committed suicide, which set in motion a painful adjustment for Dr. Schum. "You will never get over this but you can get through it," was a phrase Dr. Schum found helpful (Schum 2007). Also helpful is the support group Survivors of Suicide. Part of the recovery process is accepting that one cannot stop the suicide of those who are adamant about taking their own life and that one is not responsible for the suicide of another. Indeed, family members often harbor the belief that they could have done something to prevent the suicide. Singer Judy Collins lost her son to suicide and began to attend a support group for people who had lost a loved one to suicide. When Collins asked whether there was something she could have done, one of the members of a group she attended answered with a resounding no:

> I was sitting on his bed saying, "I love you, Jim. Don't do this. How can you do this?" I had my hand on his hand, my cheek on his cheek. He said excuse me, reached his other hand around, took the gun from under the pillow, and blew his head off. My face was inches from his. If somebody wants to kill himself or herself, there is nothing you can do to stop them. (Collins 1998, 210)

RAZORS PAIN YOU; RIVERS ARE DAMP;
ACIDS STAIN YOU; AND DRUGS CAUSE CRAMP.
GUNS AREN'T LAWFUL; NOOSES GIVE;
GAS SMELLS AWFUL; YOU MIGHT AS WELL LIVE.

—Dorothy Parker, poet

EXTRAMARITAL AFFAIRS IN OTHER COUNTRIES

Researcher Pam Druckerman (2007) wrote *Lust in Translation*, in which she reflects how affairs are viewed throughout the world. First, terms for having an affair vary; for the Dutch, it is called, "pinching the cat in the dark"; in Taiwan, it is called "a man standing in two boats"; and in England, "playing off sides." Second, how an affair is regarded differs by culture. In America, the script for discovering a partner's affair involves confronting the partner and ending the marriage. In France, the script does not involve confronting the partner and does not assume that the affair means the end of the marriage; rather, letting time pass to let a partner go through the experience without pressure or comment is the norm." In America, presidential candidate John Edwards's disclosure of his affair with a campaign employee ended his political life.

Divorce and Remarriage

SECTIONS

Just as weddings are a time of celebration and joy, divorce is a time of dismay and sadness. Divorce is the end of a dream of living happily ever after with one's mate. Whether a divorce is the beginning of the end of one's life (some never recover) or the beginning of a new life depends on the person. College students are not unacquainted with divorce. In a sample of 1,319 undergraduates, 30.9 percent reported that their parents were divorced (Knox and Zusman 2009).

14.1 Macro Factors Contributing to Divorce

Sociologists emphasize that social context creates outcomes. This is best illustrated in the statistic that the Puritans in Massachusetts, from 1639 to 1760, averaged only one divorce per year (Morgan 1944). The social context of that era involved strong pro-family values and strict divorce laws, with the result that divorce was almost nonexistent. In contrast, divorce occurs more frequently today as a result of various structural and cultural factors, also known as macro factors (Lowenstein 2005).

Increased Economic Independence of Women

In the past, an unemployed wife was dependent on her husband for food and shelter. No matter how unhappy her marriage was, she stayed married because she was economically dependent on her husband. Her husband literally represented her lifeline. Finding gainful employment outside the home made it possible for a wife to afford to leave her husband if she wanted to. Now that about three-fourths of wives are employed, fewer wives are economically trapped in unhappy marriage relationships. As we noted earlier, a wife's employment does not increase the risk of divorce in a happy marriage. However, it does provide an avenue of escape for women in unhappy or abusive marriages (Kesselring and Bremmer 2006).

Employed wives are also more likely to require an egalitarian relationship; although some husbands prefer this role relationship, others are unsettled by it. Another effect of a wife's employment is that she may meet someone new in the workplace so that she becomes aware of an alternative to her current partner. Finally, unhappy husbands may be more likely to divorce if their wives are employed and able to be financially independent (less alimony and child support).

Changing Family Functions and Structure

Many of the protective, religious, educational, and recreational functions of the family have been largely taken over by outside agencies. Family members may now look to the police for protection, the church or synagogue for meaning, the school for education, and commercial recreational facilities for fun rather than to each other within the family for fulfilling these needs. The result is that, although meeting emotional needs remains a primary function of the family, fewer reasons exist to keep a family together.

In addition to the changing functions of the family brought on by the Industrial Revolution, the family structure has changed from that of larger extended families in rural communities

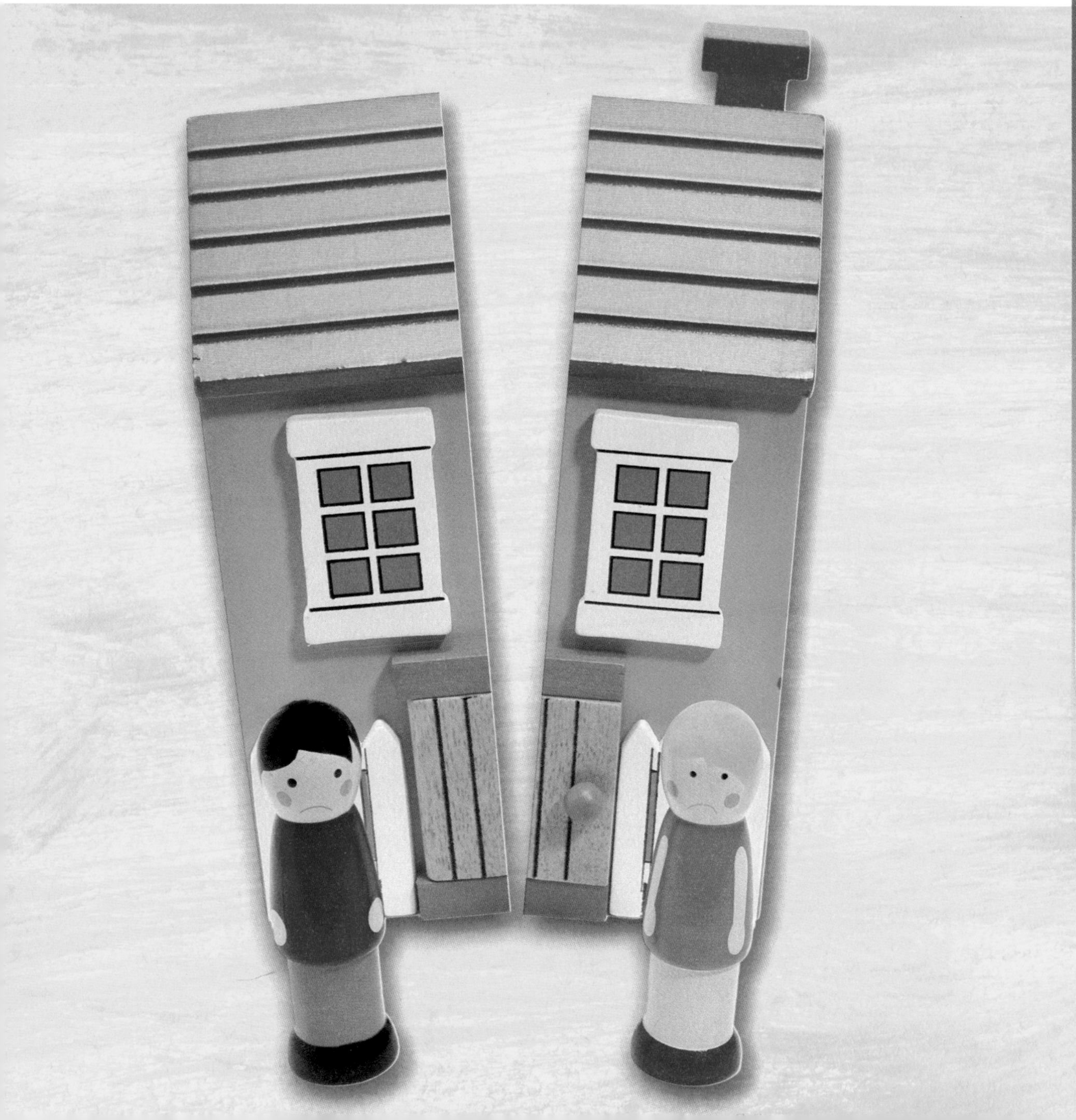

no-fault divorce a divorce in which neither party is identified as the guilty party or the cause of the divorce.

to smaller nuclear families in urban communities. In the former, individuals could turn to a lot of people in times of stress; in the latter, more stress necessarily falls on fewer shoulders. Cohen and Savaya (2003) documented an increased divorce rate among Muslim Palestinian citizens of Israel due to an increased acceptance of the modern views of marriage—increased emphasis on happiness and compatibility.

Liberal Divorce Laws

All states recognize some form of **no-fault divorce**—where neither party is identified as the guilty party or the cause of the divorce (for example, due to adultery). In effect, divorce is granted after a period of separation ranging from six weeks to twelve months. Nevada requires the shortest waiting period of six weeks. The goal of no-fault divorce has been to try to make divorce less acrimonious. However, this has not always been successful as spouses who divorce may still fight over custody of the children, child support, spouse support, and division of property. Indeed, almost 90 percent of people who go through the divorce process report it as a negative experience (Clarke-Stewart and Brentano 2006). Although researchers disagree whether no-fault divorce is associated with more divorce (Drewianka 2008; Allen et al. 2006), a backlash has occurred and there is a movement to make divorce more difficult. One divorced spouse said, "I should have stayed married when the hard times hit—it was just too easy to walk out" (personal communication).

Fewer Moral and Religious Sanctions

Many priests and clergy recognize that divorce may be the best alternative in particular marital relationships and attempt to minimize the guilt that congregational members may feel at the failure of their marriage. Churches increasingly embrace single and divorced or separated individuals, as evidenced by "divorce adjustment groups."

More Divorce Models

As the number of divorced individuals in our society increases, the probability increases that a person's friends, parents, siblings, or children will be divorced. The more divorced people a person knows, the more normal divorce will seem to that person. The less deviant the person perceives divorce to be, the greater the probability that person will divorce if his or her own marriage becomes strained. Divorce has become so common that numerous websites exclusively for divorced individuals are available (for example, www.heartchoice.com/divorce).

Mobility and Anonymity

When individuals are highly mobile, they have fewer roots in a community and greater anonymity. Spouses who move away from their respective family and friends often discover that they are surrounded by strangers who don't care if they stay married or not. Divorce thrives when pro-marriage social expectations are not operative. In addition, the factors of mobility and anonymity also result in the removal of a consistent support system to help spouses deal with the difficulties they may encounter in marriage.

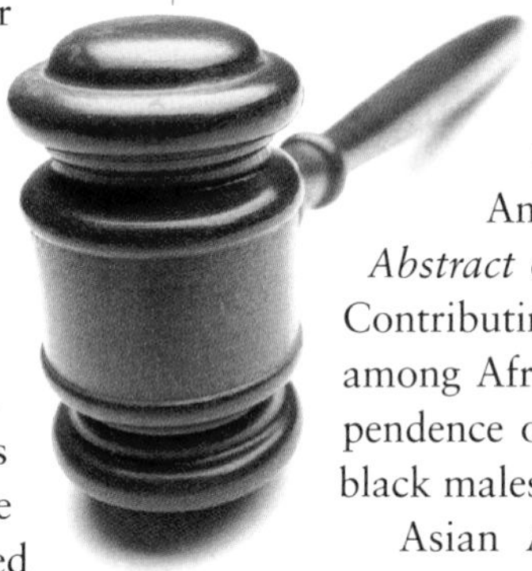

Race and Culture

A higher percentage of African Americans are divorced than European Americans (11.5 versus 10.4) (*Statistical Abstract of the United States, 2009*, Table 55). Contributing to this higher percentage of divorce among African Americans is the economic independence of black females and the difficulty for black males to find and keep stable jobs.

Asian Americans and Mexican Americans have lower divorce rates than European Americans or African Americans because they consider the family unit to be of greater value (familism) than their individual interests (individualism). Unlike familistic values in Asian cultures, individualistic values in American culture emphasize the goal of personal happiness in marriage. When spouses stop having fun (when individualistic goals are no longer met), they sometimes feel no reason to stay married. Of 1,319 undergraduates at a large southeastern university, only 4.5 percent agreed with the statement, "I would not divorce my spouse for any reason" (Knox and Zusman 2009). In two national samples, 36 percent agreed that "The personal happiness of an individual is more important than putting up with a bad marriage" (Amato et al. 2007). Reflecting an individualistic philosophy, Geraldo Rivera asked of his divorces, "Who cares if I've been married five times?"

14.2 Micro Factors Contributing to Divorce

Although macro factors may make divorce a viable cultural alternative to marital unhappiness, they are not sufficient to "cause" a divorce. One spouse must choose to divorce and initiate proceedings. Such a view is micro in that it focuses on the individual decisions and interactions within specific family units. The following subsections discuss some of the micro factors that may be operative in influencing a couple toward divorce.

> **The best single** *predictor of whether a couple is going to divorce is contempt.*
>
> *—John and Julie Gottman, marriage therapists*

Falling Out of Love

"Falling out of love" was a top reason for divorce that men reported (Enright 2004). Previti and Amato (2003) noted that the absence of love was also associated with both men and women being more likely to divorce. Indeed, almost half of 1,319 undergraduates reported that they would divorce a spouse they no longer loved (Knox and Zusman 2009).

Negative Behavior

Physical or emotional abuse and alcohol or drug abuse were among the top behavioral reasons identified by women who divorced in midlife (Enright 2004). Ostermann et al. (2005) confirmed that a discrepancy in the amount of alcohol spouses consume is associated with divorce. When the presence of negative behavior is coupled with the absence of positive behavior, the combination can be deadly. Shumway and Wampler (2002) emphasized that the absence of positive behaviors such as "small talk," "reminiscing about shared times together," and "encouragement" increases a couple's marital dissatisfaction.

© iStockphoto.com

People marry because they anticipate greater rewards from being married than from being single. During courtship, each partner engages in a high frequency of positive verbal (compliments) and nonverbal (eye contact, physical affection) behavior toward the other. The good feelings the partners experience as a result of these positive behaviors encourage them to marry to "lock in" these feelings across time. Just as love feelings are based on positive behavior from a partner, negative feelings are created when a partner engages in a high frequency of negative behavior and thoughts of divorce (to escape the negative behavior) occur.

Affair

Some extramarital affairs result in divorce (O'Leary 2005). The spouse having an affair may feel unloved at home where there is little to no sex. Involvement in an affair may bring both love and sex and speed the spouse toward divorce. Alternatively, an at-home spouse may become indignant and demand that the partner leave. Although most spouses (75 percent) do not leave their mates for a lover, an extramarital relationship may weaken the emotional tie between the spouses so that they are less inclined to stay married. Combined with a partner having an affair is the spouse who may feel betrayed and may terminate the marriage. Of 1,319 undergraduates, 67.6 percent agreed with the statement, "I would divorce a spouse who had an affair" (Knox and Zusman 2009). However, as we saw in the section on extramarital affairs in Chapter 13, Stress and Crisis in Relationships, most couples do not end a marriage because of an affair. Unmarried undergraduates say they would end a marriage, but married couples (who have invested a great deal) usually do not.

satiation the state in which a stimulus loses its value with repeated exposure.

Lack of Conflict Resolution Skills

Managing differences and conflict in a relationship helps to reduce the negative feelings that develop in a relationship. Some partners respond to conflict by withdrawing emotionally from their relationship; others respond by attacking, blaming, and failing to listen to their partner's point of view. Ways to negotiate differences and reduce conflict were discussed in Chapter 4.

Value Changes

Both spouses change throughout the marriage. "He (or she) is not the same person I married" is a frequent observation of people contemplating divorce. People may undergo radical value changes after marriage. One minister married and decided seven years later that he did not like the confines of the marriage role. He left the ministry, earned a PhD, and began to drink and have affairs. His wife, who had married him when he was a minister, now found herself married to a clinical psychologist who spent his evenings at bars with other women. The couple divorced.

Jane Fonda noted that she experienced a change in her assertiveness, that she was no longer willing to just go along with her mate but began to specify what her needs were and to negotiate their fulfillment:

> The very thing that I feared the most—that I would gain my voice and lose my man—was actually happening . . . The problem comes when what you need and what you see isn't seen or needed by your partner. It doesn't mean your partner is bad; it just means that he or she wants something else in life. . . . I could see Ted [Turner] withering before my eyes. Clearly he wasn't going to be able (or willing) to make the journey with me. We agreed to separate. (Fonda 2005, p. 545)

Because people change throughout their lives, the person who one selects at one point in life may not be the same partner one would select at another point. Margaret Mead, the famous anthropologist, noted that her first marriage was a student marriage; her second, a professional partnership; and her third, an intellectual marriage to her soul mate, with whom she had her only child. At each of several stages in her life, she experienced a different set of needs and selected a mate who fulfilled those needs.

Satiation

Satiation, also referred to as habituation, refers to the state in which a stimulus loses its value with repeated exposure. Spouses may tire of each other. Their stories are no longer new, their sex is repetitive, and their presence no longer stimulates excitement as it did in courtship. Some people who feel trapped by the boredom of constancy divorce and seek what they believe to be more excitement by returning to singlehood and, potentially, new partners. A developmental task of marriage is for couples to enjoy being together and not demand a constant state of excitement (which is difficult over a fifty-year period). The late comedian George Carlin said, "If all of your needs are not being met, drop some of your needs." If spouses did not expect so much of marriage, maybe they would not be disappointed.

Perception That One Would Be Happier If Divorced

Brinig and Allen (2000) noted that women file two-thirds of divorce applications. Women's behavior may be based on the perception that they will achieve greater power over their own life, money (in the form of child support and/or alimony) without having the liability of dealing with an unsupportive husband on a daily basis, and greater control over their children,

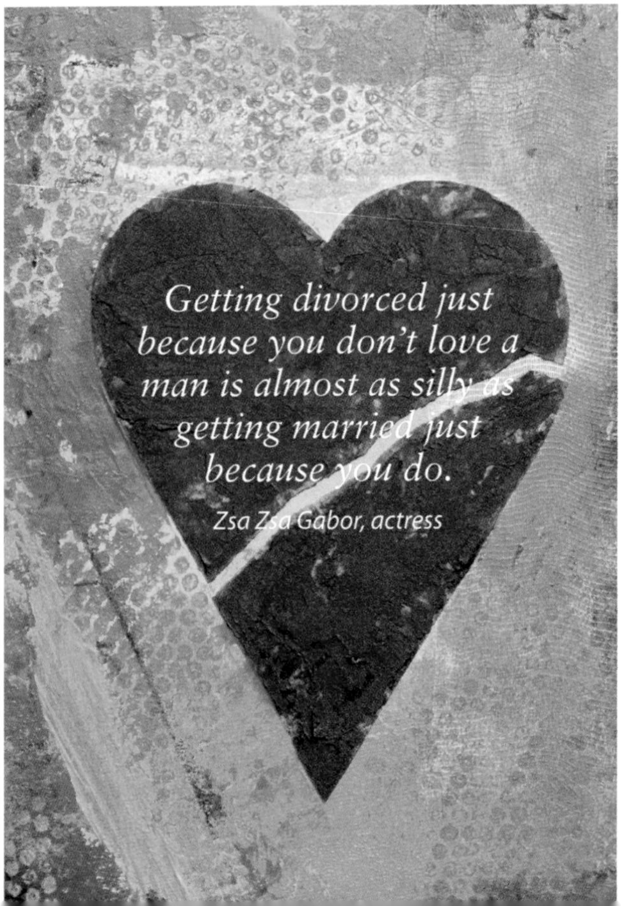
Getting divorced just because you don't love a man is almost as silly as getting married just because you do.
Zsa Zsa Gabor, actress

© Andrea Haase/iStockphoto.com

TOP TWENTY-FIVE FACTORS ASSOCIATED WITH DIVORCE

There is debate about the character of people who divorce: are they selfish, amoral people incapable of making good on a commitment to each other? Or are they individuals who care a great deal about relationships and won't settle for a bad marriage? Indeed, they may divorce precisely because they value marriage and want to rescue their children from being reared in an unhappy home. Some research does suggest that narcissistic individuals tend to leave when there is low satisfaction in a relationship. The more of the following factors that exist in a marriage, the more vulnerable a couple is to divorce.

1. Courting less than two years, which means partners know less about each other
2. Having little in common can limit the activities the couple can bond over
3. Marrying at age 17 and younger
4. Differing in race, education, religion, social class, age, values, and libido
5. Not being religiously devout
6. Having a cohabitation history or an established history of breaking social norms
7. Having been previously married, which implies being less fearful of divorce
8. Having no children or only female children
9. Having limited education
10. Living in an urban residence where there is less social control and greater anonymity
11. Engaging in infidelity
12. Growing up with divorced parents or parents who never married and never lived together
13. Having poor communication skills
14. Husband experiencing unemployment when his role as breadwinner is the norm
15. Wife obtaining employment, which may threaten her spouse
16. A spouse experiencing depression, anxiety disorder, alcoholism, physical illness, or imprisonment
17. Having seriously ill child, which impacts stress, finances, and a couple's time together
18. A spouse having low self-esteem
19. African Americans live under more oppressive conditions, which increases stress in the marital relationship (55 percent of African American marriages will end within fifteen years, in contrast to 42 percent of European American marriages, and 23 percent of Asian marriages)
20. Going through retirement
21. Experiencing rape
22. Having premarital pregnancy or unwanted child
23. Having stepchildren in the household
24. Having high debt
25. Experiencing violence or abuse

Sources: Thompson, P. 2008. Desperate housewives? Communication difficulties and the dynamics of marital (un)happiness. *The Economic Journal* 118:1640–52; Amato, P. R., and B. Hohmann-Marriott. 2007. A comparison of high- and low-distress marriages that end in divorce. *Journal of Marriage and Family* 69:621–38; Clarke-Stewart, A., and C. Brentano. 2006. *Divorce: Causes and consequences*. New Haven: Yale University Press; Lopoo, L. M., and B. Western. 2005. Incarceration and the formation and stability of marital unions. *Journal of Marriage and the Family* 67:721–35; Foster, J. D. 2008. Incorporating personality into the investment model: Probing commitment processes across individual differences in narcissism. *Journal of Social and Personal Relationships* 25:211–23.

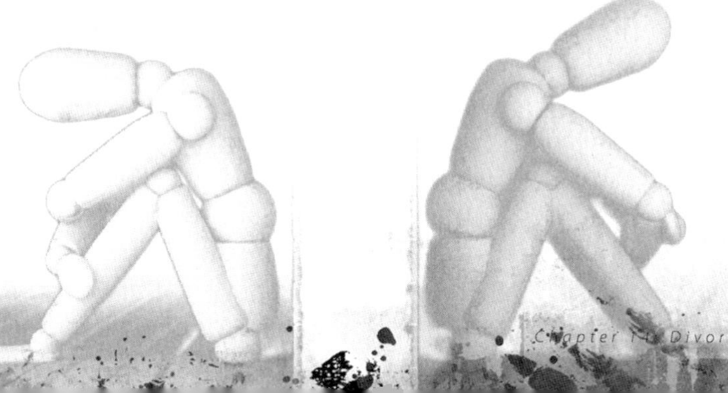

because women are awarded custody in 80 percent of cases. The researchers argue that "who gets the children" is by far the greatest predictor of who files for divorce, and they contend that if the law presumes that joint custody will follow divorce, there will be fewer women filing for divorce because women will have less to gain. The article on the previous page presents the other top characteristics of people most likely to seek a divorce.

14.3 Ending an Unsatisfactory Relationship

Breaking up is never easy. Almost a third of 279 undergraduates reported that they "sometimes" remained in relationships they thought should end (31 percent) or that became "unhappy" (32.3 percent). This finding reflects the ambivalence students sometimes feel in ending an unsatisfactory relationship and their reluctance to do so (Knox et al. 2002). All relationships have difficulties, and all necessitate careful consideration of various issues before they are ended. Before pulling the plug on a relationship, one should consider the following:

1. **Is there any desire or hope to revive and improve the relationship?** In some cases, people end relationships and later regret having done so. Setting unrealistically high standards may eliminate an array of individuals who might be superb partners, companions, and mates. If the reason for ending a relationship is conflict over an issue or set of issues, an alternative to ending the relationship is to attempt to resolve the issues through negotiating differences, compromising, and giving the relationship more time. (We do not recommend giving an abusive relationship more time, as abuse, once started, tends to increase in frequency and intensity. Nor do we recommend staying in a relationship in which there are great differences in interests, styles, personalities, and levels of attachment. It is often more prudent for an individual to find someone with whom he or she is more compatible than to try to "remake" a current partner.)
2. **Acknowledge and accept that terminating a relationship may be painful for both partners.** There may be no way to stop the hurt. One person said, "I can't live with him anymore, but I don't want to hurt him either." The two feelings are incompatible. To end a relationship with a loving partner is usually hurtful to both partners. People who end a relationship usually conclude that the pain and suffering of staying in a relationship is more than they will experience from leaving.
3. **Blame oneself for the end.** One way for an individual to end a relationship is to blame himself by giving a reason that is specific to him ("I need more freedom," "I want to go to graduate school in another state," "I'm not ready to settle down," and so on). If an individual blames his or her partner or gives him or her a way to make things better, the relationship may continue because the individual ending the relationship may feel obligated to give the partner a second chance if change is promised.
4. **Cut off the relationship completely.** It is easier for the person ending the relationship to continue to see the other person without feeling too hurt. However, the other person will probably have a more difficult time and will heal faster if the partner stays away completely. Alternatively, some people are skilled at ending love relationships and turning them into friendships. Though this is difficult and infrequent, it can be rewarding across time for the respective partners.
5. **Learn from the terminated relationship.** There are many reasons a relationship may end, but some behaviors contribute more to failed relationships. These behaviors, such as being too controlling; being oversensitive, jealous, or too picky; cheating; fearing commitment; and being unable to compromise and negotiate conflict, may not be conscious actions. Some of the benefits of terminating a relationship are recognizing one's own contribution to the breakup and working on any characteristics, such as unwanted behaviors, that might be a source of problems. Otherwise, one might repeat the process.
6. **Allow time to grieve over the end of the relationship.** Ending a love relationship is painful. It is okay to feel this pain, to hurt, to cry. Allowing oneself time to experience such grief will help one heal for the next relationship. Recovering from a serious relationship can take twelve to eighteen months.

Remaining Unhappily Married

Suppose you are unhappily married? Is your best choice to stay in this relationship or take your chances and divorce? Hawkins and Booth (2005) analyzed longitudinal data of spouses in unhappy marriages over a twelve-year period and found that those who stayed unhappily married had lower life satisfaction, self-esteem, and overall health compared to those who divorced whether or not they remarried. Similarly, Gardner and Oswald (2006) found that the psychological functioning and happiness of spouses going through divorce improved after the divorce—hence divorce was good for them. However, Waite et al. (2009) compared those who were currently unhappily married (over a five-year period) with those who divorced and remarried. The researchers did not find more positive outcomes for the latter. Hence, the data are unclear if remaining unhappily married or getting divorced and remarried has a more positive outcome for the spouses.

14.4 Gender Differences in Filing for Divorce

Enright (2004) noted that two-thirds of her 1,147 respondents who filed for divorce were women. They felt that their lives would be better without their husbands (few worried they would be separated from their children). Indeed, the primary reason for delay was financial and, once they could see a way to survive financially, they were on their way out of the marriage. The top advantage women reported on the other side of the divorce was feeling a sense of renewed "self-identity." Men are less likely to seek a divorce because they view the cost as separation from their children; women get primary custody of the children in 80 percent of divorces. Men who are not bonded with their children and/or those who are involved in a new relationship are more likely to seek a divorce.

In short, she tried to tame [Sean] Penn and he tried to domesticate her. It proved to be a recipe for emotional trauma and, ultimately divorce.

—*Andrew Morton,* Madonna

14.5 Consequences for Spouses Who Divorce

For both women and men, divorce is often an emotional and financial disaster (see Table 14.1, which identifies stages and issues of the divorce process). In addition to the death of a spouse, separation and divorce are among the most difficult of life's crisis events.

Psychological Consequences of Divorce

Sakraida (2005) interviewed women who both initiated and were the recipients of terminated relationships and found that the person being dropped (the "dumpee") was more vulnerable to depression and ruminated more about divorce. Thuen and Rise (2006) found that perceived control was associated with positive psychological adjustment to divorce (the "dumpee" typically feels less control).

How do women and men differ in their emotional and psychological adjustment to divorce? Siegler and Costa (2000) noted that women fare better emotionally after separation or divorce than do men. They note

© Laura Luongo/Getty Images

Table 14.1
Stages and Issues of Divorce Adjustment

Stage 1: Pre-separation

Issues

Personal: Spouses may consider seeing a marriage counselor to improve their marriage. If they decide to divorce, spouses should get a formal and legal separation agreement drawn up and signed before living in separate residences.

Spouse: If divorce is inevitable, each person should adopt the perspective that they will nurture a relationship that is as positive as possible with the soon-to-be former spouse. The entire family will benefit from such a relationship.

Children and Relatives: Children should be informed once a definite decision has been made and a separation agreement developed.

Finances: Spouses should anticipate a drop in income, look for alternative housing, and seek employment if unemployed.

Legal: If spouses have a civil relationship, they should contact a divorce mediator to develop the terms of the separation agreement. If mediation is not an option, each spouse needs to hire the best attorney he/she can afford.

Stage 2: Separation

Issues

Personal: To help with the emotional devastation of divorce, one should consider seeing a therapist without his or her spouse. One should not separate (move out) until signing a formal or legal agreement.

Spouse: Each spouse should endeavor to cooperate and be civil with his or her former partner.

Children: Spouses should inform children of the decision to divorce and make clear to them that they are not to blame and that the divorce will not change the parents' love for them.

Relatives: Each person should tell his or her parents and friends of the decision to divorce. By reaching out the soon-to-be former partner's parents, each spouse demonstrates the willingness to maintain a positive relationship with them, which will benefit the children.

Finances: The divorce mediator or attorney will instruct each person to develop an inventory of possessions. Each partner should open their own savings and checking accounts.

Legal: Spouses should try to complete divorce mediation with a mediator. If this is not possible, they may use an attorney to develop a legal separation agreement acceptable to both parties. The more that is agreed upon, the more time and money saved in legal fees. Mediated divorces cost about $2,000 and take two to three months. A litigated divorce costs $30,000 and takes about three years.

Stage 3: Divorce

Issues

Personal: Each person should nurture relationships with friends and family who will provide support during the divorce process.

Spouse: Each spouse should continue to nurture as civil a relationship as possible with his or her soon-to-be former partner.

Children: Ensure that children have frequent and regular contact with each parent and nurture their relationships with both parents and all grandparents.

Relatives: Spouses should continue to have regular contact with friends and family who provide emotional support.

Finances: Because of the financial changes spouses are experiencing, each should be frugal.

Legal: Spouses should follow his or her attorney's instructions.

Stage 4: Post-divorce

Issues

Personal: Each person should actively seek other relationships, moving slowly when committing to a new partner. Eighteen months is the recommended waiting period before committing to a new partner.

Former Spouse: Each person should strive to maintain as civil a relationship as possible. Each should support and be positive about the former partner's involvement in a new relationship.

Children: Both parents should spend individual time with each child. Parents shouldn't require the children to like or enjoy new partners.

Relatives: Both sets of grandparents should have access to the children.

Finances: Both former partners should continue to be frugal and move toward getting out of debt.

Legal: Each person should make a payment plan with his or her attorney to begin reducing this debt.

Adapted from The Divorce Room at Heartchoice.com. Used by permission.

that women are more likely than men to not only have a stronger network of supportive relationships but also to profit from divorce by developing a new sense of self-esteem and confidence, because they are thrust into more independent roles. On the other hand, men are more likely to have been dependent on their wives for domestic and emotional support and to have a weaker external emotional support system. As a result, divorced men are more likely than divorced women to date more partners sooner and to remarry more quickly. Hence, there are gender differences in adjustment to divorce, but these differences balance out over time.

We have been discussing the personal emotional consequences of divorce for the respective spouses, but the extended family and friends also feel the impact of a couple's divorce. Whereas some parents are happy and relieved that their offspring are divorcing, others grieve. Either way, the relationship with their grandchildren may be jeopardized. Friends often feel torn and divided in their loyalties. The courts divide the property and grant custody of the children, but who gets the friends? Indeed, an individual going through divorce will lose an average of three friends (Clarke-Stewart and Brentano 2006).

Recovering from a Broken Heart

A sample of 410 freshmen and sophomores at a large southeastern university completed a confidential survey revealing their recovery from a previous love relationship (Knox et al. 2000). Some of the findings were as follows:

1. **Sex differences in relationship termination.** Women were significantly more likely than men to report that they initiated the breakup (50 percent versus 40 percent). Sociologists suggest that women terminating relationships more often is related to their desire to select a better father for their offspring. One female student recalled, "I got tired of his lack of ambition—I just thought I could do better. He's a nice guy but living in a trailer is not my idea of a life."

2. **Sex differences in relationship recovery.** Though recovery was not traumatic for either men or women, men reported more difficulty than women did in adjusting to a breakup. When respondents were asked to rate their level of difficulty from "no problem" (0) to "complete devastation" (10), women scored 4.35 and men scored 4.96. In explaining why men might have more difficulty adjusting to terminated relationships, some of the female students said, "Men have such inflated egos, they can't believe that a woman would actually dump them." Others said, "Men are oblivious to what is happening in a relationship and may not have a clue that it is heading toward an abrupt end. When it does end, they are in shock."

3. **Time or new partner as factors in recovery.** The passage of time and involvement with a new partner were identified as the most helpful factors in getting over a love relationship that ended. Though the difference was not statistically significant, men more than women reported that "a new partner" was more helpful in relationship recovery (34 percent versus 29 percent). Similarly, women more than men reported that "time" was more helpful in relationship recovery (34 percent versus 29 percent).

4. **Other findings.** Other factors associated with recovery for women and men were "moving to a new location" (13 percent versus 10 percent) and recalling that "the previous partner lied to me" (7 percent versus 5 percent). Men were much more likely than women to use alcohol to help them get over a previous partner (9 percent versus 2 percent). Neither men nor women reported using therapy to help them get over a partner (1 percent versus 2 percent). These data suggest that breaking up was not terribly difficult for these undergraduates (but more difficult for men than women) and that both time and a new partner enabled their recovery. These undergraduates are also young and may not view relationships at this age as permanent, making it easier to move on.

Some partners seek revenge as a way of recovering from a relationship that was ended by the other partner. Breed et al. (2007) detailed one example of a former girlfriend who downloaded Internet child pornography on her partner's home computer, called the cops, and had him arrested for possession of child pornography. The penalty was fifteen months in prison per image. She loaded eighteen on his computer. He was found guilty.

Financial Consequences

Getting divorced affects one's finances. Both women and men experience a drop in income following divorce, but women may suffer more. Because men usually have greater financial resources, they may take all they can

postnuptial agreement similar to a premarital agreement, a postnuptial agreement specifies what is to be done with property and holdings at death or divorce.

shared parenting dysfunction the set of behaviors by both parents that are focused on hurting the other parent and are counterproductive for a child's well-being.

with them when they leave. The only money they may continue to give to an ex-wife is court-ordered child support or spousal support (alimony). Although alimony is rare, it is awarded. Former presidential candidate John McCain pays his former wife $17,000 in alimony annually. Some states, such as Texas, do not award alimony at all. However, most states do provide for an equitable distribution of property, whereby property is divided according to what seems fairest to each party, on the basis of a number of factors (such as ability to earn a living, fault in breaking up the marriage, and so on).

Although 56 percent of custodial mothers are awarded child support, the amount is usually inadequate, infrequent, and not dependable, and women are forced to work (sometimes at more than one job) to take financial care of their children. A comparison of divorced mothers, single mothers by choice, and married mothers revealed that economic stability and involvement of the father were more important than family structure in determining the quality of life for the mothers (Segal-Engelchin and Wozner 2005).

Job loss is also associated with divorce. Covizzi (2008) emphasized that getting divorced is associated with losing one's job, and men are more vulnerable than women. The stress of divorce takes its toll on one's ability to function, and productivity may drop, resulting in the person being fired.

How money is divided depends on whether the couple had a prenuptial agreement or a **postnuptial agreement**. Such agreements are most likely to be upheld if an attorney insists on four conditions—full disclosure by both parties, independent representation by separate counsel, absence of coercion or duress, and terms that are fair and equitable (Abut 2005).

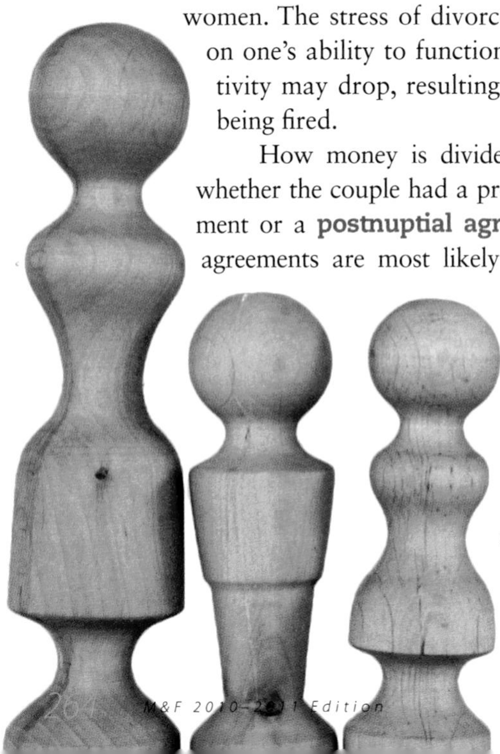

Fathers' Separation from Children

According to Finley (2004, F7), ". . . divorce transforms family power from intact patriarchy to post-divorce matriarchy," where women are typically given custody of the children and child support. Trinder (2008) emphasized that these women serve as gatekeepers for the relationship their husbands have with their children. Their patterns range from being proactive whereby they encourage such relationships and involvement to attempting to destroy the relationship.

As a result, about 5 million divorced dads wake up every morning in one apartment or home, while their children wake up in another with their separated or divorced mother. These are noncustodial fathers who may find the gate to their children shut so that they are allowed to see their children only at specified times (for example, two weekends a month).

Ahrons and Tanner (2003) found that low father involvement, not divorce, has a negative impact on the father-child relationship. Fathers who stay involved in the lives of their children emotionally, physically, and economically (in spite of being a noncustodial parent, having an adversarial former spouse, and a remarriage) mitigate any negative effects on the relationship with their children. Indeed, some relationships with the father may improve because a father may spend more one-on-one time with his children.

Juby et al. (2007) found that fathers who begin a pattern of close involvement with their children during separation and early divorce tend to maintain a close pattern of involvement. Fathers who delay such involvement, become pair-bonded with a new partner, and have a child with the partner, tend to maintain minimal relationships with their nonresident children. Some fathers reduce contact with their children when their former spouse remarries and provides a stepfather for the offspring.

Shared Parenting Dysfunction

Shared parenting dysfunction refers to the set of behaviors on the part of each parent (embroiled in a

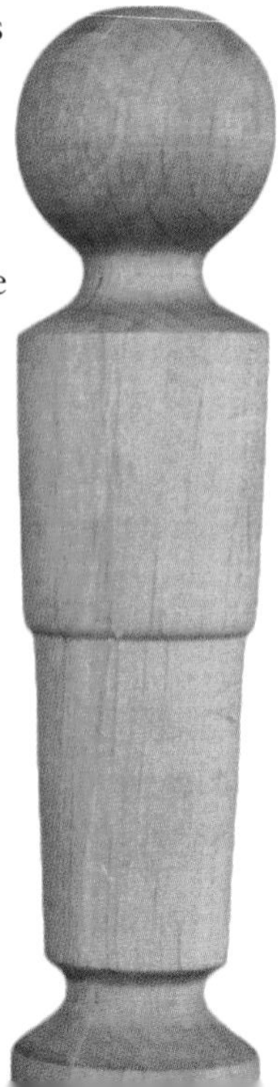

divorce) that are focused on hurting the other parent and are counterproductive for the well-being of the children. Turkat (2002) identified some examples:

- A parent who forced the children to sleep in a car to prove the other parent had bankrupted them
- A noncustodial parent who burned down the house of the primary residential parent after losing a court battle over custody of the children
- A parent who bought a cat for the children because the other parent was highly allergic to cats

Finally, some of the most destructive displays of shared parenting dysfunction may include kidnapping, physical abuse, and murder (p. 390).

Parental Alienation Syndrome

Shared parenting dysfunction may lead to parental alienation syndrome. **Parental alienation syndrome** is a disturbance in which children are obsessively preoccupied with deprecation and/or criticism of a parent, denigration that is unjustified and/or exaggerated (Gardner 1998). Celebrity Alec Baldwin (2008) experienced this phenomenon firsthand after his divorce from Kim Basinger. He noted that she systematically tried to destroy the relationship he had with his daughter and he found the court system to be of no help (he spent $3 million on the divorce and custody hearings). His plight is detailed in his book *A Promise to Ourselves: A Journey through Fatherhood and Divorce.*

Although the "alienators" are fairly evenly balanced between fathers and mothers (Gardner 1998), the custodial parent (more often the mother) has more opportunity and control to alienate the child from the other parent. A couple need not necessarily be divorcing for parental alienation syndrome to occur—this phenomenon may also occur in intact families (Baker 2006).

Several types of behavior that either parent may engage in to alienate a child from the other parent include the following:

1. Minimizing the importance of contact and the relationship with the other parent, including moving far away with the child to make regular contact difficult
2. Exhibiting excessively rigid boundaries; rudeness, or refusal to speak to or inability to tolerate the presence of the other parent, even at events important to the child; refusal to allow the other parent near the home for drop-off or pick-up visitations
3. Having no concern about missed visits with the other parent
4. Showing no positive interest in the child's activities or experiences during visits with the other parent and withholding affection if the child expresses positive feelings about the absent parent
5. Granting autonomy to the point of apparent indifference ("It's up to you if you want to see your dad, I don't care.")
6. Overtly expressing dislike of a visitation ("OK, visit, but you know how I feel about it.")
7. Refusing to discuss anything about the other parent ("I don't want to hear about . . . ") or showing selective willingness to discuss only negative matters
8. Using innuendo and accusations against the other parent, including statements that are false, and blaming the parent for the divorce
9. Portraying the child as an actual or potential victim of the other parent's behavior
10. Demanding that the child keep secrets from the other parent
11. Destroying gifts or memorabilia of the other parent
12. Promoting loyalty conflicts (such as by offering an opportunity for a desired activity that conflicts with scheduled visitation) (Schacht 2000; Teich 2007; and Baker and Darnall 2007)

parental alienation syndrome a disturbance in which children are obsessively preoccupied with deprecation or criticism of a parent.

© Joelle Gould/iStockphoto.com

The most telling sign of children who have been alienated from a parent is the irrational behavior of the children, who for no properly explained reason says that they want nothing further to do with one of the parents. Indeed such children have a lack of ambivalence toward the alienation, lack of guilt or remorse about the alienation, and take the alienating parent's side in the conflict (Baker and Darnall 2007).

Children who are alienated from one parent are sometimes unable to see through the alienation process and regard their negative feelings as natural. Such children are similar to those who have been brainwashed by cult leaders to view outsiders negatively. Baker (2005) interviewed thirty-eight adults who had experienced parental alienation as children and observed several areas of impact, including low self-esteem, depression, drug or alcohol abuse, lack of trust, alienation from one's own children, and divorce.

Sometimes parents intent on alienating their children from the other parent may discover that the children resent the custodial parent for such deprivation. In addition, the children may feel deceived if they are told negative things about the other parent and later learn that these were designed to foster a negative relationship with that parent. The result is often a strained and distanced relationship with the custodial parent when the children grow up—an unintended consequence. Alternatively, the negative socialization toward the other parent may create a lifelong bias against that parent.

14.6 Effects of Divorce on Children

The well-being of children who grow up in homes where their parents remain in low-conflict marriages is higher than children who have divorced parents. Children who grow up in homes where their parents are married but who constantly fight also suffer in terms of their subjective well-being (Sobolewski and Amato 2007).

Gordon (2005) noted that divorce may actually benefit children. When parental conflict is very high prior to divorce, children benefit by no longer being subjected to the relentless anger and emotional abuse they observe between their parents. Indeed, when comparing children whose conflicted parents divorced with children whose parents were still together, the children were very similar. In self-reports of 158 Israeli young adults whose parents divorced when they were adolescents, Sever et al. (2007) determined that, although the children had painful feelings, almost half the participants reported more positive than negative outcomes. These included maturity and growth, empowerment, empathy, and relationship savvy. Lambert (2007) identified other advantages for children whose parents divorced as learning to be resilient, developing closer relationships with siblings, having happier parents, learning lessons about what not to do in a relationship, and receiving more attention. DeCuzzi et al. (2004) also reported that, although 26 percent of the undergraduates in their study reported that the divorce of their parents had a negative effect, 32.9 percent reported a *positive* effect. Finally, when children of divorced parents become involved in their own marriage to a supportive, well-adjusted partner, the negative effects of parental divorce are mitigated (Hetherington 2003).

Kelly and Emery (2003) reviewed the literature on the effect of divorce on children and concluded the following:

> . . . [I]t is important to emphasize that approximately 75–80% of children and young adults do not suffer from major psychological problems, including depression; have achieved their education and career goals; and retain close ties to their families. They enjoy intimate relationships, have not divorced, and do not appear to be scarred with immutable negative effects from divorce. (p. 357–58)

Indeed, Coltrane and Adams (2003) emphasized that claiming that divorce seriously damages children is a "symbolic tool used to defend a specific moral vision for families and gender roles within them" (p. 369). They go on to state that "Understanding this allows us to see divorce not as the universal moral evil depicted by divorce reformers, but as a highly individualized process that engenders different experiences and reactions among various family members . . ." (p. 370).

Nevertheless, some children experience negative fallout from their parents' divorce. Kilmann et al. (2006) compared 147 college females with intact biological parents with 157 college females whose parents had divorced. Compared to those with intact parents, females whose parents had divorced had lower self-esteem and rated both their biological fathers and mothers more negatively. Other disadvantages Lambert (2007) identified include being disrupted because of visitation arrangements, stressful holidays, not having a role model for a good relationship, feeling the divorce hurt one's siblings, financial hardships, and having no dad while growing up.

Although the primary factor that determines the effect of divorce on children is the degree to which the divorcing parents are civil or adversarial, legal and physical custody are important issues.

Who Gets the Children?

Judges who are assigned to hear initial child custody cases must make a judicial determination regarding whether one or both parents will have decisional authority on major issues affecting the children (called "**legal custody**"), and the distribution of parenting time (called "visitation" or "**physical custody**"). Toward this end, judges in all states are guided by the statutory dictum called "best interests of the child." In some states (Florida, Michigan, California, New Jersey, and others), specific statutory custody factors have been enacted to guide judges in making "best interest" determinations.

In a highly contested custody case, a judge will often appoint a mental health professional to conduct a custody evaluation to assist the judge in determining what will be the best future arrangement for the child. Of course, each custody case is different because the circumstances of the children are different, but some of the frequently employed custody factors include the following:

1. The child's age, maturity, sex, and activities, including culture and religion—all relevant information about the child's life—keeping the focus of custody on the child's best interests
2. The wishes expressed by the child, particularly the older child (judges will often interview children over 6 years old in chambers)
3. Each parent's capacity to care for and provide for the emotional, intellectual, financial, religious, and other needs of the child (including the work schedules of the parents)
4. The parents' ability to agree, communicate, and cooperate in matters relating to the child
5. The nature of the child's relationship with the parents, which considers the child's relationship with other significant people, such as members of the child's extended family
6. The need to protect the child from physical and psychological harm caused by abuse or ill treatment, especially domestic violence
7. The past and present parental attitudes and behavior (dealing with issues of parenting skills and personalities)
8. The proposed plans for caring for the child (The judge will want to know how each parent proposes to raise the child, including proposed parenting times for the other parent.) (Lewis 2008)

legal custody decisional authority over major issues involving the child.

physical custody also called "visitation," refers to distribution of parenting time following divorce.

These and other custody factors, whether presented by the custody evaluator or by testimony, will become the basis for the judge's custody determination.

Minimizing Negative Effects of Divorce on Children

Researchers have identified the following conditions under which a divorce has the fewest negative consequences for children:

1. **Healthy parental psychological functioning.** Children of divorced parents benefit to the degree that the parents remain psychologically fit and positive, and socialize their children to view the divorce as a "challenge to learn from." Parents who nurture self-pity, abuse alcohol or drugs, and socialize their children to view the divorce as a tragedy from which they will never recover create negative outcomes for their children. Some divorcing parents can benefit from therapy as a method for coping with their anger or depression and for making choices in the best interest of their children.

   Some parents also enroll their children in the "New Beginnings Program"—"an empirically driven prevention program designed to promote child resilience during the post-divorce period" (Hipke et al. 2002, 121). The program focuses on improving the quality of the primary residential parent-child relationship, ensuring continued

discipline, reducing exposure to parental conflict, and providing access to the nonresidential parent. Outcome data reveal that not all children benefit, particularly those with "poor regulatory skills" and "demoralized" mothers (p. 127).

2. **A cooperative relationship between the parents.** The most important variable in a child's positive adjustment to divorce is when the child's parents continue to maintain a cooperative relationship throughout the separation, divorce, and post-divorce period. In contrast, bitter parental conflict places the children in the middle. One daughter of divorced parents said, "My father told me, 'If you love me, you would come visit me,' but my mom told me, 'If you love me, you won't visit him.'" Baum (2003) confirmed that the longer and more conflicted the legal proceedings, the worse the co-parental relationship. Bream and Buchanan (2003) noted that children of conflictual divorcing parents are "children in need." Numerous states mandate parenting classes as part of the divorce process. Whitehurst et al. (2008) confirmed that divorcing spouses who attended a court-ordered six-session (two hours each) Cooperative Parenting and Divorce Program benefited in parenting skills improvement as well as improved relationships with their children. The finding was true for both women and men. The court-approved "Positive Parenting through Divorce" program may also be taken online (http://www.positiveparentingthroughdivorce.com/).

3. **Parents' attention to the children and allowing them to grieve.** Children benefit when both the custodial and the noncustodial parent continue to spend time with them and to communicate to them that they love them and are interested in them. Parents also need to be aware that their children do not want the divorce and to allow them to grieve over the loss of their family as they knew it. Indeed, children do not want their parents to separate. Some children are devastated to the point of suicide. A team of researchers identified 15,555 suicides among 15- to 24-year-olds in thirty-four countries in a one-year period and found an association between divorce rates and suicide rates (Johnson et al. 2000).

4. **Encouragement to see noncustodial parent.** Children benefit when custodial parents (usually mothers) encourage and maintain regular and stable visitation schedules with the noncustodial parent following divorce. Cashmore et al. (2008) interviewed sixty adolescents (ages 12 through 19) and their divorced nonresident parent (usually the father) and found that overnight stays with the nonresident parent were associated with reported greater closeness and better quality relationships than was true if the parent and child had only daytime contact.

5. **Attention from the noncustodial parent.** Children benefit when they receive frequent and consistent attention from noncustodial parents, usually the fathers. Noncustodial parents who do not show up at regular intervals exacerbate their children's emotional insecurity by teaching them, once again, that parents cannot be depended on. Parents who show up often and consistently teach their children to feel loved and secure. Sometimes joint custody solves the problem of children's access to their parents. Hsu et al. (2002) noted the devastating effect on children who grow up without a father.

6. **Assertion of parental authority.** Children benefit when both parents continue to assert their parental authority and continue to support the discipline practices of each other.

7. **Regular and consistent child-support payments.** Support payments (usually from the father to the mother) are associated with economic stability for the child.

8. **Stability.** Moving to a new location causes children to be cut off from their friends, neighbors, and teachers. It is important to keep their life as stable as possible during a divorce.

9. **Children in a new marriage.** Manning and Smock (2000) found that divorced noncustodial fathers who remarried and who had children in the new marriages were more likely to shift their emotional and economic resources to the new family unit than were fathers who did not have new biological children. Fathers might be alert to this potential and consider each child, regardless of when or with whom the child was born, as worthy of a father's continued love, time, and support.

Parents who show up often and consistently teach their children to feel loved and secure.

© Blend Images/Jupiterimages / © John Solie/iStockphoto.com

WHAT DIVORCING PARENTS MIGHT TELL THEIR CHILDREN: AN EXAMPLE

The following are the words of a mother of two children (8 and 12) as she tells her children of the pending divorce. It assumes that both parents take some responsibility for the divorce and are willing to provide a united front to the children. The script should be adapted for one's own unique situation.

Daddy and I want to talk to you about a big decision that we have made. A while back we told you that we were having a really hard time getting along, and that we were having meetings with someone called a therapist who has been helping us talk about our feelings, and deciding what to do about them.

We also told you that the trouble we are having is not about either of you. Our trouble getting along is about our grown-up relationship with each other. That is still true. We both love you very much, and love being your parents. We want to be the best parents we can be.

Daddy and I have realized that we don't get along so much, and disagree about so many things all the time, that we want to live separately, and not be married to each other anymore. This is called getting divorced. Daddy and I care about each other but we don't love each other in the way that happily married people do. We are sad about that. We want to be happy, and want each other to be happy. So to be happy we have to be true to our feelings.

It is not your fault that we are going to get divorced. And it's not our fault. We tried for a very long time to get along living together but it just got too hard for both of us.

We are a family and will always be your family. Many things in your life will stay the same. Mommy will stay living at our house here, and Daddy will move to an apartment close by. You both will continue to live with mommy and daddy but in two different places. You will keep your same rooms here, and will have a room at Daddy's apartment. You will be with one of us every day, and sometimes we will all be together, like to celebrate somebody's birthday, special events at school, or scouts. You will still go to your same school, have the same friends, go to soccer, baseball, and so on. You will still be part of the same family and will see your aunts, uncles, and cousins.

The most important things we want you both to know are that we love you, and we will always be your mom and dad . . . nothing will change that. It's hard to understand sometimes why some people stop getting along and decide not to be friends anymore, or if they are married decide to get divorced. You will probably have lots of different feelings about this. While you can't do anything to change the decision that daddy and I have made, we both care very much about your feelings. Your feelings may change a lot. Sometimes you might feel happy and relieved that you don't have to see and feel daddy and me not getting along. Then sometimes you might feel sad, scared, or angry. Whatever you are feeling at any time is ok. Daddy and I hope you will tell us about your feelings, and it's OK to ask us about ours. This is going to take some time to get used to. You will have lots of questions in the days to come. You may have some right now. Please ask any question at any time.

Daddy and I are here for you. Today, tomorrow, and always. We love you with our heart and soul.

divorce mediation process in which divorcing parties make agreements with a third party (mediator) about custody, visitation, child support, property settlement, and spousal support.

10. Age and reflection on the part of children of divorce. Sometimes children whose parents are divorced benefit from growing older and reflecting on their parents' divorce as an adult rather than a child. Nielsen (2004) emphasized that daughters who feel distant from their fathers can benefit from examining the divorce from the viewpoint of the father (Was he alienated by the mother?), the cultural bias against fathers (they are maligned as "deadbeat dads" who "abandon their families for a younger woman"), and the facts about divorced dads (they are more likely to be depressed and suicidal following divorce than mothers). Adolescents also bear some responsibility for the post-divorce relationships with their parents. Menning (2008) emphasized that adolescents are not passive recipients of their parents' divorce but may actively accelerate or decelerate having a positive or negative relationship with their parents by using a variety of relationship management strategies. For example, by deciding to shut down and disclose nothing to their parents about their lives, they increase the emotional distance between them and their parents.

11. Divorce education program for children. Gilman et al. (2005) found that Kid's Turn, a San Francisco area divorce education program for children, was effective in reducing the conflict between parents and children of a group of sixty 7- to 9-year-old children. However, the children also had more reconciliation fantasies.

14.7 Conditions of a "Successful" Divorce

Although acknowledging that divorce is usually an emotional and economic disaster, it is possible to have a "successful" divorce. Indeed, most people are resilient and "are able to adapt constructively to their new life situation within two to three years following divorce, a minority being defeated by the marital breakup, and a substantial group of women being enhanced" (Hetherington 2003, 318). The following are some of the behaviors spouses can engage in to achieve this:

1. Mediate rather than litigate the divorce. Divorce mediators encourage a civil, cooperative, compromising relationship while moving the couple toward an agreement on the division of property, custody, and child support. By contrast, attorneys make their money by encouraging hostility so that spouses will prolong the conflict, thus running up higher legal bills. In addition, the couple cannot divide money spent on divorce attorneys (average is $15,000 for *each* side, so a litigated divorce cost will start at $30,000). Benton (2008) noted that the worse thing divorcing spouses can do is to respectively hire the "meanest, nastiest, most expensive yard dog lawyer in town" because doing so will only result in a protracted expensive divorce where neither spouse will "win." Because the greatest damage to children from a divorce is a continuing hostile and bitter relationship between their parents, some states require **divorce mediation** as a mechanism to encourage civility in working out

EACH DIVORCE, PAINFUL THOUGH IT MAY HAVE BEEN AT THE TIME, MARKED A STEP FORWARD, AN OPPORTUNITY FOR SELF-DEFINITION RATHER THAN FAILURE—ALMOST LIKE THE REPOTTING OF A PLANT WHEN THE ROOTS DON'T FIT ANYMORE.

—*Jane Fonda (of her three divorces)*

differences and to clear the court calendar from protracted court battles.

2. **Co-parent with the ex-spouse.** Setting aside negative feelings about one's ex-spouse so as to cooperatively co-parent not only facilitates parental adjustment but also takes children out of the line of fire. Such co-parenting translates into being cooperative when one parent needs to change a child care schedule, sitting together during a performance by the children, and showing appreciation for the other parent's skill in responding to a crisis with the children.

3. **Take some responsibility for the divorce.** Because marriage is an interaction between spouses, one person is seldom totally to blame for a divorce. Rather, both spouses share reasons for the demise of the relationship. Each spouse should take some responsibility for what went wrong.

4. **Learn from the divorce.** Individuals should view the divorce as an opportunity to improve oneself for future relationships. What could be done differently in the next relationship?

5. **Create positive thoughts.** Divorced people are susceptible to feeling as though they are failures. They see themselves as Divorced people with a capital D, a situation sometimes referred to as "hardening of the categories" disease. Improving self-esteem is important for divorced people. They can do this by systematically thinking positive thoughts about themselves. One technique is to write down twenty-one positive statements about oneself ("I am honest," "I have strong family values," "I am a good parent," and so on) and transfer them to three-by-five cards, each containing three statements. One card could be carried along each day and read at three regularly spaced intervals (for example, 7:00 A.M., 1:00 P.M., and 7:00 P.M.). This ensures positive thoughts about oneself and preventing a drift into a negative set of thoughts ("I am a failure" or "no one wants to be with me.")

6. **Avoid alcohol and other drugs.** The stress and despair that some people feel following a divorce make them particularly vulnerable to the use of alcohol or other drugs. These should be avoided because they produce an endless negative cycle. For example, stress is relieved by alcohol; alcohol produces a hangover and negative feelings; the negative feelings are relieved by more alcohol, producing more negative feelings, and so on.

7. **Relax without drugs.** Deep muscle relaxation can be achieved by systematically tensing and relaxing each of the major muscle groups in the body. Alternatively, yoga, transcendental meditation, and massage can induce a state of relaxation in some people. Whatever the form, it is important to schedule a time each day for relaxation.

8. **Engage in aerobic exercise.** Exercise helps one to not only counteract stress but also to avoid it. Jogging, swimming, riding an exercise bike, or other similar exercise for thirty minutes every day increases the oxygen to the brain and helps facilitate clear thinking. In addition, aerobic exercise produces endorphins in the brain, which create a sense of euphoria ("runner's high").

9. **Engage in fun activities.** Some divorced people sit at home and brood over their "failed" relationship. This only compounds their depression. Doing what they have previously found enjoyable—swimming, horseback riding, skiing, sporting events with friends—provides an alternative to sitting on the couch alone.

10. **Continue interpersonal connections.** Adjustment to divorce is facilitated when continuing relationships with friends and family. These individuals provide emotional support and help buffer the feeling of isolation and aloneness. First Wives World (www.firstwivesworld.com) is a new interactive site to provide an Internet social network for women transitioning through divorce.

11. **Let go of the anger for the ex-partner.** Former spouses who stay negatively attached to an ex by harboring resentment and trying to get back at the ex prolong their adjustment to divorce. The old adage that you can't get ahead by getting even is relevant to divorce adjustment.

12. **Allow time to heal.** Because self-esteem usually drops after divorce, a person is

divorcism the belief that divorce is a disaster.

covenant marriage type of marriage that permits divorce only under specific conditions.

often vulnerable to making commitments before working through feelings about the divorce. The time period most people need to adjust to divorce is between twelve and eighteen months. Although being available to others may help to repair one's self-esteem, getting remarried during this time should be considered cautiously. Two years between marriages is recommended.

13. **Progress through various psychological stages of divorce.** Reva Wiseman (1975) identified the various psychological stages a person goes through when getting a divorce. These include the following:

    a. **Denial.** Marital problems are ignored or attributed to an external cause. "We are fine" or "this is normal" are ways of coping with the emotional and physical distance from the partner.

    b. **Loss or depression.** Spouses confront the reality that they will divorce and lose their once-intimate relationship permanently, and depression sets in.

    c. **Anger or ambivalence.** Spouses turn their anger toward each other and become critical, vindictive, and even violent. Strong negative emotions ensue during the marital separation, which is a normal response to the loss of an important attachment figure (for example, a spouse). Some spouses are not capable of detaching and maintain a negative attachment; by staying bitter and resentful, they remain attached. They may also want to hold on to the dying relationship because it feels safe.

    d. **New lifestyle and identify.** Ex-spouses begin life as single adults and detach from marital identity. Men typically drink more alcohol. Women typically turn to girlfriends for support. Both may enter a period of being sexually indiscriminate.

    e. **Acceptance and integration.** Individuals recognize their new status, loss of anger, and acceptance, and move on to new relationships. Some never reach this stage but harbor resentments or blame the ex for the end of the marriage.

14.8 Divorce Prevention

Divorce remains stigmatized in our society, as evidenced by the term **divorcism**—the belief that divorce is a disaster. In view of this cultural attitude, a number of attempts have been made to reduce it. Marriage education workshops provide an opportunity for couples to meet with other couples and a leader who provides instruction in communication, conflict resolution, and parenting skills. Stanley et al. (2005) found positive outcomes in marital functioning for couples in marriage education classes provided for the U.S. Army.

Another attempt at divorce prevention is **covenant marriage** (now available in Louisiana, Arizona, and Arkansas), which emphasizes the importance of staying married and permits divorce only under certain conditions (Byrne and Carr 2005). In covenant marriages, couples agree to the following when they marry: (1) marriage preparation (meeting with a counselor who discusses marriage and their relationship); (2) full disclosure of all information that could reasonably affect a partner's decision to marry (for example, previous marriages, children, STIs, one's homosexuality); (3) an oath that their marriage is a lifelong commitment; (4) an agreement to consider divorce only for "serious" reasons such as abuse, adultery, and imprisonment for a felony or separation of more than two years; (5) an agreement to see a marriage counselor if problems threaten the marriage; and (6) not to divorce until after a two-year "cooling off" period (*Economist* 2005).

Although most of a sample of 1,324 adults in a telephone survey in Louisiana, Arizona, and Minnesota were positive about covenant marriage (Hawkins et al. 2002), fewer than 3 percent of marrying couples elected covenant marriages when given the opportunity to do so (Licata 2002). Although already married couples can convert their standard marriages to covenant marriages, there are no data on how many have done so (Hawkins et al. 2002).

14.9 Remarriage

Divorced spouses usually waste little time getting involved in new relationships. Indeed, one-fourth date someone new before the divorce is final. Within two years, 75 percent of divorced women and 80 percent of divorced men are in serious, exclusive relationships (Enright 2004). When comparing divorced individuals who have remarried against divorced individuals who have not remarried, the remarried individuals report greater personal and relationship happiness.

© Food Pix/Jupiterimages

Remarriage for Divorced Individuals

Ninety percent of remarriages are of people who are divorced rather than widowed. The principle of **homogamy** is illustrated in the selection of a new spouse (never-married individuals tend to marry never-married individuals, divorced individuals tend to marry other divorced individuals, and widowed individuals tend to marry other widowed individuals). The majority of divorced people remarry for many of the same reasons as for a first marriage—love, companionship, emotional security, and a regular sex partner. Other reasons are unique to remarriage and include financial security (particularly for a wife with children), help in rearing one's children, the desire to provide a "social" father or mother for one's children, escape from the stigma associated with the label "divorced person," and legal threats regarding the custody of one's children. With regard to the latter, the courts view a parent seeking custody of a child more favorably if the parent is married.

Regardless of the reason for remarriage, it is best to proceed slowly into a remarriage. Some may benefit from involvement in a divorce adjustment program. Vukalovich and Caltabiano (2008) assessed the effectiveness of an adjustment separation or divorce program. Twenty females and ten males completed a pre- and post-questionnaire. Overall, the results indicated that participants made significant adjustment gains following participation in the program.

For others, grieving over the loss of a first spouse through divorce or death and developing a relationship with a new partner takes time. At least two years is the recommended interval between the end of a marriage and a remarriage (Marano 2000). Older divorced women (over 40) are less likely than younger divorced women to remarry. Not only are fewer men available, but also the **mating gradient**, whereby men tend to marry women younger than themselves, is operative. In addition, older women are more likely to be economically independent, to enjoy living alone, to value the freedom of singlehood, and to want to avoid the restrictions of marriage. Divorced people getting remarried are usually about ten years older than those marrying for the first time. People in their mid-thirties who are con-

I don't think *I'll get married again. I'll just find a woman I don't like and give her a house.*

—Lewis Grizzard

homogamy tendency to select someone with similar characteristics to marry.

mating gradient the tendency for husbands to be more advanced than their wives with regard to age, education, and occupational success.

negative commitment individuals who remain emotionally invested in their relationship with their former spouse, despite remarriage.

sidering remarriage are more likely to have finished school and to be established in a job or career than individuals in their first marriages.

Courtship for previously married individuals is usually short and takes into account the individuals' respective work schedules and career commitments. Because each partner may have children, much of the couple's time together includes their children. In subsequent marriages, eating pizza at home and renting a DVD to watch with the kids often replaces the practice of going out alone on an expensive dinner date during courtship before a first marriage.

Preparation for Remarriage

Trust is a major issue for people getting remarried. Brimhall et al. (2008) interviewed sixteen remarried individuals and found that most reported that their first marriage ended over trust issues—the partner betrayed them by having an affair or by hiding or spending money without the partner's knowledge. Each was intent on ensuring a foundation of trust in the new marriage.

Some people who remarry first live together before the wedding. Teachman (2008) found that living together before a remarriage does not increase the risk of divorce in a subsequent marriage. Poortman and Lyngstad (2008) found a similar pattern.

People remarrying may also develop a prenuptial agreement. These soon-to-be remarried spouses may have assets to protect as well as wanting to ensure that their respective children get whatever portion of their estate they desire. We discussed prenuptial agreements in detail in Chapter 5, and they can be one way to ensure that children will receive their remarrying parent's estate.

Issues of Remarriage

Several issues challenge people who remarry (Swallow 2004; Ganong and Coleman 1994; Goetting 1982).

Boundary Maintenance Movement from divorce to remarriage is not a static event that is over after a brief ceremony. Rather, ghosts of the first marriage—in terms of the ex-spouse and, possibly, the children—must be dealt with. A parent must decide how to relate to an ex-spouse to maintain a good parenting relationship for the biological children while keeping an emotional distance to prevent problems from developing with a new partner. Some spouses continue to be emotionally attached to and have difficulty breaking away from an ex-spouse. These former spouses have what Masheter (1999) terms a "**negative commitment**," Masheter says such individuals "have decided to remain [emotionally] in this relationship and to invest considerable amounts of time, money, and effort in it . . . [T]hese individuals do not take responsibility for their own feelings and actions, and often remain "stuck," unable to move forward in their lives" (p. 297).

One former spouse recalled a conversation with his ex from whom he had been divorced for more than seventeen years. "Her anger coming through the phone seemed like she was still feeling the divorce as though it had happened this morning," he said.

Emotional Remarriage Remarriage involves beginning to trust and love another person in a new relationship. Such feelings may come slowly as a result of negative experiences in a previous marriage.

Psychic Remarriage Divorced individuals considering remarriage may find it difficult to give up the freedom and autonomy of being single and to develop a mental set conducive to pairing. This transition may be particularly difficult for people who sought a divorce as a means to personal growth and autonomy. These individuals may fear that getting remarried will put unwanted constraints on them.

Community Remarriage This stage involves a change in focus from single friends to a new mate and other couples with whom the new pair will interact. The bonds of friendship established during the divorce period may be particularly valuable because they have given support at a time of personal crisis. Care should be taken not to drop these friendships.

Parental Remarriage Because most remarriages involve children, people must work out the nuances of living with someone else's children. Mothers are usually awarded primary physical custody, and this translates into a new stepfather adjusting to the mother's children and vice versa. For individuals who have children from a previous marriage who do not live primarily with them, a new spouse must adjust to these children on weekends, holidays, and vacations, or at other visitation times.

Economic and Legal Remarriage A second marriage may begin with economic responsibilities to a first marriage. Alimony and child support often threaten the harmony and sometimes even the economic survival of second marriages. Although the income of a new wife is not used legally to decide the amount her new husband is required to pay in child support for his children of a former marriage, his ex-wife may petition the court for more child support. The ex-wife may do so, however, on the premise that his living expenses are reduced with a new wife and that, therefore, he should be able to afford to pay more child support. Although an ex-wife is not likely to win, she can force the new wife to court and disclose her income (all with considerable investment of time and legal fees for a newly remarried couple).

Economic issues in a remarriage may become evident in another way. A remarried woman who receives inadequate child support from an ex-spouse and needs money for her child's braces, for instance, might wrestle with how much money to ask her new husband for.

© Medioimages/Jupiterimages

Remarriage for Widowed Individuals

Only 10 percent of remarriages involved widows or widowers. Nevertheless, remarriage for widowed individuals is usually very different from remarriage for divorced people. Unlike divorced individuals, widowed individuals are usually much older and their children are grown. A widow or widower may marry someone of similar age or someone who is considerably older or younger. Marriages in which one spouse is considerably older than the other are referred to as May-December marriages (discussed in Chapter 7, Marriage Relationships). Here we will discuss only **December marriages**, in which both spouses are elderly.

A study of twenty-four elderly couples found that the primary motivation for remarriage was the need to escape loneliness or the need for companionship (Vinick 1978). Men reported a greater need to remarry than did women.

Most of the spouses (75 percent) met through a mutual friend or relative and married less than a year after their partner's death (63 percent). Increasingly, elderly individuals are meeting online. Some sites cater to older individuals seeking partners, including seniorfriendfinder.com and thirdage.com.

The children of the couples in Vinick's study had mixed reactions to their parent's remarriage. Most of the children were happy that their parent was happy and felt relieved that someone would now meet the companionship needs of their elderly parent on a more regular basis. However, some children disapproved of the marriage out of concern for their inheritance rights. "If that woman marries Dad," said a woman with two children, "she'll get everything when he dies. I love him and hope he lives forever, but when he's gone, I want the house I grew up in." Though children may be less than approving of the remarriage of their widowed parent, adult friends of the couple, including the kin of the deceased spouses, are usually very approving (Ganong and Coleman 1994).

December marriage a new marriage in which both spouses are elderly.

Stages of Involvement with a New Partner

After a legal separation or divorce (or being widowed), a parent who becomes involved in a new relationship passes through various transitions (see Table 14.2)

Table 14.2
Stages of Parental Repartnering

Relationship Transition	Definition
Dating initiation	The parent begins to date.
Child introduction	The children and new dating partner meet.
Serious involvement	The parent begins to present the relationship as "serious" to the children.
Sleepover	The parent and the partner begin to spend nights together when the children are present.
Cohabitation	The parent and the partner combine households.
Breakup of a serious relationship	The relationship experiences a temporary or permanent disruption.
Pregnancy in the new relationship	A planned or unexpected pregnancy occurs.
Engagement	The parent announces plans to remarry.
Remarriage	The parent and partner create a legal or civil union.

Source: E. R. Anderson, and S. M. Greene. 2005. Transitions in parental repartnering after divorce. *Journal of Divorce & Remarriage* 43:49 (http://www.haworthpress.com/web/jdr/).

FAMILIES DON'T BLEND. NOT LIKE ICE CREAM OR FRUIT DRINKS, ANYWAY. BUT MORE LIKE PATCHWORK QUILTS. TAKE THE PIECES AND SEW THEM TENDERLY TOGETHER. DON'T DENY THAT YOU'RE TWO FAMILIES; REJOICE IN THOSE DIFFERENCES.

—*Van Chapman, researcher*

blended family a family where new spouses blend their children from previous marriages.

binuclear a family that spans two households, often because of divorce.

stepfamily (step relationships) a family in which partners bring children from previous marriages into the new home, where they may also have a child of their own.

(Anderson and Greene 2005). These not only affect the individuals and their relationship but any children and/or extended family.

Stability of Remarriages National data reflect that remarriages are more likely than first marriages to end in divorce in the early years of remarriage (Clarke and Wilson 1994). Remarriages most vulnerable to divorce are those that involve a woman bringing a child into the new marriage. Teachman (2008) analyzed data on women (N = 655) from National Survey of Family Growth to examine the correlates of second marital dissolution. He found that women who brought stepchildren into their second marriage experienced an elevated risk of marital disruption. Premarital cohabitation or giving birth while cohabiting with a second husband did not raise the risk of marital dissolution, however. In addition, marrying a man who brought a child to the marriage did not increase the risk of marital disruption. One possible explanation for why a woman bringing a child into a second marriage is related to greater instability is that she may be less attentive to the new husband and more of a mother than a wife.

Second marriages, in general, are more susceptible to divorce than first marriages because divorced individuals are less fearful of divorce than individuals who have never divorced. So, rather than stay in an unhappy second marriage, the spouses leave.

Though remarried people are more vulnerable to divorce in the early years of their subsequent marriage, they are less likely to divorce after fifteen years of staying in the second marriage than those in first marriages (Clarke and Wilson 1994). Hence, these spouses are likely to remain married because they want to, not because they fear divorce. McCarthy and Ginsberg (2007) also noted that couples in functional and stable second marriages take greater pride and report higher satisfaction in their marriage than couples in their first marriage. Not only are they relieved at not having to live in their former relationships, but they are older and more skilled in working out the nuances of a life together.

14.10 Stepfamilies

Stepfamilies, also known as blended, binuclear, remarried, or reconstituted families, represent the fastest-growing type of family in the United States. Indeed, more than half of Americans are members of a blended family (Christian 2005). A **blended family** is one in which spouses in a new marriage relationship blend their children from at least one previous marriage. The term **binuclear** refers to a family that spans two households; when a married couple with children divorce, their family unit typically spreads into two households.

There is a movement away from the use of the term blended because stepfamilies really do not blend. The term **stepfamily** (sometimes referred to as **step relationships**) is the term currently in vogue. Leon and Angst (2005) reviewed American films over a thirteen-year period and found that stepfamilies and remarriages were depicted in negative or mixed ways. This section examines how stepfamilies differ from nuclear families; how they are experienced from the viewpoints of women, men, and children; and the developmental tasks that must be accomplished to make a successful stepfamily.

© Donna Elkow-Nash/iStockphoto.com

Definition and Types of Stepfamilies

Although there are various types of stepfamilies (Prather 2008), the most common is a family in which the partners bring children from previous relationships into the new home, where they may also have a child of their own. The couple may be married or living together, heterosexual or homosexual, and of any race. Although a stepfamily can be created when an individual who has never married or a widowed parent with children marries a person with or without children, most stepfamilies today are composed of spouses who are divorced and who bring children into a new marriage. This is different from stepfamilies characteristic of the early twentieth century, which more often were composed of spouses who had been widowed.

As noted, stepfamilies may be both heterosexual and homosexual. Lesbian stepfamilies model gender flexibility in that a biological mother and a stepmother tend to share parenting (in contrast to a traditional family, in which the mother may take primary responsibility for parenting and the father is less involved). This allows a biological mother some freedom from motherhood as well as support in it. In gay male stepfamilies, the gay men may also share equally in the work of parenting.

Myths of Stepfamilies

Various myths abound regarding stepfamilies, including that new family members will instantly bond emotionally, that children in stepfamilies are damaged and do not recover, that stepfamilies are not "real" families, and that stepmothers are "wicked" and "home-wreckers." The Stepfamily Association of America has identified other myths, such that part-time (weekend) stepfamilies are easier than full-time stepfamilies, that it is easier or better if the biological parents withdraw, and that stepfamilies formed after the death of a parent rather than the divorce of a parent are "easier" (Stepfamily Association, 2009).

© Donna Elkow-Nash/iStockphoto.com

Unique Aspects of Stepfamilies

Stepfamilies differ from nuclear families in a number of ways (see Table 14.3). To begin with, children in nuclear families are biologically related to both parents, whereas children in stepfamilies are biologically related to only one parent. Also, in nuclear families, both biological parents live with their children, whereas only one biological parent in stepfamilies lives with the children. In some cases, the children alternate living with each parent.

Though nuclear families are not immune to loss, everyone in a stepfamily has experienced the loss of a love partner, which results in grief. About 70 percent of children whose parents have divorced are living without their biological father (who some children desperately hope will reappear and reunite the family). The respective spouses may also have experienced emotional disengagement and physical separation from a once loved partner. Stepfamily members may also experience losses because of having moved away from the house in which they lived, their familiar neighborhood, and their circle of friends.

Stepfamily members are also connected psychologically to others outside their unit. Bray and Kelly (1998) referred to these relationships as the "ghosts at the table":

> Children are bound to absent parents, adults to past lives and past marriages. These invisible psychological bonds are the ghosts at the table and because they play on the most elemental emotions—emotions like love and loyalty and guilt and fear—they have the power to tear a marriage and stepfamily apart. (p. 4)

Children in nuclear families have also been exposed to a relatively consistent set of beliefs, values, and behavior patterns. When children enter a stepfamily, they "inherit" a new parent, who may bring into the family unit a new set of values and beliefs and a new way of living. Likewise, a stepparent now lives with children who may rear differently from the way in which the biological parents reared the child. "His kids had never been to church," reported one frustrated stepparent.

Another unique aspect of stepfamilies is that the relationship between the biological parent and the children has existed longer than the relationship between the adults in the remarriage. Jane and her twin children have a nine-year relationship and are emotionally bonded to each other. However, Jane has known her new partner only a year, and although her children like their new stepfather, they hardly know him.

In addition, the relationship between biological parents and their children is of longer duration than that of stepparents and stepchildren. The short history of the relationship between children and stepparents is one factor that may contribute to increased conflict between these two during children's adolescence. Children may also become confused and wonder whether they are

stepism the assumption that stepfamilies are inferior to biological families.

disloyal to their biological parent if they become friends with a stepparent.

Stepfamilies are also unique, in that stepchildren have two homes they may regard as theirs, unlike children in a nuclear family, who have one home they regard as theirs. In some cases of joint custody, children spend part of each week with one parent and part with the other; they live with two sets of adult parents in two separate homes.

Money, or lack of it, from an ex-spouse may be a source of conflict. In some stepfamilies, an ex-spouse is expected to send child support payments to the parent who has custody of the children. Fewer than one-half of these fathers send any money; those who do may be irregular in their payments. Fathers who pay regular child support tend to have higher incomes, to have remarried, to live close to their children, and to visit them regularly. They are also more likely to have legal shared or joint custody, which helps to ensure that they will have access to their children.

Fathers who do not voluntarily pay child support and are delinquent by more than one month may have their wages garnished by the state. Some fathers change jobs frequently and move around to make it difficult for the government to keep up with them. Such dodging of the law is frustrating to custodial mothers who need the child support money. Added to the frustration is the fact that fathers are legally entitled to see their children even when they do not pay court-ordered child support. This angers mothers who must give up their children on weekends and holidays to a man who is not supporting his children financially. Such distress on the part of mothers is probably conveyed to the children.

New relationships in stepfamilies experience almost constant flux. Each member of a new stepfamily has many adjustments to make. Issues that must be dealt with include how the mate feels about the partner's children from a former marriage, how the children feel about the new stepparent, and how the newly married spouse feels about the spouse's sending alimony and child support payments to an ex-spouse. In general, families in the Bray and Kelly (1998) study did not begin to think and act like a family until the end of the second or third year. These early years are the most vulnerable, with a quarter of stepfamilies ending during that period.

Stepfamilies are also stigmatized. **Stepism** is the assumption that stepfamilies are inferior to biological families. Stepism, like racism, heterosexism, sexism, and ageism, involves prejudice and discrimination. Social changes need to be made to give support to stepfamilies. For example, if there is a graduation banquet, is this an opportunity for children to invite all four of their parents? More often, children are forced to choose, which usually results in selecting the biological parent and ignoring the stepparents. The more adults children have who love and support them (for example, both biological parents and stepparents), the better for the children, and our society should support this.

Stepparents have no childfree period. Unlike newly married couples in nuclear families, who typically have their first child about two and one-half years after their wedding, many remarried couples begin their marriage with children in the house.

Profound legal differences exist between nuclear and blended families. Whereas biological parents in nuclear families are required in all states to support their children, only five states require stepparents to provide financial support for their stepchildren. Thus, when stepparents divorce, this and other discretionary types of economic support usually stop.

Other legal matters with regard to nuclear families versus stepfamilies involve inheritance rights and child custody. Stepchildren do not automatically inherit from their stepparents, and courts have been reluctant to give stepparents legal access to stepchildren in the event of a divorce. In general, U.S. law does not consistently recognize stepparents' roles, rights, and obligations regarding their stepchildren (Malia 2005). Without legal support to ensure such access, these relationships tend to become more distant and nonfunctional.

THE BOND THAT LINKS YOUR TRUE FAMILY IS NOT ONE OF BLOOD, BUT OF RESPECT AND JOY IN EACH OTHER'S LIFE.

— Richard Bach, American writer

Finally, extended family networks in nuclear families are smooth and comfortable, whereas those in stepfamilies often become complex and strained. Table 14.3 summarizes the differences between nuclear families and stepfamilies.

Table 14.3
Differences between Nuclear Families and Stepfamilies

Nuclear Families	Stepfamilies
1. Children are (usually) biologically related to both parents.	1. Children are biologically related to only one parent.
2. Both biological parents live together with children.	2. As a result of divorce or death, one biological parent does not live with the children. In the case of joint physical custody, children may live with both parents, alternating between them.
3. Beliefs and values of members tend to be similar.	3. Beliefs and values of members are more likely to be different because of different backgrounds.
4. The relationship between adults has existed longer than relationship between children and parents.	4. The relationship between children and parents has existed longer than the relationship between adults.
5. Children have one home they regard as theirs.	5. Children may have two homes they regard as theirs.
6. The family's economic resources come from within the family unit.	6. Some economic resources may come from an ex-spouse.
7. All money generated stays in the family.	7. Some money generated may leave the family in the form of alimony or child support.
8. Relationships are relatively stable.	8. Relationships are in flux: new adults adjusting to each other; children adjusting to a stepparent; a stepparent adjusting to stepchildren; stepchildren adjusting to each other.
9. No stigma is attached to nuclear family.	9. Stepfamilies are stigmatized.
10. Spouses had a childfree period.	10. Spouses had no childfree period.
11. Inheritance rights are automatic.	11. Stepchildren do not automatically inherit from stepparents.
12. Rights to custody of children are assumed if divorce occurs.	12. Rights to custody of stepchildren are usually not considered.
13. Extended family networks are smooth and comfortable.	13. Extended family networks become complex and strained.
14. Nuclear family may not have experienced loss.	14. Stepfamily has experienced loss.

Stepfamilies in Theoretical Perspective

Structural functionalists, conflict theorists, and symbolic interactionists view stepfamilies from the following different points of view:

1. **Structural-functional perspective.** To the structural functionalist, integration or stability of the system is highly valued. The very structure of the stepfamily system can be a threat to the integration and stability of a family system. The social structure of stepfamilies consists of a stepparent, a biological parent, biological children, and stepchildren. Functionalists view the stepfamily system as vulnerable to an alliance between the biological parent and the biological children who have a history together.

   In 75 percent of the cases, the mother and children create an alliance. The stepfather, as an outsider, may view this alliance between the mother and her children as the mother's giving the children too much status or power in the family. Whereas a mother may relate to her children as equals, a stepfather may relate to the children as unequals whom he attempts to discipline. The result is a fragmented parental subsystem whereby the stepfather accuses the mother of being too soft and she accuses him of being too harsh.

   Structural family therapists suggest that parents should have more power than children and that they should align themselves with each other. Not to do so is to give children family power, which they may use to splinter the parents off from each other and create another divorce.

2. **Conflict perspective.** Conflict theorists view conflict as normal, natural, and inevitable as well as functional in that it leads to change. Conflict in a stepfamily system is seen as desirable in that it leads to equality and individual autonomy.

Conflict is a normal part of stepfamily living. The spouses, parents, children, and stepchildren are constantly in conflict for the limited resources of space, time, and money. Space refers to territory (rooms) or property (television, speakers, or electronic games) in the house that the stepchildren may fight over. Time refers to the amount of time that the parents will spend with each other, with their biological children, and with their stepchildren. Money must be allocated in a reasonably equitable way so that each member of the family has a sense of being treated fairly.

Problems arise when space, time, and money are limited. Two new spouses who each bring a child from a former marriage into the house have a situation fraught with potential conflict. Who sleeps in which room? Who gets to watch which channel on television?

To further complicate the situation, suppose the couple have a baby. Where does the baby sleep? Because both parents may have full-time jobs, the time they have for the three children is scarce, not to speak of the fact that a baby will require a major portion of their available time. As for money, the cost of the baby's needs, such as formula and disposable diapers, will compete with the economic needs of the older children. Meanwhile, the spouses may need to spend time alone and may want to spend money as they wish. All these conflicts are functional, because they increase the chance that a greater range of needs will be met within the stepfamily.

3. **Interactionist perspective.** Symbolic interactionists emphasize the meanings and interpretations that members of a stepfamily develop for events and interactions in the family. Children may blame themselves for their parents' divorce and feel that they and their stepfamily are stigmatized; parents may view stepchildren as spoiled.

Stepfamily members also nurture certain myths. Stepchildren sometimes hope that their parents will reconcile and that their nightmare of divorce and stepfamily living will end. This is the myth of reconciliation. Another is the myth of instant love, usually held by stepparents, who hope that the new partner's children will instantly love them. Although this does happen, particularly if the child is young and has no negative influences from the other parent, it is unlikely.

Stages in Becoming a Stepfamily

Just as a person must pass through various developmental stages in becoming an adult, a stepfamily goes through a number of stages as it overcomes various obstacles. Researchers such as Bray and Kelly (1998) and Papernow (1988) have identified various stages of development in stepfamilies. These stages include the following:

STAGE 1 **Fantasy.** Both spouses and children bring rich fantasies into a new marriage. Spouses fantasize that their new marriage will be better than the previous one. If the new spouse has adult children, they assume that these children will be open to a rewarding relationship with them. Young children have their own fantasy—they hope that their biological parents will somehow get back together and that the stepfamily will be temporary.

STAGE 2 **Reality.** Instead of realizing their fantasies, new spouses may find that stepchildren ignore or are rude to them. Indeed, stepparents may feel that they are outsiders in an already-functioning unit (the biological parent and child).

STAGE 3 **Being assertive.** Initially a stepparent assumes a passive role and accepts the frustrations and tensions of stepfamily life. Eventually, however, resentment can reach a level where the stepparent is driven to make changes. The stepparent may make the partner aware of the frustrations and suggest that the marital relationship should have priority some of the time. The stepparent may also make specific requests, such as reducing the number of conversations the partner has with the ex-spouse, not allowing the dog on the furniture, or requiring the stepchildren to use better table manners. This stage is successful to the degree that the partner supports the recommendations for change. A crisis may ensue.

STAGE 4 **Strengthening pair ties.** During this stage, the remarried couple solidify their relationship by making it a priority. At the same time, the biological parent must back away somewhat from the parent-child relationship so that the new partner can have the opportunity to establish a relationship with the stepchildren.

This relationship is the product of small units of interaction and develops slowly across time. Many day-to-day activities, such as watching television, eating meals, and riding in the car together, provide opportunities for the stepparent-stepchild relationship to develop. It is important that the stepparent not attempt to replace the relationship that the stepchildren have with their biological parents.

STAGE 5 **Recurring change.** A hallmark of all families is change, but this is even more true of stepfamilies. Bray and Kelly (1998) note that, even though a stepfamily may function well when the children are preadolescent, a new era can begin when the children become teenagers and begin to question how the family is organized and run. Such questioning by adolescents is not unique to stepchildren.

Michaels (2000) noted that spouses who become aware of the stages through which stepfamilies pass report that they feel less isolated and unique. Involvement in stepfamily discussion groups such as the Stepfamily Enrichment Program provides enormous benefits.

A hallmark of all families is change.

14.11 Strengths of Stepfamilies

Stepfamilies have both strengths and weaknesses. Strengths include children's exposure to a variety of behavior patterns, their observation of a happy remarriage, adaptation to stepsibling relationships inside the family unit, and greater objectivity on the part of the stepparent.

Exposure to a Variety of Behavior Patterns

Children in stepfamilies experience a variety of behaviors, values, and lifestyles. They have the advantage of living on the inside of two families. Although this may be confusing and challenging for children or adolescents, they are learning early how different family patterns can be. For example, one 12-year-old had never seen a couple pray at the dinner table until his mom remarried a man who was accustomed to a "blessing" before meals.

Happier Parents

Although children may want their parents to get back together, they often come to observe that their parents are happier alone than married. If a parent remarries, the children may see the parent in a new and happier relationship. Such happiness may spill over into the context of family living so that there is less tension and conflict.

Opportunity for a New Relationship with Stepsiblings

Though some children reject their new stepsiblings, others are enriched by the opportunity to live with a new person to whom they are now "related." One 14-year-old remarked, "I have never had an older brother to do things with. We both like to do the same things and I couldn't be happier about the new situation." Some stepsibling relationships are maintained throughout adulthood.

Nevertheless, there is stress associated with having stepsiblings. Tillman (2008) studied sibling relationships

© PhotoAlto/Alamy

developmental task a skill that allows a family to grow as a cohesive unit.

in stepfamilies (referring to them as "nontraditional siblings") and noted that adolescents with stepsiblings living in the same house made lower grades and had more school-related behavior problems. The researcher noted that these outcomes are related to the complexity and stress of living with stepsiblings.

More Objective Stepparents

Because of the emotional tie between a parent and a child, some parents have difficulty discussing certain issues or topics. A stepparent often has the advantage of being less emotionally involved and can relate to a child at a different level. One 13-year-old said of the relationship with his new stepmom, "She went through her own parents' divorce and knew what it was like for me to be going through my dad's divorce. She was the only one I could really talk to about this issue. Both my dad and mom were too angry about the subject to be able to talk about it."

14.12 Developmental Tasks for Stepfamilies

A **developmental task** is a skill that, if mastered, allows a family to grow as a cohesive unit. Developmental tasks that are not mastered will edge the family closer to the point of disintegration. Some of the more important developmental tasks for stepfamilies are discussed in this section.

Acknowledge Losses and Changes

As noted previously, each stepfamily member has experienced the loss of a spouse or a biological parent in the home. Family members experience significant losses of an attachment figure (Marano 2000). These losses are sometimes compounded by home, school, neighborhood, and job changes. Feelings about these losses and changes should be acknowledged as important and consequential. In addition, children should not be required to love their new stepparent or stepsiblings (and vice versa). Such feelings will develop only as a consequence of positive interaction over an extended period of time. Resiliency in stepfamilies is also facilitated by having a strong marriage, support from family and friends, and good communication (Greeff and Du Toit 2009).

Preserving original relationships is helpful in reducing a child's grief over loss. It is sometimes helpful for the biological parent and child to take time to nurture their relationship apart from stepfamily activities. This will reduce the child's sense of loss and any feelings of jealousy toward new stepsiblings.

Nurture the New Marriage Relationship

It is critical to the healthy functioning of a new stepfamily that the new spouses nurture each other and form a strong unit. Indeed, the adult dyad is vulnerable because the couple had no childfree time upon which to build a common base (Pacey 2005). Once a couple develops a core relationship, they can communicate, cooperate, and compromise with regard to the various issues in their

The principal challenge *of stepfamily life is building an emotionally satisfying marriage.*

—James Bray and John Kelly, Stepfamilies: Love, Marriage and Parenting in the First Decade

new blended family. Too often spouses become child-focused and neglect the relationship on which the rest of the family depends. Such nurturing translates into spending time alone with each other, sharing each other's lives, and having fun with each other. One remarried couple goes out to dinner at least once a week without the children. "If you don't spend time alone with your spouse, you won't have one," says one stepparent. Two researchers studied 115 stepfamily couples and found that the husbands tended to rank "spouse" as their number one role (over parent or employee), whereas wives tended to rank "parent" as their number one role (Degarmo and Forgatch 2002).

Integrate the Stepfather into the Child's Life

Stepfathers who become interested in what their stepchild does and who spend time alone with the stepchild report greater integration into the life of the stepchild and the stepfamily. The benefits to both the stepfather and the stepchild in terms of emotional bonding are enormous. In addition, the mother of the child feels closer to her new husband if he has bonded with her offspring.

Allow Time for Relationship between Partner and Children to Develop

In an effort to escape single parenthood and to live with one's beloved, some individuals rush into remarriage without getting to know each other. Not only do they have limited information about each other, but their respective children may also have spent little or no time with their future stepparent. One stepdaughter remarked, "I came home one afternoon to find a bunch of plastic bags in the living room with my soon-to-be stepdad's clothes in them. I had no idea he was moving in. It hasn't been easy." Both adults and children should have had meals together and spent some time in the same house before becoming bonded by marriage as a family. Schrodt et al. (2008) reported the positive effects of daily communication between stepparents and stepchildren. Both reported higher satisfaction when they had frequent daily communication. As noted previously, stepparents are encouraged to discipline their own children because often the stepparent and stepchild have had insufficient time to experience a relationship that allows for discipline from the stepparent.

Have Realistic Expectations

Because of the complexity of meshing the numerous relationships involved in a stepfamily, it is important to be realistic. Dreams of one big happy family often set up stepparents for disappointment, bitterness, jealousy, and guilt. As noted previously, stepfamily members often do not begin to feel comfortable with each other until the third year (Bray and Kelly 1998). Just as nuclear and single-parent families do not always run smoothly, neither do stepfamilies.

Accept Your Stepchildren

Rather than wishing your stepchildren were different, accepting them is more productive. All children have positive qualities; find them and make them the focus of your thinking. Stepparents may communicate acceptance of their stepchildren through verbal praise and positive or affectionate statements and gestures. In addition, stepparents may communicate acceptance by engaging in pleasurable activities with their stepchildren and participating in daily activities such as homework, bedtime preparation, and transportation to after-school activities.

Funder (1991) studied 313 parents who had been separated for five to eight years and who had become involved with new partners. In general, the new partners were very willing to be involved in the parenting of their new spouses' children. Such involvement was highest when the children lived in the household.

Establish Your Own Family Rituals

Rituals are one of the bonding elements of nuclear families. Stepfamilies may integrate the various family members by establishing common rituals, such as summer vacations, visits to and from extended kin, and religious celebrations. These rituals are most effective if they are new and unique, not mirrors of rituals in the previous marriages and families.

Decide about Money

Money is an issue of potential conflict in stepfamilies because it can be a scarce resource, and several people may want to use it for their respective needs. For example, a father may want a new computer; the mother may want a new car; the mother's children may want bunk beds, dance lessons, and a satellite dish; the father's children may want a larger room, clothes, and a phone.

How do the newly married couple and their children decide how money should be spent?

Some stepfamilies put all their resources into one bank and draw out money as necessary without regard for whose money it is or for whose child the money is being spent. Others keep their money separate; the parents have separate incomes and spend them on their respective biological children. Although no one pattern is superior to another, it is important for remarried spouses to agree on whatever financial arrangements they live by.

In addition to deciding how to allocate resources fairly in a stepfamily, remarried couples may face decisions regarding sending the children and stepchildren to college. Remarried couples may also make a will that is fair to all family members.

Give Parental Authority to Your Spouse

Adults should discuss how much authority a stepparent exercises over children before they get married. Some couples divide the authority—both spouses discipline their own children. This is the strategy we recommend. The downside is that stepchildren may test the stepparent in such an arrangement when the biological parent is not around. One stepmother said, "Joe's kids are wild when he isn't here because I'm not supposed to discipline them."

Support the Children's Relationship with Their Absent Parent

A continued relationship with both biological parents is critical to the emotional well-being of children. Ex-spouses and stepparents should encourage children to have a positive relationship with both biological parents. Respect should also be shown for the biological parent's values. Bray and Kelly (1998) note that this is "particularly difficult" to exercise. "But asking a child about an absent parent's policy on movies or curfews shows the child that, despite their differences, his/her mother and father still respect each other" (p. 92).

Cooperate with the Children's Biological Parent and Co-Parent

A cooperative, supportive, and amicable co-parenting relationship between the biological parents and stepparents is a win-win situation for the children and parents. Otherwise, children are continually caught between the cross fire of the conflicted parental sets.

Support the Children's Relationship with Grandparents

It is important to support children's continued relationships with their natural grandparents on both sides of the family. This is one of the more stable relationships in the child's changing world of adult relationships. Regardless of how ex-spouses feel about their former in-laws, they should encourage their children to have positive feelings for their grandparents. One mother said, "Although I am uncomfortable around my ex-in-laws, I know they are good to my children, so I encourage and support my children spending time with them."

Anticipate Great Diversity

Stepfamilies are as diverse as nuclear families. It is important to let each family develop its own uniqueness. One remarried spouse said, "The kids were grown when the divorces and remarriages occurred, and none of the kids seem particularly interested in getting involved with the others" (Rutter 1994, 68). Ganong et al. (1998) also emphasized the importance of resisting the notion that stepfamilies are not "real" families and that adoption makes them a real family. Indeed, social policies need to be developed that allow for "the establishment of some legal ties between the stepparent and stepchild without relinquishing the biological parent's legal ties" (p. 69).

© bilderlounge/Jupiterimages

Parental Status Inventory*

The Parental Status Inventory (PSI) is a fourteen-item inventory that measures the degree to which respondents consider their stepfather to be a parent on an 11-point scale from 0 percent to 100 percent. Read each of the following statements and circle the percentage indicating the degree to which you regard the statement as true.

1. **I think of my stepfather as my father. (0%, 10, 20, 30, 40, 50%, 60, 70, 80, 90, 100%)**
2. **I am comfortable when someone else refers to my stepfather as my father or dad. (0%, 10, 20, 30, 40, 50%, 60, 70, 80, 90, 100%)**
3. **I think of myself as his daughter/son. (0%, 10, 20, 30, 40, 50%, 60, 70, 80, 90, 100%)**
4. **I refer to him as my father or dad. (0%, 10, 20, 30, 40, 50%, 60, 70, 80, 90, 100%)**
5. **He introduces me as his son/daughter. (0%, 10, 20, 30, 40, 50%, 60, 70, 80, 90, 100%)**
6. **I introduce my mother and him as my parents. (0%, 10, 20, 30, 40, 50%, 60, 70, 80, 90, 100%)**
7. **He and I are just like father and son/daughter. (0%, 10, 20, 30, 40, 50%, 60, 70, 80, 90, 100%)**
8. **I introduce him as "my father" or "my dad." (0%, 10, 20, 30, 40, 50%, 60, 70, 80, 90, 100%)**
9. **I would feel comfortable if he and I were to attend a father-daughter/father-son function, such as a banquet, baseball game, or cookout, alone together. (0%, 10, 20, 30, 40, 50%, 60, 70, 80, 90, 100%)**
10. **I introduce him as "my mother's husband" or "my mother's partner." (0%, 10, 20, 30, 40, 50%, 60, 70, 80, 90, 100%)**
11. **When I think of my mother's house, I consider him and my mother to be parents to the same degree. (0%, 10, 20, 30, 40, 50%, 60, 70, 80, 90, 100%)**
12. **I consider him to be a father to me. (0%, 10, 20, 30, 40, 50%, 60, 70, 80, 90, 100%)**
13. **I address him by his first name. (0%, 10, 20, 30, 40, 50%, 60, 70, 80, 90, 100%)**
14. **If I were choosing a greeting card for him, the inclusion of the words *father* or *dad* in the inscription would prevent me from choosing the card. (0%, 10, 20, 30, 40, 50%, 60, 70, 80, 90, 100%)**

Scoring

First, reverse the scores for items 10, 13, and 14. For example, if you circled a 90, change the number to 10; if you circled a 60, change the number to 40; and so on. Add the percentages and divide by 14. A 0 percent reflects that you do not regard your stepdad as your parent at all. A 100 percent reflects that you totally regard your stepdad as your parent. The percentages between 0 percent and 100 percent show the gradations from no regard to total regard of your stepdad as parent.

Norms

Respondents in two studies (one in Canada and one in America) completed the scale. The numbers of respondents in the studies were 159 and 156, respectively, and the average score in the respective studies was 45.66 percent. Between 40 percent and 50 percent of both Canadians and Americans viewed their stepfather as their parent.

*Developed by Dr. Susan Gamache. 2000. Hycroft Medical Centre, Vancouver, B.C., Canada. Details on construction of the scale, including validity and reliability, are available from Dr. Gamache (gamache @interchange.ubc.ca). The PSI is used in this text with the permission of Dr. Gamache and may not be used otherwise (except as an in-class student exercise) without written permission.

Relationships in the Later Years

Legendary screen actress Katharine Hepburn died at age 97. Her biographer, Scott Berg (2003), spent the last twenty years of Hepburn's life as a close friend. During one of their numerous after-dinner conversations, he asked her, "So what do you think it's all about? Life, I mean. What's the purpose?" Her reply . . .

> To work hard and to love someone. And to have some fun. And if you're lucky, you keep your health and somebody loves you back. (p. 366)

Work, loving, and being loved . . . not a bad recipe for a full life. However, the end of life comes for us all and we have challenges as we age, including the cost of even a brief stay in the hospital, the cost of prescription medication, and care in a retirement or long-term care facility. Although most young couples marrying today rarely give a thought to their own aging, the care of their aging parents is an issue that looms ahead.

SECTIONS

15.1 Age and Ageism

About 13 percent of the 310 million individuals in the United States are age 65 and older (*Statistical Abstract of the United States, 2009*, Table 11). This represents about 40 million "elderly." By 2030, this percentage will grow to 20 percent of the population in the United States (Willson 2007). In this chapter, we focus on the factors that confront individuals and couples as they age and the dilemma of how to care for aging parents. We begin by looking at the concept of age.

All societies have a way to categorize their members by age. And all societies provide social definitions for particular ages.

age term defined chronologically, physiologically, psychologically, sociologically, and culturally.

The Concept of Age

A person's **age** may be defined chronologically, physiologically, psychologically, sociologically, and culturally. Chronologically, an "old" person is defined as one who has lived a certain number of years. How many years it takes to be regarded as old varies with one's own age. Children of 12 may regard siblings of 18 as old—and their parents as "ancient." Teenagers and parents may regard themselves as "young" and reserve the label "old" for their grandparents' generation.

Chronological age has obvious practical significance in everyday life. Bureaucratic organizations and social programs identify chronological age as a criterion of certain social rights and responsibilities. One's age determines the right to drive, vote, buy alcohol or cigarettes, and receive Social Security and Medicare benefits.

Age has meaning in reference to the society and culture of the individual. In ancient Greece and Rome, where the average life expectancy was 20 years, one was old at 18; similarly, one was old at 30 in medieval Europe and at age 40 in the United States in 1850. In the United States today, however, people are usually not considered old until they reach age 65. However, our society is moving toward new chronological definitions of "old." Three groups of the elderly are the "young-old," the "middle-old," and the "old-old." The young-old are typically between the ages of 65 and 74; the middle-old, 75 to 84, and

Table 15.1
Life Expectancy

Year	White Males	Black Males	White Females	Black Females
2010	76.5	70.2	81.3	77.2
2015	77.1	71.4	81.8	78.2
2020	77.7	72.6	82.4	79.2

Source: *Statistical Abstract of the United States, 2009*, 128th ed. Washington, DC: U.S. Bureau of the Census, 2009, Table 100.

the old-old, 85 and beyond. Current life expectancy is shown in Table 15.1.

The age of a person influences that person's view of when a person becomes "old." Individuals aged 18 to 35 identify 50 as the age when the average man or woman becomes "old." However, those between the ages of 65 and 74 define "old" as 80 (Cutler 2002). Tanner (2005) noted that views of the elderly are changing from people who are needy and dependent to people who are active and resourceful. Research is underway to extend life. Dr. Aubrey de Grey of the Department of Genetics, University of Cambridge, predicted that continued research will make adding *hundreds* of years to one's life possible. Maher and Mercer (2009) noted that the most optimistic predictions are that we are two or three decades away from significant breakthroughs (Ray Kurzweil says we are forty-nine years away). Maher and Mercer (2009) also note the paths are known via genetic engineering, tissue or organ replacement, and the merging of computer technology with human biology. Should the human lifespan be extended hundreds of years, imagine the impact on marriage—"till death do us part"?

Some individuals have been successful in delaying the aging process. Frank Lloyd Wright, the famous architect, enjoyed the most productive period of his life between the ages 80 and 92. George Burns and Bob Hope were active into their nineties. Fitness guru Jack LaLanne remains active in his mid-nineties. Boulton-Lewis et al. (2006) found that continuing to learn was associated with the elderly continuing to feel healthy.

Physiologically, people are old when their auditory, visual, respiratory, and cognitive capabilities decline significantly. Indeed, Siedlecki (2007) confirmed that increased age is associated with increased difficulty in retrieving information. Becoming "disabled" is associated with being "old." Individuals tend to see themselves as disabled when their driver's licenses are taken away and when home health care workers come to their home to care for them (Kelley-Moore et al. 2006). Sleep changes occur for the elderly, including going to bed earlier, waking up during the night, and waking up earlier in the morning, as well as issues like restless legs syndrome, snoring, and obstructive sleep apnea (Wolkove et al. 2007).

People who need full-time nursing care for eating, bathing, and taking medication properly and who are placed in nursing homes are thought of as being old. Failing health is the criterion the elderly use to define themselves as old (O'Reilly 1997), and successful aging is typically defined as maintaining one's health, independence, and cognitive ability. Jorm et al. (1998) observed that the prevalence of successful aging declines steeply from age 70 to age 80. Garrett and Martini (2007) noted the impact on the U.S. health care system as increasing numbers of baby boomers age.

People who have certain diseases are also regarded as old. Although younger individuals may suffer from Alzheimer's, arthritis, and heart problems, these ailments are more often associated with aging. As medical science conquers more diseases, the physiological definition of aging changes so that it takes longer for people to be defined as "old."

Psychologically, a person's self-concept is important in defining how old that person is. As individuals begin to fulfill the roles associated with the elderly—retiree, grandparent, nursing home resident—they begin to see themselves as aging. Sociologically, once they occupy these roles, others begin to see them as "old." Barrett (2005) analyzed national data of 2,681 midlife respondents to assess how women and men differed in their age identity. Because women were typically pair-bonded with older partners, they typically had younger age identities than men.

Culturally, the society in which an individual lives defines when and if a person becomes old and what being old means. In U.S. society, the period from age 18 through 64 is generally subdivided into young adulthood, adulthood, and middle age. Cultures also differ in terms of how they view and take care of their elderly.

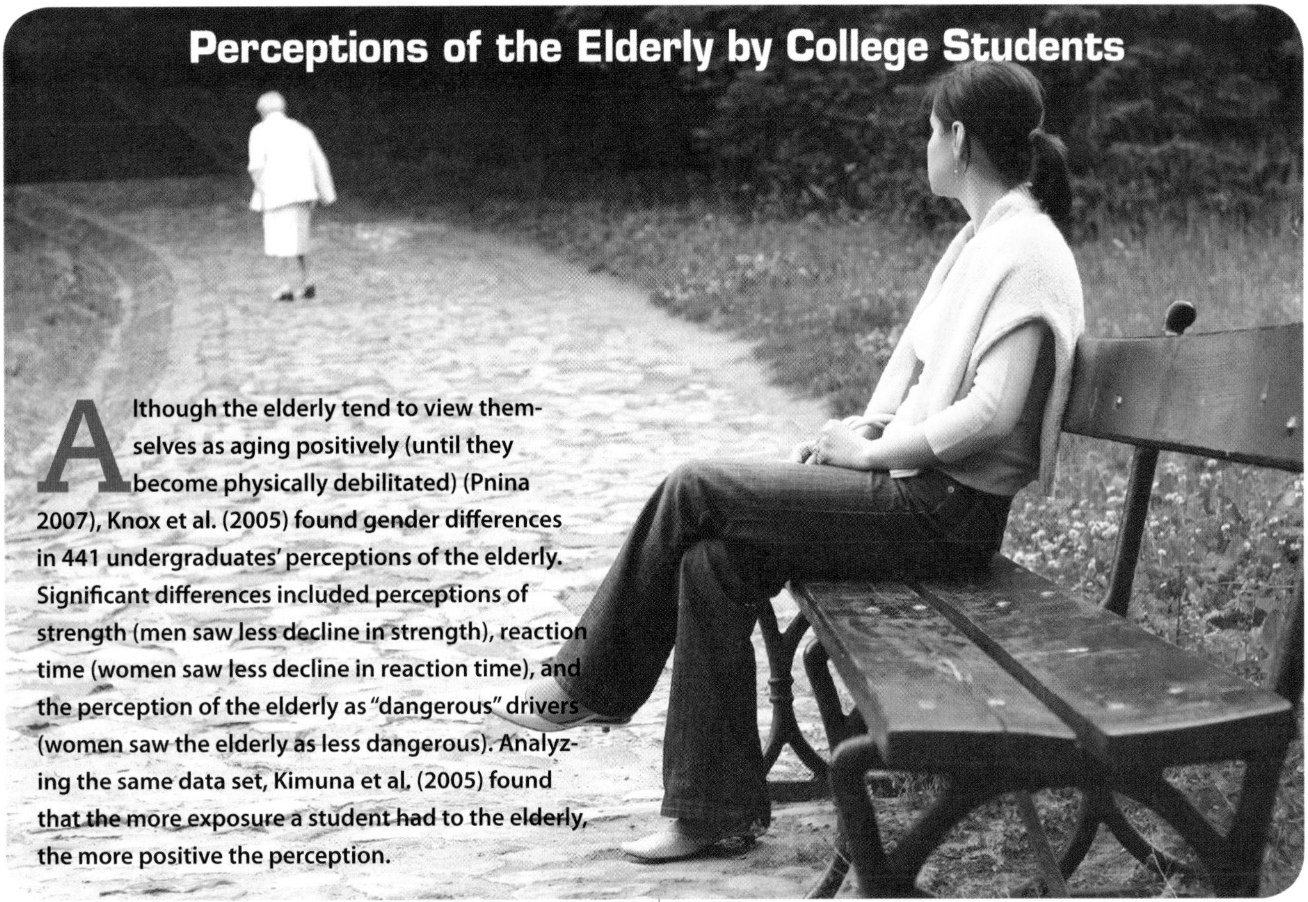

© Wojciech Krusinski/iStockphoto.com

Perceptions of the Elderly by College Students

Although the elderly tend to view themselves as aging positively (until they become physically debilitated) (Pnina 2007), Knox et al. (2005) found gender differences in 441 undergraduates' perceptions of the elderly. Significant differences included perceptions of strength (men saw less decline in strength), reaction time (women saw less decline in reaction time), and the perception of the elderly as "dangerous" drivers (women saw the elderly as less dangerous). Analyzing the same data set, Kimuna et al. (2005) found that the more exposure a student had to the elderly, the more positive the perception.

Spain is particularly noteworthy in terms of care for the elderly, with eight of ten elderly people receiving care from family members and other relatives. The elderly in Spain report very high levels of satisfaction in the relationships with their children, grandchildren, and friends (Fernandez-Ballesteros 2003).

Ageism

Every society has some form of **ageism**—the systematic persecution and degradation of people because they are old. Ageism is similar to sexism, racism, and heterosexism. The elderly are shunned, discriminated against in employment, and sometimes victims of abuse. Media portrayals contribute to the negative image of the elderly. They are portrayed as difficult, complaining, and burdensome and are often underrepresented in commercials and comic strips.

Negative stereotypes and media images of the elderly engender **gerontophobia**—a shared fear or dread of the elderly, which may create a self-fulfilling prophecy. For example, an elderly person forgets something and attributes the behavior to age. A younger person, however, engaging in the same behavior, is unlikely to attribute forgetfulness to age, given cultural definitions surrounding the age of the onset of senility. Individuals are also thought to become more inflexible and more conservative as they age. Analysis of national data suggests this is not true. In fact, the elderly become more tolerant (Danigelis et al. 2007).

ageism systematic persecution and degradation of people because they are old.

gerontophobia fear or dread of the elderly.

The negative meanings associated with aging underlie the obsession of many Americans to conceal their age by altering their appearance. With the hope of holding on to youth a little bit longer, aging Americans spend billions of dollars each year on plastic surgery, exercise equipment, hair products, facial creams, and Botox injections.

The latest attempt to reset the aging clock is to have regular injections of human growth hormone (HGH), which promises to lower blood pressure, build muscles without extra exercise, increase the skin's elasticity,

gerontology the study of aging.

thicken hair, and heighten sexual potency. It is part of the regimen of clinics such as Lifespan (in Beverly Hills), which costs $1,000 a month after an initial workup of $5,000. In the absence of long-term data, many physicians remain skeptical—there is no fountain of youth.

Theories of Aging

Gerontology is the study of aging. Table 15.2 identifies several theories, the level (macro or micro) of the theory, the theorists typically associated with the theory, assumptions, and criticisms. As noted, there are diverse ways of conceptualizing the elderly. Currently popu-

Table 15.2
Theories of Aging

Name of Theory	Level of Theory	Theorists	Basic Assumptions	Criticisms
Disengagement	Macro	Elaine Cumming, William Henry	The gradual and mutual withdrawal of the elderly and society from each other is a natural process. It is also necessary and functional for society that the elderly disengage so that new people can be phased in to replace them in an orderly transition.	Not all people want to disengage; some want to stay active and involved. Disengagement does not specify what happens when the elderly stay involved.
Activity	Macro	Robert Havighurst	People continue the level of activity they had in middle age into their later years. Though high levels of activity are unrelated to living longer, they are related to reporting high levels of life satisfaction.	Ill health may force people to curtail their level of activity. The older a person, the more likely the person is to curtail activity.
Conflict	Macro	Karl Marx, Max Weber	The elderly compete with youth for jobs and social resources such as government programs (Medicare).	The elderly are presented as disadvantaged. Their power to organize and mobilize political resources such as the American Association of Retired Persons is underestimated.
Age stratification	Macro	M. W. Riley	The elderly represent a powerful cohort of individuals passing through the social system that both affect and are affected by social change.	Too much emphasis is put on age, and little recognition is given to other variables within a cohort such as gender, race, and socioeconomic differences.
Modernization	Macro	Donald Cowgill	The status of the elderly is in reference to the evolution of the society toward modernization. The elderly in premodern societies have more status because what they have to offer in the form of cultural wisdom is more valued. The elderly in modern technologically advanced societies have low status because they have little to offer.	Cultural values for the elderly, not level of modernization, dictate the status of the elderly. Japan has high respect for the elderly and yet is highly technological and modernized.
Symbolic	Micro	Arlie Hochschild	The elderly socially construct meaning in their interactions with others and society. Developing social bonds with other elderly can ward off being isolated and abandoned. Meaning is in the interpretation, not in the event.	The power of the larger social system and larger social structures to affect the lives of the elderly is minimized.
Continuity	Micro	Bernice Neugarten	The earlier habit patterns, values, and attitudes of the individual are carried forward as a person ages. The only personality change that occurs with aging is the tendency to turn one's attention and interest on the self.	Other factors than one's personality affect aging outcomes. The social structure influences the life of the elderly rather than vice versa.

lar in sociology is the life-course perspective (Willson 2007). This approach examines differences in aging across cohorts by emphasizing that "individual biography is situated within the context of social structure and historical circumstance" (p. 150).

15.2 Caregiving for the Frail Elderly—the "Sandwich Generation"

Elderly people are defined as **frail** if they have difficulty with at least one personal care activity or other activities related to independent living; the severely disabled are unable to complete three or more personal care activities. These personal care activities include bathing, dressing, getting in and out of bed, shopping for groceries, and taking medications. About 6.1 percent of the U.S. adult population over the age of 65 are defined as being severely disabled and are not living in nursing homes.

Only 6.8 percent of the frail elderly have long-term health care insurance (Johnson and Wiener 2006). Hence, the bulk of their care falls to the children of these elderly. The term *children* typically means female adult children. Indeed, women account for about two-thirds of all unpaid caregivers (ibid.). The adults who provide **family caregiving** to these elderly parents (and their own children simultaneously) are known as the "**sandwich generation**" because they are in the middle of taking care of the needs of both their parents and children.

Caregiving for an elderly parent has two meanings. One, caregiving refers to providing personal help with the basics of daily living such as helping the parent get in and out of bed, bathing, toileting, and eating. A second form of caregiving refers to instrumental activities such as grocery shopping, money management (including paying bills), and driving the parent to the doctor.

The typical caregiver is a middle-aged married woman who works outside the home. High levels of stress and fatigue may accompany caring for one's elders. Martire and Stephens (2003) noted even higher levels of fatigue and competing demands among women who were both employed and caring for an aging parent. Of a sample of women at midlife (most of whom had children), 55 percent reported that they were providing care to their mothers, and 34 percent were caring for their fathers (Peterson 2002). The number of individuals in the sandwich generation will increase for the following reasons:

frail elderly person who has difficulty with at least one personal care activity or other activities related to independent living.

family caregiving care provided to the elderly by family members.

sandwich generation individuals who attempt to meet the needs of their children and elderly parents at the same time.

1. **Longevity.** The over-85 age group, the segment of the population most in need of care, is the fastest-growing segment of our population.
2. **Chronic disease.** In the past, diseases took the elderly quickly. Today, diseases such as arthritis and Alzheimer's are associated not with an immediate death sentence but a lifetime of managing the illness and being cared for by others. Family caregivers of parents with Alzheimer's note the difficulty of the role: "He's not the man I married," lamented one wife. Hilgeman et al. (2007) noted that having a positive view of taking care of an Alzheimer's patient was associated with experiencing stress in doing so.
3. **Fewer siblings to help.** The current generation of elderly had fewer children than the elderly in previous generations. Hence, the number of adult siblings to help look after parents is more limited. Only children are more likely to feel the weight of caring for elderly parents alone.
4. **Commitment to parental care.** Contrary to the myth that adult children in the United States abrogate responsibility for taking care of their elderly parents, most children institutionalize their parents only as a last resort. Furthermore, Wells (2000) identified some benefits of caregiving, including a closer relationship to the dependent person and a feeling of enhanced self-esteem. Most of Peterson's (2002) sample of women caring for their parents did not view doing so as a burden. Asian children, specifically Chinese children, are socialized to expect to take care of their elderly. Zhan and Montgomery (2003) observed, "Children were raised for the security of old age" (p. 209).
5. **Lack of support for the caregiver.** Caring for a dependent, aging parent requires a great deal of

age discrimination discriminating against a person because of age.

effort, sacrifice, and decision making on the part of more than 14 million adults in the United States who are challenged with this situation. The emotional toll on the caregiver may be heavy. Guilt (over not doing enough), resentment (over feeling burdened), and exhaustion (over the relentless care demands) are common feelings that are sometimes mixed. One caregiver adult child said, "I must be an awful person to begrudge taking my mother supper, but I feel that my life is consumed by the demands she makes on me, and I have no time for myself, my children, or my husband." Marks et al. (2002) noted an increase in symptoms of depression among a national sample of caregivers (of a child, parent, or spouse). Older caregivers also risk their own health in caring for their aging parents (Wallsten 2000). Caregiving can also be expensive and can devastate a family budget.

Some reduce the strain of caring for an elderly parent by arranging for home health care. This involves having a nurse go to the home of a parent and provide such services as bathing the parent and giving medication. Other services may include taking meals to the elderly (for example, through Meals on Wheels). The National Family Caregiver Support Program, enacted in 2000, provides support services for individuals (including grandparents) who provide family caregiving services. Such services might include eldercare resource and referral services, caregiver support groups, and classes on how to care for an aging parent. In addition, states are increasingly providing family caregivers a tax credit or deduction.

© Thomas Flügge/iStockphoto.com

Offspring who have no help may become overwhelmed and frustrated. Elder abuse, an expression of such frustration, is not unheard of. Many wrestle with the decision to put their parents in a nursing home or other long-term care facility.

15.3 Issues Confronting the Elderly

Numerous issues become concerns as people age. In middle age, the issues are early retirement (sometimes forced), job layoffs (recession-related cutbacks), **age discrimination** (discrimination against a person because of age; for example, older people are often not hired), separation or divorce from a spouse, and adjustment to children leaving home. For some in middle age, grandparenting is an issue if they become the primary caregiver for their grandchildren. As a couple moves from the middle to the later years, the issues become more focused on income, housing, health, retirement, and sexuality.

Income

For most individuals, the end of life is characterized by reduced income. Social Security and pension benefits, when they exist, are rarely equal to the income a retired person formerly earned.

Financial planning to provide end-of-life income is important. Kemp et al. (2005) interviewed fifty-one mid- and later-life individuals about their financial planning to identify the conditions under which such planning is initiated. They found catalysts for planning were employer programs and the offering of retirement seminars. Constraints on such planning were losing one's job or unforeseen expenses (in both cases, no resources were available to plan around). Other life events that could be either catalysts or constraints were health changes, death of a spouse, divorce, or remarriage. Some adults buy long-term health care insurance. Such insurance can be costly and does not always cover needed expenses.

To be caught at the end of life without adequate resources is not unusual. Women are particularly disadvantaged because their work history has often been discontinuous, part-time, and low-paying. Social Security and private pension plans favor those with continuous full-time work histories.

Housing

Of those over the age of 75, 78.7 percent live in and own their own home—hence, most do not live in a nursing home (*Statistical Abstract of the United States, 2009*, Table 950). Home is typically where the elderly have lived in the same neighborhood for years. Even those who do not live in their own home are likely to live in a family setting.

For the most part, the physical housing of the elderly is adequate. Indeed, only 6 percent of the housing units inhabited by the elderly are inadequate. Where deficiencies exist, the most common are inadequate plumbing and heating. However, as individuals age, they find themselves living in homes that are not "elder-friendly." Such elder-friendly homes feature bathroom doors wide enough for a wheelchair, grab bars in the bathrooms, and the absence of stairs (Nicholson 2003). The most recent trend in housing for the elderly is home health care (mentioned previously), as an alternative to nursing home care. In this situation, the elderly person lives in a single-family dwelling, and other people are hired to come in regularly to help with various needs.

Other elderly individuals live in group living or shared housing arrangements, referred to as **cohousing**. Residents essentially plan the communities in which they own a unit and have a common area where they meet for meals several times a week (they typically rotate responsibilities of cooking these meals). The upside of cohousing is companionship with others. The downside is the need to make decisions by consensus on what to do when members can no longer take care of themselves. Silver Sage Village (www.silversagevillage.com) and ElderSpirit Community (www.elderspirit.net) are two such cohousing arrangements.

Some elderly singles choose to maintain their own home even though they are in a partnered relationship. Karlsson and Borell (2005) interviewed 116 elderly men and women living in an increasingly popular arrangement: living apart together (LAT), in which partners retain their own homes although they are involved in a long-term intimate relationship. These researchers focused on the motives of the women who sought to use their homes as a way of establishing boundaries to influence their interaction with partners, friends, and kin. All the women studied seem to prioritize the possibility of keeping their various social relations separate from one another. For example, one widow involved in a new relationship insisted on keeping her house and living apart from her new boyfriend so that she could have her children visit when she wanted and for as long as she wanted. "If I lived with him, he would get upset every time they came around. I wouldn't like that," she said.

cohousing an arrangement where elderly individuals live in group housing or shared living arrangements.

Only
6 percent of the
housing units inhabited
by the elderly are
inadequate.

Physical Health

Our current cultural value for health is in great contrast to previous times. Film legend Clark Gable drank and smoked heavily and was told by his physicians he was on the way to a coronary. He paid no attention and died of a coronary at age 59. Dean Martin also drank and smoked himself to death.

Good physical health is the single most important determinant of an elderly person's reported happiness (Smith et al. 2002). Though health varies by social class and race, with higher social class whites reporting greater health (Willson 2007), most elderly individuals, even those of advanced years, continue to define themselves as being in good health. Ostbye et al. (2006) studied an elderly population in Cache County, Utah, and found that 80 percent to 90 percent of those aged 65 to 75 were healthy on ten dimensions of health (independent living, vision, hearing, activities of daily living, instrumental activities of daily living, absence of physical illness, cognition, healthy mood, social support and participation, and religious participation and spirituality). Prevalence of excellent and good self-reported health decreased with age, to approximately 60 percent among those aged 85 and older. Although most (over 90 percent) of the elderly do not exercise, Morey et al. (2008) studied the elderly ages 65 to 94 and found that the greater their physical activity, the greater their ability to function physically.

dementia the mental disorder most associated with aging, whereby the normal cognitive functions are slowly lost.

Even individuals who have a chronic, debilitating illness maintain a perception of good health as long as they are able to function relatively well. However, when elders' vision, hearing, physical mobility, and strength are markedly diminished, their sense of well-being is often significantly impacted (Smith et al. 2002). For many, the ability to experience the positive side of life seems to become compromised after age 80 (ibid.). Driving accidents also increase with aging. However, unlike teenagers, who also have a high percentage of automobile accidents, the elderly are less likely to die in an accident because they are usually driving at a much slower speed (Pope 2003).

Some elderly become so physically debilitated that questions about the quality of life sometimes lead to consideration of physician-assisted suicide. Useda et al. (2007) compared a group of adults over 50 who attempted suicide versus those who had been successful in killing themselves and found that those in the latter group were more focused, planned, and organized in their movement toward ending their own life. Some debilitated elderly ask their physicians to end their lives. The debilitated elderly may also ask spouses or adult children for help in ending their life.

© iStockphoto.com / © Comstock Images/Jupiterimages

Mental Health

Aging also affects mental processes. Elderly people (particularly those 85 and older) more often have a reduced capacity for processing information quickly, for cognitive attention to a specific task, for retention, and for motivation to focus on a task. However, judgment may not be affected, and experience and perspective are benefits to decision making. A team of researchers (Vance et al. 2005) noted that being socially active, which often leads to physical activity, is associated with keeping one's cognitive functioning. Silverstein et al. (2006) also emphasized that the elderly in rural China who live with their children and grandchildren report greater psychological well-being than the elderly who live with just their children. Hispanic elderly are also likely to live with their children, and some elderly grandparents take care of the grandchildren while the parents work outside the home.

Mental health may worsen for some elderly. Mood disorders, with depression being the most frequent, are more common among the elderly. Ryan et al. (2008) analyzed detailed reproductive histories of 1,013 women aged 65 years and over and found that the prevalence of depressive symptoms was 17 percent. Women who reported menopause at an earlier age had an increased risk. Women who had taken the oral contraceptive pill for at least ten years were less likely to report depression. The elderly who abuse alcohol and who do not exercise are also more likely to report being depressed (Van Gool et al. 2007).

Regrets may also be related to depression. Elderly women who have not had children and who have not accepted their childlessness also report more mental distress than those who have had children or those who have accepted their childfree status (Wu and Hart 2002). The mental health of elderly men seems unaffected by their parental status.

Regardless of the source of depression, it has a negative impact on an elderly couple's relationship. Sandberg et al. (2002) studied depression in twenty-six elderly couples and concluded, "The most striking finding was the frequent mention of marital conflict and confrontation among the depressed couples and the almost complete absence of it among the nondepressed couples" (p. 261).

Dementia, which includes Alzheimer's disease, is the mental disorder most associated with aging. Its presence is assessed in numerous ways, including clinician rating methods, questionnaire-based methods, and performance-based methods (Clare et al. 2005). In spite of the association, only 3 percent of the aged population experience severe cognitive impairment—the most common symptom is loss of memory. It can be devastating

I WOULDN'T SWAP ONE WRINKLE OF MY FACE FOR ALL THE ELIXIRS OF YOUTH. ALL THESE WRINKLES REPRESENT A SMILE, A GRIMACE OF PAIN AND DISAPPOINTMENT . . . SOME PART OF BEING FULLY ALIVE.

— *Helen Hayes, actress*

to an individual and his or her partner. An 87-year-old woman, who was caring for her 97-year-old demented husband, said, "After 56 years of marriage, I am waiting for him to die, so I can follow him. At this point, I feel like he'd be better off dead. I can't go before him and abandon him" (Johnson and Barer 1997, 47).

Retirement

Retirement represents a rite of passage through which most elderly pass. Some are "young" retirees who make their fortune early in life and simply stop working. More often individuals are "old" retirees who have worked thirty to forty years and retire. In 1983, Congress increased the retirement age at which individuals can receive full Social Security benefits, from 65 (for those born before 1938) to 67 (for those born after 1960). People can take early retirement at age 62, with reduced benefits. Retirement affects an individual's status, income, privileges, power, and prestige. Fabian (2007) noted that, for most of our history, the concept of retirement did not exist—older individuals were viewed as a source of wisdom and they continued to work. In the twentieth century, retirement was developed in reference to the economy, which was faced with an aging population and surplus labor (Willson 2007). Indeed, retirement is a socially programmed stage of life that is being reevaluated.

People least likely to retire are unmarried, widowed, single-parent women who need to continue working because they have no pension or even Social Security benefits—if they don't work or continue to work, they will have no income, so retirement is not an option. Some workers experience what is called **blurred retirement** rather than a clear-cut one. A blurred retirement means the individual works part-time before completely retiring or takes a "bridge job" that provides a transition between a lifelong career and full retirement. About half of all people between ages 55 and 64 who quit their primary careers take some kind of bridge job (Dychtwald and Kadlec 2005).

Individuals who have a positive attitude toward retirement are those who have a pension waiting for them, are married (and thus have social support for the transition), have planned for retirement, are in good health, and have high self-esteem. Those who regard retirement negatively have no pension waiting for them, have no spouse, gave no thought to retirement, have bad health, and have negative self-esteem (Mutran et al. 1997). Nordenmark and Stattin (2009) analyzed national data in Sweden and found that better psychosocial retirement adjustment outcomes resulted from voluntary rather than forced retirement. Also, men whose skills were no longer needed reported a more difficult adjustment than women.

blurred retirement the process of retiring gradually so that the individual works part-time before completely retiring or takes a bridge job that provides a transition between a lifelong career and full retirement.

In general, most retired individuals enjoy it. Few regard continuing to work as a privilege. Although they still prefer to be active, they want to pick their own activity and pace, as they could not do when they were working. Collins and Smyer (2005) found that the loss of one's job via retirement was not a major issue, and most were very resilient in their feelings about life and self.

Retirement may have positive consequences for a marriage. Szinovacz and Schaffer (2000) analyzed national data and concluded that husbands viewed their wife's retirement as associated with a reduction in heated arguments. If both spouses were retired, they had a perceived reduction of disagreements. Webber et al. (2000) found that spouses had an easier time adjusting to retirement if they had similar retirement goals or

Levitra, Cialis, Viagra medications taken by aging men to help them get and maintain an erection.

expectations in the areas of leisure, family relationships, friendships, and finances.

Some individuals experience disenchantment during retirement; it is not enjoyable and does not live up to their expectations. Lee Iacocca, former president of Chrysler Corporation, reported that he was bored with retirement, missed "the action," and warned others "never to retire" but to stay busy and involved. Some heed his advice and either "die in harness" or go back to work. Dychtwald and Kadlec (2005) observed that some retired people benefit from volunteering—giving back their time and money to attack poverty, illiteracy, oppression, crime, and so on. Examples of volunteer organizations are the Service Corps of Retired Executives, known as SCORE (www.score.org), Experience Works (www.experienceworks.org), and Generations United (www.gu.org).

Sexuality

Levitra, **Cialis**, and **Viagra** (prescription drugs that help a man obtain and maintain an erection) are advertised regularly on television and have given cultural visibility to the issue of sexuality among the elderly. Though the elderly (both men and women) experience physiological changes that impact sexuality, older adults often continue their interest in sexual activity. Indeed, there is a new era of opportunity to continue sexual functioning into one's later years, which has replaced the earlier script of decline and end to sexual functioning as a natural part of aging (Potts et al. 2006).

Table 15.3 describes the physiological changes that elderly men experience during the sexual response cycle. Table 15.4 describes the physical changes elderly women experience during the sexual response cycle.

The National Institute on Aging surveyed 3,005 men and women ages 57 to 85 and found that sexual activity decreases with age (Lindau et al. 2007). Almost three-fourths (73 percent) of those 57 to 64 reported being sexually active in the last twelve months. This percentage declined to about half (53 percent) for those 65 to 74 and to about a fourth (26 percent) for those ages 75 to 80. An easy way to remember these percentages is three-fourths of those around 60, a half of those about 70 and a fourth of those around 80 report being sexually active. In general, women reported less sexual activity due to the absence of a sexual partner. Those most sexually healthy were also in good health. Diabetes and hypertension were major causes of sexual dysfunction. Only one out of seven reported using Viagra or other such medications. The most frequent sexual problem for men was erectile dysfunction; for women, the problems were low sexual desire (43 percent reporting), less vaginal lubrication (39 percent) and inability to climax (34 percent). Beckman et al. (2006) also reported that sexual interest continued for 95 percent of their sample of

Table 15.3
Physiological Sexual Changes in Elderly Men

Phases of Sexual Response	Changes in Men
Excitement phase	As men age, getting an erection can take them longer. Although a young man may get an erection within 10 seconds, elderly men may take several minutes (10 to 30). During this time, they usually need intense stimulation (manual or oral). Unaware that the greater delay in getting erect is a normal consequence of aging, men who experience this for the first time may panic and have erectile dysfunction.
Plateau phase	The erection may be less rigid than when the man was younger, and there is usually a longer delay before ejaculation. This latter change is usually regarded as an advantage by both the man and his partner.
Orgasm phase	Orgasm in the elderly male is usually less intense, with fewer contractions and less fluid. However, orgasm remains an enjoyable experience, as over 70 percent of older men in one study reported that having a climax was very important when having a sexual experience.
Resolution phase	The elderly man loses his erection rather quickly after ejaculation. In some cases, the erection will be lost while the penis is still in the woman's vagina and she is thrusting to cause her orgasm. The refractory period is also increased. Whereas a young male needs only a short time after ejaculation to get an erection, the elderly man may need considerably longer.

Source: Adapted from W. Boskin, G. Graf, and V. Kreisworth. 1990. *Health dynamics: Attitudes and behaviors*, p. 209. St. Paul: West. Used by permission.

Table 15.4
Physiological Sexual Changes in Elderly Women

Phases of Sexual Response	Changes in Women
Excitement phase	Vaginal lubrication takes several minutes or longer, as opposed to 10 to 30 seconds when younger. Both the length and the width of the vagina decrease. Considerable decreased lubrication and vaginal size are associated with pain during intercourse. Some women report decreased sexual desire and unusual sensitivity of the clitoris.
Plateau phase	Little change occurs as the woman ages. During this phase, the vaginal orgasmic platform is formed and the uterus elevates.
Orgasm phase	Elderly women continue to experience and enjoy orgasm. Of women aged 60 to 91, almost 70 percent reported that having an orgasm made for a good sexual experience. With regard to their frequency of orgasm now as opposed to when they were younger, 65 percent said "unchanged," 20 percent "increased," and 14 percent "decreased."
Resolution phase	Defined as a return to the preexcitement state, the resolution phase of the sexual response phase happens more quickly in elderly than in younger women. Clitoral retraction and orgasmic platform disappear quickly after orgasm. This is most likely a result of less pelvic vasocongestion to begin with during the arousal phase.

Source: Adapted from W. Boskin, G. Graf, and V. Kreisworth. 1990. *Health dynamics: Attitudes and behaviors*, p. 210. St. Paul: West. Used by permission.

more than 500 people in their seventies. Alford-Cooper (2006) studied the sexuality of couples married for more than fifty years and found that sexual interest, activity, and capacity declined with age but that marital satisfaction did not decrease with these changes.

Although sex may occur less often, it remains satisfying for most elderly. Winterich (2003) found that, in spite of vaginal, libido, and orgasm changes past menopause, women of both sexual orientations reported that they continued to enjoy active, enjoyable sex lives due to open communication with their partners.

The debate continues about whether menopausal women should become involved in estrogen replacement therapy and estrogen-progestin replacement therapy (collectively referred to as HRT—hormone replacement therapy). Schairer et al. (2000) studied 2,082 cases of breast cancer and concluded that the estrogen-progestin regimen increased the risk of breast cancer beyond that associated with estrogen alone. Beginning in 2003, women were no longer routinely encouraged to take HRT, and the effects of their not doing so on their physical and emotional health were minimal. A major study (Hays et al. 2003) of more than 16,608 postmenopausal women aged 50 to 79 found no significant benefits from HRT in terms of quality of life. Those with severe symptoms (for example, hot flashes, sleep disturbances, irritability) do seem to benefit without negative outcomes. Indeed, one woman reported, "I'd rather be on estrogen so my husband can stand me (and I can stand myself)." New data suggest that women who have had a hysterectomy can benefit from estrogen-alone therapy without raising their breast cancer risk.

15.4 Successful Aging

Researchers who worked on the Landmark Harvard Study of Adult Development (Valliant 2002) followed 824 men and women from their teens into their eighties and identified those factors associated with successful aging. These include not smoking (or quitting early), developing a positive view of life and life's crises, avoiding alcohol and substance abuse, maintaining healthy weight, exercising daily, continuing to educate oneself, and having a happy marriage. Indeed, those who were identified as "happy and well" were six times more likely to be in a good marriage than those who were identified as "sad and sick." Conversely, spouses with higher levels of negative marital behavior had more chronic health problems, physical disability, and poorer perceived health (Bookwala 2005). Those with negative marital interaction (particularly from lower socioeconomic backgrounds) were more likely to die of a heart attack than spouses for whom no such marital interaction existed (Krause 2005).

Not smoking is "probably the single most significant factor in terms of health" according to Valliant (2002), of the Landmark Harvard Study of Adult Development. Smokers who quit before age 50 were as healthy at 70 as those who had never smoked.

Exercise is one of the most beneficial activities the elderly can engage in to help them maintain good health. Yet the occurrence of exercise activities significantly decreases as a person ages. A team of researchers (Lees et al. 2005) interviewed fifty-seven adults over the age of 65 about their exercise behavior. Those who did not exercise noted fear of falling and inertia for avoiding exercise. Those who did exercise identified inertia, time constraints, and physical ailments as being the most significant barriers to exercise. Wrosch et al. (2007) studied the exercise behavior of 172 elderly adults and found that the greatest predictor of whether elderly people exercised was their having done so previously. Hence, people who exercised at year 2 of the study were likely to still be exercising at year 5 of the study. Stephenson et al. (2007) noted that community-sponsored walking programs at malls were beneficial in getting the elderly to exercise. Jack LaLanne, the health fitness guru, remarked at age 90, "I hate to exercise; I love the results." At his 90th birthday party, he challenged his well-wishers to be at his 115th birthday party.

Agahi (2008) also noted a strong positive effect of involvement in social leisure activities and successful aging among a representative sample of 1,246 men and women ages 65 to 95. Participating in only a few activities doubled mortality risk compared to those with the highest participation levels, even after controlling for age, education, walking ability, and other health indicators. Strongest benefits were found for engagement in organizational activities and study circles among women and hobby activities and gardening among men.

The elderly also benefit from a credible role model for successful aging. Horton and Deakin (2007) note that such a model is older than themselves and is "active, vigorous and illustrative of the high quality of life that is possible into very late age." Superathletic models are often not effective because they are sometimes viewed as intimidating and not realistic.

15.5 Relationships and the Elderly

Relationships continue into old age. Here we examine relationships with one's spouse, siblings, and children.

Relationships among Elderly Spouses

Marriages that survive into late life are characterized by little conflict, considerable companionship, and mutual supportiveness. All but one of the thirty-one spouses over age 85 in the Johnson and Barer (1997) study reported "high expressive rewards" from their mate. Walker and Luszcz (2009) reviewed the literature on elderly couples and found marital satisfaction related to equality of roles and marital communication. Health may be both improved by positive relationships and decreased by negative relationships.

Field and Weishaus (1992) reported interview data on seventeen couples who had been married an average of fifty-nine years and found that the husbands and wives viewed their marriages very differently. Men tended to report more marital satisfaction, more pleasure in the way their relationships had been across time, more pleasure in shared activities, and closer affectional ties. Even though club activities, including church attendance, were related to marital satisfaction, this study found that financial stability, amount of education, health, and intelligence were not. Sex was also more important to the husbands. Every man in the study reported that sex was always an important part of the relationship with his wife, but only four of the seventeen wives reported the same.

The wives, according to the researchers, presented a much more realistic view of their long-term marriage. For this generation of women, "it was never as important for them to put the best face on things" (Field and Weishaus 1992, 273). Hence, many of these wives were not unhappy; they were just more willing to report disagreements and changes in their marriages across time.

Only a small percentage (8 percent) of individuals older than 100 are married. Most married centenarians are men in their second or third marriage. Many

Married 84 Years, and Still Loving*

Herbert and Zelmyra Fisher of the Brownsville community in North Carolina have been married for more than eighty-four years. They have the world record of the longest marriage for a living couple (they have a certificate from the Guinness Book of Records) (Sawyer 2008).

Herbert was born June 10, 1905. His hearing is going but his mind is sharp. Zelmyra was born December 10, 1907. She uses a walker to get around the house and yard. The two of them can still give their reasons for marrying on May 13, 1924. "He was not mean; he was not a fighter," Zelmrya said. "He was quiet and kind. He was not much to look at but he was sweet." Herbert said Zelmyra never gave him any trouble. "No, no trouble at all. We never argued, but we might have disagreed," he said.

Norma Godette, one of the couple's five children, said her parents have gotten along well through the years. "One time, mama wanted to work. Daddy told her she could not work, that he could take care of the family. She slipped down to Cherry Point and got a job as a caretaker there," Godette said. "Well, it was done; she got the job. I had to let it be," Herbert said.

© Courtesy of Liz Bowles/*New Bern Sun Journal* / © Nicholas Belton/iStockphoto.com / © Eliza Snow/iStockphoto.com

Different religions did not tear the two apart. He is a member of Pilgrim Chapel Missionary Baptist Church. She is a member of Jones Chapel African Methodist Episcopal Zion Church. The churches are in James City, where they both grew up. For all of their married life, they have attended their own churches. They go their own ways on Sunday morning. She reads the Bible daily.

The two watch television together. "We separate when the baseball comes on," Zelmyra said. Herbert loves baseball, especially the Atlanta Braves. He also enjoys golf, because one of his sons-in-law plays the game.

They have no secret or sage advice as to why their marriage has lasted so long. "I didn't know I would be married this long," Herbert said. "But I lived a nice holy life and go to church every Sunday. Yes sir, anything for her."

Zelmyra said Herbert was the only boyfriend she ever had. "We got along good," she said. "There was no trouble." She said she is not tired of seeing him. "I didn't think I'd be married this long. He is quiet," she said. Zelmyra said her husband had no annoying habits. They both said they shared the title of "boss." They worked hard and put all five of the children through college.

The two sit on the porch, and as a train goes by, they count the cars. They also watch the neighbors who walk by. Herbert makes his bed each day and sweeps the floor. He also checks on his wife as she rests. Between the rests, they enjoy their children, ten grandchildren, nine great-grandchildren, and nieces and nephews. Both say that if they had it to do over, they would not change their life.

* This article appeared in The *New Bern Sun Journal* (New Bern, NC), Editor John Huff. Used by permission.

thanatology examination of the social dimensions of death, dying, and bereavement.

have outlived some of their children. Marital satisfaction in these elderly marriages is related to a high frequency of expressing love feelings to one's partner. Though it is assumed that spouses who have been married for a long time should know how their partners feel, this is often not the case. Telling each other "I love you" is very important to these elderly spouses.

Renewing an Old Love Relationship

Some widowed or divorced elderly try to find and renew an earlier love relationship. Researcher Nancy Kalish (1997) surveyed 1,001 individuals who reported that they had renewed an old love relationship. Two-thirds of those who had contacted a previous love (mostly by phone or letter) were female; one-third were male. Almost two-thirds (62 percent) reported that the person they made contact with was their first love, and 72 percent reported that they were still in love with and together with the person with whom they had renewed their relationship. An example of one reunited couple (interviewed by Kalish) follows:

> He was my first boyfriend. It was all very innocent. Mostly we were good friends. We walked home from school together, and went out a couple of times. My parents didn't like him for some reason. Neither of us remembers why we broke up. We went on to long marriages with other people. Sixty-three years passed, and we were both widowed when we went to our high school reunion. I knew as soon as we saw each other. We got to talking and nothing else mattered. It seemed like something that was destined to happen.
>
> We were married on my eightieth birthday. He's so loving and kind and caring and understanding and peaceful. He doesn't argue. He's very respectful, a perfect gentlemen. I just love everything about him. He's all I ever wanted. (p. 11)

Weintraub (2006) suggested some cautions in renewing an old love relationship. These include (1) discussing the original breakup or what went wrong—if one partner was hurt badly, the other must take responsibility or make amends; and (2) going slow in reviving the relationship. She also identified some successful couples who reunited after a long interval. Harry Kullijian contacted Carol Channing seventy years after they left high school. Carol's mother had broken up the relationship. Kullijian and Channing married in 2003.

Relationship with Siblings at Age 85 and Beyond

Relationships that the elderly have with their siblings are primarily emotional (enjoying time together) rather than functional (in which a sibling provides money or services). Earlier in the text, we noted that sibling relationships, particularly those between sisters, are the most enduring of all relationships.

Relationship with One's Own Children at Age 85 and Beyond

In regard to relationships of the elderly with their children, emotional and expressive rewards are high. Actual caregiving is rare. Only 12 percent of the Johnson and Barer (1997) sample of adults older than 85 lived with their children. Most preferred to be independent and to live in their own residence. "This independent stance is carried over to social supports; many prefer to hire help rather than bother their children. When hired help is used, children function more as mediators than regular helpers, but most are very attentive in filling the gaps in the service network" (p. 86).

Relationships among multiple generations will increase. Whereas three-generation families have been the norm, four- and five-generation families will increasingly become the norm. These changes have already become visible.

15.6 The End of One's Life

Thanatology is the examination of the social dimensions of death, dying, and bereavement (Bryant 2007). The end of one's life sometimes involves the death of one's spouse.

Death of One's Spouse

The death of one's spouse is one of the most stressful life events a person ever experiences. Because women tend to live longer than men, and because women are

IF YOU WERE GOING TO DIE SOON AND HAD ONLY ONE PHONE CALL TO MAKE, WHO WOULD YOU CALL AND WHAT WOULD YOU SAY? AND WHY ARE YOU WAITING?

—Stephen Levine

often younger than their husbands, women are more likely than men to experience the death of their marital partner.

Although individual reactions and coping mechanisms for dealing with the death of a loved one vary, several reactions to death are common. These include shock, disbelief and denial, confusion and disorientation, grief and sadness, anger, numbness, physiological symptoms such as insomnia or lack of appetite, withdrawal from activities, immersion in activities, depression, and guilt. Eventually, surviving the death of a loved one involves the recognition that life must go on, the need to make sense out of the loss, and the establishment of a new identity. However, research is not consistent on the degree to which individuals are best served by continuing or relinquishing the emotional bonds with the deceased (Stroebe and Schut 2005).

Women and men tend to have different ways of reacting to and coping with the death of a loved one. Women are more likely than men to express and share feelings with family and friends and are also more likely to seek and accept help, such as attending support groups of other grievers. Initial responses of men are often cognitive rather than emotional. From early childhood, males are taught to be in control, to be strong and courageous under adversity, and to be able to take charge and fix things. Showing emotions is labeled as weak.

Men sometimes respond to the death of their spouse in behavioral rather than emotional ways. Sometimes they immerse themselves in work or become involved in physical action in response to the loss. For example, a widower immersed himself in repairing a beach cottage he and his wife had recently bought. Later, he described this activity as crucial to getting him through those first two months. Another coping mechanism for men is the increased use of alcohol and other drugs.

Women's response to the death of their husbands may necessarily involve practical considerations. Johnson and Barer (1997) identified two major problems of widows—the economic effects of losing a spouse and the practical problems of maintaining a home alone. The latter involves such practical issues as cleaning the gutters, painting the house, and changing the filters in the furnace.

Whether a spouse dies suddenly or after a prolonged illness has an impact on the reaction of the remaining spouse. The sudden rather than prolonged death of one's spouse is associated with being less at peace with death and being more angry. The suddenness of the death provides no cognitive preparation time.

The age at which one experiences the death of a spouse is also a factor in one's adjustment. People in their eighties may be so consumed with their own health and disability concerns that they have little emotional energy left to grieve. On the other hand, a team of researchers (Hansson et al. 1999) noted that death does not end the relationship with the deceased. Some widows and widowers report a feeling that their spouses are with them at times and are watching out for them a year after the death of their beloved. They may also dream of them, talk to

their photographs, and remain interested in carrying out their wishes. Such continuation of the relationship may be adaptive by providing meaning and purpose for the living or maladaptive in that it may prevent one from establishing new relationships.

Involvement with New Partners at Age 80 and Beyond

Most women who live to age 80 have lost their husbands. At age 80, only 53 men are available for every 100 women. Patterns women use to adjust to this lopsided man-woman ratio include dating younger men, romance without marriage, and "share-a-man" relationships. Most individuals in their eighties who have lost a partner do not remarry (Stevens 2002). Those who do report a need for companionship and to provide meaning in life as the primary reasons. Stevens (2002) noted "evidence of continuing loyalty to the deceased" in all the late-life partnerships she studied, suggesting that new partners do not simply "replace" former partners.

More than 3 million women are married to men who are at least ten years younger than they are. Although traditionally women were socialized to seek older, financially established men, the sheer shortage of men has encouraged many women to seek younger partners. These age-discrepant relationships were discussed in Chapter 7, Marriage Relationships.

Faced with a shortage of men but reluctant to marry those who are available, some elderly women are willing to share a man. In elderly retirement communities such as Palm Beach, Florida, women count themselves lucky to have a man who will come for lunch, take them to a movie, or be an escort to a dance. They accept the fact that the man may also have lunch, go to a movie, or go dancing with other women.

However, finding a new spouse is usually not a goal. Women in their later years have also moved away from the idea that they must remarry and have become more accepting of the idea that they can enjoy the romance of a relationship without the obligations of a marriage. Many enjoy their economic independence, their control over their life space, and their freedom not to be a nurse to an aging partner.

To avoid marriage, some elderly couples live together. Parenthood is no longer a goal, and many do not want to entangle their assets. For some, marriage would mean the end of their Social Security benefits or other pension moneys.

Preparing for One's Own Death

What is it like for those near the end of life to think about death? To what degree do they go about actually "preparing" for death? Johnson and Barer (1997) interviewed forty-eight individuals with an average age of 93 to find out their perspective on death. Most interviewees were women (77 percent); of them 56 percent lived alone, but 73 percent had some sort of support in terms of children or one or more social support services. The following findings are specific to those who died within a year after the interview.

Thoughts in the Last Year of Life Most had thought about death and saw their life as one that would soon end. Most did so without remorse or anxiety. With their spouses and friends dead and their health failing, they accepted death as the next stage in life. Some comments follow:

> If I die tomorrow, it would be all right. I've had a beautiful life, but I'm ready to go.
>
> My husband is gone, my children are gone, and my friends are gone.
>
> That's what is so wonderful about living to be so old. You know death is near and you don't even care.
>
> I've just been diagnosed with cancer, but it's no big deal. At my age, I have to die of something. (Johnson and Barer 1997, 205)

The major fear these respondents expressed was not the fear of death but of the dying process. Dying in a nursing home after a long illness is a dreaded fear. Sadly, almost 60 percent of the respondents died after a long, progressive illness. They had become frail, fatigued, and burdened by living. They identified dying in their sleep as the ideal way to die. Some hasten their death by no longer taking their medications or wish they could terminate their own life. "I'm feeling kind of useless. I don't enjoy anything anymore. . . . What the heck am I living for? I'm ready to go anytime—straight to hell. I'd take lots of sleeping pills if I could get them" (Johnson and Barer 1997, 204). Nakashima and Canda (2005) studied sixteen hospice patients and found that having a positive death was associated with coming to terms with one's own mortality, finding meaning in death and dying, and spiritual well-being.

Behaviors in the Last Year of Life Aware that they are going to die, most simplify their life, disengage from social relationships, and leave final instructions. In simplifying their life, they sell their home and belongings and move to smaller quarters. One 81-year-old

© Sally Llanes/iStockphoto.com

woman sold her home, gave her car away to a friend, and moved into a nursing home. The extent of her belongings became a chair, lamp, and television.

Disengaging from social relationships is selective. Some maintain close relationships with children and friends, but others "let go." They may no longer send out Christmas cards, stop sending letters, and phone calls become the source of social connections. Some leave final instructions in the form of a will or handwritten note expressing wishes of where to be buried, handling costs associated with disposal of the body, and what to do about pets. One of Johnson and Barer's (1997) respondents left $30,000 to specific caregivers to take care of each of several pets (p. 204).

Elderly who may have counted on children to take care of them may discover that they have too few children who may have scattered because of job changes, divorce, or both (Kutza 2005). The result is that the elderly may have to fend for themselves.

One of the last legal acts of the elderly is to have a will drawn up. Stone (2008) emphasized how wills may stir up sibling rivalry (e.g., one sibling may be left more than another), be used as a weapon against a second spouse (e.g., by leaving all of one's possessions to one's children), or reveal a toxic secret (e.g., reveal a mistress and several children who are left money).

> **For in all the world** *there are no people so piteous and forlorn as those who are forced to eat the bitter bread of dependency in their old age, and find how steep are the stairs of another man's house.*
>
> —*Dorothy Dix, mental health advocate*

Diversity in Elder Care

Whereas female children in the United States have the most frequent contact and are more involved in the caregiving of their elderly parents than male children, the daughters-in-law in Japan offer the most help to elderly individuals. The female child who is married gives her attention to the parents of her husband (Ikegami 1998).

Eastern cultures emphasize filial piety, which is love and respect toward their parents. Filial piety involves respecting parents, bringing no dishonor to parents, and taking good care of parents (Jang and Detzner 1998). Western cultures are characterized by filial responsibility emphasizing duty, protection, care, and financial support.

Part of filial piety does involve financial support, in evidence in Chinese culture. There, loss of income is minimized among the elderly Chinese, who receive money from their children. Logan and Bian (2003) noted that money from "children accounts for nearly a third of parents' incomes" (p. 85).

© C Squared Studios/Getty Images

Test coming up? Now what?

With M&F you have a multitude of study aids at your fingertips.
After reading the chapters, check out these ideas for further help.

Review cards include all learning outcomes, definitions, and summaries for each chapter.

Flash Cards give you three additional ways to check your comprehension of marriage and family concepts.

Other great ways to help you study include **Interactive Quizzing, Games, Videos, Study Worksheets, Note Taking Outlines** and **Internet Activities.**

You can find it all at **4ltrpress.cengage.com/mf.**

References

Chapter 1

Allen, K. R., E. K. Husser, D. J. Stone, and C. E. Jordal. 2008. Agency and error in young adults' stories of sexual decision making. *Family Relations* 57:517–29.

Amato, P. R., A. Booth, D. R. Johnson, and S. F. Rogers. 2007. *Alone together: How marriage in America is changing*. Cambridge, Massachusetts: Harvard University Press.

Blumer, H. G. 1969. The methodological position of symbolic interaction. In *Symbolic interactionism: Perspective and method*. Englewood Cliffs, NJ: Prentice-Hall.

Busby, D. M., B. Gardner, C. Brandt, and N. Taniguchi. 2005. The family of origin parachute model: Landing safely in adult romantic relationships. *Family Relations* 54:254–64.

Cooley, C. H. 1964. *Human nature and the social order.* New York: Schocken.

Corra, M., S. Carter, J. S. Carter, and D. Knox. 2009. Trends in marital happiness by sex and race, 1973-2006. *Journal of Family Issues* (in press).

DeCuzzi, A., D. Knox, and M. Zusman. 2006. Racial differences in perceptions of women and men. *College Student Journal* 40:343–49.

Dotson-Blake, K., D. Knox, and A. Holman. Forthcoming. College student rank/gender/race and attitudes toward therapy.

Bristol, K., and B. Farmer 2005. Sexuality among southeastern university students: A survey. Unpublished data, East Carolina University, Greenville, NC.

Brown, G. L., S. C. Mangelsdorf, C. Neff, S. J. Schoppe-Sullivan, and C. A. Frosch. 2009. Young children's self-concepts: Associations with child temperament, mothers' and fathers' parenting, and triadic family interaction. *Merrill-Palmer Quarterly* 55:207.

Foster, J. D. 2008. Incorporating personality into the investment model: Probing commitment processes across individual differences in narcissism. *Journal of Social and Personal Relationships* 25:211–23.

Gavin, J. 2003. *Deep in a dream: The long night of Chet Baker.* New York: Welcome Rain.

Generation Y Data. 2007. http://www.docuticker.com/wp-content/uploads/2007/09/factsheet_geny.htm.

Hall, S. 2009. Personal communication.

Harris Poll. 2007. Pets as family members. www.harrispollonline.com (retrieved December 11, 2007).

Henley-Walters, L., W. Warzywoda-Kruszynska, and T. Gurko. 2002. Cross-cultural studies of families: Hidden differences. *Journal of Comparative Family Studies* 33:433–50.

Howard, Margo. 2003. *A life in letters: Ann Landers's letters to her only child.* New York: Warner.

James, S. D. 2008. Wild child speechless after tortured life. *ABC News,* May 7.

Knox, D., and M. E. Zusman. 2009. Relationship and sexual behaviors of a sample of 1319 university students. Unpublished data collected for this text. Department of Sociology, East Carolina University, Greenville, NC.

Lehmiller, J. J., and C. R. Agnew. 2007. Perceived marginalization and the prediction of romantic relationship stability. *Journal of Marriage and Family* 69:1036–49.

Lemonick, M. D. 2006. The rise and fall of the cloning king. *Time,* January 6, 40–43.

Lorber, J. 1998. *Gender inequality: Feminist theories and politics.* Los Angeles, CA: Roxbury.

Mead, G. H. 1934. *Mind, self, and society.* Chicago: University of Chicago Press.

Meinhold, J. L., A. Acock, and A. Walker. 2006. The influence of life transition statuses on sibling intimacy and contact in early adulthood. Paper presented at the Annual Meeting of the National Council on Family Relations in Orlando in November 2005.

Murdock, G. P. 1949. *Social structure.* New York: Free Press.

Newton, N. 2002. *Savage girls and wild boys: A history of feral children.* New York: Thomas Dunne Books/St. Martin's Press.

Neyer, F. J., and F. R. Lang. 2003. Blood is thicker than water: Kinship orientation across adulthood. *Journal of Personality and Social Psychology* 84:310–21.

Novilla, M., B. Lelinneth, M. D. Barnes, N. G. De La Cruz, P. N. Williams, and J. Rogers. 2006. Public health perspectives on the family. *Family and Community Health* 29:28–42.

Parker, L. 2005. When pets die at the vet, grieving owners call lawyers. *USA Today,* March 15, A1.

Pescosolido, B. A., J. K. Martin, A. Lang, and S. Olafsdottir. 2008. Rethinking theoretical approaches to stigma: A Framework Integrating Normative Influences on Stigma (FINIS). *Social Science & Medicine* 67:431–51.

Pryor, J. H., S. Hurtado, L. DeAngelo, J. Sharkness, L. C. Romero, W. K. Korn, and S. Trans. 2008. *The American freshmen: National norms for fall 2008.* Los Angeles: Higher Education Research Institute, UCLA.

Richman, J. A., L. Cloninger, and K. M. Rospenda. 2008. Macrolevel stressors, terrorism, and mental health outcomes: Broadening the stress paradigm. *American Journal of Public Health* 98:323–30.

Silverstein, L. B., and C. F. Auerbach. 2005. (Post) modern families. In *Families in global perspective,* ed. Jaipaul L. Roopnarine and U. P. Gielen, 33–48. Boston, MA. Pearson Education.

Statistical Abstract of the United States, 2009, 128th ed. Washington, DC: U.S. Bureau of the Census.

Taylor, A. C., and A. Bagd. 2005. The lack of explicit theory in family research: The case analysis of the *Journal of Marriage and the Family 1990–1999.* In *Sourcebook of family theory & research,* ed. Vern L. Bengtson, Alan C. Acock, Katherine R. Allen, Peggye Dilworth-Anderson, and David M. Klein, 22–25. Thousand Oaks, CA: Sage Publications.

Testa, M., J. A. Livingston, and C. VanZile-Tamsen. 2005. The impact of questionnaire administration mode on response rate and reporting of consensual and nonconsensual sexual behavior. *Psychology of Women Quarterly* 29:345–52.

Thomas, S. G. 2007. *Buy buy baby: How consumer culture manipulates parents and harms young minds.* Boston: Houghton Mifflin.

Tupelo, A., and E. Freeman. 2008. Polyamory. Presentation to Sociology of Human Sexuality class, Department of Sociology, East Carolina University, Greenville, NC. October.

Turner, J. H. 2007. Self, emotions, and extreme violence: Extending symbolic interactionist theorizing. *Symbolic Interaction* 30:501–31.

Veenhoven, R. 2007. Quality-of-life-research. In *21st century sociology: A reference handbook,* ed. Clifton D. Bryant and Dennis L. Peck, 54–62. Thousand Oaks, California: Sage Publications.

Vogel, D. L., D. A. Gentile, and S. A. Kaplan. 2008. The influence of television on willingness to seek therapy. *Journal of Clinical Psychology* 64:276–81.

White, J. M., and D. M. Klein. 2002. *Family theories,* 2d ed. Thousand Oaks, CA: Sage Publications.

Zeitzen, M. K. 2008. *Polygamy: A cross-cultural analysis.* Oxford: Berg.

Chapter 2

Abowitz, D. A., D. Knox, M. Zusman, and A. McNeely, 2009. Beliefs about romantic relationships: Gender differences among undergraduates. *College Student Journal* 43:276–284.

Aissen, K., and S. Houvouras. 2006. Family first: Negotiating motherhood and doctoral studies. Southern Sociological Society, New Orleans, March 24.

Bacik, I., and E. Drew. 2006. Struggling with juggling: Gender and work/life balance in the legal professions. *Women's Studies in International Forum* 29:136–46.
Bem, S. L. 1983. Gender schema theory and its implications for child development: Raising gender-aschematic children in a gender-schematic society. *Signs* 8:596–616.
Cheng, C. 2005. Processes underlying gender-role flexibility: Do androgynous individuals know more or know how to cope? *Journal of Personality* 73:645–74.
Cohen-Kettenis, P. T. 2005. Gender change in 46, XY persons with 5[alpha]-reductase-2 deficiency and 17[beta]-hydroxysteroid dehydrogenase-3 deficiency. *Archives of Sexual Behavior* 34:399–411.
Colapinto, J. 2000. *As nature made him: The boy who was raised as a girl.* New York: Harper Collins.
Consolatore, D. 2002. What next for the women of Afghanistan? *The Humanist* 62:10–15.
Cordova, J. V., C. B. Gee, and L. Z. Warren. 2005. Emotional skillfulness in marriage: Intimacy as a mediator of the relationship between emotional skillfulness and marital satisfaction. *Journal of Social & Clinical Psychology* 24:218–35.
Corra, M., J. S. Carter, and D. Knox. 2006. Marital happiness by sex and race: A second look. Paper, Annual Meeting of the American Sociological Association, New York, August.
Crawley, S. L., L. J. Foley, and C. L. Shehan. 2008. *Gendering bodies.* Boston: Rowman and Littlefield.
Denton, M. L. 2004. Gender and marital decision making: Negotiating religious ideology and practice. *Social Forces* 82:1151–80.
Dotson-Blake, K., D. Knox, and A. Holman 2008. College student attitudes toward marriage, family, and sex therapy. Unpublished data from 288 undergraduate/graduate students. East Carolina University, Greenville, NC.
Faulkner, R., A. M. Davey, and A. Davey. 2005. Gender-related predictors of change in marital satisfaction and marital conflict. *American Journal of Family Therapy* 33:61–83.
Fortune 500 in 2008 http://money.cnn.com/galleries/2008/fortune/0804/gallery.500_women_ceos.fortune/index.html.
Georgas, J., T. Bafiti, K. Mylonas, and L. Papademou. 2005. Families in Greece. In *Families in global perspective,* ed. J. L. Roopnaraine and U. P. Gielen, 207–24. Boston: Pearson Allyn & Bacon.
Giordano, P. C., W. D. Manning, and M. A. Longmore. 2005. The romantic relationships of African-American and white adolescents. *The Sociological Quarterly* 46:545–68.
Grief, G. L. 2006. Male friendships: Implications from research for family therapy. *Family Therapy* 33:1–15.
Haynie, D. L., and D. W. Osgood. 2005. Reconsidering peers and delinquency: How do peers matters? *Social Forces* 84:1109–30.
Heller, N. 2008. Will the transgender dad be a father? What goes on the birth certificate. http://www.slate.com/id/2193475/ (accessed June 13, 2008).
Henry, R. G., R. B. Miller, and R. Giarrusso. 2005. Difficulties, disagreements, and disappointments in late-life marriages. *International Journal of Aging & Human Development* 61:243–65.
Hepp, U., A. Spindler, and G. Milos. 2005. Eating disorder symptomalogy and gender role orientation. *International Journal of Eating Disorders* 37:227–33.
Hill, D. B. 2007. Differences and similarities in men's and women's sexual self-schemas. *Journal of Sex Research* 44:135–44.
Ingram, S., J. L. Ringle, K. Hallstrom, D. E. Schill, et al. 2008. Coping with crisis across the lifespan: The role of a telephone hotline. *Journal of Child and Family Studies* 17:663–75.
Irvolino, A. C., M. Hines, S. E. Golombok, J. Rust, and R. Plomin. 2005. Genetic and environmental influences on sex-typed behavior during the preschool years. *Child Development* 76:826–40.
Kennedy, M. 2007. Gender role observations of East Africa. Written exclusively for this text.
Kilmartin, C., T. Smith, A. Green, H. Heinzen, M. Kuchler, and D. Kolar. 2008. A real time social norms intervention to reduce male sexism. *Journal Sex Roles* 59: 264–73.
Kim, J. L., C. L. Sorsoli, K. Collins, B. A. Zylbergold, D. Schooler, and D. L. Tolman. 2007. From sex to sexuality: Exposing the heterosexual script on primetime network television. *Journal of Sex Research* 44:145–57.
Kimmel, M. S. 2001. Masculinity as homophobia: Fear, shame, and silence in the construction of gender identity. In *Men and masculinity: A text reader,* ed. T. F. Cohen, 29–41. Belmont, CA: Wadsworth.
Kimmel, M. S. 2006. *Manhood in America: A cultural history.* New York: Oxford University Press.
Knox, D., S. Hatfield, and M. E. Zusman. 1998. College student discussion of relationship problems. *College Student Journal* 32:19–21.
Knox, D., and M. E. Zusman. 2009. Relationship and sexual behaviors of a sample of 1,319 university students. Unpublished data collected for this text. Department of Sociology, East Carolina University, Greenville, NC.
Knox, D., M. E. Zusman, and H. R. Thompson. 2004. Emotional perceptions of self and others: Stereotypes and data. *College Student Journal* 38:130–42.
Kohlberg, L. 1966. A cognitive-developmental analysis of children's sex-role concepts and attitudes. In *The development of sex differences,* ed. E. E. Macoby. Stanford, CA: Stanford University Press.
———. 1969. State and sequence: The cognitive developmental approach to socialization. In *Handbook of socialization theory and research,* ed. D. A. Goslin, 347–480. Chicago: Rand McNally.
Krafchick, J. L., T. S. Zimmerman, S. A. Haddock, and J. H. Banning. 2005. Best-selling books advising parents about gender: A feminist analysis. *Family Relations* 54:84–101.
Kurson, R. 2007. *Crashing through: A true story of risk, adventure, and the man who dared to see.* New York: Random House.
Lareau, A., and E. B. Weininger. 2008. Time, work, and family life: Reconceptualizing gendered time patterns through the case of children's organized activities. *Sociological Forum* 23:419–54.
Lipsitz, L. A. 2005. The elderly people of post-soviet Ukraine: Medical, social, and economic challenges. *Journal of the American Geriatrics Society* 53:2,216–20.
Ma, M.K. 2005. The relation of gender-role classifications to the prosocial and antisocial behavior of Chinese adolescents. *Journal of Genetic Psychology* 166:189–201.
Maume, D. J. 2006. Gender differences in taking vacation time. *Work and Occupations* 33:161–90.
McGinty, K., D. Knox, and M. E. Zusman. 2006. Research report on undergraduate women who prefer a traditional man. Unpublished research created for this text.
McNeely, A., D. Knox, and M. E. Zusman. 2004. Beliefs about men: Gender differences among college students. Poster for Annual Meeting of the Southern Sociological Society, Atlanta, April 16–17.
_____. 2005. College student beliefs about women: Some gender differences. *College Student Journal* 39:769–74.
McPherson, M., L. Smith-Lovin, and M. E. Brashears. 2006. Social isolation in America, 1985–2004. *American Sociological Review* 71:353–75.
Mead, M. 1935. *Sex and temperament in three primitive societies.* New York: William Morrow.
Meinhold, J. L., A. Acock, and A. Walker. 2006. The influence of life transition statuses on sibling intimacy and contact in early adulthood. Presented at the National Council on Family Relations Annual meeting in Orlando in 2005.
Meyer-Bahlburg, H. F. L. 2005. Introduction: Gender dysphoria and gender change in persons with intersexuality. *Archives of Sexual Behavior* 34:371–74.
Miller, A. S., and R. Stark. 2002. Gender and religiousness: Can socialization explanations be saved? *American Journal of Sociology* 107:1399–423.

Mirchandani, R. 2005. Postmodernism and sociology: From the epistemological to the empirical. *Sociological Theory* 23:86–115.
Moghadam, V. M. 2002. Patriarchy, the Taliban, and the politics of public space in Afghanistan. *Women's Studies International Forum* 25:19–31.
Monin, J. K., M. S. Clark, and E. P. Lemay. 2008. Communal responsiveness in relationships with female versus male family members. *Journal Sex Roles* 59:176–88.
Monro, S. 2000. Theorizing transgender diversity: Towards a social model of health. *Sexual and Relationship Therapy* 15:33–42.
Moore, O., S. Kreitler, M. Ehrenfeld, and N. Giladi. 2005. Quality of life and gender identity in Parkinson's disease. *Journal of Neural Transmission* 112:1511–22.
Neighbors, L., J. Sobal, C. Liff, and D. Amiraian. 2008. Weighing weight: Trends in body weight evaluation among young adults, 1990 and 2005. *Journal of Sex Roles* 59:68–80.
Parra-Cardona, J. R., D. Córdova Jr., K. Holtrop, F. A. Villarruel, and E. Wieling. 2008. Shared ancestry, evolving stories: Similar and contrasting life experiences described by foreign born and U.S. born Latino parents. *Family Process* 47:157–73.
Peoples, J. G. 2001.The cultural construction of gender and manhood. In *Men and masculinity: A text reader*, ed. T. F. Cohen, 9–18. Belmont, CA: Wadsworth.
Peters, J. K. 2005. Gender remembered: The ghost of "unisex" past, present, and future. *Women's Studies* 34:67–83.
Pew Research. 2008. The U.S. religious landscape survey. Pew Forum on Religion & Public Life. http://pewresearch.org/pubs/743/united-states-religion.
Pollack, W. S. (with T. Shuster). 2001. *Real boys' voices*. New York: Penguin Books.
Pretorius, E. 2005. Family life in South Africa. In *Families in global perspective*, ed. J. L. Roopnaraine and U. P. Gielen, 363–80. Boston: Pearson Allyn & Bacon.
Probert, B. 2005. "I Just Couldn't Fit It In": Gender and unequal outcomes in academic careers. *Gender, Work & Organization* 12:50–73.
Raj, A., C. Gomez, and J. G. Silverman. 2008. Driven to a fiery death: The tragedy of self-immolation in Afghanistan. *The New England Journal of Medicine* 358:2201–17.
Rivadeneyraa, R., and M. J. Lebob. 2008. The association between television-viewing behaviors and adolescent dating role attitudes and behaviors. *Journal of Adolescence* 31:291–305.
Roopnarine, J. L., P. Bynoe, R. Singh, and R. Simon. 2005. Caribbean families in English-speaking countries. In *Families in global perspective*, ed. J. L. Roopnaraine and U. P. Gielen, 311–29. Boston: Pearson Allyn & Bacon.
Ross, C., D. Knox, and M. Zusman. Forthcoming. "Hey Big Boy": Characteristics of university women who initiate relationships with men. *College Student Journal*.
Royo-Vela, M., J. Aldas-Manzano, I. Kuster, and N. Vila. 2008. Adaptation of marketing activities to cultural and social context: Gender role portrayals and sexism in Spanish commercials. *Sex Roles* 58:379–91.
Skaine, R. 2002. *The women of Afghanistan under the Taliban*. Jefferson, NC: McFarland.
Statistical Abstract of the United States, 2009. 128th ed. Washington, DC: U.S. Bureau of the Census.
Stone, P. 2007. *Opting out?* Berkeley: University of California Press.
Sumsion, J. 2005. Male teachers in early childhood education: Issues and case study. *Early Childhood Research Quarterly* 20:109–23.
Taylor, R. L. 2002. Black American families. In *Minority families in the United States: A multicultural perspective*, ed. Ronald L. Taylor, 19–47. Upper Saddle River, NJ: Prentice Hall.
Thompson, M. 2008. America's medicated army. *Time*, July 16, 38–42.
Vail-Smith, K., D. Knox, and M. Zusman. 2007. The lonely college male. *International Journal of Men's Health* 6:273–79.
Walker, R. B., and M. A. Luszcz. 2009. The health and relationship dynamics of late-life couples: a systematic review of the literature. *Ageing and Society* 29: 455–81.
Walzer, S. 2008. Redoing gender through divorce. *Journal of Social and Personal Relationships* 25:5–21.
Ward, C. A. 2001. Models and measurement of psychological androgyny: A cross-cultural extension of theory and research. *Sex Roles: A Journal of Research* 43:529–52.
Welch, V. 2008. *Doctorate recipients from United States universities: Selected Tables 2007*. Chicago: National Opinion Research Center.
Wilcox, W. B., and S. L. Nock. 2006. What's love got to do with it? Equality, equity, commitment and marital quality. *Social Forces* 84:1321–45.
Woodhill, B. M., and C. A. Samuels. 2003. Positive and negative androgyny and their relationship with psychological health and well-being. *Sex Roles* 48:555–65.

Chapter 3

Adams, M., et al. 2008. Rhetoric of alternative dating: Investment versus exploration. Southern Sociological Society, Richmond, VA, April.
Albright, J. M. 2007. How do I love thee and thee and thee?: Self-presentation, deception, and multiple relationships online. In *Online Matchmaking*, ed. M. T. Whitty, A. J. Baker, and J. A. Inman, 81–93. New York: Palgrave Macmillan.
Avellar, S., and P. J. Smock. 2005. The economic consequences of the dissolution of cohabiting unions. *Journal of Marriage and the Family* 67:315–27.
Baker, A. J. 2007. Expressing emotion in text: Email communication of online couples. In *Online Matchmaking*, ed. M. T. Whitty, A. J. Baker, and J. A. Inman, 97–111. New York: Palgrave Macmillan.
Baxter, J. 2005. To marry or not to marry: Marital status and the household division of labor. *Journal of Family Issues* 26:300–21.
Bogle, K. A. 2002. From dating to hooking up: Sexual behavior on the college campus. Paper presented at the Annual Meeting of the Society for the Study of Social Problems, Summer.
_____. 2008. *Hooking up: Sex, dating, and relationships on campus*. New York: New York University Press.
Booth, A., E. Rustenbach, and S. McHale. 2008. Early family transitions and depressive symptom changes from adolescence to early adulthood. *Journal of Marriage and Family* 70:3–14.
Brewster, C. D. D. 2006. African American male perspective on sex, dating, and marriage. *The Journal of Sex Research* 43:32–33.
Brown, S. L., J. R. Bulanda, and G. R. Lee. 2005. The significance of nonmarital cohabitation: Marital status and mental health benefits among middle-aged and older adults. *Journals of Gerontology Series B—Psychological Sciences and Social Sciences* 60(1):S21–S29.
Cohan, C. L., and S. Kleinbaum. 2002. Toward a greater understanding of the cohabitation effect: Premarital cohabitation and marital communication. *Journal of Marriage and the Family* 64:180–92.
Davis, S. N., T. N. Greenstein, and J. P. Gerteisen Marks. 2007. Effects of union type on division of household labor. *Journal of Family Issues* 28:1246–72.
DePaulo, B. 2006. *Singled out: How singles are stereotyped, stigmatized, and ignored, and still live happily ever after.* New York: St. Martin's Press.
Donn, J. 2005. Adult development and well-being of midlife never-married singles. Unpublished dissertation. Miami University, Oxford, Ohio.
Dush, C., M. Kamp, and P. R. Amato. 2005. Consequences of relationship status and quality for subjective well-being. *Journal of Social and Personal Relationships* 22:607–27.

Eggebeen, D. J. 2005. Cohabitation and exchanges of support. *Social Forces* 83:1097–110.

England, P., and R. J. Thomas. 2006. The decline of the date and the rise of the college hook up. In *Family in transition,* 14th ed., ed. A. S. Skolnick and J. H. Skolnick, 151–62. Boston: Pearson Allyn & Bacon.

Eshbaugh, E. M., and G. Gute. 2008. Hookups and sexual regret among college women. *The Journal of Social Psychology* 148:77–87.

Gibbs, J. L., N. B. Ellison, and R. D. Heino. 2006. Self-presentation in online personals: The role of anticipated future interaction, self-disclosure, and perceived success in Internet dating. *Communication Research* 33:152–77.

Ha, J. H. 2008. Changes in support from confidants, children, and friends following widowhood. *Journal of Marriage and Family* 70:306–29.

Hansen, T., T. Moum, and A. Shapiro. 2007. Relational and individual well-being among cohabitors and married individuals in midlife. *Journal of Family Issues* 28:910–33.

Harcourt, W. 2005. Gender and community in the social construction of the Internet/Hanging out in the virtual pub: Masculinities and relationships online. *Signs: Journal of Women in Culture and Society* 30:1,981–84.

Hemstrom, O. 1996. Is marriage dissolution linked to differences in mortality risks for men and women? *Journal of Marriage and the Family* 58:366–78.

Hess, J. 2009. Personal communication. Appreciation is expressed to Judye Hess for the development of this section. For more information about Judye Hess, see http://www.psychotherapist.com/judyehess/

Hodge, A. 2003. Video chatting and the males who do it. Paper presented at the 73rd Annual Meeting of the Eastern Sociological Association, Philadelphia, February 28.

Jagger, E. 2005. Is thirty the new sixty? Dating, age and gender in postmodern consumer society. *Sociology* 9:89–106.

Jamieson, L., M. Anderson, D. McCrone, F. Bechhofer, R. Stewart, and Y. Li. 2002. Cohabitation and commitment: Partnership plans of young men and women. *The Sociological Review* 50:356–77.

Jerin, R. A., and B. Dolinsky. 2007. Cyber-victimization and online dating. In *Online Matchmaking,* ed. M. T. Whitty, A. J. Baker, and J. A. Inman, 147–56. New York: Palgrave Macmillan.

Jones, J. 2006. Marriage is for white people. *The Washington Post* March 26, B3–B4.

Kalmijn, M. 2007. Gender differences in the effects of divorce, widowhood and remarriage on intergenerational support: Does marriage protect fathers? *Social Forces* 85:1079–85.

Knox, D., and U. Corte. 2007. "Work it out/See a counselor": Advice from spouses in the separation process. *Journal of Divorce and Remarriage* 48:79–90.

Knox, D., and M. E. Zusman. 2009. Relationship and sexual behaviors of a sample of 1319 university students. Unpublished data collected for this text. Department of Sociology, East Carolina University, Greenville, NC.

Knox, D., M. Zusman, V. Daniels, and A. Brantley. 2002. Absence makes the heart grow fonder? Long-distance dating relationships among college students. *College Student Journal* 36:365–67.

Lara, A. 2005. "One for the price of two: Some couples find their marriages thrive when they share separate quarters." *San Francisco Chronicle,* June 29. http://www.sfgate.com/cgi-bin/article.cgi?file=/c/a/2005/06/29/HOG7HDEB7B1.DTL.

Lucas, R. E., A. E. Clark, Y. Georgellis, and E. Diener. 2003. Reexamining adaptation and the set point model of happiness: Reactions to changes in marital status. *Journal of Personality and Social Psychology* 84:527–39.

Luff, T., and K. Hoffman. 2006. College dating patterns: Cultural and structural influences. Roundtable presentation, Southern Sociological Society, New Orleans, LA, March 24.

Madden, M., and A. Lenhart 2006. *Online dating.* Washington, DC: Pew Internet & American Life Project.

Mahoney, S. 2006. The secret lives of single women—lifestyles, dating and romance: A study of midlife singles. *AARP: The Magazine* May/June, 62–69.

Manning, W. D., M. A. Longmore, and P. C. Giordano. 2007. The changing institution of marriage: Adolescents' expectations to cohabit and to marry. *Journal of Marriage and Family* 69:559–75.

McGinty, C. 2006. Internet dating. Presentation to courtship and marriage class, East Carolina University, Greenville, NC.

McKenna, K. Y. A. 2007. A progressive affair: Online dating to real world mating. In *Online Matchmaking,* ed. M. T. Whitty, A. J. Baker, and J. A. Inman, 112–24. New York: Palgrave Macmillan.

Ohayon, M. 2005. *Cowboy del Amor* (video released in 2005).

Oppenheimer, V. K. 2003. Cohabiting and marriage during young men's career-development process. *Demography* 40:127–49.

Pryor, J. H., S. Hurtado, L. DeAngelo, J. Sharkness, L. C. Romero, W. K. Korn, and S. Trans. 2008. *The American Freshmen: National Norms for Fall 2008.* Los Angeles: Higher Education Research Institute, UCLA.

Raley, R. K., M. L. Frisco, and E. Wildsmith. 2005. Maternal cohabitation and educational success. *Sociology of Education* 78:144–64.

Renshaw, S. W. 2005. "Swing Dance" and "Closing Time": Two ethnographies in popular culture. Dissertation Abstracts International. A: *The Humanities and Social Sciences* 25:4355A–56A.

Rhoades, G. K., S. M. Stanley, and H. J. Markman. 2009. Couples' reasons for cohabitation: Associations with individual well-being and relationship quality. *Journal of Family Issues* 30:233–46.

Sassler, S., and A. Cunningham. 2008. How cohabitors view childrearing. *Sociological Perspectives* 51:3–29.

Schoen, R., N. S. Landale, and K. Daniels. 2007. Family transitions in young adulthood. *Demography* 44:807–30.

Sharp, E. A., and L. Ganong. 2007. Living in the gray: Women's experiences of missing the marital transition. *Journal of Marriage and Family* 69:831–44.

Skinner, K. B., S. J. Bahr, D. R. Crane, and V. R. A. Call. 2002. Cohabitation, marriage, and remarriage. *Journal of Family Issues* 23:74–90.

Smock, P. J. 2000. Cohabitation in the United States: An appraisal of research themes, findings, and implications. *Annual Review of Sociology* 26:1–20.

Spitzberg, B. H., and W. R. Cupach. 2007. Cyberstalking as (mis) matchmaking. In *Online matchmaking,* ed. M. T. Whitty, A. J. Baker, and J. A. Inman, 127–46. New York: Palgrave Macmillan.

Statistical Abstract of the United States, 2009. 128th ed. Washington, DC: U.S. Bureau of the Census.

Stepp, L. S. 2007. *Unhooked: How young women pursue sex, delay love, and lose at both.* Riverhead, New York: Riverhead Publishing Co.

Stringfield, M. D. 2008. Online dating. Unpublished paper, East Carolina University.

Thomas, M. E. 2005. Girls, consumption space and the contradictions of hanging out in the city. *Social and Cultural Geography* 6:587–605.

Twenge, J. 2006. *Generation me.* New York: Free Press.

Vanderkam, L. 2006. Love (or not) in an iPod world. *USA Today* February 14, 13A.

Weden, M., and R. T. Kimbro. 2007. Racial and ethnic differences in the timing of first marriage and smoking cessation. *Journal of Marriage and Family* 69:878–87.

Whitty, M. T. 2007. The art of selling one's "self" on an online dating site: The BAR approach. In *Online matchmaking,* ed. M. T. Whitty, A. J. Baker, and J. A. Inman, 37–69. New York: Palgrave Macmillan.

Wienke, C. and G. J. Hill. 2009. Does the "marriage benefit" extend to partners in gay and lesbian relationships?: Evidence from a random sample of sexually active adults. *Journal of Family Issues* 30:259–73.

Wilson, G. D., J. M. Cousins, and B. Fink. 2006. The CQ as a predictor of speed-date outcomes. *Sexual & Relationship Therapy* 21:163–69.

Yeoman, B. 2006. Rethinking the commune. *AARP: The Magazine* March/April, 88–97.

Yurchisin, J., K. Watchravesrighkan, and D. M. Brown. 2005. An exploration of identity recreation in the context of Internet dating. *Social Behavior and Personality: An International Journal* 33:735–50.

Chapter 4

Alexandrov, E. Q. 2005. Couple attachment and the quality of marital relationships. *Attachment and Human Development* 7:123–52.

Behringer, A. M. 2005. Bridging the gap between Mars and Venus: A study of communication meanings in marriage. *Dissertation Abstracts International, A: The Humanities and Social Sciences* 65:4007A–8A.

Berle, W., with B. Lewis. 1999. *My father uncle Miltie.* New York: Barricade Books, Inc.

Bos, E. H., A. L. Bouhuys, E. Geerts, T. W. D. P. Van Os, and J. Ormel. 2007. Stressful life events as a link between problems in nonverbal communication and recurrence of depression. *Journal of Affective Disorders* 97:161–69.

Campbell, S. 2005. *Seven keys to authentic communication and relationship satisfaction.* New York: New World Library.

Caughlin, J. P., and M. B. Ramey. 2005. The demand/withdraw pattern of communication in parent-adolescent dyads. *Personal Relationships* 12:337–55.

Day, L., and J. Maltby. 2005. Forgiveness. *Journal of Psychology* 139:553–55.

DeMaria, R. M. 2005. Distressed couples and marriage education. *Family Relations* 54:242–53.

Dunbar, N. E., and J. K. Burgoon. 2005. Perceptions of power and interactional dominance in interpersonal relationships. *Journal of Social and Personal Relationships* 22:207–33.

Ennis, E., A. Vrij, and C. Chance. 2008. Individual differences and lying in everyday life. *Journal of Social and Personal Relationships* 25:105–118.

Finkenauer, C., and H. Hazam. 2000. Disclosure and secrecy in marriage: Do both contribute to marital satisfaction? *Journal of Social and Personal Relationships* 17:245–63.

Fonda, J. 2005. *My life, so far.* New York: Random House.

Gable, S. L., H. T. Reis, and G. Downey. 2003. He said, she said: A Quasi-signal detection analysis of daily interactions between close relationship partners. *Psychological Science* 14:100–05.

Gallmeier, C. P., M. E. Zusman, D. Knox, and L. Gibson. 1997. Can we talk? Gender differences in disclosure patterns and expectations. *Free Inquiry in Creative Sociology* 25:129–225.

Gibbs, J. L., N. B. Ellison, and R. D. Heino. 2006. Self-presentation in online personals: The role of anticipated future interaction, self-disclosure, and perceived success in Internet dating. *Communication Research* 33:152–77.

Gordon, K. C., D. H. Baucom, and D. K. Snyder. 2005. Treating couples recovering from infidelity: An integrative approach. *Journal of Clinical Psychology* 61:1393–405.

Gottman, J. 1994. *Why marriages succeed or fail.* New York: Simon & Schuster.

Greeff, A. P., and T. De Bruyne. 2000. Conflict management style and marital satisfaction. *Journal of Sex and Marital Satisfaction* 26:321–34.

Haas, S. M., and L. Stafford. 2005. Maintenance behaviors in same-sex and marital relationships: A matched sample comparison. *Journal of Family Communication* 5:43–60.

Henry, R. G., R. B. Miller, and R. Giarrusso. 2005. Difficulties, disagreements, and disappointments in late-life marriages. *International Journal of Aging & Human Development* 61:243–65.

Hertenstein, M. J., M. J. Hertenstein, J. M. Verkamp, A. M. Kerestes, and R. M. Holmes. 2007. The communicative functions of touch in humans, nonhuman primates, and rats: A review and synthesis of the empirical research. *Genetic Social and General Psychology Monographs* 132:5–94.

Hetherington, E. M. 2003. Intimate pathways: Changing patterns in close personal relationships across time. *Family Relations* 52:318–31.

Ingoldsby, B. B., G. T. Horlacher, P. L. Schvaneveldt, and M. Matthews. 2005. Emotional expressiveness and marital adjustment in Ecuador. *Marriage and Family Review* 38:25–44.

Knobloch, L. K. 2008. The content of relational uncertainty within marriage. *Journal of Social and Personal Relationships* 25:467–95.

Knox, D., and Zusman, M. E. 2009. Relationship and sexual behaviors of a sample of 1,319 university students. Unpublished data collected for this text. Department of Sociology, East Carolina University, Greenville, NC.

Knox, D., C. Schacht, J. Turner, and P. Norris. 1995. College students' preference for win-win relationships. *College Student Journal* 29:44–6.

Kurdek, L. A. 1994. Areas of conflict for gay, lesbian, and heterosexual couples: What couples argue about influences relationship satisfaction. *Journal of Marriage and the Family* 56:923–34.

———. 1995. Predicting change in marital satisfaction from husbands' and wives' conflict resolution styles. *Journal of Marriage and the Family* 57:153–64.

Marano, H. E. 1992. The reinvention of marriage. *Psychology Today* January/February, 49.

Marchand, J. F., and E. Hock. 2000. Avoidance and attacking conflict-resolution strategies among married couples: Relations to depressive symptoms and marital satisfaction. *Family Relations* 49:201–6.

McAlister, A. R., N. Pachana, and C. J. Jackson. 2005. Predictors of young dating adults' inclination to engage in extra dyadic sexual activities: A multi-perspective study. *British Journal of Psychology* 96:331–50.

McKenna, K. Y. A., A. S. Green, and M. E. J. Gleason. 2002. Relationship formation on the Internet: What's the big attraction? *Journal of Social Issues* 58:9–22.

Patford, J. L. 2000. Partners and cross-sex friends: A preliminary study of the way marital and de facto partnerships affect verbal intimacy with cross-sex friends. *Journal of Family Studies* 6:106–19.

Preston, P. 2005. Nonverbal communication: Do you really say what you mean? *Journal of Healthcare Management* 50:83–7.

Punyanunt-Carter, N. N. 2006. An analysis of college students' self-disclosure behaviors. *College Student Journal* 40:329–31.

Purkett, T. 2009. Sexually transmitted infections. Presentation to Courtship and Marriage class, Spring.

Rayer, A. J., and B. L. Volling. 2005. The role of husbands' and wives' emotional expressivity in the marital relationship. *Sex Roles: A Journal of Research* 52:577–88.

Rhoades, G. K., S. M. Stanley, and H. J. Markman. 2009. Couples' reasons for cohabitation: Associations with individual well-being and relationship quality. *Journal of Family Issues* 30:233–46.

Robbins, C. A. 2005. ADHD couple and family relationships: Enhancing communication and understanding through Imago Relationship Therapy. *Journal of Clinical Psychology* 61:565–78.

Rosof, F. 2005. An investigation of the therapeutic role of communication in couple relationships. Dissertation, Union Institute US. Dissertation Abstract International, Section B, Vol. 25 (8-B):4302.

Sanford, K. 2007. Hard and soft emotion during conflict: Investigating married couples and other relationships *Personal Relationships* 14:65–90.

Seguin-Levesque, C., M. L. N. Laliberte, L. G. Pelletier, C. Blanchard, and R. J. Vallerand. 2003. Harmonious and obsessive passion for the Internet: Their associations with the couple's relationship. *Journal of Applied Social Psychology* 33:197–221.

Shinn, L. K., and M. O'Brien 2008. Parent-child conversational styles in middle childhood: Gender and social class differences. *Sex Roles* 59:61–9.

Stanley, S. M., H. J. Markman, and S. W. Whitton. 2002. Communication, conflict, and commitment: Insights on the foundations of relationship success from a national survey. *Interpersonal Relations* 41:659–66.

Tannen, D. 1990. *You just don't understand: Women and men in conversation.* London: Virago.

_____. 2006. *You're wearing that? Understanding mothers and daughters in conversation.* New York: Random House.

Toussaint, L., and J. R. Webb. 2005. Gender differences in the relationship between empathy and forgiveness. *Journal of Social Psychology* 145:673–85.

Turner, A. J. 2005. Communication basics. Personal communication.

Vail-Smith, K., L. MacKenzie, and D. Knox. Forthcoming. The illusion of safety in "monogamous" undergraduates. *American Journal of Health Behavior* 34.

Walther, J. B., T. Loh, and L. Granka. 2005. Let me count the ways: The interchange of verbal and nonverbal cues in computer-mediated and face-to-face affinity. *Journal of Language and Social Psychology* 34:36–65.

Chapter 5

Ackerman, D. 1994. *A natural history of love.* New York: Random House.

Amato, P. R., A. Booth, D. R. Johnson, and S. F. Rogers. 2007. *Alone together: How marriage in America is changing.* Cambridge, Massachusetts: Harvard University Press.

Ambwani, S., and J. Strauss. 2007. Love thyself before loving others? A qualitative and quantitative analysis of gender differences in body image and romantic love. *Sex Roles* 56:13–22.

Assad, K. K., M. B. Donnellan, and R. D. Conger. 2007. Optimism: An enduring resource for romantic relationships. *Journal of Personality and Social Psychology* 93:285–96.

Barelds, D. P. H., and P. Barelds-Dijkstra. 2007. Love at first sight or friends first? Ties among partner personality trait similarity, relationship onset, relationship quality, and love. *Journal of Social and Personal Relationships* 24:479–96.

_____. 2007. Relations between different types of jealousy and self and partner perceptions of relationship quality. *Clinical Psychology & Psychotherapy* 14:176–88.

Bianchi, A. J., and L. Povilavicius. 2006. Cultural meanings versus academic meaning of self-esteem. Paper presented for Southern Sociological Society, New Orleans, LA, March 24.

Billingham, R. E., P. B. Perera, and N. A. Ehlers. 2005. College women's rankings of the most undesirable marriage and family forms. *College Student Journal* 39:749–53.

Bogg, R. A., and J. M. Ray. 2006. The heterosexual appeal of socially marginal men. *Deviant Behavior* 27:457–77.

Brantley, A., D. Knox, and M. E. Zusman. 2002. When and why gender differences in saying "I love you" among college students. *College Student Journal* 36:614–15.

Bratter, J. L., and R. B. King. 2008. "But Will It Last?": Marital instability among interracial and same-race couples. *Family Relations* 57:160–71.

Brehm, S. S. 1992. *Intimate relationships,* 2nd ed. New York: McGraw-Hill.

Bulcroft, R., K. Bulcroft, K. Bradley, and C. Simpson. 2000. The management and production of risk in romantic relationships: A postmodern paradox. *Journal of Family History* 25:63–92.

Burford, M. 2005. The un-Hollywood wife. In *The Oprah Magazine Book*, 174. Birmingham, Alabama: Oxmoor House.

Busby, D. M., D. C. Ivey, S. M. Harris, and C. Ates. 2007. Self-directed, therapist-directed, and assessment-based interventions for premarital couples. *Family Relations* 56: 279–90.

Buss, D. M. 2000. Prescription for passion. *Psychology Today* May/June, 54–61.

Carroll, J. S., and W. J. Doherty. 2003. Evaluating the effectiveness of premarital prevention programs: A meta-analytic review of outcome research. *Family Relations* 52:105–18.

Cassidy, M. L., and G. Lee. 1989. The study of polyandry: A critique and synthesis. *Journal of Comparative Family Studies* 20:1–11.

Chu, K. 2007. As higher education costs rise, so do debt loads. *USA Today*, May 25, 3B.

Clarkwest, A. 2007. Spousal dissimilarity, race, and marital dissolution. *Journal of Marriage and the Family* 69:639–53.

Crowell, J. A., D. Treboux, and E. Waters. 2002. Stability of attachment representations: The transition to marriage. *Developmental Psychology* 38:467–79.

DeCuzzi, A., D. Knox, and M. Zusman. 2006. Racial differences in perceptions of women and men. *College Student Journal* 40:343–49.

DeHart, T., S. L. Murray, B. W. Pelham, and P. Rose. 2002. The regulation of dependency in parent-child relationships. *Journal of Experimental Social Psychology* 39:59–67.

Demir, M. 2008. Sweetheart, you really make me happy: Romantic relationship quality and personality as predictors of happiness among emerging adults. *Journal of Happiness Studies* 9:257–77.

Diamond, L. M. 2003. What does sexual orientation orient? A biobehavioral model distinguishing romantic love and sexual desire. *Psychological Review* 110:173–92.

Dotson-Blake, K., D. Knox, and A. Holman. 2008. College student attitudes toward marriage, family, and sex therapy. Unpublished data from 288 undergraduate/graduate students. East Carolina University, Greenville, NC.

Dugan, J. 2005. Colonial America. http://www.suite101.com/article.cfm/colonial_america_retired/61531/1 (retrieved on December 12).

Duncan, S. F., T. B. Holman, and C. Yang. 2007. Factors associated with involvement in marriage preparation programs. *Family Relations* 56:270–78.

East, L., D. Jackson, L. O'Brien, and K. Peters. 2007. Use of the male condom by heterosexual adolescents and young people: Literature review. *Journal of Advanced Nursing* 59:103–10.

Edwards, T. M. 2000. Flying solo. *Time*, August 28, 47–53.

Fehr, B., and C. Harasymchuk. 2005. The experience of emotion in close relationships: Toward an integration of the emotion-in-relationships and interpersonal script models. *Personal Relationships* 12:181–96.

Fisher, H. 2009. *Why him? Why her?* New York: Henry Holt and Company.

Foster, J. D. 2008. Incorporating personality into the investment model: Probing commitment processes across individual differences in narcissism. *Journal of Social and Personal Relationships* 25:211–23.

Fuller, J. A., and R. M. Warner. 2000. Family stressors as predictors of codependency. *Genetic, Social, and General Psychology Monographs* 126:5–22.

Gallmeier, C. P., M. E. Zusman, D. Knox, and L. Gibson. 1997. Can we talk? Gender differences in disclosure patterns and expectations. *Free Inquiry in Creative Sociology* 25:219–25.

Gattis, K. S., S. Berns, L. E. Simpson, and A. Christensen. 2004. Birds of a feather or strange birds? Ties among personality dimensions, similarity, and marital quality. *Journal of Family Psychology* 18:564–78.

Gavin, J. 2003. *Deep in a dream: The long night of Chet Baker.* New York: Welcome Rain.

Guerrero, L. K., M. R. Trost, and S. M. Yoshimura. 2005. Romantic jealousy: Emotions and communicative responses. *Personal Relationships* 12:233–52.

Halpern, C. T., R. B. King, S. G. Oslak, and J. R. Udry. 2005. Body mass index, dieting, romance, and sexual activity in adolescent girls: Relationships over time. *Journal of Research on Adolescence* 15:535–59.

Haring, M., P. L. Hewitt, and G. L. Flett. 2003. Perfectionism, coping, and quality of relationships. *Journal of Marriage and the Family* 65:143–59.
Hart, K. 2007. Love by arrangement: The ambiguity of "spousal choice" in a Turkish village. *Journal of the Royal Anthropological Institute* 13:345–63.
Haynie, D. L., P. C. Giordano, W. D. Manning, and M. A. Longmore. 2005. Adolescent romantic relationships and delinquency involvement. *Criminology* 43:177–210.
Hendrick, S. S., C. Hendrick, and N. L. Adler. 1988. Romantic relationships: Love, satisfaction, and staying together. *Journal of Personality and Social Psychology* 54:980–88.
Hohmann-Marriott, B. E., and P. Amato. 2008. Relationship quality in interethnic marriages and cohabitation. *Social Forces* 87: 825–55.
Huston, T. L., J. P. Caughlin, R. M. Houts, S. E. Smith, and L. J. George. 2001. The connubial crucible: Newlywed years as predictors of marital delight, distress, and divorce. *Journal of Personality and Social Psychology* 80:237–52.
Ingoldsby, B., P. Schvaneveldt, and C. Uribe. 2003. Perceptions of acceptable mate attributes in Ecuador. *Journal of Comparative Family Studies* 34:171–86.
Jefson, C. 2006. Candy hearts: Messages about love, lust, and infatuation. *Journal of School Health* 76:117–22.
Jones, D. 2006. One of USA's exports: Love, American style. *USA Today*, February 14, 1B.
Kalmijn, M., and H. Flap. 2001. Assortative meeting and mating: Unintended consequences of organized settings for partner choices. *Social Forces* 79:1289–312.
Khanchandani, L. 2005. Jealousy during dating among college women. Paper presented at Third Annual East Carolina University Undergraduate Research and Creative Activities Symposium, Greenville, NC, April 8.
Kito, M. 2005. Self-disclosure in romantic relationships and friendships among American and Japanese college students. *Journal of Social Psychology* 145:127–40.
Knox, D., R. Breed, and M. Zusman. 2007. College men and jealousy. *College Student Journal* 41:494–98.
Knox, D., and K. McGinty. Forthcoming. Searching for homogamy: An in-class exercise. *College Student Journal.*
Knox, D., and M. E. Zusman. 2009. Relationship and sexual behaviors of a sample of 1,319 university students. Unpublished data. Department of Sociology, East Carolina University, Greenville, NC.
Knox, D., M. E. Zusman, L. Mabon, and L. Shivar. 1999. Jealousy in college student relationships. *College Student Journal* 33:328–29.
Knox, D., M. Zusman, and W. Nieves. 1998. What I did for love: Risky behavior of college students in love. *College Student Journal* 32:203–05.
Knox, D., M. E. Zusman, and W. Nieves. 1997. College students' homogamous preferences for a date and mate. *College Student Journal* 31:445–48.
Lee, J. A. 1973. *The colors of love: An exploration of the ways of loving.* Don Mills, Ontario: New Press.
———. 1988. Love-styles. In *The psychology of love*, ed. R. Sternberg and M. Barnes, 38–67. New Haven, CT: Yale University Press.
Lewis, S. K., and V. K. Oppenheimer. 2000. Educational assortative mating across marriage markets: Non-Hispanic whites in the United States. *Demography* 37:29–40.
Luo, S. H., and E. C. Klohnen. 2005. Assortative mating and marital quality in newlyweds: A couple-centered approach. *Journal of Personality and Social Psychology* 88:304–26.
Medora, N. P., J. H. Larson, N. Hortacsu, and P. Dave. 2002. Perceived attitudes towards romanticism: A cross-cultural study of American, Asian-Indian, and Turkish young adults. *Journal of Comparative Family Studies* 33:155–78.
Meehan, D., and C. Negy. 2003. Undergraduate students' adaptation to college: Does being married make a difference? *Journal of College Student Development* 44:670–90.
Meyer, J. P., and S. Pepper. 1977. Need compatibility and marital adjustment in young married couples. *Journal of Personality and Social Psychology* 35:331–42.
Paul, E. L., B. McManus, and A. Hayes. 2000. "Hookups": Characteristics and correlates of college students' spontaneous and anonymous sexual experiences. *Journal of Sex Research* 37:76–88.
Petrie, J., J. A. Giordano, and C. S. Roberts. 1992. Characteristics of women who love too much. *Affilia: Journal of Women and Social Work* 7:7–20.
Pew Research Center. 2007. Trends in political values and core attitudes: 1987–2007. Washington DC: Pew Research Center for the People & the Press.
Picca, L. H., and J. R. Feagin. 2007. *Two-faced racism.* New York: Routledge.
Pimentel, E. E. 2000. Just how do I love thee? Marital relations in urban China. *Journal of Marriage and the Family* 62:32–47.
Pines, A. M. 1992. *Romantic jealousy: Understanding and conquering the shadow of love.* New York: St. Martin's Press.
Radmacher, K., and M. Azmitia. 2006. Are there gendered pathways to intimacy in early adolescents' and emerging adults' friendships? *Journal of Adolescent Research* 21:415–48.
Redbook. 2005. Survey on "Description of Relationship with a Partner." August, 14.
Regan, P. C., and A. Joshi. 2003. Ideal partner preferences among adolescents. *Social Behavior and Personality* 31:13–20.
Ridley, C. 2009. Personal communication with Dr. Ridley who retired from University of Arizona in 2009.
Ross, C. B. 2006. An exploration of eight dimensions of self-disclosure on relationship. Paper for Southern Sociological Society, New Orleans, LA. March 24.
Sack, K. 2008. Health benefits inspire rush to marry, or divorce. *The New York Times*, August 12.
Saint, D. J. 1994. Complementarity in marital relationships. *Journal of Social Psychology* 134:701–4.
Schoebi, D. 2008. The coregulation of daily affect in marital relationships. *Journal of Family Psychology* 22:595–604.
Skowron, E. A. 2000. The role of differentiation of self in marital adjustment. *Journal of Counseling Psychology* 47:229–37.
Snyder, D. K., and J. M. Regts. 1990. Personality correlates of marital dissatisfaction: A comparison of psychiatric, maritally distressed, and nonclinic samples. *Journal of Sex and Marital Therapy* 90:34–43.
Soriano, C. G. 2005. Prince Charles and Camilla to wed. *USA Today*, A1.
Sprecher, S. 2002. Sexual satisfaction in premarital relationships: Associations with satisfaction, love, commitment, and stability. *Journal of Sex Research* 39:190–96.
Sprecher, S., and P. C. Regan. 2002. Liking some things (in some people) more than others: Partner preferences in romantic relationships and friendships. *Journal of Social and Personal Relationships* 19:463–81.
Statistical Abstract of the United States, 2009, 128th ed. Washington, DC: U.S. Bureau of the Census.
Sternberg, R. J. 1986. A triangular theory of love. *Psychological Review* 93:119–35.
Strassberg, D. S., and S. Holty. 2003. An experimental study of women's Internet personal ads. *Archives of Sexual Behavior* 32:253–61.
Swenson, D., J. G. Pankhurst, and S. K. Houseknecht. 2005. Links between families and religion. In *Sourcebook of family theory & research*, ed. V. L. Bengtson, A. C. Acock, K. R. Allen, P. Dilworth-Anderson, and D. M. Klein, 530–33. Thousand Oaks, California: Sage Publications.
Teachman, J. 2003. Premarital sex, premarital cohabitation, and the risk of subsequent marital disruption among women. *Journal of Marriage and the Family* 65:444–55.

Tilley, D. S., and M. Brackley. 2005. Men who batter intimate partners: A grounded theory study of the development of male violence in intimate partner relationships. *Issues in Mental Health Nursing* 26:281–97.

Toro-Morn, M., and S. Sprecher. 2003. A cross-cultural comparison of mate preferences among university students: The United States versus the People's Republic of China (PRC). *Journal of Comparative Family Studies* 34:151–62.

Trachman, M., and C. Bluestone. 2005. What's love got to do with it? *College Teaching* 53:131–36.

Twenge, J. 2006. *Generation me*. New York: Free Press.

Tzeng, O. C. S., K. Wooldridge, and K. Campbell. 2003. Faith love: A psychological construct in intimate relations. *Journal of the Indiana Academy of the Social Sciences* 7:11–20.

Vesselinov, E. 2008. Members only: Gated communities and residential segregation in the metropolitan United States. *Sociological Forum* 23:536–55.

Waller, W., and R. Hill. 1951. *The family: A dynamic interpretation*. New York: Holt, Rinehart and Winston.

Walster, E., and G. W. Walster. 1978. *A new look at love*. Reading, MA: Addison-Wesley.

West III, J. L. W. 2005. *The perfect hour*. New York: Random House.

White, T. 1990. *Rock lives*. New York: Henry Holt and Co.

Wilson, G. D., and J. M. Cousins. 2005. Measurement of partner compatibility: Further validation and refinement of the CQ test. *Sexual and Relationship Therapy* 20:421–29.

Winch, R. F. 1955. The theory of complementary needs in mate selection: Final results on the test of the general hypothesis. *American Sociological Review* 20:552–55.

Xie, Y., J. M. Raymo, K. Govette, and A. Thornton. 2003. Economic potential and entry into marriage and cohabitation. *Demography* 40:351–64.

Zusman, M. E., J. Gescheidler, D. Knox, and K. McGinty. 2003. Dating manners among college students. *Journal of Indiana Academy of Social Sciences* 7:28–32.

Chapter 6

Amato, P. R., A. Booth, D. R. Johnson, and S. F. Rogers. 2007. *Alone together: How marriage in America is changing*. Cambridge, MA: Harvard University Press.

Barnes, K., and J. Patrick. 2004. Examining age-congruency and marital satisfaction. *The Gerontologist* 44:185–87.

Billingsley, S., M. Lim, and G. Jennings. 1995. Themes of long-term, satisfied marriages consummated between 1952–1967. *Family Perspective* 29:283–95.

Blakely, K. 2008. Busy brides and the business of family life. *Journal of Family Issues* 29:639–43.

Brunsma, D. L. 2005. Interracial families and the racial identification of mixed-race children: Evidence from the early childhood longitudinal study. *Social Forces* 84:1131–57.

Bulanda, J. R., and S. L. Brown. 2007. Race-ethnic differences in marital quality and divorce. *Social Science Research* 36:945–59.

Burford, M. 2005. The un-Hollywood wife. In *The Oprah Magazine Book*, 174. Birmingham, AL: Oxmoor House.

Cadden, M., and D. Merrill 2007. What married people miss most (based on *Reader's Digest* study of 1,001 married adults). *USA Today*, D1.

Caforio, G. 2003. *Handbook on the sociology of the military*. New York: Kluwer Academic.

Coontz, S. 2000. Marriage: Then and now. *Phi Kappa Phi Journal* 80:16–20.

Corra, M., S. Carter, J. S. Carter, and D. Knox. Forthcoming. Trends in marital happiness by sex and race, 1973–2006. *Journal of Family Issues*.

Cozza, S. J., R. S. Chun, and J. A. Polo. 2005. Military families and children during operation Iraqi freedom. *Psychiatric Quarterly* 76:371–78.

DeMaria, R. M. 2005. Distressed couples and marriage education. *Family Relations* 54:242–53.

DeOllos, I. Y. 2005. Predicting marital success or failure: Burgess and beyond. In *Sourcebook of family theory and research*, ed. Vern L. Bengtson, Alan C. Acock, Katherine R. Allen, Peggye Dilworth-Anderson, and David M. Klein, 134–36. Thousand Oaks, CA: Sage Publications.

Dew, J. 2008. Debt change and marital satisfaction change in recently married couples. *Family Relations* 57:60–71.

Dowd, D. A., M. J. Means, J. F. Pope, and J. H. Humphries. 2005. Attributions and marital satisfaction: The mediated effects of self-disclosure. *Journal of Family and Consumer Sciences* 97:22–27.

Easterling, B. A. 2005. *The invisible side of military careers: An examination of employment and well-being among military spouses*. PhD diss., University of North Florida.

Fisher, T. D., and J. K. McNulty. 2008. Neuroticism and marital satisfaction: The mediating role played by the sexual relationship. *Journal of Family Psychology* 22:112–23.

Fu, X. 2006. Impact of socioeconomic status on inter-racial mate selection and divorce. *Social Science Journal* 43:239–58.

Gaines, S. O., Jr., and J. Leaver. 2002. Interracial relationships. In *Inappropriate relationships: The unconventional, the disapproved, and the forbidden*, ed. R. Goodwin and D. Cramer, 65–78. Mahwah, NJ: Lawrence Erlbaum Associates.

Gibson, V. 2002. *Cougar: A guide for older women dating younger men*. Boston, MA: Firefly Books.

Glover, K. R., M. Phelps, A. Sean Burleson, and A. Dessie. 2006. Friends or lovers: Understanding the cognitive underpinnings of the transition from affiliation to care-giving in romantic relationships. 4th Annual ECU Research and Creative Activities Symposium, April 21, East Carolina University, Greenville, NC.

Gottman, J., and S. Carrere. 2000. Welcome to the love lab. *Psychology Today* September/October, 42.

Hill, M. R., and V. Thomas. 2000. Strategies for racial identity development: Narratives of black and white women in interracial partner relationships. *Family Relations* 49:193–200.

Huyck, M. H., and D. L. Gutmann. 1992. Thirty 'something years of marriage: Understanding experiences of women and men in enduring family relationships. *Family Perspective* 26:249–65.

Karney, B. R., and J. S. Crown. 2007. *Families under stress: An assessment of data, theory, and research on marriage and divorce in the military*. Rand Corporation: National Defense Research Institute; Prepared for the Office of the Secretary of Defense. http://www.rand.org/pubs/monographs/2007/RAND_MG599.pdf (accessed November 18, 2008).

Kennedy, R. 2003. *Interracial intimacies*. New York: Pantheon.

Knox, D., and M. E. Zusman. 2009. Relationship and sexual behaviors of a sample of 1,319 university students. Data collected for this text. Department of Sociology, East Carolina University, Greenville, NC.

Kreaer, D. A. 2008. Guarded borders: Adolescent interracial romance and peer trouble at school. *Social Forces* 87: 887–910.

Lindquist, J. H. 2004. When race makes no difference: Marriage and the military. *Social Forces* 83:731–57.

Macomber, J. E., J. Murray, and M. Stagner. 2005. Investigation of programs to strengthen and support healthy marriages. http://www.urban.org/url.cfm?ID/411141.

Marklein, M. B. 2008. High mark for foreign students here. *USA Today*, November 17, 4D.

Martin, J. A., and P. McClure. 2000. Today's active duty military family: The evolving challenges of military family life. In *The military family: A practice guide for human service providers*, ed. J. A. Martin, L. N. Rosen, and L. R. Sparacino, 3–24. Connecticut: Praeger Publishers.

McNulty, J. K. 2008. Forgiveness in marriage: Putting the benefits into context. *Journal of Family Psychology* 22:171–83.

Michael, R. T., J. H. Gagnon, E. O. Laumann, and G. Kolata. 1994. *Sex in America: A definitive survey.* Boston, MA: Little, Brown.
Military Family Resource Center. 2003. Active Duty Families in 2003 Demographics Report. http://www.mfrc-dodqol.org/pdffiles/demo2003/SectionIIIActiveDutyFamilies.pdf (accessed January 30, 2005).
Montemurro, B. 2006. *Something old, something bold.* New Brunswick, NJ: Rutgers University Press.
Morr Serewicz, M. C., and D. J. Canary. 2008. Assessments of disclosure from the in-laws: Links among disclosure topics, family privacy orientations, and relational quality. *Journal of Social and Personal Relationships* 25:333–57.
Murdock, G. P. 1949. *Social structure.* New York: Free Press.
NCFR Policy Brief. 2004. *Building strong communities for military families.* Minneapolis, MN: National Council on Family Relations.
North, R. J., C. J. Holahan, R. H. Moos, and R. C. Cronkite. 2008. Family support, family income, and happiness: A 10-year perspective. *Journal of Family Psychology* 22:475–83.
Nuner, J. E. 2004. A qualitative study of mother-in-law/daughter-in-law relationships. *Dissertation Abstracts International, A: The Humanities and Social Sciences,* 65(August): 712A–13A.
Page, S. 2006. War has hurt USA. *USA Today*, March 17, A1.
Plagnol, A. C., and R. A. Easterlin. 2008. Aspirations, attainments, and satisfaction: Life cycle differences between American women and men. *Journal of Happiness Studies.* Published online July 2008.
Pollard, M., B. Karney, and D. Loughran. n.d. Comparing Rates of Marriage and Divorce in Civilian, Military, and Veteran Populations. http://paa2008.princeton.edu/download.aspx?submissionId-81696 (accessed November 18, 2008).
Rowland, I. 2006. Choosing to have children or choosing to be child-free: Australian students' attitudes towards the decisions of heterosexual and lesbian women. *Australian Psychologist* 41:55–59.
Saleska, S. 2004. Exploratory study of problems and stresses dependent military spouses experience. Paper presented at the Second Annual East Carolina University Research and Scholarship Day, March 26, Greenville, NC.
Sayer, L. C., S. M. Bianchi, and J. P. Robinson. 2004. Are parents investing less in children? Trends in mothers' and fathers' time with children. *American Journal of Sociology* 110:1–43.
Sherif-Trask, B. 2003. Love, courtship, and marriage from a cross-cultural perspective: The upper middle class Egyptian example. In *Mate Selection Across Cultures* ed. R. R. Hamon and B. B. Ingoldsby, 121–36. Thousand Oaks, CA: Sage Publications.
Sobal, J., C. F. Bove, and B. S. Rauschenbach. 2002. Commensal careers at entry into marriage: Establishing commensal units and managing commensal circles. *Sociological Review* 50:378–97.
Stuckey, D., and A. Gonzalez. 2006. Princeton Survey Research Associates Poll on marriage happiness. *USA Today*, March 7, 1A.
Treas, J., and D. Giesen. 2000. Sexual infidelity among married and cohabiting Americans. *Journal of Marriage and the Family* 62:48–60.
Vaillant, C. O., and G. E. Vaillant. 1993. Is the U-curve of marital satisfaction an illusion? A 40-year study of marriage. *Journal of Marriage and the Family* 55:230–39.
Wallerstein, J., and S. Blakeslee. 1995. *The good marriage.* Boston, MA: Houghton-Mifflin.
Weigel, D. J., and D. S. Ballard-Reisch. 2002. Investigating the behavioral indicators of relational commitment. *Journal of Social and Personal Relationships* 19:403–23.
Whiteman, S. D., S. M. McHale, and A. C. Crouter. 2007. Longitudinal changes in marital relationships: The role of offspring's pubertal development. *Journal of Marriage and Family* 69:1005–20.
Wilmoth, J., and G. Koso. 2002. Does marital history matter? Marital status and wealth outcomes among preretirement adults. *Journal of Marriage and the Family* 64:254–68.
Wilson, G., and J. Cousins. 2005. Measurement of partner compatibility; further validation and refinement of the CQ test. *Sexual and Relationship Therapy* 20:421–29.
Wilson, S. M., L. W. Ngige, and L. J. Trollinger. 2003. Kamba and Maasai paths to marriage in Kenya. In *Mate selection across cultures,* ed. R. R. Hamon and B. B. Ingoldsby, 95–117. Thousand Oaks, CA: Sage Publications.
Wolfe, D. 2006. Personal communication. Jacksonville, NC: Camp Lejeune Marine Corp Base.

Chapter 7

Amato, P. R. 2004. Tension between institutional and individual views of marriage. *Journal of Marriage and Family* 66:959–65.
American Psychological Association. 2004. *Sexual Orientation, Parents, & Children.* APA Policy Statement on Sexual Orientation, Parents, & Children. APA Online. http://www.apa.org.
Asch-Goodkin, J. 2006. An unsuccessful attempt to adopt a constitutional amendment that bans gay marriage. *Contemporary Pediatrics* 23:14–15.
Avery, A., J. Chase, L. Johansson, S. Litvak, D. Montero, and M. Wydra. 2007. America's changing attitudes toward homosexuality, civil unions, and same-gender marriages. *Social Work* 52:71–79.
Bartholomew, K., K. V. Regan, M. A. White, and D. Oram. 2008. Patterns of abuse in male same-sex relationships. *Violence and Victims* 23:617–37.
Berg, N., and D. Lein. 2006. Same-sex behaviour: U.S. frequency estimates from survey data with simultaneous misreporting and non-response. *Applied Economics* 39:757–70.
Besen, W. 2000. Introduction. In *Feeling free: Personal stories—How love and self-acceptance saved us from "ex-gay" ministries.* Washington, DC: Human Rights Campaign Foundation, 7.
Black, D., G. Gates, S. Sanders, and L. Taylor. 2000. Demographics of the gay and lesbian population in the United States: Evidence from available systematic data sources. *Demography* 37:139–54.
Bobbe, J. 2002. Treatment with lesbian alcoholics: Healing shame and internalized homophobia for ongoing sobriety. *Health and Social Work* 27:218–23.
Bontempo, D. E., and A. R. D'Augelli. 2002. Effects of at-school victimization and sexual orientation on lesbian, gay, or bisexual youths' health risk behavior. *Journal of Adolescent Health* 30:364–74.
Bradford, J., K. Barrett, and J. A. Honnold. 2002. *The 2000 census and same-sex households: A user's guide.* New York: National Gay and Lesbian Task Force Policy Institute, Survey and Evaluation Research Laboratory, and Fenway Institute. http://www.thetaskforce.org.
Buxton, A. P. 2004. Paths and pitfalls: How heterosexual spouses cope when their husbands or wives come out. *Journal of Couple and Relationship Therapy* 3:95–109.
Buxton, A. 2005. A family matter: When a spouse comes out as gay, lesbian, or bisexual. *Journal of GLBT Family Studies* 1:49–70.
Cahill, S., and S. Slater. 2004. *Marriage: Legal protections for families and children. Policy brief.* Washington, DC: National Gay and Lesbian Task Force Policy Institute.
Carpenter, C., and G. J. Gates. 2008. Gay and lesbian partnership: Evidence from California. *Demography* 45:573–691.
Centers for Disease Control and Prevention. 2005. *HIV/AIDS Surveillance Report* 2004, Vol. 16. http://www.cdc.gov.
Chase, B. 2000. NEA president Bob Chase's historic speech from 2000 GLSEN Conference. http://www.glsen.org.
Cianciotto, J., and S. Cahill. 2006. Youth in the crosshairs: The third wave of ex-gay activism. National Gay and Lesbian Task Force Policy Institute. http:www.thetaskforce.org.
Clunis, D. M., and G. Dorsey Green. 2003. *The lesbian parenting book,* 2nd ed. Emeryville, CA: Seal Press.

Crowl, A., S. Ahn, and J. Baker. 2008. A meta-analysis of developmental outcomes for children of same-sex and heterosexual parents. *Journal of GLBT Family Studies* 4:385–407.

Curtis, C. 2003. Poll: U.S. public is 50–50 on gay marriage. *PlanetOut* (October 7). http://www.planetout.com.

_____. 2004. Poll: 1 in 20 High school students is gay. *PlanetOut*. http://www.planetout.com.

Custody and Visitation. 2000. *Human Rights Campaign FamilyNet*. http://familynet.hrc.org.

Diamond, L. M. 2003. What does sexual orientation orient? A biobehavioral model distinguishing romantic love and sexual desire. *Psychological Review* 110:173–92.

Dozetos, B. 2001. School shooter taunted as 'gay.' *PlanetOut* (March 7). http:www.planetout.com.

Fone, B. 2000. *Homophobia: A history*. New York: Henry Holt.

Garnets, L., G. M. Herek, and B. Levy. 1990. Violence and victimization of lesbians and gay men: Mental health consequences. *Journal of Interpersonal Violence* 5:366–83.

Gay, Lesbian, and Straight Education Network. 2000. Homophobia 101: Teaching respect for all. Gay, Lesbian, and Straight Education Network. http:www.glsen.org.

Gilman, S. E., S. D. Cochran, V. M. Mays, M. Hughes, D. Ostrow, and R. C. Kessler. 2001. Risk of psychiatric disorders among individuals reporting same-sex sexual partners in the National Comorbidity Survey. *American Journal of Public Health* 91:933–39.

Golombok, S., B. Perry, A. Burston, C. Murray, J. Money-Somers, M. Stevens, and J. Golding. 2003. Children with lesbian parents: A community study. *Developmental Psychology* 39:20–33.

Gottman, J. M., R. W. Levenson, C. Swanson, K. Swanson, R. Tyson, and D. Yoshimoto. 2003. Observing gay, lesbian, and heterosexual couples' relationships: Mathematical modeling of conflict interaction. *Journal of Homosexuality* 45:65–91.

Green, J. C. 2004. The American religious landscape and political attitudes: a baseline for 2004. Pew Forum on Religion and Public Life. http://pewforum.org.

Green, R. J., J. Bettinger, and E. Sacks. 1996. Are lesbian couples fused and gay male couples disengaged? In *Lesbians and gays in couples and families*, ed. J. Laird and R. J. Green, 185–230. San Francisco: Jossey-Bass.

Held, Myka. 2005. Mix it up: T-shirts and activism (March 16). http://www.tolerance.org/teens.

Herek, G. M. 2002. Heterosexuals' attitudes toward bisexual men and women in the United States. *The Journal of Sex Research* 39:264–74.

Holthouse, D. 2005. Curious cures. *Intelligence Report* 117 (Spring):14.

Howard, J. 2006. *Expanding resources for children: Is adoption by gays and lesbians part of the answer for boys and girls who need homes?* New York: Evan B. Donaldson Adoption Institute.

Human Rights Campaign. 2000. *Feeling free: Personal stories—How love and self-acceptance saved us from "ex-gay" ministries*. Washington, DC: Human Rights Campaign Foundation.

_____. 2005. *The state of the workplace for lesbian, gay, bisexual, and transgendered Americans*, 2004. Washington, DC: Human Rights Campaign. http://www.hrc.org.

Israel, T., and J. J. Mohr. 2004. Attitudes toward bisexual women and men: Current research, future directions. In *Current research on bisexuality*, ed. R. C. Fox, 117–34. New York: Harrington Park Press.

Jenkins, M., E. G. Lambert, and D. N. Baker. 2009. The attitudes of black and white college students toward gays and lesbians. *Journal of Black Studies* 39:589–601.

Johnston, E. 2005. Massachusetts releases data on same-sex marriages. *PlanetOut* (May 5). http://www.planetout.com.

Kinsey, A. C., W. B. Pomeroy, and C. E. Martin. 1948. *Sexual behavior in the human male*. Philadelphia: Saunders.

Kinsey, A. C., W. B. Pomeroy, C. E. Martin, and P. H. Gebhard. 1953. *Sexual behavior in the human female*. Philadelphia: Saunders.

Kirkpatrick, R. C. 2000. The evolution of human sexual behavior. *Current Anthropology* 41:385–414.

Knox, D., and M. E. Zusman. 2009. Relationship and sexual behaviors of a sample of 1,319 university students. Unpublished data collected for this text. Department of Sociology, East Carolina University, Greenville, NC.

Kurdek, L. A. 2008. Change in relationship quality for partners from lesbian, gay male, and heterosexual couples. *Journal of Family Psychology* 22:701–11.

______. 2005. What do we know about gay and lesbian couples? Current Directions in *Psychological Science* 14:251–54.

______. 2004. Gay men and lesbians: The family context. In *Handbook of contemporary families: Considering the past, contemplating the future*, ed. M. Coleman and L. H. Ganong, 96–115. Thousand Oaks, CA: Sage Publications.

______. 1994. Conflict resolution styles in gay, lesbian, heterosexual nonparent, and heterosexual parent couples. *Journal of Marriage and the Family* 56:705–22.

Landis, D. 1999. Mississippi Supreme Court made a tragic mistake in denying custody to gay father, experts say. *American Civil Liberties Union News*, February 17. http://www.aclu.org.

LAWbriefs. 2005. Recent developments in sexual orientation and gender identity law. LAWbriefs 7(April):1.

Lever, J. 1994. The 1994 *Advocate* survey of sexuality and relationships: The men. *The Advocate* August 23, 16–24.

Loftus, J. 2001. America's liberalization in attitudes toward homosexuality, 1973 to 1998. *American Sociological Review* 66:762–82.

Luther, S. 2006. Domestic partner benefits. Human Rights Campaign (March). http://www.hrc.org.

Mathy, R. M. 2007. Sexual orientation moderates online sexual activity. In *Online Matchmaking* ed. M. T. Whitty, A. J. Baker, and J. A. Inman, 159–77. New York: Paulgrave Macmillan.

McKinney, J. 2004. *The Christian case for gay marriage*. Pullen Memorial Baptist Church (February 8). http://www.pullen.org.

McLean, K. 2004. Negotiating (non)monogamy: Bisexuality and intimate relationships. In *Current research on bisexuality*, ed. R. C. Fox, 82–97. New York: Harrington Park Press.

Meyer, I. H., J. Dietrich, and S. Schwartz. 2008. Lifetime prevalence of mental disorders and suicide attempts in diverse lesbian, gay, and bisexual populations. *American Journal of Public Health* 98:1004–03.

Michael, R. T., J. H. Gagnon, E. O. Laumann, and G. Kolata. 1994. *Sex in America: A definitive survey*. Boston: Little, Brown.

Mohipp, C., and M. M. Morry. 2004. Relationship of symbolic beliefs and prior contact to heterosexuals' attitudes toward gay men and lesbian women. *Canadian Journal of Behavioral Science* 36:36–44.

Moltz, K. 2005. Testimony of Kathleen Moltz, given before the United States Senate Judiciary Committee, Subcommittee on the Constitution, Civil Rights and Property Rights. Human Rights Campaign (April 13). http://www.hrc.org.

National Center for Lesbian Rights. 2003. Second-parent adoptions: A snapshot of current law. http://www.nclrights.org.

National Coalition of Anti-Violence Programs. 2005. 2004 National hate crimes report: Anti-lesbian, gay, bisexual and transgender violence in 2004. *New York: National Coalition of Anti-Violence Programs*.

National Gay and Lesbian Task Force. 2004. Anti-Gay parenting laws in the U.S. National Gay and Lesbian Task Force (June). http://www.thetaskforce.org.

_____. 2005. Second-parent adoption in the U.S. National Gay and Lesbian Task Force. http://www.thetaskforce.org.

_____. 2005–2006. Marriage and partnership recognition. http://www.thetaskforce.org.

Ochs, R. 1996. Biphobia: It goes more than two ways. In *Bisexuality: The psychology and politics of an invisible minority,* ed. B. A. Firestein, 217–39. Thousand Oaks, CA: Sage Publications.

Ogilvie, G. S., D. L. Taylor, T. Trussler, R. Marchand, M. Gilbert, A. Moniruzzaman, and M. L. Rekart. 2008. Seeking sexual partners on the Internet: A marker for risky sexual behaviour in men who have sex with men. *Canadian Journal of Public Health* 99:185–89.

Otis, M. D., S. S. Rostosky, E. D. B. Riggle, and R. Hamrin. 2006. Stress and relationship quality in same-sex couples. *Journal of Social and Personal Relationships* 23:81–99.

Page, S. 2003. Gay rights tough to sharpen into political "wedge issue." *USA Today*, July 28, 10A.

Palmer, R., and R. Bor. 2001. The challenges to intimacy and sexual relationships for gay men in HIV serodiscordant relationships: A pilot study. *Journal of Marital and Family Therapy* 27:419–31.

Parelli, S. 2007. Why ex-gay therapy doesn't work. *The Gay & Lesbian Review Worldwide* 14:29–32.

Paul, J. P. 1996. Bisexuality: Exploring/exploding the boundaries. In *The lives of lesbians, gays, and bisexuals: Children to adults,* ed. R. Savin-Williams and K. M. Cohen, 436–61. Fort Worth, TX: Harcourt Brace.

Pearcey, M. 2004. Gay and bisexual married men's attitudes and experiences: Homophobia, reasons for marriage, and self-identity. *Journal of GLBT Family Studies* 1:21–42.

Peplau, L. A., R. C. Veniegas, and S. N. Campbell. 1996. Gay and lesbian relationships. In *The lives of lesbians, gays, and bisexuals: Children to adults,* ed. R. C. Savin-Williams and K. M. Cohen, 250–73. Fort Worth, TX: Harcourt Brace.

Pew Research Center. 2008. Gay Marriage Opposed. http://pewforum.org/docs/index.php?DocID=39 (accessed November 16).

_____. 2006. Less opposition to gay marriage, adoption and military service (March 22). http://people-press.org.

Pinello, D. R. 2008. Gay marriage: For better or for worse? What we've learned from the evidence. *Law & Society Review* 42:227–30.

Platt, L. 2001. Not your father's high school club. *American Prospect* 12:A37–39.

Porche, M. V., and D. M. Purvin. 2008. "Never in Our Lifetime": Legal marriage for same-sex couples in long-term relationships. *Family Relations* 57:144–59.

Potok, M. 2005. Vilification and violence. *Intelligence Report* 117(Spring):1.

Pryor, J. H., S. Hurtado, L. DeAngelo, J. Sharkness, L. C. Romero, W. K. Korn, and S. Trans. 2008. *The American freshmen: National Norms for Fall 2008.* Los Angeles: Higher Education Research Institute, UCLA.

Puccinelli, M. 2005. Students support, decry gays with t-shirts. CBS 2 Chicago (April 19). http://cbs2chicago.com.

Rosin, H., and R. Morin. 1999. In one area, Americans still draw a line on acceptability. *Washington Post National Weekly Edition* 16(January 11):8.

Rosario, M., E. W. Schrimshaw, J. Hunter, and L. Braun. 2006. Sexual identity development among lesbian, gay, and bisexual youth: Consistency and change over time. *Journal of Sex Research* 43:46–58.

Saad, L. 2005. Gay rights attitudes a mixed bag. Gallup Organization (May 20). http://www.gallup.com.

Sanday, P. R. 1995. Pulling train. In *Race, class, and gender in the United States,* 3rd ed., ed. P. S. Rothenberg, 396–402. New York: St. Martin's Press.

Savin-Williams, R. C. 2006. Who's gay? Does it matter? Current Directions in *Psychological Science* 15:40–44.

Serovich, J. M., S. M. Craft, P. Toviessi, R. Gangamma, et al. 2008. A systematic review of the research base on sexual reorientation therapies. *Journal of Marital and Family Therapy* 34:227–39.

SIECUS (Sexuality Information and Education Council of the United States). 2009. Facts. Retrieved March 23. http://www.dianedew.com/siecus.htm

Statistical Abstract of the United States, 2009, 128th ed. Washington, DC: U.S. Bureau of the Census.

Sullivan, A. 1997. The conservative case. In *Same sex marriage: Pro and con,* ed. A. Sullivan, 146–54. New York: Vintage Books.

Tao, G. 2008. Sexual orientation and related viral sexually transmitted disease rates among U.S. women aged 15 to 44 years. *American Journal of Public Health* 98:1007–10.

Tobias, S., and S. Cahill. 2003. School lunches, the Wright brothers, and gay families. National Gay and Lesbian Task Force. http://www.thetaskforce.org.

Tyagart, C. E. 2002. Legal rights to homosexuals in areas of domestic partnerships and marriages: Public support and genetic causation attribution. *Educational Research Quarterly* 25:20–29.

Wagner, C. G. 2006. Homosexual relationships. *Futurist* 40:6.

Wainwright, J., S. T. Russell, and C. J. Patterson. 2004. Psychosocial adjustment, school outcomes, and romantic relationships of adolescents with same-sex parents. *Child Development* 75:1886–98.

Walsh, K. T. 2006. And now it's her turn. *U.S. News & World Report* (May 15):27.

Zaleski, R. M. 2007. What's in a name? (equal treatment for gay couples) (New Jersey) *New Jersey Law Journal,* March 30.

Chapter 8

Beckman, N. M., M. Waern, I. Skoog, and The Sahlgrenska Academy at Göteborg University, Sweden. 2006. Determinants of sexuality in 70 year olds. *The Journal of Sex Research* 43:2–3.

Bersamin, M., M. Todd, D. A. Fisher, D. L. Hill, J. W. Grube, and S. Walker. 2008. Parenting practices and adolescent sexual behavior: A longitudinal study. *Journal of Marriage and Family* 70:97–112.

Bristol, K., and B. Farmer. 2005. *Sexuality among Southeastern university students: A survey.* Unpublished data. Greenville, NC: East Carolina University.

Brucker, H., and P. Bearman 2005. After the promise: The STD consequences of adolescent virginity pledges. *Journal of Adolescent Health* 36:271–78.

Carpenter, L. M. 2003. Like a virgin . . . again? Understanding secondary virginity in context. Paper presented at the 73rd Annual Meeting of the Eastern Sociological Society, Philadelphia, February 28.

Colson, M., A. Lemaire, P. Pinton, K. Hamidi, and P. Klein. 2006. Sexual behaviors and mental perception, satisfaction, and expectations of sex life in men and women in France. *The Journal of Sexual Medicine* 3:121–31.

Cutler, W. B., E. Friedmann, and N. L. McCoy. 1998. Pheromonal influences on sociosexual behavior in men. *Archives of Sexual Behavior* 27:1–13.

Davidson, J. K., Sr., N. B. Moore, J. R. Earle, and R. Davis. 2008. Sexual attitudes and behavior at four universities: Do region, race, and/or religion matter? *Adolescence* 43:189–223.

DeLamater, J., and M. Hasday. 2007. The sociology of sexuality. In *21st century sociology: A reference handbook,* ed. C. D. Bryant and D. L. Peck, 254–64. Thousand Oaks, California: Sage Publications.

Eisenman, R., and M. L. Dantzker. 2006. Gender and ethnic differences in sexual attitudes at a Hispanic-serving university. *The Journal of General Psychology* 133:153–63.

England, P., and R. J. Thomas. 2006. The decline of the date and the rise of the college hook up. In *Family in transition,* 14th ed., ed. A. S. Skolnick and J. H. Skolnick, 151–62. Boston: Pearson Allyn & Bacon.

Gibbs, N. 2008. The pursuit of purity. *Time*, July 28, 46–49.

Graham, C. A., S. A. Sanders, R. R. Milhausen, and K. R. McBride. 2004. Turning on and turning off: A focus group study of the factors that affect women's sexual arousal. *Archives of Sexual Behavior* 33:527–38.

Grammer, K., F. Bernard, and N. Neave. 2005. Human pheromones and sexual attraction. *European Journal of Obstetrics & Gynecology and Reproductive Biology* 118:135–42.

Greene, K. and S. Faulkner. 2005. Gender, belief in the sexual double standard, and sexual talk in heterosexual dating relationships. *Sex Roles* 53:239–51.

Halstead, M. J. 2005. Teaching about love. *British Journal of Educational Studies* 53:290–305.

Hattori, M. K., and F. N. Dodoo. 2007. Cohabitation, marriage and sexual monogamy in Nairobi's slums. *Social Science and Medicine* 64:1067–72.

Heintz, A. J., and R. M. Melendez. 2006. Intimate partner violence and HIV/STD risk among lesbian, gay, bisexual and transgender individuals. *Journal of Interpersonal Violence* 21:193–208.

Hollander, D. 2006. Many teenagers who say they have taken a Virginity Pledge retract that statement after having intercourse. *Perspectives on Sexual and Reproductive Health* 38:168–73.

Hughes, M., K. Morrison, and K. J. Asada. 2005. What's love got to do with it? Exploring the impact of maintenance rules, love attitudes, and network support on friends with benefits relationships. *Western Journal of Communication* 69:49–66.

Impett, E. A., L. A. Peplau, and S. L. Gable. 2005. Approach and avoidance sexual motives: Implications for personal and interpersonal well-being. *Personal Relationships* 12:465–82.

Kim, J. L., C. L. Sorsoli, K. Collins, B. A. Zylbergold, D. Schooler, and D. L. Tolman. 2007. From sex to sexuality: Exposing the heterosexual script on primetime network television. *Journal of Sex Research* 44:145–57.

Knox, D., and M. E. Zusman. 2009. Relationship and sexual behaviors of a sample of 1,319 university students. Unpublished data collected for this text. Department of Sociology, East Carolina University, Greenville, NC.

Knox, D., M. Zusman, and A. McNeely. 2008. University student beliefs about sex: Men vs. women *College Student Journal* 42:181–85.

Kornreich, J. L., K. D. Hern, G. Rodriguez, and L. F. O'Sullivan. 2003. Sibling influence, gender roles, and the sexual socialization of urban early adolescent girls. *Journal of Sex Research* 40:101–10.

Laumann, E. O., A. Paik, D. B. Glasser, J.-H. Kang, T. Wang, B. Levinson, E. D. Moreira, Jr., A. Nicolosi, and C. Gingell. 2006. A cross-national study of subjective sexual well-being among older women and men: Findings from the global study of sexual attitudes and behaviors. *Archives of Sexual Behavior* (April).

Lenton, A. P., and A. Bryan. 2005. An affair to remember: The role of sexual scripts in perceptions of sexual intent. *Personal Relationships* 12:483–98.

Levin, R. 2004. Smells and tastes: their putative influence on sexual activity in humans. *Sexual & Relationship Therapy* 19:451–62.

Liu, C. 2003. Does quality of marital sex decline with duration? *Archives of Sexual Behavior* 32:55–60.

Lykins, A. D., E. Janssen, and C. A. Graham. 2006. The relationship between negative mood and sexuality in heterosexual college women and men. *The Journal of Sex Research* 43:136–44.

Masters, W. H., and V. E. Johnson. 1970. *Human sexual inadequacy.* Boston: Little, Brown.

Mathy, R. M. 2007. Sexual orientation moderates online sexual activity. In *Online Matchmaking,* ed. M. T. Whitty, A. J. Baker, and J. A. Inman, 159–77. New York: Paulgrave Macmillan.

McGinty, K., D. Knox, and M. Zusman. 2007. Friends with benefits: Women want "friends," men want "benefits." *College Student Journal* 41:1128–31.

Michael, R. T., J. H. Gagnon, E. O. Laumann, and G. Kolata. 1994. *Sex in America.* Boston: Little, Brown.

Miller, S. A., and E. S. Byers. 2004. Actual and desired duration of foreplay and intercourse: Discordant and misperceptions within heterosexual couples. *The Journal of Sex Research* 41:301–09.

Nobre, P. J., and J. Pinto-Gouveia. 2006. Dysfunctional sexual beliefs as vulnerability factors for sexual dysfunction. *The Journal of Sex Research* 43:68–74.

O'Reilly, S., D. Knox, and M. Zusman. 2007. College student attitudes toward pornography use. *College Student Journal* 41:402–06.

Puentes, J., D. Knox, and M. Zusman. 2008. Participants in "Friends with Benefits" relationships. *College Student Journal* 42:176–80.

Raley, R. K. 2000. Recent trends and differentials in marriage and cohabitation: The United States. In *The ties that bind,* ed. L. J. Waite, 19–39. New York: Aldine de Gruyter.

Richey, E., D. Knox, and M. Zusman. 2009. Sexual values of 783 undergraduates. *College Student Journal* 43:175–80.

Simon, W., and J. Gagnon. 1998. Psychosexual development. *Society* 35:60–68.

True Love Waits. 2006. http://www.lifeway.com/tlw/students/join.asp (accessed January 14, 2006).

Uecker, J. E. 2008. Religion, pledging, and the premarital sexual behavior of married young adults. *Journal of Marriage and the Family* 70:728–44.

Vail-Smith, K., D. Knox, and L. M. Whetstone. Forthcoming. The illusion of safety in "monogamous" relationships. *American Journal of Health Behavior.*

Chapter 9

Baden, A. L., and M. O. Wiley. 2007. Counseling adopted persons in adulthood: Integrating practice and research. *Counseling Psychologist* 35:868–79.

Begue, L. 2001. Social judgment of abortion: A black-sheep effect in a Catholic sheepfold. *Journal of Social Psychology* 141:640–50.

Berger, R., and M. Paul. 2008. Family secrets and family functioning: The case of donor assistance. *Family Process* 47:553–66.

Block, S. 2008. Adopting domestically can lower hurdles to claiming tax credit. *USA Today*, August 19, 3B.

Bristol, K., and B. Farmer. 2005. *Sexuality among southeastern university students: a survey*. Unpublished data, East Carolina University, Greenville, NC.

Brown, J. D. 2008. Foster parents' perceptions of factors needed for successful foster placements. *Journal of Child and Family Studies* 17:538–55.

Finer, L. B., L. F. Frohwirth, L. A. Dauphinne, S. Singh, and A. M. Moore. 2005. Reasons U.S. women have abortions: quantitative and qualitative reasons. *Perspectives on Sexual and Reproductive Health* 37:110–18.

Flower Kim, K. M. 2003. We are family. Paper presented at the 73rd Annual Meeting of the Eastern Sociological Society, February 27, Philadelphia.

Garrett, T. M., H. W. Baillie, and R. M. Garrett. 2001. *Health care ethics*, 4th ed. Upper Saddle River, NJ: Prentice Hall.

Ge, X., M. N. Natsuaki, D. Martin, J. M. Neiderhiser, D. S. Shaw, L. Scaramella, J. B. Reid, and D. Reiss. 2008. Bridging the divide: Openness in adoption and postadoption psychosocial adjustment among birth and adoptive parents issue: Public health perspectives on family interventions. American Psychological Association. Sage Periodicals Press.

Gross, J., and W. Conners. 2007. Surge in adoptions raises concern in Ethiopia. *The New York Times*, June 4, A16.

Hammarberg, K., J. R. W. Fisher, and H. J. Rowe. 2008. Women's experiences of childbirth and post-natal healthcare after assisted conception. *Human Reproduction* 23:1567–74.

Hollingsworth, L. D. 1997. Same race adoption among African Americans: A ten-year empirical review. *African American Research Perspectives* 13:44–49.

Huh, N. S., and W. J. Reid. 2000. Intercountry, transracial adoption and ethnic identity: a Korean example. *International Social Work* 43:75–87.

Kennedy, R. 2003. *Interracial intimacies*. New York: Pantheon.
Kero, A., and A. Lalos. 2004. Reactions and reflections in men, 4 and 12 months post-abortion. *Journal of Psychosomatic Obstetrics and Gynecology* 25:135–43.
Knox, D., and M. E. Zusman. 2009. Relationship and sexual behaviors of a sample of 1,319 university students. Unpublished data collected for this text. Department of Sociology, East Carolina University, Greenville, NC.
Koropeckyj-Cox, T., and G. Pendell. 2007a. Attitudes about childlessness in the United States: Correlates of positive, neutral, and negative responses. *Journal of Family Issues* 28:1054–82.
_____. 2007b. The gender gap in attitudes about childlessness in the United States. *Journal of Marriage and Family* 69:899–915.
Lee, D. 2006. Device brings hope for fertility clinics. http://www.indystar.com/apps/pbcs.dll/article?AID/20060221/BUSINESS/602210365/1003 (retrieved February 22, 2006).
Leung, P., S. Erich, and H. Kanenberg. 2005. A comparison of family functioning in gay/lesbian, heterosexual and special needs adoptions. *Children and Youth Services Review* 27:1031–44.
Levine, J. A., C. R. Emery, and H. Pollack. 2007. The well-being of children born to teen mothers. *Journal of Marriage and Family* 69:105–22.
Livermore, M. M., and R. S. Powers. 2006. Unfulfilled plans and financial stress: unwed mothers and unemployment. *Journal of Human Behavior in the Social Environment* 13:1–17.
Maill, C. E., and K. March. 2005. Social support for changes in adoption practice: gay adoption, open adoption, birth reunions, and the release of confidential identifying information. *Families in Society* 86:83–92.
Major, B., M. Appelbaum, and C. West. 2008. Report of the APA task force on mental health and abortion. Retrieved August 23, 2008, from http://www.apa.org/releases/abortion-report.pdf.
Martin, S. 2008. Recent changes in fertility rates in the United States: What do they tell us about American's changing families? Report for Council on Contemporary Families. http://www.contemporaryfamilies.org/subtemplate.php?t=briefingPapers&ext=changesinfertility (retrieved August 10, 2008).
McDermott, E., and H. Graham. 2005. Resilient young mothering: social inequalities, late modernity and the problem of teenage motherhood. *Journal of Youth Studies* 8:59–79.
Mollborn, S. 2007. Making the best of a bad situation: Material resources and teenage parenthood. *Journal of Marriage and Family* 69:92–104.
Nation's Health. Teen birth rate rises, 38(1):7.
Nickman, S. L., A. A. Rosenfeld, P. Fine, J. C. MacIntyre, D. J. Pilowsky, R. Howe, A. Dereyn, M. Gonzales, L. Forsythe, and S. A. Sveda. 2005. Children in adoptive families: overview and update. *Journal of the American Academy of Child and Adolescent Psychiatry* 44:987–95.
Pew Research. 2008. The U.S. religious landscape survey. The Pew Forum on Religion & Public Life. http://pewresearch.org/pubs/743/united-states-religion.
Rochman, B. 2009. The ethics of octuplets. *Time*, February 16, p. 43–44.
Rolfe, A. 2008. "You've got to grow up when you've got a kid": Marginalized young women's accounts of motherhood. *Journal of Community & Applied Social Psychology* 18:299–312.
Ross, R., D. Knox, M. Whatley, and J. N. Jahangardi. 2003. Transracial adoption: some college student data. Paper presented at the 73rd Annual Meeting of the Eastern Sociological Society, Philadelphia.
Scheib, J. E., M. Riordan, and S. Rubin. 2005. Adolescents with open-identity sperm donors: reports from 12–17 year olds. *Human Reproduction* 20:239–52.
Schwarz, E. B., R. Smith, J. Steinauer, M. F. Reeves, and A. B. Caughey. 2008. Measuring the effects of unintended pregnancy on women's quality of life. *Contraception* 78:204–10.
Simon, R. J., and R. M. Roorda. 2000. *In their own voices: Transracial adoptees tell their stories*. New York: Columbia University Press.
Smith, T. 2007. Four kids is the new standard. National Public Radio, August 5.
Statistical Abstract of the United States, 2009. 128th ed. Washington, DC. U.S. Bureau of the Census.
Steinberg, J. R., and N. F. Russo. 2008. Abortion and anxiety: What's the relationship? *Social Science & Medicine* 67:238–42.
Stone, A. 2006. Drives to ban gay adoption heat up. *USA Today*, February 21, A1.
Thomas, K. A., and R. C. Tessler. 2007. Bicultural socialization among adoptive families: Where there is a will, there is a way. *Journal of Family Issues* 28:1189–1219.
Wang, Y. A., D. Healy, D. Black, and E. A. Sullivan. 2008. Age-specific success rate for women undertaking their first assisted reproduction technology treatment using their own oocytes in Australia, 2002–2005. *Human Reproduction* 23:1533–1639.
Wilson, H., and A. Huntington. 2006. Deviant mothers: the construction of teenage motherhood in contemporary discourse. *Journal of Social Policy* 35:59–76.
Wirtberg I., A. Möller, L. Hogström, S. E. Tronstad, and A. Lalos. 2007. Life 20 years after unsuccessful infertility treatment. *Human Reproduction* 22:598–604.
Wolters, J., D. Knox, and M. Zusman. Forthcoming. Male and female attitudes toward transracial adoption. *Journal of Indiana Academy of Social Sciences*.
Zachry, E. M. 2005. Getting my education: Teen mothers' experiences in school before and after motherhood. *Teachers College Record* 107:2566–98.

Chapter 10

Baumrind, D. 1966. Effects of authoritative parental control on child behavior. *Child Development* 37:887–907.
Bock, J. D. 2000. Doing the right thing? Single mothers by choice and the struggle for legitimacy. *Gender and Society* 14:62–86.
Booth, C. L., K. A. Clarke-Stewart, D. L. Vandell, K. McCartney, and M. T. Owen. 2002. Child-care usage and mother-infant "quality time." *Journal of Marriage and the Family* 64:16–26.
Bost, K. K., M. J. Cox, M. R. Burchinal, and C. Payne. 2002. Structural and supporting changes in couples' family and friendships networks across the transition to parenthood. *Journal of Marriage and the Family* 64:517–31.
British Columbia Reproductive Mental Health Program. 2005. Reproductive mental health: Psychosis. http://www.bcrmh.com/disorders/psychosis.htm (retrieved June 15, 2005).
Bronte-Tinkew, J., J. Carrano, A. Horowitz, and A. Kinukawa. 2008. Involvement among resident fathers and links to infant cognitive outcomes. *Journal of Family Issues* 29:1211–31.
Brook, J. S., K. Pahl, and P. Cohen. 2008. Associations between marijuana use during emerging adulthood and aspects of the significant other relationship in young adulthood. *Journal of Child and Family Studies* 17:1–12.
Bushman, B. J., and J. Cantor. 2003. Media ratings for violence and sex. *American Psychologist* 58:130–41.
Castrucci, B. C., J. F. Culhane, E. K. Chung, I. Bennett, and K. F. McCollum. 2006. Smoking in pregnancy: Patient and provider risk reduction behavior. *Journal of Public Health Management & Practice* 12:68–76.
Clarke, J. I. 2004. The overindulgence research literature: Implications for family life educators. Poster at the National Council on Family Relations, Annual Meeting, November. Orlando, Florida.
Claxton, A., and M. Perry-Jenkins. 2008. No fun anymore: Leisure and marital quality across the transition to parenthood. *Journal of Marriage and the Family* 70:28–43.

Cornelius-Cozzi, T. 2002. Effects of parenthood on the relationships of lesbian couples. *PROGRESS: Family Systems Research and Therapy* 11:85–94.

Cui, M., F. D. Fincham, and B. Kay Pasley. 2008. Young adult romantic relationships: The role of parents' marital problems and relationship efficacy. *Personality and Social Psychology Bulletin* 34:1226–35.

Diamond, A., J. Bowes, and G. Robertson. 2006. Mothers' safety intervention strategies with toddlers and their relationship to child characteristics. *Early Child Development and Care* 176:271–84.

Facer, J., and R. Day. 2004. Explaining diminished marital satisfaction when parenting adolescents. Poster at an Annual Meeting National Council on Family Relations, Orlando, Florida.

Fadiman, C., ed. 1985. *The Little, Brown book of anecdotes.* Boston: Little, Brown and Co.

Flouri, E., and A. Buchanan. 2003. The role of father involvement and mother involvement in adolescents' psychological well-being. *British Journal of Social Work* 33:399–406.

Galambos, N. L., E. T. Barker, and D. M. Almeida. 2003. Parents *do* matter: Trajectories of change in externalizing and internalizing problems in early adolescence. *Child Development* 74:578–94.

Gavin, L. E., M. M. Black, S. Minor, Y. Abel, and M. E. Bentley. 2002. Young, disadvantaged fathers' involvement with their infants: An ecological perspective. *Journal of Adolescent Health* 31:266–76.

Green, S. E. 2003. "What do you mean 'what's wrong with her?'" Stigma and the lives of families of children with disabilities. *Social Science & Medicine* 57:1361–74.

Hammarberg, K., J. R. Fisher, and K. H. Wynter. 2008. Psychological and social aspects of pregnancy, childbirth and early parenting after assisted conception: A systematic review. *Human Reproduction* 14:395–415.

Hira, N. A. 2007. The baby boomers' kids are marching into the workplace and look out. This crop of twentysomethings really is different. *Fortune Magazine*, May.

Kim-Cohen, J., T. E. Moffitt, A. Taylor, S. J. Pawlby, and A. Caspi. 2005. Maternal depression and children's antisocial behavior: Nature and nurture effects. *Archives of General Psychiatry* 62:173–82.

Knox, D., and K. Leggett. 2000. *The divorced dad's survival book: How to stay connected with your kids.* Reading, MA: Perseus Books.

Kolko, D. J., L. D. Dorn, O. Bukstein, and J. D. Burke. 2008. Clinically referred ODD children with or without CD and healthy controls: Comparisons across contextual domains. *Journal of Child and Family Studies* 17: 714–34.

Kouros, C. D., C. E. Merrilees, and E. M. Cummings. 2008. Marital conflict and children's emotional security in the context of parental depression. *Journal of Marriage and Family* 70:684 –97.

Lee, J. 2008. "A Kotex and a smile": Mothers and daughters at menarche. *Journal of Family Issues* 29:1325–47.

Lengua, L. J., S. A. Wolchik, I. N. Sandler, and S. G. West. 2000. The additive and interactive effects of parenting and temperament in predicting problems of children of divorce. *Journal of Clinical Child Psychology* 29:232–44.

Louv, R. 2006. *Last child in the woods.* Chapel Hill: Algonquin Books.

Mayall, B. 2002. *Toward a sociology of childhood.* Philadelphia, PA: Open University Press.

McBride, B. A., S. J. Schoppe, and T. R. Rane. 2002. Child characteristics, parenting stress, and parental involvement: Fathers versus mothers. *Journal of Marriage and Family* 64:998–1011.

McKinney, C., and K. Renk 2008. Differential parenting between mothers and fathers: Implications for late adolescents. *Journal of Family Issues* 29:806–27.

McLanahan, S. S. 1991. The long term effects of family dissolution. In *When families fail: The social costs*, ed. Brice J. Christensen, 5–26. New York: University Press of America for the Rockford Institute.

McLanahan, S. S., and K. Booth. 1989. Mother-only families: Problems, prospects, and politics. *Journal of Marriage and the Family* 51:557–80.

Morton, A. 2003. *Madonna.* New York: St. Martin's Press.

Pew Research Center. 2007. Motherhood today: Tougher challenges, less success. May 2. http://pewresearch.org/pubs/468/motherhood.

Pinheiro, R. T., R. A. da Silva, P. V. S. Magalhaes, B. L. Hortam, and K. A. T. Pinheiro. 2008. Two studies on suicidality in the postpartum. *Acta Psychiatrica Scandinavica* 118:160–62.

Pinquart, M., and R. K. Silbereisen. 2002. Changes in adolescents' and mothers' autonomy and connectedness in conflict discussions: An observation study. *Journal of Adolescence* 25:509–22.

Pong, S. L., and B. Dong. 2000. The effects of change in family structure and income on dropping of out of middle and high school. *Journal of Family Issues* 21:147–69.

Rapoport, B., and C. Le Bourdais. 2008. Parental time and working schedules. *Journal of Population Economics* 21:903–33.

Rhea, D. Personal Communication.

Sammons, L. 2008. Personal communication, Grand Junction, Colorado.

Schoppe-Sullivan, S. J., G. L. Brown, E. A. Cannon, S. C. Mangelsdorf, and M. S. Sokolowski. 2008. Maternal gatekeeping, coparenting quality, and fathering behavior in families with infants. *Journal of Family Psychology* 22:389–97.

Shellenbarger, S. 2006. Helicopter parents go to work: Moms and dads are now hovering at the office. *The Wall Street Journal*, March 16, D1.

Shields, B. 2005. *Down came the rain: My journey through postpartum depression.* New York: Hyperion.

Stanley, S. M., and H. J. Markman. 1992. Assessing commitment in personal relationships. *Journal of Marriage and the Family* 54:595–608.

Sugarman, S. D. 2003. Single-parent families. In *All our families: New policies for a new century,* 2nd ed., ed. M. A. Mason, A. Skolnick, and S. D. Sugarman, 14–39. New York: Oxford University Press.

Suitor, J. J., and K. Pillemer. 2007. Mothers' favoritism in later life: The role of children's birth order. *Research on Aging* 29:32–42.

Sulloway, F. J. 1996. *Born to rebel: Birth order, family dynamics, and creative lives.* New York: Vintage Books.

_____. 2007. Birth order and intelligence. *Age and Intelligence* 316:1711–21.

Talwar, V., and K. Lee. 2008. Social and cognitive correlates of children's lying behavior. *Child Development* 79:866–81.

Thomas, P. A., E. M. Krampe, and R. R. Newton. 2008. Father presence, family structure, and feelings of closeness to the father among adult African American children. *Journal of Black Studies* 38:529–41.

Tucker, C. J., S. M. McHale, and A. C. Crouter. 2003. Dimensions of mothers' and fathers' differential treatment of siblings: Links with adolescents' sex-typed personal qualities. *Family Relations* 52:82–89.

Twenge, J. M., W. K. Campbell, and C. A. Foster. 2003. Parenthood and marital satisfaction: A meta-analytic review. *Journal of Marriage and Family* 65:574–83.

Usher-Seriki, K. K., M. S. Bynum, and T. A. Callands. 2008. Mother–daughter communication about sex and sexual intercourse among middle- to upper-class African American girls. *Journal of Family Issues* 29:901–17.

Ward, R. A., and G. D. Spitze. 2007. Nestleaving and coresidence by young adult children: The role of family relations. *Research on Aging* 29:257–71.

Webb, F. J. 2005. The new demographics of families. In *Sourcebook of family theory & research,* ed. V. L. Bengtson, A. C. Acock, K. R.

Allen, P. Dilworth-Anderson, and D. M. Klein, 101–02. Thousand Oaks, California: Sage Publications.
Wyckoff, S. C., K. S. Miller, R. Forehand, J. J. Bau, A. Fasula, N. Long, and L. Armistead. 2008. Patterns of sexuality communication between preadolescents and their mothers and fathers. *Journal of Child and Family Studies* 17:649–53.

Chapter 11

Ahnert, L., and M. E. Lamb. 2003. Shared care: Establishing a balance between home and child care settings. *Child Development* 74:1044–49.
Amato, P. R., A. Booth, D. R. Johnson, and S. F. Rogers. 2007. *Alone together: How marriage in America is changing.* Cambridge, Massachusetts: Harvard University Press.
Boushey, H., and C. E. Weller. 2008. Has growing inequality contributed to rising household economic distress? *Review of Political Economy* 20:1–2.
Bryant, C. M., R. J. Taylor, K. D. Lincoln, L. M. Chatters, and J. S. Jackson. 2008. Marital satisfaction among African Americans and Black Caribbeans: Findings from the National Survey of American Life. *Family Relations* 57:239–354.
Cinamon, R. G. 2006. Anticipated work-family conflict: effects of gender, self-efficacy, and family background. *Career Development Quarterly* 54:202–16.
Davies, J. B., S. Sandstrom, A. Shorrocks, and E. N. Wolff. 2006. (Dec. 5). *The World Distribution of Household Wealth.* United Nations University—World Institute for Development Economics Research.
De Schipper, J. C., L. W. C. Tavecchio, and M. H. Van IJzendoorn. 2008. Children's attachment relationships with day care caregivers: Associations with positive caregiving and the child's temperament. *Social Development* 17:454–65.
Deutsch, F. M., A. P. Kokot, and K. S. Binder. 2007. College women's plans for different types of egalitarian marriages. *Journal of Marriage and Family* 69:916–29.
Dew, J. 2008. Debt change and marital satisfaction change in recently married couples. *Family Relations* 57:60–72.
Gordon, R. A., R. Kaestner, and S. Korenman. 2007. The effects of maternal employment on child injuries and infectious disease. *Demography* 44:307–26.
Gordon, R. A., and R. S. Högnäs. 2006. The best laid plans: Expectations, preferences, and stability of child-care arrangements. *Journal of Marriage and Family* 68:373–93.
Gordon, J. R., and K. S. Whelan-Berry. 2005. Contributions to family and household activities by the husbands of midlife professional women. *Journal of Family Studies* 26:899–923.
Hochschild, A. R. 1989. *The second shift.* New York: Viking.
———. 1997. *The time bind.* New York: Metropolitan Books.
Keller, E. G. 2008. *The comeback: Seven stories of women who went from career to family and back again.* New York: Bloomsbury.
Kiecolt, K. J. 2003. Satisfaction with work and family life: No evidence of a cultural reversal. *Journal of Marriage and the Family* 65:23–35.
Lavee, Y., and A. Ben-Ari. 2007. Relationship of dyadic closeness with work-related stress: A daily diary study. *Journal of Marriage and Family* 69:1021–35.
Mason, M. A., and M. Goulden. 2004. Do babies matter? The effect of family formation on the lifelong careers of academic men and women. Annual Conference of the National Council on Family Relations, November. Orlando, Florida.
Morin, R., and D. Cohn. 2008. Women call the shots at home; Public mixed on gender roles in jobs, gender and power. Pew Research Center, September 25.
Noonan, M. C., and M. E. Corcoran. 2004. The mommy track and partnership: Temporary delay or dead end? *The Annals of the American Academy of Political and Social Science* 596:130–50.
Pepper, T. 2006. Fatherhood: Trying to do it all. *Newsweek*, International ed., February 27.
Presser, H. B. 2000. Nonstandard work schedules and marital instability. *Journal of Marriage and the Family* 62:93–110.
Rose, K., and K. J. Elicker. 2008. Parental decision making about child care. *Journal of Family Issues* 29:1161–79.
Rosen, E. 2006. Derailed on the mommy track? There's help to get going again. *New York Times* 155(10):1–3.
Schoen, R., N. M. Astone, K. Rothert, N. J. Standish, and Y. J. Kim. 2002. Women's employment, marital happiness, and divorce. *Social Forces* 81:643–62.
Sefton, B. W. 1998. The market value of the stay-at-home mother. *Mothering* 86:26–29.
Shapiro, M. 2007. Money: A therapeutic tool for couples therapy. *Family Process* 46:279–91.
Snyder, K. A. 2007. A vocabulary of motives: Understanding how parents define quality time. *Journal of Marriage and Family* 69:320–40.
Society for the Advancement of Education. 2005. Mr. Mom nation reaches new peak. *USA Today*, August 13.
Stanfield, J. B. 1998. Couples coping with dual careers: A description of flexible and rigid coping styles. *Social Science Journal* 35:53–62.
Stanley, S. M., and L. A. Einhorn. 2007. Hitting pay dirt: Comment on "Money: A therapeutic tool for couples therapy." *Family Process* 46:293–99.
Stone, P. 2007. *Opting out?* Berkeley: University of California Press.
Statistical Abstract of the United States, 2009. 128th ed. Washington, DC: U.S. Bureau of the Census.
Tucker, P. 2005. Stay-at-home dads. *The Futurist* 39:12–15.
Walker, S. K. 2000. Making home work: Family factors related to stress in family child care providers. Poster at the Annual Conference of the National Council on Family Relations, November. Minneapolis.
Warash, B. G., C. A. Markstrom, and B. Lucci. 2005. The early childhood environment rating scale—revised as a tool to improve child care centers. *Education* 126:240–50.
Whitehead, B. D., and D. Popenoe. 2004. The state of our union: The social health of marriage in America. The National Marriage Project. Rutgers University. http://www.marriage.rutgers.edu/.
Zibel, A. 2008. Home loan trouble break records again. *Associated Press*, September 5, http://news.yahoo.com/s/ap/20080905/ap_on_bi_ge/home_foreclosures.

Chapter 12

Brecklin, L. R., and S. E. Ullman. 2005. Self-defense or assertiveness training and women's responses to sexual attacks. *Journal of Interpersonal Violence* 20:738–62.
Buddie, A. M., and M. Testa. 2005. Rates and predictors of sexual aggression among students and nonstudents. *Journal of Interpersonal Violence* 20:713–24.
Busby, D. M., T. B. Holman, and E. Walker. 2008. Pathways to relationship aggression between adult partners. *Family Relations* 57:72–83.
Campbell, R., and S. M. Wasco. 2005. Understanding rape and sexual assault: 20 years of progress and future directions. *Journal of Interpersonal Violence* 20:127–31.
Chapleau, K. M., D. L. Oswald, and B. L. Russell. 2008. Male rape myths: The role of gender, violence, and sexism. *Journal of Interpersonal Violence* 23:600–15.
Daigle, L. E., B. S. Fisher, and F. T. Cullen. 2008. The violent and sexual victimization of college women: Is repeat victimization a problem? *Journal of Interpersonal Violence* 23:1296–1313.
Edleson, J. L., L. F. Mbilinyi, S. K. Beeman, and A. K. Hagemeister. 2003. How children are involved in adult domestic violence. *Journal of Interpersonal Violence* 18:18–32.

Erwin, M. J., R. R. M. Gershon, M. Tiburzi, and S. Lin. 2005. Reports of intimate partner violence made against police officers. *Journal of Family Violence* 20:13–20.

Few, A. L., and K. H. Rosen 2005. Victims of chronic dating violence: How women's vulnerabilities link to their decisions to stay. *Family Relations* 54:265–79.

Flack, Jr., W. F., M. L. Caron, S. J. Leinen, K. G. Breitenbach, A. M. Barber, E. N. Brown, C. T. Gilbert, T. F. Harchak, M. M. Hendricks, C. E. Rector, H. T. Schatten, and H. C. Stein. 2008. "The Red Zone": Temporal risk for unwanted sex among college students. *Journal of Interpersonal Violence* 23:1177–96.

Gidycz, C. A., A. V. Wynsberghe, and K. M. Edwards. 2008. Prediction of women's utilization of resistance strategies in a sexual assault situation: A prospective study. *Journal of Interpersonal Violence* 23:571–88.

Gottman, J. 2007. The mathematics of love. http://www.edge.org/3rd_culture/gottman05/gottman05_index.html (accessed August 23).

Grych, J. H., G. T. Harold, and C. J. Miles. 2003. A prospective investigation of appraisals as mediators of the link between interparental conflict and child adjustment. *Child Development* 74:1176–96.

Ham-Rowbottom, K. A., E. E. Gordon, K. L. Jarvis, and R. W. Novaco. 2005. Life constraints and psychological well-being of domestic violence shelter graduates. *Journal of Family Violence* 20:109–22.

Haskett, M. E., S. S. Scott, R. Grant, C. S. Ward, and C. Robinson. 2003. Child-related cognitions and affective functioning of physically abusive and comparison parents. *Child Abuse and Neglect* 27:663–86.

Howard, D. E., S. Feigelman, X. Li, S. Gross, and L. Rachuba. 2002. The relationship among violence victimization, witnessing violence, and youth distress. *Journal of Adolescent Health* 31:455–62.

Jerin, R. A., and B. Dolinsky. 2007. Cyber victimization and online dating. In *Online Matchmaking*, ed. M. T. Whitty, A. J. Baker, and J. A. Inman,147–56. New York: Palgrave Macmillan.

Johnson, L., M. Todd, and G. Subramanian. 2005. Violence in police families: Work-family spillover. *Journal of Family Violence* 20:3–13.

Katz, J., J. Moore, and P. May. 2008. Physical and sexual covictimization from dating partners: A distinct type of intimate abuse? *Violence Against Women* 14:961–73.

Katz, J., and L. Myhr. 2008. Perceived conflict patterns and relationship quality associated with verbal sexual coercion by male dating partners. *Journal of Interpersonal Violence* 23:798–804.

Kernic, M. A., M. E. Wolfe, V. L. Holt, B. McKnight, C. E. Huebner, and F. P. Rivara. 2003. Behavioral problems among children whose mothers are abused by an intimate partner. *Child Abuse & Neglect* 27:1231–46.

Kitzmann, K. M., N. K. Gaylord, A. R. Holt, and E. D. Kenny. 2003. Child witnesses to domestic violence: A meta-analytic review. *Journal of Clinical and Consulting Psychology* 71:339–52.

Knox, D., and M. E. Zusman. 2009. Relationship and sexual behaviors of a sample of 1,319 university students. Unpublished data collected for this text. Department of Sociology, East Carolina University, Greenville, NC.

Komarow, S. 2005. Report: Military women devalued. *USA Today*, August 26.

Kreager, D. A. 2007. Unnecessary roughness? School sports, peer networks, and male adolescent violence. *American Sociological Review* 72:705–24.

Kress, V. E., J. J. Protivnak, and L. Sadlak. 2008. Counseling clients involved with violent intimate partners: The mental health counselor's role in promoting client safety. *Journal of Mental Health Counseling* 30:200–11.

Meloy, J. R., and H. Fisher. 2005. Some thoughts on the neurobiology of stalking. *Journal of Forensic Science* 50:1472–80.

Nayak, M. B., C. A. Byrne, M. K. Martin, and A. G. Abraham. 2003. Attitudes toward violence against women: A cross-nation study. *Sex Roles: A Journal of Research* 49:333–43.

Oswald, D. L., and B. L. Russell. 2006. Perceptions of sexual coercion in heterosexual dating relationships: the role of aggressor gender and tactics. *The Journal of Sex Research* 43:87–98.

Rand, M. R. 2003. The nature and extent of recurring intimate partner violence against women in the United States. *Journal of Comparative Family Studies* 34:137–46.

Rosenbaum, A., and P. A. Leisring. 2003. Beyond power and control: Towards an understanding of partner abusive men. *Journal of Comparative Family Studies* 34:7–21.

Rothman, E., and J. Silverman. 2007. The effect of a college sexual assault prevention program on first year students' victimization rates. *Journal of American College Health* 55:283–90.

Rousseve, A. 2005. Domestic violence in the United States. *Georgetown Journal of Gender & the Law* 6:431–58.

Sarkar, N. N. 2008. The impact of intimate partner violence on women's reproductive health and pregnancy outcome. *Journal of Obstetrics and Gynecology* 28:266–78.

Schaeffer, C. M., P. C. Alexander, K. Bethke, and L. S. Kretz. 2005. Predictors of child abuse potential among military parents: Comparing mothers and fathers. *Journal of Family Violence* 20:123–30.

Shen, H., and S. B. Sorenson. 2005. Restraining orders in California: A look at statewide data. *Violence Against Women* 11:912–33.

Silvergleid, C., and E. S. Mankowski. 2006. How batterer intervention programs work: Participant and facilitator accounts of processes of change. *Journal of Interpersonal Violence* 21:139–59.

Spitzberg, B. H., and W. R. Cupach. 2007. Cyberstalking as (mis) matchmaking. In *Online Matchmaking*, ed. M. T. Whitty, A J. Baker, and J. A. Inman, 127–46. New York: Palgrave Macmillan.

Straus, M. A. 2000. Corporal punishment and primary prevention of physical abuse. *Child Abuse and Neglect* 24:1109–14.

Struckman-Johnson, C., D. Struckman-Johnson, and P. B. Anderson. 2003. Tactics of sexual coercion: When men and women won't take no for an answer. *Journal of Sex Research* 40:76–86.

Stuart, G. L. 2005. Improving violence intervention outcomes by integrating alcohol treatment. *Journal of Interpersonal Violence* 20:388–93.

Swan, S. C., L. J. Gambone, J. E. Caldwell, T. P. Sullivan, and D. L. Snow. 2008. A review of research on women's use of violence with male intimate partners. *Violence and Victims* 23:301–15.

Ulman, A., and M. A. Straus. 2003. Violence by children against mothers in relation to violence between parents and corporal punishment by parents. *Journal of Comparative Family Studies* 34:41–56.

Vandello, J. A., and D. Cohen. 2003. Male honor and female fidelity: Implicit cultural scripts that perpetuate domestic violence. *Journal of Personality and Social Psychology* 84:997–1010.

Vazquez, S., M. K. Stohr, K. Skow, and M. Purkiss. 2005. Why is a woman still not safe when she's home? Seven years of NIBRS data on victims and offenders of intimate partner violence. *Criminal Justice Studies: A Critical Journal of Crime, Law and Society* 18:125–46.

Verma, R. K. 2003. Wife beating and the link with poor sexual health and risk behavior among men in urban slums in India. *Journal of Comparative Family Studies* 34:1–61.

Walsh, W. 2002. Spankers and nonspankers: Where they get information on spanking. *Family Relations* 51:81–88.

Whatley, M. 2005. The effect of participant sex, victim dress, and traditional attitudes on casual judgments for marital rape victims. *Journal of Family Violence* 20:191–201.

Chapter 13

Allen, E. S., G. K. Rhoades, S. M. Stanley, H. J. Markman, et al. 2008. Premarital precursors of marital infidelity. *Family Process* 47:243–60.

Bagarozzi, D. A. 2008. Understanding and treating marital infidelity: A multidimensional model. *The American Journal of Family Therapy* 36:1–17.

Baggerly, J., and H. A Exum. 2008. Counseling children after natural disasters: Guidance for family therapists. *The American Journal of Family Therapy* 36:79–93.

Bermant, G. 1976. Sexual behavior: Hard times with the Coolidge effect. In *Psychological research: The inside story*, ed. M. H. Siegel and H. P. Zeigler. New York: Harper and Row.

Black, K., and M. Lobo. 2008. A conceptual review of family resilience factors. *Journal of Family Nursing* 14:1–33.

Browder, B. S. 2005. *On the up and up: A survival guide for women living with men on the down low.* New York: Kensington Publishers Corp.

Burke, M. L., G. G. Eakes, and M. A. Hainsworth. 1999. Milestones of chronic sorrow: Perspectives of chronically ill and bereaved persons and family caregivers. *Journal of Family Nursing* 5:387–94.

Burr, W. R., and S. R. Klein. 1994. *Reexamining family stress: New theory and research.* Thousand Oaks, CA: Sage Publications.

Butterworth, P., and B. Rodgers. 2008. Mental health problems and marital disruption: Is it the combination of husbands' and wives' mental health problems that predicts later divorce? *Social Psychiatry and Psychiatric Epidemiology* 43:758–64.

Carlson-Catalano, J. 2003. Director of Clinical Biofeedback Services. *Health innovations*. Greenville, NC. Personal communication, June 9.

Collins, J. 1998. *Singing lessons: A memoir of love, loss, hope, and healing.* New York: Pocket Books.

Cramer, R. E., R. E. Lipinski, J. D. Meteer, and J. A. Houska. 2008. Sex differences in subjective distress to unfaithfulness: Testing competing evolutionary and violation of infidelity expectations hypotheses. *The Journal of Social Psychology* 148:389–406.

De Castro, S., and J. T. Guterman. 2008. Solution-focused therapy for families with suicide. *Journal of Marital and Family Therapy* 34:93–107.

Djamba, Y. K., M. J. Crump, and A. G. Jackson. 2005. Levels and determinants of extramarital sex. Paper presented at the Southern Sociological Society, March. Charlotte, NC.

Dotson-Blake, K., D. Knox, and A. Holman. 2009. College student attitudes toward marriage, family, and sex therapy. Unpublished data from 288 undergraduate/graduate students. East Carolina University, Greenville, NC.

Druckerman, P. 2007. *Lust in translation.* New York: Penguin Group.

Duparcq, E. 2008. US soldiers in Iraq can find stress deadlier than enemy. Associated Press, October 14. http://news.yahoo.com/s/afp/20081015/wl_mideast_afp/iraqunrestus.

Ellis, R. T., and J. M. Granger. 2002. African American adults' perceptions of the effects of parental loss during adolescence. *Child and Adolescent Social Work Journal* 19:271–86.

Elmslie, B., and E. Tebaldi. 2008. So, what did you do last night? The economics of infidelity. *Kyklos* 61:391–406.

Enright, E. 2004. A house divided. *AARP The Magazine*, July/August, 60.

Falba, T. M. T., J. L. Sindelar, and W. T. Gallo. 2005. The effect of involuntary job loss on smoking intensity and relapse. *Addiction* 100:1330–39.

Feather, William. Personal Communication.

Field, N. P., E. Gal-Oz, and G. A. Bananno. 2003. Continuing bonds and adjustment at 5 years after the death of a spouse. *Journal of Consulting and Clinical Psychology* 71:110–17.

Flynn, M. A. T., D. A. McNeil, B. Maloff, D. Matasingwa, M. Wu, C. Ford, and S. C. Tough. 2006. Reducing obesity and related chronic disease risk in children and youth: a synthesis of evidence with 'best practice' recommendations. *Obesity Reviews* 7:7–66.

Friedrich, R. M., S. Lively, and L. M. Rubenstein. 2008. Siblings' coping strategies and mental health services: A national study of siblings of persons with schizophrenia. *Psychiatric Services* 59:261–73.

Goode, E. 1999. New study finds middle age is prime of life. *New York Times*, July 17, D6.

Hall, J. H., W. Fals-Stewart, and F. D. Fincham. 2008. Risky sexual behavior among married alcoholic men. *Journal of Family Psychology* 22:287–99.

Haugland, B. S. M. 2005. Recurrent disruptions of rituals and routines in families with paternal alcohol abuse. *Family Relations* 54:225–41.

Ingram, S., J. L. Ringle, K. Hallstrom, D. E. Schill, V. M. Gohr, and R. W. Thompson. 2008. Coping with crisis across the lifespan: The role of a telephone hotline. *Journal of Child and Family Studies* 17:663–75.

Insel, T. R. 2008. Assessing the economic costs of serious mental illness. *The American Journal of Psychiatry* 165:663–66.

Knox, D., and M. E. Zusman. 2009. Relationship and sexual behaviors of a sample of 1,319 university students. Unpublished data collected for this text. Department of Sociology, East Carolina University, Greenville, NC.

Kottke, J. 2008. The Eliot Spitzer affair and the business of sex. http://kottke.org/08/03/the-eliot-spitzer-affair-and-the-business-of-sex (retrieved December 6, 2008).

Legerski, E. M., M. Cornwall, and B. O'Neil. 2006. Changing locus of control: Steelworkers adjusting to forced unemployment. *Social Forces* 84:1521–37.

Leonard, R., and A. Burns. 2006. Turning points in the lives of midlife and older women: Five-year follow-up. *Australian Psychologist* 41:28–36.

Levitt, M. J., J. Levitt, G. L. Bustos, N. A. Crooks, J. D. Santos, P. Telan, J. Hodgetts, and A. Milevsky. 2005. Patterns of social support in middle childhood to early adolescent transition: Implications for adjustment. *Social Development* 14:398–420.

Linquist, L., and C. Negy. 2005. Maximizing the experiences of an extrarelational affair: An unconventional approach to a common social convention. *Journal of Clinical Psychology/In Session* 61:1421–28.

Macmillan, R., and R. Gartner. 2000. When she brings home the bacon: Labor-force participation and the risk of spousal violence against women. *Journal of Marriage and the Family* 61:947–58.

Mahoney, D. 2005. Mental illness prevalence high, despite advances. *Clinical Psychiatry News* 33:1–2.

Mattson, M. E., and R. Z. Litten. 2005. Combining treatments for alcoholism: Why and how? *Journal of Studies on Alcohol* July:8–16.

Melhem, N. M., D. A. Brent, M. Ziegler, S. Iyengar, et al. 2007. Familial pathways to early-onset suicidal behavior: Familial and individual antecedents of suicidal behavior. *American Journal of Psychiatry* 164:1364–71.

Merline, A. C., J. E. Schulenberg, P. M. O'Malley, J. G. Bachman, and L. D. Johnston. 2008. Substance use in marital dyads: Premarital assortment and change over time. *Journal of Studies on Alcohol and Drugs* 69:352–65.

Michael, S. T., and C. R. Snyder. 2005. Getting unstuck: The roles of hope, finding meaning, and rumination in the adjustment to bereavement among college students. *Death Studies* 29:435–59.

Miley, W. M., and W. Frank. 2006. Binge and non-binge college students' perceptions of other students' drinking habits. *College Student Journal* 40:259–62.

Miller, M. 2005. Where's the outrage? *Social Policy* 35:5–8.

Mordoch, E., and W. A. Hall. 2008. Children's perceptions of living with a parent with a mental illness: Finding the rhythm and maintaining the frame. *Qualitative Health Research* 18:1127–35.

Morell, V. 1998. A new look at monogamy. *Science* 281:1982.

Mostaghimi, L., W. H. Obermeyer, B. Ballamudi, D. M. Gonzalez, and R. M. Benca. 2005. Effects of sleep deprivation on wound healing. *Journal of Sleep Research* 12:213–19.

Neff, L. A., and B. R. Karney. 2007. Stress crossover in newlywed marriage: A longitudinal and dyadic perspective. *Journal of Marriage and Family* 69:594–607.

Neuman, M. G. 2008. *The truth about cheating: Why men stray and what you can do to prevent it.* New York: John Wiley & Sons.

Olson, M. M., C. S. Russell, M. Higgins-Kessler, and R. B. Miller. 2002. Emotional processes following disclosure of an extramarital affair. *Journal of Marital and Family Therapy* 28:423–34.

Ostbye, T., K. M. Krause, M. C. Norton, J. Tschanz, L. Sanders, K. Hayden, C. Pieper, and K. A. Welsh-Bohmer. 2006. Ten dimensions of health and their relationships with overall self-reported health and survival in a predominately religiously active elderly population: The Cache County Memory Study. *Journal of the American Geriatrics Society* 54:199–209.

Ozer, E. J., S. R. Best, T. L. Lipsey, and D. S. Weiss. 2003. Predictors of posttrauamatic stress disorder and symptoms in adults: A meta-analysis. *Psychological Bulletin* 129:52–73.

Park, C. 2006. Exploring relations among religiousness, meaning, and adjustment to lifetime and current stressful encounters in later life. *Anxiety Stress & Coping* 19:33–45.

Pratt, L. A., and D. J. Brody. 2008. Depression in the United States household population, 2005–2006. NCHS Data Brief, no. 7, September.

Pyle, S. A., J. Sharkey, G. Yetter, E. Felix, M. J. Furlog, and W. S. Poston. 2006. Fighting an epidemic: The role of schools in reducing childhood obesity. *Psychology in the Schools* 43:361–76.

Roberto, K. A. 2005. Families and policy: Health issues of older women. In *Sourcebook of family theory & research*, ed. V. L. Bengtson, A. C. Acock, K. R. Allen, P. Dilworth-Anderson, and D. M. Klein, 547–48. Thousand Oaks, CA: Sage Publications.

Routh, K., and J. N. Rao. 2006. A simple, and potentially low-cost method for measuring the presence of childhood obesity. *Child: Care, Health & Development* 32:239–45.

Schabus, M., K. Hodlmoser, T. Pecherstorfer, and G. Klosch. 2005. Influence of midday nap on declarative memory performance and motivation. *Somnologie* 9:148–53.

Schneider, J. P. 2000. Effects of cybersex addiction on the family: Results of a survey. *Sexual Addiction and Compulsivity* 7:31–58.

_____. 2003. The impact of compulsive cybersex behaviors on the family. *Sexual and Relationship Therapy* 18:329–55.

Schum, T. R. 2007. Dave's dead! Personal tragedy leading a call to action in preventing suicide. *Ambulatory Pediatrics* 7:410–12.

Sharpe, L., and L. Curran. 2006. Understanding the process of adjustment to illness. *Social Science and Medicine* 62:1153–66.

Smith, L. R. 2005. Infidelity and emotionally focused therapy: A program design. Dissertation Abstracts. International, Section B, The Sciences and Engineering, 65(10-B):5423.

Statistical Abstract of the United States, 2009, 128th ed. Washington, DC: U.S. Bureau of the Census.

Szabo, A., S. E. Ainsworth, and P. K. Danks. 2005. Experimental comparison of the psychological benefits of aerobic exercise, humor, and music. *International Journal of Humor Research* 18:235–46.

Taliaferro, L. A., B. A. Rienzo, M. D. Miller, R. M. Pigg, and V. J. Dodd. 2008. High school youth and suicide risk: Exploring protection afforded through physical activity and sport participation. *The Journal of School Health* 78:545–56.

Teitler, J. O., and N. E Reichman. 2008. Mental illness as a barrier to marriage among unmarried mothers. *Journal of Marriage and Family* 70:772–83.

Termini, K. A. 2006. Reducing the negative psychological and physiological effects of chronic stress. 4th Annual ECU Research and Creative. Activities Symposium, April 21, East Carolina University, Greenville, NC.

Tetlie, T., N. Eik-Nes, T. Palmstierna, P. Callaghan, and J. A Nøttestad. 2008. The effect of exercise on psychological and physical health outcomes: Preliminary results from a Norwegian forensic hospital. *Journal of Psychosocial Nursing & Mental Health Services* 46:38–44.

Unnever, J. D., F. T. Cullen, and B. K. Applegate. 2005. Turning the other cheek: Reassessing the impact of religion on punitive ideology. *Justice Quarterly* 22:304–39.

Waller, M. R. 2008. How do disadvantaged parents view tensions in their relationships? Insights for relationship longevity among at-risk couples. *Family Relations* 57:128–43.

Weckwerth, A. C., and D. M. Flynn. 2006. Effect of sex on perceived support and burnout in university students. *College Student Journal* 40:237–49.

White, A. M., D. W. Jamieson-Drake, and H. S. Swartzwelder. 2002. Prevalence and correlates of alcohol-induced blackouts among college students: Results of an e-mail survey. *Journal of American College Health* 51:117–32.

Chapter 14

Abut, C. C. 2005. Ten common questions about postnuptial agreements. *New Jersey Law Journal*, August 15.

Ahrons, C. R., and J. L. Tanner. 2003. Adult children and their fathers: Relationship changes 20 years after parental divorce. *Family Relations* 52:340–51.

Allen, D. W., K. Pendakur, and W. Suen. 2006. No-fault divorce and the compression of marriage ages. *Economic Inquiry* 44:547–59.

Amato, P. R., A. Booth, D. R. Johnson, and S. F. Rogers. 2007. *Alone together: How marriage in America is changing*. Cambridge, Massachusetts: Harvard University Press.

Anderson, E. R., and S. M. Greene. 2005. Transitions in parental repartnering after divorce. *Journal of Divorce & Remarriage* 43:47–62.

Baker, A. J. L. 2005. The long-term effects of parental alienation on adult children: A qualitative research study. *American Journal of Family Therapy* 33:289–302.

_____. 2006. Patterns of Parental Alienation Syndrome: A qualitative study of adults who were alienated from a parent as a child. *American Journal of Family Therapy* 34:63–78.

Baker, A. J. L., and D. Darnall. 2007. A construct study of the eight symptoms of severe parental alienation syndrome: A survey of parental experiences. *Journal of Divorce & Remarriage* 47:55–62.

Baldwin, A. 2008. *A promise to ourselves: A journey through fatherhood and divorce*. New York: St. Martin's Press.

Baum, N. 2003. Divorce process variables and the co-parental relationship and parental role fulfillment of divorced parents. *Family Process* 42:117–31.

Benton, S. D. 2008. Divorce mediation. Lecture. November 10, East Carolina University.

Bray, J. H., and J. Kelly. 1998. *Stepfamilies: Love, marriage and parenting in the first decade*. New York: Broadway Books.

Bream, V., and A. Buchanan. 2003. Distress among children whose separated or divorced parents cannot agree on arrangements for them. *British Journal of Social Work* 33:227–38.

Breed, R., D. Knox, and M. Zusman. 2007. "Hell hath no fury" . . . Legal consequences of having Internet child pornography on one's computer. Poster, Southern Sociological Society, Atlanta, March.

Brimhall, A., K. Wampler, and T. Kimball. 2008. Learning from the past, altering the future: A tentative theory of the effect of past relationships on couples who remarry. *Family Process* 47:373–408.

Brinig, M. F., and D. W. Allen. 2000. "These boots are made for walking": Why most divorce filers are women. *American Law and Economic Association* 2:126–69.

Byrne, A., and D. Carr. 2005. Commentaries on: Singles in society and science. *Psychological Inquiry* 16:84–141.

Cashmore, J., P. Parkinson, and A. Taylor. 2008. Overnight stays and children's relationships with resident and nonresident parents after divorce. *Journal of Family Issues* 29:707–14.

Christian, A. 2005. Contesting the myth of the 'Wicked stepmother': Narrative analysis of an online stepfamily support group. *Western Journal of Communication* 69:27–48.

Clarke, S. C., and B. F. Wilson. 1994. The relative stability of remarriages: A cohort approach using vital statistics. *Family Relations* 43:305–10.

Clarke-Stewart, A., and C. Brentano. 2006. Divorce: *Causes and consequences.* New Haven: Yale University Press.

Cohen, O., and R. Savaya. 2003. Lifestyle differences in traditionalism and modernity and reasons for divorce among Muslim Palestinian citizens of Israel. *Journal of Comparative Family Studies* 34:283–94.

Coltrane, S., and M. Adams. 2003. The social construction of the divorce "problem": Morality, child victims, and the politics of gender. *Family Relations* 52:363–72.

Covizzi, I. 2008. Does union dissolution lead to unemployment? A longitudinal study of health and risk of unemployment for women and men undergoing separation. *European Sociological Review* 24:347–62.

Decuzzi, A., D. Knox, and M. Zusman. 2004. The effect of parental divorce on relationships with parents and romantic partners of college students. Roundtable, Southern Sociological Society, Atlanta, April 17.

Degarmo, D. S., and M. S. Forgatch. 2002. Identity salience as a moderator of psychological and marital distress in stepfather families. *Social Psychology Quarterly* 65:266–84.

Drewianka, S. 2008. Divorce law and family formation. *Journal of Population Economics* 21:19–25.

Economist. 2005. "Yes, I really do."374:31–32.

Enright, E. 2004. A house divided. *AARP The Magazine,* July/August, 60.

Finley, G. E. 2004. Divorce inequities. NCFR *Family Focus Report* 49(3):F7.

Fonda, J. 2005. *My life so far.* New York: Random House.

Funder, K. 1991. New partners as co-parents. *Family Matters* April:44–46.

Ganong, L. H., and M. Coleman. 1994. *Remarried family relationships.* Thousand Oaks, CA: Sage Publications.

Ganong, L. H., M. Coleman, M. Fine, and A. K. McDaniel. 1998. Issues considered in contemplating stepchild adoption. *Family Relations* 47:63–71.

Gardner, J., and A. J. Oswald. 2006. Do divorcing couples become happier by breaking up? *Journal of the Royal Statistical Society: Series A (Statistics and Society)* 169:319–36.

Gardner, R. A. 1998. *The parental alienation syndrome.* 2nd ed. Cresskill, NJ: Creative Therapeutics.

Gilman, J., D. Schneider, and R. Shulak. 2005. Children's ability to cope post-divorce: The effects of Kids' Turn intervention program on 7- to 9-year-olds. *Journal of Divorce and Remarriage* 42:109–26.

Goetting, A. 1982. The six stations of remarriage: The developmental tasks of remarriage after divorce. *The Family Coordinator* 31:213–22.

Gordon, R. M. 2005. The doom and gloom of divorce research: Comment on Wallerstein and Lewis (2004) *Psychoanalytic Psychology* 22:450–51.

Gottman, J. M., J. S. Gottman, and J. de Claire. 2006. *Ten lessons to transform your marriage: America's lab experts share their strategies for strengthening your relationship.* New York: Random House.

Greeff, A. P., and C. Du Toit. 2009. Resilience in remarried families. *The American Journal of Family Therapy* 37:114–26.

Hawkins, A. J., S. L. Nock, J. C. Wilson, L. Sanchez, and J. D. Wright. 2002. Attitudes about covenant marriage and divorce: Policy implications from a three-state comparison. *Family Relations* 51:166–75.

Hawkins, D. N., and A. Booth. 2005. Unhappily ever after: Effects of long-term, low-quality marriages on well-being. *Social Forces* 84:445–65.

Hetherington, E. M. 2003. Intimate pathways: Changing patterns in close personal relationships across time. *Family Relations* 52:318–31.

Hipke, K. N., S. A. Wolchik, I. N. Sandler, and S. L. Braver. 2002. Predictors of children's intervention-induced resilience in a parenting program for divorced mothers. *Family Relations* 51:121–29.

Hsu, M., D. L. Kahn, and C. Huang. 2002. No more the same: The lives of adolescents in Taiwan who have lost fathers. *Family Community Health* 25:43–56.

Johnson, G. R., E. G. Krug, and L. B. Potter. 2000. Suicide among adolescents and young adults. A cross-national comparison of 34 countries. *Suicide and Life-Threatening Behavior* 30:74–82.

Juby, H., J. Michel Billette, B. Laplante, and C. Le Bourdais. 2007. Nonresident fathers and children: Parents' new unions and frequency of contact. *Journal of Family Issues* 28:1220–45.

Kelly, J. B., and R. E. Emery. 2003. Children's adjustment following divorce: Risk and resilience perspectives. *Family Relations* 52:352–62.

Kesselring, R. G., and D. Bremmer. 2006. Female income and the divorce decision: Evidence from micro data. *Applied Economics* 38:1605–17.

Kilmann, P. R., L. V. Carranza, and J. M. C. Vendemia. 2006. Recollections of parent characteristics and attachment patterns for college women of intact vs. non-intact families. *Journal of Adolescence* 29:89–102.

Knox, D., and M. E. Zusman. 2009. Relationship and sexual behaviors of a sample of 1,319 university students. Unpublished data collected for this text. Department of Sociology, East Carolina University, Greenville, NC.

Knox, D., M. E. Zusman, M. Kaluzny, and C. Cooper. 2000. College student recovery from a broken heart. *College Student Journal* 34:322–24.

Knox, D., M. E. Zusman, K. McGinty, and B. Davis. 2002. College student attitudes and behaviors toward ending an unsatisfactory relationship. *College Student Journal* 36:630–34.

Lambert, A. N. 2007. Perceptions of divorce advantages and disadvantages: A comparison of adult children experiencing one parental divorce versus multiple parental divorces. *Journal of Divorce & Remarriage* 48:55–77.

Leon, K., and E. Angst. 2005. Portrayals of stepfamilies in film: Using media images in remarriage education. *Family Relations* 54:3–23.

Lewis, K. 2008. Personal communication. Dr. Lewis is also the author of *Five Stages of Child Custody.* Glenside, PA: CCES Press.

Licata, N. 2002. Should premarital counseling be mandatory as a requisite to obtaining a marriage license? *Family Court Review* 40:518–32.

Lopoo, L. M., and B. Western. 2005. Incarceration and the formation and stability of marital unions. *Journal of Marriage and the Family* 67:721–35.

Lowenstein, L. F. 2005. Causes and associated factors of divorce as seen by recent research. *Journal of Divorce & Remarriage* 42:153–71.

Malia, S. E. C. 2005. Balancing family members' interests regarding stepparent rights and obligations: A social policy challenge. *Family Relations* 54:298–319.

Manning, W. D., and P. J. Smock. 2000. "Swapping" families: Serial parenting and economic support for children. *Journal of Marriage and the Family* 62:111–22.

Marano, H. E. 2000. Divorced? Don't even think of remarrying until you read this. *Psychology Today*, March/April, 56–64.

Masheter, C. 1999. Examples of commitment in postdivorce relationships between spouses. In *Handbook of interpersonal commitment and relationship stability,* ed. J. M. Adams and W. H. Jones, 293–306. New York: Academic/Plenum Publishers.

McCarthy, B. W., and R. L. Ginsberg. 2007. Second marriages: Challenges and risks. *The Family Journal* 15:119–23.

Menning, C. L. 2008. "I've Kept It That Way on Purpose": Adolescents' management of negative parental relationship traits after divorce and separation. *Journal of Contemporary Ethnography* 37:586–97.

Michaels, M. L. 2000. The stepfamily enrichment program: A preliminary evaluation using focus groups. *American Journal of Family Therapy* 28:61–73.
Morgan, E. S. 1944. *The Puritan family.* Boston: Public Library.
Nielsen, L. 2004. *Embracing your father: How to build the relationship you always wanted with your dad.* New York: McGraw-Hill.
O'Leary, K. D. 2005. Commentary on intrapersonal, interpersonal, and contextual factors in extramarital involvement. *Clinical Psychology: Science and Practice* 12:131–33.
Ostermann, J., F. A. Sloan, and D. H. Taylor. 2005. Heavy alcohol use and marital dissolution in the USA. *Social Science and Medicine* 61:2304–20.
Pacey, S. 2005. Step change: The interplay of sexual and parenting problems when couples form stepfamilies. *Sexual & Relationship Therapy* 20:359–69.
Papernow, P. L. 1988. Stepparent role development: From outsider to intimate. In *Relative strangers*, ed. William R. Beer, 54–82. Lanham, MD: Rowman and Littlefield.
Poortman, A., and T. H. Lyngstad. 2008. Dissolution risks in first and higher order marital and cohabiting unions. *Social Science Research* 36:1431–47.
Prather, J. E. 2008. Brave new stepfamilies: Diverse paths toward stepfamily living. *Contemporary Sociology* 37:33–35.
Previti, D., and P. R. Amato. 2003. Why stay married? Rewards, barriers, and marital stability. *Journal of Marriage and Family* 65:561–73.
Rutter, V. 1994. Lessons from stepfamilies. *Psychology Today*, May/June 27:68.
Sakraida, T. 2005. Divorce transition differences of midlife women. *Issues in Mental Health Nursing* 26:225–49.
Schacht, T. E. 2000. Protection strategies to protect professionals and families involved in high-conflict divorce. *UALR Law Review* 22(3):565–92.
Schrodt, P., J. Soliz, and D. O. Braithwaite. 2008. A social relations model of everyday talk and relational satisfaction in stepfamilies. *Communication Monographs* 75:190–202.
Segal-Engelchin, D., and Y. Wozner. 2005. Quality of life to single mothers in Israel: A comparison to single mothers and divorced mothers. *Marriage and Family Review* 37:7–28.
Sever, I., J. Guttmann, and A. Lazar. 2007. Positive consequences of parental divorce among Israeli young adults: A long-term effect model. *Marriage and Family Review* 42:7–21.
Shumway, S. T., and R. S. Wampler. 2002. A behaviorally focused measure for relationships: The couple behavior report (CBR). *The American Journal of Family Therapy* 30:311–21.
Siegler, I., and P. Costa. 2000. Divorce in midlife. Paper presented at the Annual Meeting of the American Psychological Association, Boston.
Sobolewski, J. M., and P. R. Amato. 2007. Parents' discord and divorce, parent-child relationships and subjective well-being in early adulthood: Is feeling close to two parents always better than feeling close to one? *Social Forces* 85:1105–25.
Stanley, S. M., E. S. Allen, H. J. Markman, C. Saiz, G. Bloomstrom, R. Thomas, and W. R. Schumm. 2005. Dissemination and evaluation of marriage education in the Army. *Family Process* 44:187–201.
Statistical Abstract of the United States, 2009. 128th ed. Washington, DC: U.S. Bureau of the Census.
The Stepfamily Association. Retrieved 2009, http://www.saafamilies.org/faqs/myths.html.
Swallow, W. 2004. *The triumph of love over experience: A memoir of remarriage.* New York: Hyperion/Theia.
Teachman, J. 2008. Complex life course patterns and the risk of divorce in second marriages. *Journal of Marriage and Family* 70:294–306.
Teich, M. 2007. A divided house. *Psychology Today* 40:96–102.
Thuen, F., and J. Rise. 2006. Psychological adaptation after marital disruption: The effects of optimism and perceived control. *Scandinavian Journal of Psychology* 47:121–28.
Tillman, K. H. 2008. "Non-traditional" siblings and the academic outcomes of adolescents. *Social Science Research* 37:88–101.
Trinder, L. 2008. Maternal gate closing and gate opening in post-divorce families. *Journal of Family Issues* 29:1298–2011.
Turkat, I. D. 2002. Shared parenting dysfunction. *The American Journal of Family Therapy* 30:385–93.
Vinick, B. 1978. Remarriage in old age. *The Family Coordinator* 27:359–63.
Vukalovich, D., and N. Caltabiano. 2008. The effectiveness of a community group intervention program on adjustment to separation and divorce. *Journal of Divorce & Remarriage* 48:145–68.
Waite, L. J., Y. Luo, and A. C. Lewin. 2009. Marital happiness and marital stability: Consequences for psychological well-being. *Social Science Research* 28:201–17.
Whitehurst, D. H., S. O'Keefe, and R. A. Wilson. 2008. Divorced and separated parents in conflict: Results from a true experiment effect of a court mandated parenting education program. *Journal of Divorce & Remarriage* 48:127–44.
Wiseman, R. S. 1975. Crisis theory and the process of divorce. *Social Casework* 56:205–12.

Chapter 15

Agahi, N. 2008. Leisure activities and mortality: Does gender matter? *Journal of Aging and Health* 20:855–71.
Alford-Cooper, F. 2006. Where has all the sex gone? Sexual activity in lifetime marriage. Paper presented at the Southern Sociological Society, New Orleans, March 23–26.
Barrett, A. E. 2005. Gendered experiences in midlife: Implications for age identity. *Journal of Aging Studies* 19:163–83.
Beckman, N. M., M. Waern, I. Skoog, and The Sahlgrenska Academy at Göteborg University, Sweden. 2006. Determinants of sexuality in 70-year-olds. *The Journal of Sex Research* 43:2–3.
Berg, A. S. 2003. *Kate remembered.* New York: G. P. Putnam's Sons.
Bookwala, J. 2005. The role of marital quality in physical health during the mature years. *Journal of Aging and Health* 17:85–97.
Boulton-Lewis, G. M., L. K. Buys, and J. Kitchin. 2006. Learning and active aging. *Educational Gerontology* 32:271–82.
Bryant, C. D. 2007. The sociology of death and dying. In *21st century sociology: A reference handbook*, ed. C. D. Bryant and D. L. Peck, 156–66. Thousand Oaks, CA: Sage Publications.
Clare, L., I. Markova, F. Verhey, and G. Kenny. 2005. Awareness in dementia: A review of assessment methods and measures. *Aging and Mental Health* 9:394–404.
Collins, A. L., and M. A. Smyer. 2005. The resilience of self-esteem in late adulthood. *Journal of Aging and Health* 17:471–90.
Cutler, N. E. 2002. *Advising mature clients.* New York: Wiley.
Danigelis, N. L., M. Hardy, and S. J. Cutler. 2007. Population aging, intracohort aging, and sociopolitical attitudes. *American Sociological Review* 72:812–30.
Dychtwald, K., and D. J. Kadlec. 2005. *The power years: A user's guide to the rest of your life.* New York: Wiley.
Fabian, N. 2007. Rethinking retirement—And a footnote on diversity. *Journal of Environmental Health* 69:86–95.
Fernandez-Ballesteros, R. 2003. Social support and quality of life among older people in Spain. *Journal of Social Issues* 58:645–60.
Field, D., and S. Weishaus. 1992. Marriage over half a century: A longitudinal study. In *Changing lives*, ed. M. Bloom, 269–73. Columbia, SC: University of South Carolina Press.
Gallanis, T. P. 2002. Aging and the nontraditional family. *The University of Memphis Law Review* 32:607–42.
Garrett, N., and E. M. Martini. 2007. The boomers are coming: A total cost of care model of the impact of population aging on the cost of chronic conditions in the United States. *Disease Management* 10:51–60.

Hansson, R. O., J. O. Berry, and M. E. Berry. 1999. The bereavement experience: Continuing commitment after the loss of a loved one. In *Handbook of interpersonal commitment and relationship stability*, ed. J. M. Adams and W. H. Jones, 281–91. New York: Academic/Plenum Publishers.

Hays, J., J. K. Ockene, R. L. Brunner, J. M. Kotchen, J. E. Manson, R. E. Patterson, A. K. Aragki, S. A. Shumaker, R. G. Bryzyski, et al. 2003. Effects of estrogen plus progestin on health-related quality of life. *The New England Journal of Medicine* 348:1839–54.

Hilgeman, M. M., R. S. Allen, J. DeCoster, and L. D. Burgio. 2007. Positive aspects of caregiving as a moderator of treatment outcome over 12 months. *Psychology and Aging* 22:361–71.

Horton, S., and J. Deakin. 2007. Role models for seniors and society: Seniors' perceptions of aging successfully. *Journal of Sport & Exercise Psychology* 29:14–15.

Ikegami, N. 1998. Growing old in Japan. *Age and Ageing* 27:277–78.

Jang, S., and D. F. Detzner. 1998. Filial responsibility in cross-cultural context. Poster at the Annual Conference of the National Council on Family Relations, Milwaukee, Wisconsin.

Johnson, C. L., and B. M. Barer. 1997. *Life beyond 85 years: The aura of survivorship*. New York: Springer Publishing.

Johnson, R. W., and J. M. Wiener. 2006. A profile of frail older Americans and their caregivers. Urban Institute Report. http://www.urban.org/url.cfm?ID?311284 (posted March 1).

Jorm, A. F., H. Christensen, A. S. Henderson, P. A. Jacomb, A. E. Korten, and A. Mackinnon. 1998. Factors associated with successful ageing. *Australian Journal of Ageing* 17:33–37.

Kalish, N. 1997. *Lost & found lovers: Facts and fantasies of rekindled romances*. New York: William Morrow and Company.

Karlsson, S. G., and K. Borell. 2005. A home of their own. Women's boundary work in LAT-relationships. *Journal of Aging Studies* 19:73–84.

Kelley-Moore, J. A., J. G. Schumacher, E. Kahana, and B. Kahana. 2006. When do older adults become 'disabled'? Social and health antecedents of perceived disability in a panel study of the oldest old. *Journal of Health and Social Behavior* 47:126–42.

Kemp, C. L., C. J. Rosenthal, and M. Denton. 2005. Financial planning for later life: Subjective understandings of catalysts and constraints. *Journal of Aging Studies* 19:273–90.

Kimuna, S., D. Knox, and M. Zusman. 2005. College students' perceptions about older people and aging. *Educational Gerontology* 31:563–72.

Knox, D., S. Kimuna, and M. Zusman. 2005. College student views of the elderly: Some gender differences. *College Student Journal* 39:14–16.

Krause, N. 2005. Negative interaction and heart disease in late life: Exploring variations by socioeconomic status. *Journal of Aging and Health* 17:28–35.

Kutza, E. A. 2005. The intersection of economics and family status in later life: Implications for the future. *Marriage and Family Review* 37:3–8.

Lees, F. D., P. G. Clark, C. R. Nigg, and P. Newman. 2005. Barriers to exercise behavior among older adults: A focus-group study. *Journal of Aging and Physical Activity* 13:23–34.

Lindau, S. T., L. P. Schumm, E. O. Laumann, W. Levinson, C. A. O'Muircheartaigh, and L. J. Waite. 2007. A study of sexuality and health among older adults in the United States. *The New England Journal of Medicine* 357:762–74.

Logan, J. R., and F. Bian. 2003. Parents' needs, family structure, and regular international financial exchange in Chinese cities. *Sociological Forum* 18:85–101.

Maher, D., and C. Mercer (editors). 2009. *Introduction to religion and the implications of radical life extension*. New York: Palgrave Macmillan.

Marks, N. F., J. D. Lambert, and H. Choi. 2002. Transitions to caregiving, gender, and psychological well-being: A prospective U.S. national study. *Journal of Marriage and Family* 64:657–67.

Martire, L. M., and M. A. P. Stephens. 2003. Juggling parent care and employment responsibilities: The dilemmas of adult daughter caregivers in the workforce. *Sex Roles: A Journal of Research* 48:167–74.

Morey, M. C., R. Sloane, C. F. Pieper, and M. J. Peterson. 2008. Effect of physical activity guidelines on physical function in older adults. *Journal of the American Geriatrics Society* 56:1873–85.

Mutran, E. J., D. Reitzes, and M. E. Fernandez. 1997. Factors that influence attitudes toward retirement. *Research on Aging* 19:251–73.

Nakashima, M., and E. R. Canda. 2005. Positive dying and resiliency in later life: A qualitative study. *Journal of Aging Studies* 19:109–22.

Nicholson, T. 2003. Homeowners fail to prepare for aging. *AARP Bulletin* 44:7.

Nordenmark, M., and M. Stattin. 2009. Psychosocial wellbeing and reasons for retirement in Sweden. *Ageing and Society* 29:413–41.

O'Reilly, E. M. 1997. *Decoding the cultural stereotypes about aging: New perspectives on aging talk and aging issues*. New York: Garland.

Ostbye, T., K. M. Krause, M. C. Norton, J. Tschanz, L. Sanders, K. Hayden, C. Pieper, and K. A. Welsh-Bohmer. 2006. Ten dimensions of health and their relationships with overall self-reported health and survival in a predominately religiously active elderly population: The Cache County Memory Study. *Journal of the American Geriatrics Society* 54:199–209.

Peterson, B. E. 2002. Longitudinal analysis of midlife generativity, intergenerational roles, and caregiving. *Psychology and Aging* 17:161–68.

Pnina, R. 2007. Elderly people's attitudes and perceptions of aging and old age: the role of cognitive dissonance? *International Journal of Geriatric Psychiatry* 22:656–72.

Pope, E. 2003. MIT study: Older drivers know when to slow down. *AARP Bulletin* 44:11–12.

Potts, A., V. M. Grace, T. Vares, and N. Gavey. 2006. "Sex for life"? Men's counter-stories on "erectile dysfunction," male sexuality and ageing. *Sociology and Health and Illness* 28:306–29.

Ryan, J., I. Carrière, J. Scali, K. Ritchie, and M. Ancelin. 2008. Lifetime hormonal factors may predict late-life depression in women. *International Psychogeriatrics* 20:1203–29.

Sandberg, J. G., R. B. Miller, and J. M. Harper. 2002. A qualitative study of marital process and depression in older couples. *Family Relations* 51:256–64.

Sawyer, F. 2008. Article is abridged from that which appeared in the *Sun Journal* of New Bern, NC, September 13, 2008. Used by permission.

Schairer, C., J. Lubin, R. Troisi, S. Sturgeon, L. Brinton, and R. Hoover. 2000. Menopausal estrogen and estrogen-progestin replacement therapy and breast cancer risk. *Journal of the American Medical Association* 283:485–91.

Siedlecki, K. L. 2007. Investigating the structure and age invariance of episodic memory across the adult lifespan. *Psychology & Aging* 22:251–68.

Silverstein, M., Z. Cong, and S. Li. 2006. Intergenerational transfers and living arrangements of older people in rural China: Consequences for psychological well-being. 2006. *The Journals of Gerontology, Series B: Psychological Sciences and Social Sciences*, vol. 61B:S256–67.

Smith, J., M. Borchelt, H. Maier, and D. Jopp. 2002. Health and well-being in the young old and oldest old. *Journal of Social Issues* 58:715–33.

Statistical Abstract of the United States, 2009. 128th ed. Washington, DC: U.S. Bureau of the Census.

Stephenson, L. E., S. N. Culos-Reed, P. K. Doyle-Baker, J. A. Devonish, and J. A. Dickinson. 2007. Walking for wellness: Results from a mall walking program for the elderly. *Journal of Sport & Exercise Psychology* 29:204–14.

Stevens, N. 2002. Re-engaging: New partnerships in late-life widowhood. *Ageing International* 27:27–42.

Stone, E. 2008. The last will and testament in literature: Rupture, rivalry, and sometimes approachement from Middlemarch to Lemony Sniket. *Family Process* 47:425–39.

Stroebe, M., and H. Schut. 2005. To continue or relinquish bonds: A review of consequences for the bereaved. *Death Studies* 29:477–95.

Szinovacz, M. E., and A. M. Schaffer. 2000. Effects of retirement on marital tactics. *Journal of Family Issues* 21:367–89.

Tanner, D. 2005. Promoting the well-being of older people: Messages for social workers. *Practice* 17:191–205.

Useda, J. D., K. R. Conner, A. Beckman, N. Franus, Z. Tu, and Y. Conwell. 2007. Personality differences in attempted suicide versus suicide in adults 50 years of age or older. *Journal of Consulting and Clinical Psychology* 75:126–33.

Valliant, G. E. 2002. *Aging well: Surprising guideposts to a happier life from the Landmark Harvard study of adult development*. New York: Little, Brown.

Vance, D. E., V. G. Wadley, K. K. Ball, D. L. Roenker, and M. Rizzo. 2005. The effects of physical activity and sedentary behavior on cognitive health in older adults. *Journal of Aging & Physical Activity* 13:294–314.

Van Gool, C. H., G. Kempen, H. Bosma, J. Van Eijk, M. P. J. Van Boxtel, and J. Jolles. 2007. Associations between lifestyle and depressed mood: Longitudinal results from the Maastricht Aging Study. *American Journal of Public Health* 97:887–94.

Walker, R. B., and M. A. Luszcz. 2009. The health and relationship dynamics of late-life couples: a systematic review of the literature. *Ageing and Society* 29:455–81.

Wallsten, S. S. 2000. Effects of care giving, gender, and race on the health, mutuality, and social supports of older couples. *Journal of Aging and Health* 12:90–111.

Webber, S., J. P. Scott, and R. Wampler. 2000. Perceived congruency of goals as a predictor of marital satisfaction and adjustment in retirement. Poster at the 62nd Annual Conference of the National Council on Family Relations, Minneapolis, November 12.

Weintraub, P. 2006. Guess who's back? *Psychology Today* 39:79–84.

Wells, Y. D. 2000. Intentions to care for spouse: Gender differences in anticipated willingness to care and expected burden. *Journal of Family Studies* 5:220–34.

Willson, A. E. 2007. The sociology of aging. In *21st century sociology: A reference handbook*, ed. C. D. Bryant and D. L. Peck, 148–55. Thousand Oaks, CA: Sage Publications.

Winterich, J. A. 2003. Sex, menopause, and culture: Sexual orientation and the meaning of menopause for women's sex lives. *Gender and Society* 17:627–42.

Wolkove, N., O. Elkholy, M. Baltzan, and M. Palayew. 2007. Sleep and aging. *Canadian Medical Association Journal* 176:1299–1304.

Wrosch, C., R. Schulz, G. E. Miller, S. Lupien, and E. Dunne. 2007. Physical health problems, depressive mood, and cortisol secretion in old age: Buffer effects of health engagement control strategies *Health Psychology* 26:341–49.

Wu, Z., and R. Hart. 2002. The mental health of the childless elderly. *Sociological Inquiry* 72:21–42.

Zhan, H. J., and R. J. V. Montgomery. 2003. Gender and elder care in China: The influence of filial piety and structural constraints. *Gender and Society* 17:209–29.

Name Index

A

B

C

O

P

R

S

T

U

V

W

X

Y

Z

Subject Index

D

E

reviewcard CHAPTER 2 GENDER

Learning Outcomes

To help you succeed, we've designed a review card for each chapter.

2.1 What are the important terms

Sex refers to the biological distinction between females a
is identified on the basis of one's chromosomes, gonads, ho
and external genitals, and exists on a continuum rather than being a dichotomy. *Gender* refers to the social and psychological characteristics often associated with being female or male. Other terms related to gender include *gender identity, gender role, gender role ideology, transgender,* and *transgenderism.*

In this column, you'll find summary points supported by diagrams or tables when relevant to help you better understand important concepts.

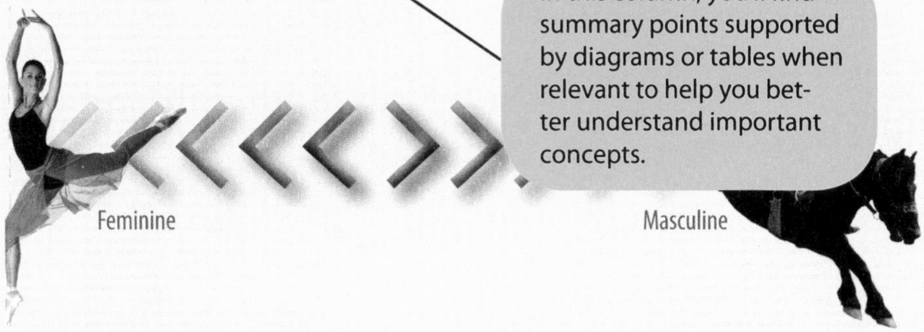

2.2 What theories explain gender role development?

Biosocial theory emphasizes that social behaviors (for example
biologically based and have an evolutionary survival function.
stayed in the nest or gathered food nearby, whereas men trave
Such a conceptualization focuses on the division of labor betw
functional for the survival of the species. Social learning theory e
of reward and punishment in explaining how children learn gender role behavior. Identification theory says that children acquire the characteristics and behaviors of their same-sex parent through a process of identification. Boys identify with their fathers; girls identify with their mothers. Cognitive-developmental theory emphasizes biological readiness, in terms of cognitive development, of the child's responses to gender cues in the environment. Once children learn the concept of gender permanence, they seek to becor
group.

Here you'll find key terms and definitions in the order in which they appear in the chapter.

How to Use This Card

1. Look over the card to preview the new concepts you'll be introduced to in the chapter.
2. Read your chapter to fully understand the material.
3. Go to class (and pay attention).
4. Review the card one more time to make sure you've registered the key concepts.
5. Don't forget, this card is only one of many M&F learning tools available to help you succeed in your marriage and family course.

2. of socialization?

Vari
lings (representing differe
race
ucation, and mass media. T
shap
uence what people think, fe
and
e family is a gendered institu
with
r. The names parents assign to
thei
oys they buy them all reflect
gen
, determining the age they are
allo
d directives such as "call when
you

When it's time to prepare for exams, use the review card and the technique to the left to ensure successful study sessions.

2.4 What are the consequences of traditional gender role socialization?

Traditional female role socialization may result in negative outcomes such as less education, less income, negative body image, and lower marital satisfaction but positive

Key Terms

androgyny a blend of traits that are stereotypically associated with masculinity and femininity.

biosocial theory (sociobiology) emphasizes the interaction of one's biological or genetic inheritance with one's social environment to explain and predict human behavior.

cross-dresser a generic term for individuals who may dress or present themselves in the gender of the opposite sex.

feminization of poverty the idea that women disproportionately experience poverty.

gender the social and psychological behaviors associated with being female or male.

gender dysphoria the condition in which one's gender identity does not match one's biological sex.

gender identity the psychological state of viewing oneself as a girl or a boy, and later as a woman or a man.

gender role ideology the proper role relationships between women and men in a society.

gender role transcendence
eworks and
pendent
ries.
ssigned
society.

hermaphrodites (intersexed individuals) people with mixed or ambiguous genitals.

intersex development refers to congenital variations in the reproductive system, sometimes resulting in ambiguous genitals.

occupational sex segregation the concentration of women in certain occupations and men in other occupations.

parental investment any investment by a parent that increases the chance that the offspring will survive and thrive.

positive androgyny a view of androgyny that is devoid of the negative traits associated with masculinity and femininity.

pseudohermaphroditism refers to a condition in which an individual is born with gonads matching the sex chromosomes, but with genitals either ambiguous or resembling those of the opposite sex.

sex the biological distinction between being female and being male.

sex roles behaviors defined by biological constraints.

sexism an attitude, action, or institutional structure that subordinates or discriminates against an individual or group because of their sex.

socialization the process through which we learn attitudes, values, beliefs, and behaviors appropriate to the social positions we occupy.

transgender a generic term for a person of one biological sex who displays characteristics of the opposite sex.

transgenderist an individual who lives in a gender role that does not match his or her biological sex, but has no desire to surgically alter his or her genitalia.

outcomes such as a longer life, a stronger relationship focus, keeping relationships on track, and a closer emotional bond with children. Traditional male role socialization may result in the fusion of self and occupation, a more limited expression of emotion, disadvantages in child custody disputes, and a shorter life but higher income, greater freedom of movement, a greater available pool of potential partners, and greater acceptance in initiating relationships. The "research application" for the chapter revealed that about 30 percent of college men in one study reported their preference for marrying a traditional wife (one who would stay at home to take care of children). These me weakens the marriage.

For additional study tools such as flashcards, interactive quizzing, games, videos, study worksheets, note taking outlines, and Internet activities, visit **4ltrpress.cengage.com/mf**

zation

	ive Consequences
	ger life
Feminization of poverty	Stronger relationship focus
Higher STD/HIV infection risk	Keep relationships on track
Negative body image	Bonding with children
Less marital satisfaction	Identity not tied to job

Consequences of Traditional Male Role Socialization

Negative Consequences	Positive Consequences
Identity tied to work role	Higher income and occupational status
Limited emotionality	More positive self-concept
Fear of intimacy; more lonely	Less job discrimination
Disadvantaged in getting custody	Freedom of movement; more partners to select from; more normative to initiate relationships
Shorter life	Happier marriage

2.5 How are gender roles changing?

Androgyny refers to a blend of traits that are stereotypically associated with both masculinity and femininity. It may also imply flexibility of traits; for example, an androgynous individual may be emotional in one situation, logical in another, assertive in another, and so forth. The concept of gender role transcendence involves abandoning gender schema (for example, becoming "gender aschematic"), so that personality traits, social and occupational roles, and other aspects of our lives become divorced from gender categories. However, such transcendence is not equal for women and men. Although females are becoming more masculine, partly because our society values whatever is masculine, men are not becoming more feminine.

transsexual an individual who has the anatomical and genetic characteristics of one sex but the self-concept of the other.

transvestite term commonly associated with homosexual men who dress provocatively as women to attract men.

true hermaphroditism an extremely rare condition in which individuals are born with both ovarian and testicular tissue.

reviewcard CHAPTER 1 RELATIONSHIPS, MARRIAGES, AND FAMILIES TODAY

Learning Outcomes

1.1 What is the view/theme of this text?

A central theme of this text is to encourage you to be proactive—to make conscious, deliberate relationship choices to enhance your own well-being and the well-being of those in your intimate groups. Though global, structural, cultural, and media influences are operative, a choices framework emphasizes that individuals have some control over their relationships. Important issues to keep in mind about a choices framework for viewing marriage and the family are that (1) not to decide is to decide, (2) some choices require correcting, (3) all choices involve trade-offs, (4) choices include selecting a positive or negative view, (5) making choices produces ambivalence, and (6) some choices are not revocable. Generation Yers (born in the early 1980s) are relaxed about relationship choices. Rather than pair bond, they "hang out," "hook up," and "live together." They are in no hurry to find "the one," to marry, and to begin a family.

1.2 What is marriage?

Marriage is a system of binding a man and a woman together for the reproduction, care (physical and emotional), and socialization of offspring. Marriage in the United States is a legal contract between a couple and their state that regulates their economic and sexual relationship. The federal government supports marriage education in the public school system with the intention of reducing divorce (which is costly to both individuals and society). The various types of marriage are polygyny, polyandry, polyamory, pantagamy, and domestic partnerships.

1.3 What is family?

In recognition of the diversity of families, the definition of family is increasing beyond the U.S. Census Bureau's definition to include two adult partners whose interdependent relationship is long-term and characterized by an emotional and financial commitment. Types of family include nuclear, extended, and blended. There are also traditional, modern, and postmodern families. See the table to the right for differences between marriage and family.

Marriage	Family
Usually initiated by a formal ceremony	Formal ceremony not essential
Involves two people	Usually involves more than two people
Ages of the individuals tend to be similar	Individuals represent more than one generation
Individuals usually choose each other	Members are born or adopted into the family
Ends when spouse dies or is divorced	Continues beyond the life of the individual
Sex between spouses is expected and approved	Sex between near kin is neither expected nor approved
Requires a license	No license needed to become a parent
Procreation expected	Consequence of procreation
Spouses are focused on each other	Focus changes with addition of children
Spouses can voluntarily withdraw from marriage with obligations to children	Spouses/parents cannot easily withdraw voluntarily with approval of state
Money in unit is spent on the couple	Money is used for the needs of children
Recreation revolves around adults	Recreation revolves around children

1.4 How have marriage and the family changed?

The advent of industrialization, urbanization, and mobility involved the demise of familism and the rise of individualism. When family members functioned together as an economic unit, they were dependent on one another for survival and were concerned about what was good for the family. The shift from familism to individualism is only one change; others include divorce replacing death as the endpoint for the majority of marriages, marriage and relationships emerging as legitimate objects of scientific study, the rise of feminism, changes in gender roles, increasing marriage age, and the acceptance of singlehood, cohabitation, and childfree marriages. Marriages and families today must also deal with the additional stresses of terrorism.

1.5 What are the theoretical frameworks for viewing marriage and the family?

Theoretical frameworks provide a set of interrelated principles designed to explain a particular phenomenon and provide a point of view. The following table gives an overview of the frameworks used in this text.

Theoretical Frameworks for Marriage and the Family

THEORY	DESCRIPTION	CONCEPTS	LEVEL OF ANALYSIS	STRENGTHS	WEAKNESSES
Social Exchange	In their relationships, individuals seek to maximize their benefits and minimize their costs.	Benefits, Costs, Profit, Loss	Individual, Couple, Family	Provides explanations of human behavior based on outcome.	Assumes that people always act rationally and all behavior is calculated.
Family Life Course Development	All families have a life course that is composed of all the stages and events that have occurred within the family.	Stages, Transitions, Timing	Institution, Individual, Couple, Family	Families are seen as dynamic rather than static. Useful in working with families who are facing transitions in their life courses.	Difficult to adequately test the theory through research.
Structural-Function	The family has several important functions within society; within the family, individual members have certain functions.	Structure, Function	Institution	Emphasizes the relation of family to society, noting how families affect and are affected by the larger society.	Families with nontraditional structures (single-parent, same-sex couples) are seen as dysfunctional.
Conflict	Conflict in relationships is inevitable, due to competition over resources and power.	Conflict, Resources, Power	Institution	Views conflict as a normal part of relationships and as necessary for change and growth.	Sees all relationships as conflictual, and does not acknowledge cooperation.
Symbolic Interaction	People communicate through symbols and interpret the words and actions of others.	Definition of the situation, Looking-glass self, Self-fulfilling prophecy	Couple	Emphasizes the perceptions of individuals, not just objective reality or the viewpoint of outsiders.	Ignores the larger social interaction context and minimizes the influence of external forces.
Family Systems	The family is a system of interrelated parts that function together to maintain the unit.	Subsystem, Roles, Rules, Boundaries, Open system, Closed system	Couple, Family	Very useful in working with families who are having serious problems (violence, alcoholism). Describes the effect family members have on each other.	Based on work with systems, troubled families, and may not apply to nonproblem families.
Feminism	Women's experience is central and different from man's experience of social reality.	Inequality, Power, Oppression	Institution, Individual, Couple, Family	Exposes inequality and oppression as explanations for frustrations women experience.	Multiple branches of feminism may inhibit central accomplishment of increased equality.

1.6 What are some factors to keep in mind when evaluating research?

Caveats that are factors to be used in evaluating research include a random sample (the respondents providing the data reflect those who were not in the sample), a control group (the group not subjected to the experimental design for a basis of comparison), terminology (the phenomenon being studied should be objectively defined), researcher bias (present in all studies), time lag (takes two years from study to print), and distortion or deception (although rare, some researchers distort their data). Few studies avoid all research problems.

Potential Research Problems in Marriage and Family

Weakness	Consequences	Example
Sample not random	Cannot generalize findings	Opinions of college students do not reflect opinions of other adults.
No control group	Inaccurate conclusions	Study on the effect of divorce on children needs control group of children whose parents are still together.
Age differences between groups of respondents	Inaccurate conclusions	Effect may be due to passage of time or to cohort differences.
Unclear terminology	Inability to measure what is not clearly defined	What is living together, marital happiness, sexual fulfillment, good communication, quality time?
Researcher bias	Slanted conclusions	Male researcher may assume that, because men usually ejaculate each time they have intercourse, women should have an orgasm each time they have intercourse.
Time lag	Outdated conclusions	Often-quoted Kinsey sex research is more than fifty years old.
Distortion	Invalid conclusions	Research subjects exaggerate, omit information, and/or recall facts or events inaccurately. Respondents may remember what they wish had happened.

Key Terms

beliefs definitions and explanations about what is thought to be true.

binuclear family family in which the members live in two households.

blended family (stepfamily) a family created when two individuals marry and at least one of them brings a child or children from a previous relationship or marriage.

civil union a pair-bonded relationship given legal significance in terms of rights and privileges.

collectivism pattern that one regards group values and goals as more important than one's own values and goals.

common-law marriage a marriage by mutual agreement between cohabitants without a marriage license or ceremony (recognized in some, but not all, states).

conflict framework view that individuals in relationships compete for valuable resources.

control group group used to compare with the experimental group that is not exposed to the independent variable being studied.

domestic partnership a relationship in which individuals who live together are emotionally and financially interdependent and are given some kind of official recognition by a city or corporation so as to receive partner benefits.

experimental group the group exposed to the independent variable.

extended family the nuclear family or parts of it plus other relatives.

familism philosophy in which decisions are made in reference to what is best for the family as a collective unit.

family a group of two or more people related by blood, marriage, or adoption.

family life course development the stages and process of how families change over time.

family life cycle stages which identify the various challenges faced by members of a family across time.

family of orientation the family of origin into which a person is born.

family of origin the family into which an individual is born or reared, usually including a mother, father, and children.

family of procreation the family a person begins by getting married and having children.

family systems framework views each member of the family as part of a system and the family as a unit that develops norms of interaction.

feminist framework views marriage and the family as contexts for inequality and oppression.

feral children wild, undomesticated children who are thought to have been reared by animals.

functionalists structural functionalist theorists who view the family as an institution with values, norms, and activities meant to provide stability for the larger society.

Generation Y children of the baby boomers, typically born between 1979 and 1984. Also known as the Millennial or Internet Generation.

hypothesis a suggested explanation for a phenomenon.

individualism philosophy in which decisions are made on the basis of what is best for the individual.

institution established and enduring patterns of social relationships.

IRB approval Institutional Review Board approval is the OK by one's college, university, or institution that the proposed research is consistent with research ethics standards and poses no undo harm to participants.

marriage a legal contract signed by a couple with the state in which they reside that regulates their economic and sexual relationship.

marriage-resilience perspective the view that changes in the institution of marriage are not indicative of a decline and do not have negative effects.

mating gradient the tendency for husbands to marry wives who are younger and have less education and less occupational success.

modern family the dual-earner family, in which both spouses work outside the home.

nuclear family family consisting of an individual, his or her spouse, and his or her children, or of an individual and his or her parents and siblings.

open relationship a stable relationship in which the partners regard their own relationship as primary but agree that each may have emotional and physical relationships with others.

pantagamy a group marriage in which each member of the group is "married" to the others.

polyamory a term meaning "many loves," whereby three or more men and women have a committed emotional and sexual relationship.

polyandry a form of polygamy in which one wife has two or more husbands.

polygamy a generic term referring to a marriage involving more than two spouses.

polygyny a form of polygamy in which one husband has two or more wives.

postmodern family non-traditional families emphasizing that a healthy family need not be heterosexual or have two parents.

primary group small, intimate, informal group.

random sample sample in which each person in the population being studied has an equal chance of being included in the sample.

role the behavior with which individuals in certain status positions are expected to engage.

secondary group large or small group characterized by impersonal and formal interaction.

sequential ambivalence the individual experiences one wish and then another.

simultaneous ambivalence the person experiences two conflicting wishes at the same time.

social exchange framework spouses exchange resources, and decisions are made on the basis of perceived profit and loss.

sociological imagination the perspective of how powerful social structure and culture are in influencing personal decision making.

status a social position a person occupies within a social group.

structural-function framework emphasizes how marriage and family contribute to the larger society.

symbolic interaction framework views marriage and families as symbolic worlds in which the various members give meaning to each other's behavior.

theoretical framework a set of interrelated principles designed to explain a particular phenomenon and to provide a point of view.

traditional family the two-parent nuclear family with the husband as breadwinner and wife as homemaker.

utilitarianism the doctrine holding that individuals rationally weigh the rewards and costs associated with behavioral choices.

values standards regarding what is good and bad, right and wrong, desirable and undesirable.

reviewcard CHAPTER 2 GENDER

Learning Outcomes

2.1 What are the important terms related to gender?

Sex refers to the biological distinction between females and males. One's biological sex is identified on the basis of one's chromosomes, gonads, hormones, internal sex organs, and external genitals, and exists on a continuum rather than being a dichotomy. *Gender* refers to the social and psychological characteristics often associated with being female or male. Other terms related to gender include *gender identity, gender role, gender role ideology, transgender,* and *transgenderism.*

2.2 What theories explain gender role development?

Biosocial theory emphasizes that social behaviors (for example, gender roles) are biologically based and have an evolutionary survival function. Traditionally, women stayed in the nest or gathered food nearby, whereas men traveled far to find food. Such a conceptualization focuses on the division of labor between women and men as functional for the survival of the species. Social learning theory emphasizes the roles of reward and punishment in explaining how children learn gender role behavior. Identification theory says that children acquire the characteristics and behaviors of their same-sex parent through a process of identification. Boys identify with their fathers; girls identify with their mothers. Cognitive-developmental theory emphasizes biological readiness, in terms of cognitive development, of the child's responses to gender cues in the environment. Once children learn the concept of gender permanence, they seek to become competent, proper members of their gender group.

2.3 What are the various agents of socialization?

Various socialization influences include parents and siblings (representing different races and ethnicities), peers, religion, the economy, education, and mass media. These shape individuals toward various gender roles and influence what people think, feel, and do in their roles as woman or man. For example, the family is a gendered institution with female and male roles highly structured by gender. The names parents assign to their children, the clothes they dress them in, and the toys they buy them all reflect gender. Parents may also be stricter on female children, determining the age they are allowed to leave the house at night, time of curfew, and directives such as "call when you get to the party."

2.4 What are the consequences of traditional gender role socialization?

Traditional female role socialization may result in negative outcomes such as less education, less income, negative body image, and lower marital satisfaction but positive

Key Terms

androgyny a blend of traits that are stereotypically associated with masculinity and femininity.

biosocial theory (sociobiology) emphasizes the interaction of one's biological or genetic inheritance with one's social environment to explain and predict human behavior.

cross-dresser a generic term for individuals who may dress or present themselves in the gender of the opposite sex.

feminization of poverty the idea that women disproportionately experience poverty.

gender the social and psychological behaviors associated with being female or male.

gender dysphoria the condition in which one's gender identity does not match one's biological sex.

gender identity the psychological state of viewing oneself as a girl or a boy, and later as a woman or a man.

gender role ideology the proper role relationships between women and men in a society.

gender role transcendence abandoning gender frameworks and looking at phenomena independent of traditional gender categories.

gender roles behaviors assigned to women and men in a society.

hermaphrodites (intersexed individuals) people with mixed or ambiguous genitals.

intersex development refers to congenital variations in the reproductive system, sometimes resulting in ambiguous genitals.

occupational sex segregation the concentration of women in certain occupations and men in other occupations.

parental investment any investment by a parent that increases the chance that the offspring will survive and thrive.

positive androgyny a view of androgyny that is devoid of the negative traits associated with masculinity and femininity.

pseudohermaphroditism refers to a condition in which an individual is born with gonads matching the sex chromosomes, but with genitals either ambiguous or resembling those of the opposite sex.

sex the biological distinction between being female and being male.

sex roles behaviors defined by biological constraints.

sexism an attitude, action, or institutional structure that subordinates or discriminates against an individual or group because of their sex.

socialization the process through which we learn attitudes, values, beliefs, and behaviors appropriate to the social positions we occupy.

transgender a generic term for a person of one biological sex who displays characteristics of the opposite sex.

transgenderist an individual who lives in a gender role that does not match his or her biological sex, but has no desire to surgically alter his or her genitalia.

transsexual an individual who has the anatomical and genetic characteristics of one sex but the self-concept of the other.

outcomes such as a longer life, a stronger relationship focus, keeping relationships on track, and a closer emotional bond with children. Traditional male role socialization may result in the fusion of self and occupation, a more limited expression of emotion, disadvantages in child custody disputes, and a shorter life but higher income, greater freedom of movement, a greater available pool of potential partners, and greater acceptance in initiating relationships. The "research application" for the chapter revealed that about 30 percent of college men in one study reported their preference for marrying a traditional wife (one who would stay at home to take care of children). These men believe that a wife's working outside the home weakens the marriage.

Consequences of Traditional Female Role Socialization

Negative Consequences	Positive Consequences
Less education/income (more dependent)	Longer life
Feminization of poverty	Stronger relationship focus
Higher STD/HIV infection risk	Keep relationships on track
Negative body image	Bonding with children
Less marital satisfaction	Identity not tied to job

Consequences of Traditional Male Role Socialization

Negative Consequences	Positive Consequences
Identity tied to work role	Higher income and occupational status
Limited emotionality	More positive self-concept
Fear of intimacy; more lonely	Less job discrimination
Disadvantaged in getting custody	Freedom of movement; more partners to select from; more normative to initiate relationships
Shorter life	Happier marriage

2.5 How are gender roles changing?

Androgyny refers to a blend of traits that are stereotypically associated with both masculinity and femininity. It may also imply flexibility of traits; for example, an androgynous individual may be emotional in one situation, logical in another, assertive in another, and so forth. The concept of gender role transcendence involves abandoning gender schema (for example, becoming "gender aschematic"), so that personality traits, social and occupational roles, and other aspects of our lives become divorced from gender categories. However, such transcendence is not equal for women and men. Although females are becoming more masculine, partly because our society values whatever is masculine, men are not becoming more feminine.

transvestite term commonly associated with homosexual men who dress provocatively as women to attract men.

true hermaphroditism an extremely rare condition in which individuals are born with both ovarian and testicular tissue.

reviewcard

CHAPTER 3
SINGLEHOOD, HANGING OUT, HOOKING UP, AND COHABITATION

Learning Outcomes

3.1 What are the attractions of singlehood?

An increasing percentage of people are delaying marriage. Between the ages of 25 and 29, 57.6 percent of males and 43.4 percent of females are not married. The Alternatives to Marriage Project is designed to give visibility and credibility to the status of being unmarried. The primary attraction of singlehood is the freedom to do as one chooses. As a result of the sexual revolution, the women's movement, and the gay liberation movement, there is increased social approval of being unmarried. Single people are those who never married as well as those who are divorced or widowed.

Reasons to Remain Single

Benefits of Singlehood	Limitations of Marriage
Freedom to do as one wishes	Restricted by spouse or children
Variety of lovers	One sexual partner
Spontaneous lifestyle	Routine, predictable lifestyle
Close friends of both sexes	Pressure to avoid close other-sex friendships
Responsible for one person only	Responsible for spouse and children
Spend money as one wishes	Expenditures influenced by needs of spouse and children
Freedom to move as career dictates	Restrictions on career mobility
Avoid being controlled by spouse	Potential to be controlled by spouse
Avoid emotional and financial stress of divorce	Possibility of divorce

3.2 What are some categories of singles?

The term *singlehood* is most often associated with young unmarried individuals. However, there are three categories of single people: the never-married, the divorced, and the widowed.

U.S. Adult Population by Relationship Status
N=222,557,000

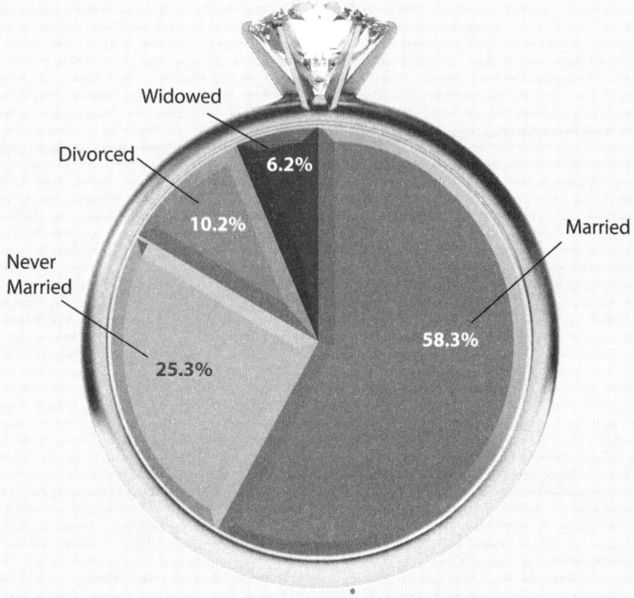

Source: *Statistical Abstract of the United States, 2009.* 128th ed. Washington, DC: U.S. Bureau of the Census, Table 56.

Key Terms

cohabitation (living together) two unrelated adults (by blood or by law) involved in an emotional and sexual relationship who sleep in the same residence at least four nights a week.

common-law marriage a marriage by mutual agreement between a cohabiting man and woman without a marriage license or ceremony.

domestic partnership a relationship in which individuals who live together are emotionally and financially interdependent and are given some kind of official recognition by a city or corporation so as to receive partner benefits.

hanging out refers to going out in groups where the agenda is to meet others and have fun.

hooking up a one-time sexual encounter in which there are generally no expectations of seeing one another again.

intentional community (commune) group of people living together on the basis of shared values and worldview.

living apart together (LAT) a committed couple who does not live in the same home.

palimony referring to the amount of money one "pal" who lives with another "pal" may have to pay if the partners terminate their relationship.

POSSLQ an acronym used by the U.S. Census Bureau that stands for "people of the opposite sex sharing living quarters."

satiation the state in which a stimulus loses its value with repeated exposure.

singlehood being unmarried.

3.3 What are some ways to find a partner?

Besides the traditional way of meeting people at work or school or through friends and going on a date, couples today may also "hang out," which may lead to "hooking up." Internet dating, video dating, speed-dating, and international dating are new forms for finding each other.

ONLINE MEETING

Pros:

- Highly efficient
- Develop a relationship without visual distraction
- Avoid crowded, loud, uncomfortable locations, like bars
- The opportunity to try on new identities

Cons:

- Ease of deception
- Relationship involvement escalates too quickly.
- Inability to assess "chemistry" through the computer
- A lot of competition

3.4 What are the functions of involvement with a partner?

There are at least seven functions of becoming involved with someone: confirmation of a social self, recreation, companionship, anticipatory socialization, status achievement, mate selection, and health enhancement.

3.5 What is cohabitation like among today's youth?

Cohabitation, also known as living together, is becoming a "normative life experience," with almost 60 percent of American women reporting that they had cohabited before marriage. Reasons for an increase in living together include delaying marriage, fear of marriage, increased social tolerance for living together, and a desire to avoid the legal entanglements of marriage. Types of relationships in which couples live together include the here-and-now, testers, engaged couples, and cohabitants forever. Most people who live together eventually marry.

Although living together before marriage does not ensure a happy, stable marriage, it has some potential advantages. These include a sense of well-being, delayed marriage, learning about yourself and your partner, and being able to disengage with minimal legal hassle. Disadvantages include feeling exploited, feeling guilty about lying to parents, and not having the same economic benefits as those who are married. Social Security and retirement benefits are paid to spouses, not live-in partners.

3.6 What are intentional communities?

An intentional community is the newer term for what was formerly known as a commune. Intentional communities encompass a wide range of ages, races, ethnicities, and social class backgrounds, and are an alternative to other types of single living that people choose to explore.

3.7 What are the pros and cons of "living apart together"?

A new lifestyle and family form is living apart together (LAT), which means that monogamous committed partners—whether married or not—carve out varying degrees of physical space between them. People living apart together exist on a continuum from partners who have separate bedrooms and baths in the same house to those who live in separate places (apartment, condo, house) in the same or different cities. Couples choose this pattern for a number of reasons, including the desire to maintain some level of independence, to enjoy their time alone, to keep their relationship exciting, and so on.

Advantages to involvement in an LAT relationship include space and privacy, sleeping without being cramped or dealing with snoring, not living with animals if there is an allergy, having family or friends over without interfering with a partner's life space, and keeping the relationship exciting. Disadvantages include being confronted with stigma or disapproval, cost, inconvenience, and waking up alone.

reviewcard CHAPTER 4 COMMUNICATION

Learning Outcomes

4.1 What is the nature of interpersonal communication?

Communication is the exchange of information and feelings by two individuals. It involves both verbal and nonverbal messages. The nonverbal part of a message often carries more weight than the verbal part.

4.2 What are various issues related to conflict in relationships?

Conflict is both inevitable and desirable. Unless individuals confront and resolve issues over which they disagree, one or both may become resentful and withdraw from the relationship. Conflict may result from one partner's doing something the other does not like, having different perceptions, or having different values. Sometimes it is easier for one partner to view a situation differently or alter a value than for the other partner to change the behavior causing the distress.

4.3 What are some principles and techniques of effective communication?

Some basic principles and techniques of effective communication include making communication a priority, maintaining eye contact, asking open-ended questions, using reflective listening, using "I" statements, complimenting each other, and sharing power. Partners must also be alert to keeping the dialogue (process) going even when they don't like what is being said (content).

Judgmental and Nonjudgmental Responses to a Partner's Saying, "I'd Like to Spend One Evening a Week with My Friends"

Nonjudgmental, Reflective Statements	Judgmental Statements
You value your friends and want to maintain good relationships with them.	You only think about what you want.
You think it is healthy for us to be with our friends some of the time.	Your friends are more important to you than I am.
You really enjoy your friends and want to spend some time with them.	You just want a night out so that you can meet someone new.
You think it is important that we not abandon our friends just because we are involved.	You just want to get away so you can drink.
You think that our being apart one night each week will make us even closer.	You are selfish.

4.4 How are relationships affected by self-disclosure, dishonesty, and lying?

The levels of self-disclosure and honesty influence intimacy in relationships. High levels of self-disclosure are associated with increased intimacy. Despite the importance of honesty in relationships, deception occurs frequently in interpersonal relationships. Partners sometimes lie to each other about previous sexual relationships, how they feel about each other, and how they experience each other sexually. Telling lies is not the only form of dishonesty. People exaggerate, minimize, tell partial truths, pretend, and engage in self-deception. Partners may withhold information or keep secrets to protect themselves and/or to preserve the relationship. However, the more intimate the relationship, the greater our desire to share our most personal and private selves with

Key Terms

accommodating style of conflict conflict style in which the respective partners are not assertive in their positions but are cooperative. Each attempts to soothe the other and to seek a harmonious solution.

avoiding style of conflict conflict style in which partners are neither assertive nor cooperative, so as to avoid confrontation.

brainstorming suggesting as many alternatives as possible without evaluating them.

branching in communication, going out on different limbs of an issue rather than staying focused on the issue.

closed-ended question question that allows for a one-word answer and does not elicit much information.

collaborating style of conflict conflict style in which partners are both assertive and cooperative. Each expresses views and cooperates to find a solution.

communication the process of exchanging information and feelings between two or more people.

competing style of conflict conflict style in which partners are both assertive and uncooperative. Each tries to force a way on the other so that there is a winner and a loser.

compromising style of conflict conflict style in which there is an intermediate solution in which both partners find a middle ground they can live with.

conflict the interaction that occurs when the behavior or desires of one person interfere with the behavior or desires of another.

congruent message one in which verbal and nonverbal behaviors match.

defense mechanisms unconscious techniques that function to protect individuals from anxiety and minimize emotional hurt.

displacement shifting one's feelings, thoughts, or behaviors from the person who evokes them onto someone else.

escapism the simultaneous denial of and withdrawal from a problem.

"I" statements statements that focus on the feelings and thoughts of the communicator without making a judgment on others.

lose-lose solution a solution to a conflict in which neither partner benefits.

nonverbal communication the "message about the message," using gestures, eye contact, body posture, tone, volume, and rapidity of speech.

open-ended question question that encourages answers that contain a great deal of information.

parallel style of conflict style of conflict whereby both partners deny, ignore, and retreat from addressing a problem issue.

power the ability to impose one's will on one's partner and to avoid being influenced by the partner.

projection attributing one's own feelings, attitudes, or desires to one's partner while avoiding recognition that these are one's own thoughts, feelings, and desires.

rationalization the cognitive justification for one's own behavior that unconsciously conceals one's true motives.

our partner and the greater the emotional consequences of not sharing. In intimate relationships, keeping secrets can block opportunities for healing, resolution, self-acceptance, and a deeper intimacy with your partner.

4.5 How are interactionist and exchange theories applied to relationship communication?

Symbolic interactionists examine the process of communication between two actors in terms of the meanings each attaches to the actions of the other. Definition of the situation, the looking-glass self, and taking the role of the other are all relevant to understanding how partners communicate.

Exchange theorists suggest that the partners' communication can be described as a ratio of rewards to costs. Rewards are positive exchanges, such as compliments, compromises, and agreements. Costs refer to negative exchanges, such as critical remarks, complaints, and attacks. When the rewards are high and the costs are low, the outcome is likely to be positive for both partners (profit). When the costs are high and the rewards low, neither may be satisfied with the outcome (loss).

4.6 What are examples of fighting fair to resolve conflict?

The sequence of resolving conflict includes deciding to address recurring issues rather than suppressing them, asking the partner for help in resolving issues, finding out the partner's point of view, summarizing in a nonjudgmental way the partner's perspective, and finding alternative win-win solutions. Defense mechanisms that interfere with conflict resolution include escapism, rationalization, projection, and displacement.

The Seven Steps

The seven steps for fair fighting and resolution of interpersonal conflict:

1. Address recurring, disturbing issues
2. Identify new desired behaviors
3. Identify perceptions to change
4. Summarize you partner's perspective
5. Generate alternative win-win solutions
6. Forgive
7. Be alert to defense mechanisms

reflective listening paraphrasing or restating what a person has said to indicate that the listener understands.

win-lose solution a solution to a conflict in which one partner benefits at the expense of the other.

win-win relationship a relationship in which conflict is resolved so that each partner derives benefits from the resolution.

"you" statements statements that blame or criticize the listener and often result in increasing negative feelings and behavior in the relationship.

Learning Outcomes

5.1 What are some ways that love has been described?

Love remains an elusive and variable phenomenon. Researchers have conceptualized love as a continuum from romanticism to realism, as a triangle consisting of three basic elements (intimacy, passion, and commitment), and as a style (from playful ludic love to obsessive and dangerous manic love).

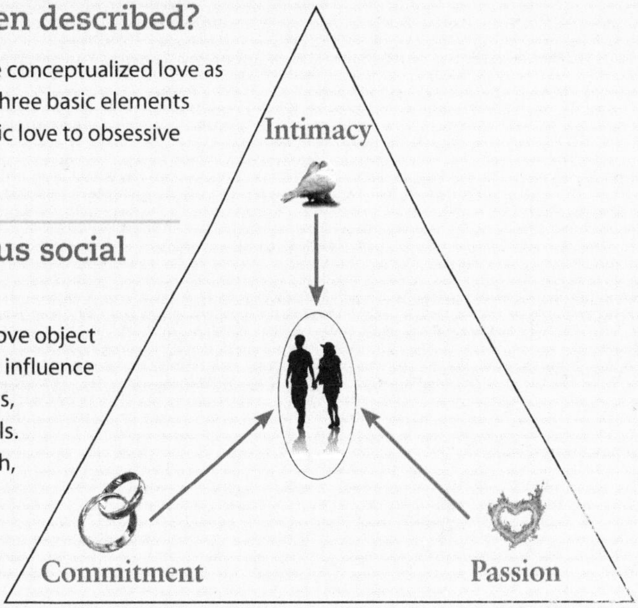

5.2 How has love expressed itself in various social and historical contexts?

The society in which we live exercises considerable control over our love object or choice and conceptualizes it in various ways. Parents inadvertently influence the mate choice of their children by moving to certain neighborhoods, joining certain churches, and enrolling their children in certain schools. Doing so increases the chance that their offspring will "hang out" with, fall in love with, and marry people who are similar in race, education, and social class.

In the 1100s in Europe, marriage was an economic and political arrangement that linked two families. As aristocratic families declined after the French Revolution, love bound a woman and man together. Previously, Buddhists, Greeks, and Hebrews had their own views of love. Love in colonial America was also tightly controlled.

5.3 How does love develop in a new relationship?

Love occurs under certain conditions. Social conditions include a society that promotes the pursuit of love, peers who enjoy it, and a set of norms that link love and marriage. Psychological conditions involve high self-esteem and a willingness to disclose one's self to others. Physiological and cognitive conditions imply that the individual experiences a stirred-up state and labels it "love." Love sometimes provides a context for problems in that a young person in love will lie to parents and become distant from them so as to be with the lover. Also, lovers experience problems such as being in love with two people at the same time, being in love with someone who is abusive, and making risky, dangerous, or questionable choices while in love (for example, not using a condom) or reacting to a former lover who has become a stalker.

5.4 How do jealousy and love interface?

Jealousy is an emotional response to a perceived or real threat to a valued relationship. Types of jealousy are reactive, anxious, and possessive. Jealous feelings may have both internal and external causes and may have both positive and negative consequences for a couple's relationship.

5.5 What are the cultural factors that influence partnering?

Two types of cultural influences in mate selection are endogamy (to marry someone inside one's own social group—race, religion, social class) and exogamy (to marry someone outside one's own family).

5.6 What are the sociological factors that influence partnering?

Sociological aspects of mate selection involve homogamy. Variables include race, age, religion, education, social class, personal appearance, attachment, personality, and open-mindedness. Couples who have a lot in common are more likely to have a happy and durable relationship.

5.7 What are the psychological factors operative in partnering?

Psychological aspects of mate selection include complementary needs, exchange theory, and parental characteristics. Personality characteristics of a potential mate desired by both men and women include being warm, kind, and open and having a sense of humor. Negative personality characteristics to avoid in a potential mate include disagreeableness or expressing few positives, poor impulse control, hypersensitivity to criticism, inflated ego, neurosis (perfectionism) or insecurity, and control by someone else (for example, parents). Paranoid, schizoid, and borderline personalities are also to be avoided.

Personality Types Problematic in a Potential Partner

Type	Characteristics	Impact on Partner
Paranoid	Suspicious, distrustful, thin-skinned, defensive	Partners may be accused of everything.
Schizoid	Cold, aloof, solitary, reclusive	Partners may feel that they can never "connect" and that the person is not capable of returning love.
Borderline	Moody, unstable, volatile, unreliable, suicidal, impulsive	Partners will never know what their Jekyll-and-Hyde partner will be like, which could be dangerous.
Antisocial	Deceptive, untrustworthy, conscienceless, remorseless	Such a partner could cheat on, lie, or steal from a partner and not feel guilty.
Narcissistic	Egotistical, demanding, greedy, selfish	Such a person views partners only in terms of their value. Don't expect such a person to see anything from a partner's point of view; expect such a person to bail in tough times.
Dependent	Helpless, weak, clingy, insecure	Such a person will demand a partner's full time and attention, and other interests will incite jealousy.
Obsessive-compulsive	Rigid, inflexible	Such a person has rigid ideas about how a partner should think and behave and may try to impose them on the partner.

5.8 What are the sociobiological factors operative in partnering?

The sociobiological view of mate selection suggests that men and women select each other on the basis of their biological capacity to produce and support healthy offspring. Men seek young women with healthy bodies, and women seek ambitious men who will provide economic support for their offspring. There is considerable controversy about the validity of this theory.

5.9 What factors should be considered when becoming engaged?

The engagement period is the time to ask specific questions about one's partner's values, goals, and marital agenda, to visit each other's parents to assess parental models, and to consider involvement in premarital educational programs and/or counseling. Negative reasons for getting married include being on the rebound, escaping from an unhappy home life, psychological blackmail, and pity.

Some couples (particularly those with children from previous marriages) decide to write a prenuptial agreement to specify who gets what and the extent of spousal support in the event of a divorce. To be valid, the document should be developed by an attorney in accordance with the laws of the state in which the partners reside. Last-minute prenuptial agreements put enormous emotional strain on the couple and are often considered invalid by the courts. Discussing a prenuptial agreement six months in advance is recommended.

5.10 What factors suggest you might consider calling off the wedding?

Factors suggesting that a couple may not be ready for marriage include being in their teens, having known each other less than two years, and having a relationship characterized by significant differences and/or dramatic parental disapproval. Some research suggests that partners with the greatest number of similarities in values, goals, and common interests are most likely to have happy and durable marriages.

reviewcard CHAPTER 5 LOVE AND BECOMING PARTNERS

Key Terms

agape love style love style characterized by a focus on the well-being of the love object, with little regard for reciprocation. The love of parents for their children is agape love.

anxious jealousy obsessive ruminations about the partner's alleged infidelity make one's life a miserable emotional torment.

arranged marriage mate selection pattern whereby parents select the spouse of their offspring.

complementary-needs theory tendency to select mates whose needs are opposite and complementary to one's own needs.

conjugal (married) love the love between married people characterized by companionship, calmness, comfort, and security.

dowry (trousseau) the amount of money or valuables a woman's father pays a man's father for the man to marry his daughter. It functioned to entice the man to marry the woman, because an unmarried daughter stigmatized the family of the woman's father.

endogamous pressures cultural attitudes reflecting approval for selecting a partner within one's social group and disapproval for selecting a partner outside one's social group.

endogamy the cultural expectation to select a marriage partner within one's social group.

engagement period of time during which committed, monogamous partners focus on wedding preparations and systematically examine their relationship.

eros love style love style characterized by passion and romance.

exchange theory theory that emphasizes that relations are formed and maintained between individuals offering the greatest rewards and least costs to each other.

exogamous pressures cultural attitudes reflecting approval for selecting a partner outside one's family group.

exogamy the cultural expectation that one will marry outside the family group.

extradyadic relationship emotional or sexual involvement between a member of a couple and someone other than the partner.

homogamy tendency to select someone with similar characteristics.

homogamy theory of mate selection theory that individuals tend to be attracted to and become involved with those who have similar characteristics.

infatuation emotional feelings based on little actual exposure to the love object.

jealousy an emotional response to a perceived or real threat to an important or valued relationship.

ludic love style love style in which love is viewed as a game whereby the love interest is one of several partners, is never seen too often, and is kept at an emotional distance.

lust sexual desire.

manic love style an out-of-control love whereby the person "must have" the love object. Obsessive jealousy and controlling behavior are symptoms of manic love.

mating gradient the tendency for husbands to be more advanced than their wives with regard to age, education, and occupational success.

open-minded being open to understanding alternative points of view, values, and behaviors.

pool of eligibles the population from which a person selects an appropriate mate.

possessive jealousy attacking the partner who is perceived as being unfaithful.

pragma love style love style that is logical and rational. The love partner is evaluated in terms of assets and liabilities.

premarital education programs formal systematized experiences designed to provide information to individuals and couples about how to have a good relationship.

prenuptial agreement formal document specifying how property will be divided if the marriage ends in divorce or by one partner's death.

principle of least interest principle stating that the person who has the least interest in a relationship controls the relationship.

racial homogamy tendency for individuals to marry someone of the same race.

reactive jealousy feelings that the partner may be straying.

religion specific fundamental set of beliefs and practices generally agreed upon by a number of people or sects.

religious homogamy tendency for people of similar religious or spiritual philosophies to seek out each other.

role (modeling) theory of partnering emphasizes that children select partners similar to the one their same sex parent selected.

romantic love an intense love whereby the lover believes in love at first sight, only one true love, and that love conquers all.

sociobiology suggests a biological basis for all social behavior.

storge love style a love consisting of friendship that is calm and nonsexual.

reviewcard CHAPTER 6 MARRIAGE RELATIONSHIPS

Learning Outcomes

6.1 What are individual motivations of marriage?

Individuals' motives for marriage include
- personal fulfillment,
- companionship,
- legitimacy of parenthood, and
- emotional and financial security.

Traditional versus Egalitarian Marriages

Traditional Marriage	Egalitarian Marriage
There is limited expectation of husband to meet emotional needs of wife and children.	Husband is expected to meet emotional needs of wife and to be involved with children.
Wife is not expected to earn income.	Wife is expected to earn income.
Emphasis is on ritual and roles.	Emphasis is on companionship.
Couples do not live together before marriage.	Couples may live together before marriage.
Wife takes husband's last name.	Wife may keep her maiden name.
Husband is dominant; wife is submissive.	Neither spouse is dominant.
Roles for husband and wife are rigid.	Roles for spouses are flexible.
Husband initiates sex; wife complies.	Either spouse initiates sex.
Wife takes care of children.	Parents share child rearing.
Education is important for husband, not for wife.	Education is important for both spouses.
Husband's career decides family residence.	Career of either spouse determines family residence.

6.2 What are social functions of marriage?

Social functions include
- continuing to provide society with socialized members,
- regulating sexual behavior, and
- stabilizing adult personalities.

6.3 What are three levels of commitment in marriage?

Marriage involves three levels of commitment
1. person-to-person,
2. family-to-family, and
3. couple-to-state.

6.4 What are two rites of passage associated with marriage?

The wedding is a rite of passage signifying the change from the role of fiancé to the role of spouse. In general women, more than men, are invested in preparation for the

Key Terms

artifact concrete symbol that an event exists; in game theory, evidence that a game exists.

"blue" wedding artifact blue artifact worn by bride that is symbolic of fidelity.

"borrowed" wedding artifact artifact worn by bride that may be a garment owned by a currently happy bride.

bride wealth also known as bride price or bride payment, the amount of money or payment in goods (e.g., cows) by the groom or his family to the wife's family for giving her up.

commensality eating with others. Most spouses eat together and negotiate who joins them.

commitment an intent to maintain a relationship.

cougar a woman, usually in her 30s or 40s, who is financially stable and mentally independent and looking for a younger man with whom to have fun.

disenchantment the change in a relationship from a state of newness and high expectation to a state of mundaneness and boredom in the face of reality.

honeymoon the time following the wedding whereby the couple become isolated to recover from the wedding and to solidify their new status change from lovers to spouses.

marital success relationship in which the partners have spent many years together and define themselves as happy and in love (hence, the factors of time and emotionality).

May-December marriage an age-discrepant marriage, usually in which the younger woman is in the spring of her life (May) and the man is in his later years (December).

"new" wedding artifact artifact worn by a bride that symbolizes the new life she is to begin (for example, a new, unlaundered undergarment).

"old" wedding artifact artifact worn by a bride that symbolizes durability of the impending marriage (for example old gold locket).

rite of passage event that marks the transition from one status to another.

wedding; the wedding is perceived to be more for the bride's family, and many women prefer a traditional wedding. The honeymoon is a time of personal recuperation and making the transition to the new role of spouse.

6.5 What changes might a person anticipate after marriage?

Changes after the wedding are

- legal,
- personal,
- social,
- economic,
- sexual, and
- parental.

6.6 What are examples of diversity in marriage relationships?

Mixed marriages include interracial, interreligious, and age-discrepant. When age-discrepant and age-similar marriages are compared, there are no differences in regard to marital happiness. There are three main types of military marriages: (1) those in which the soldier falls in love with a high school sweetheart, marries the person, and subsequently joins the military; (2) those in which the partners meet and marry after one of them has signed up for the military; and (3) the contract marriage in which a soldier will marry a civilian to get more money and benefits from the government. Military families cope with deployment, the double standard, and limited income.

6.7 What are the characteristics associated with successful marriages?

Marital success is defined in terms of both quality and durability. Characteristics associated with marital success include commitment, common interests, communication, religiosity, trust, and nonmaterialism, and having positive role models, low stress levels, and sexual desire. Marriage education programs are designed to improve marriage. There are few empirical studies on marriage education programs.

Figure 6.1

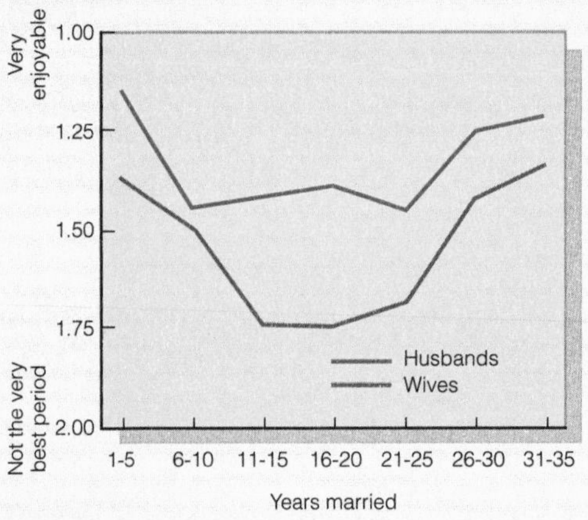

Source: Caroline O. Vaillant and George E. Vaillant. 1993. Is the U-curve of marital satisfaction an illusion? A 40-year study of marriage. *Journal of Marriage and the Family* 55:237 (Figure 6, copyrighted 1993 by the National Council on Family Relations, 3989 Central Avenue N. E., Suite 550, Minneapolis, MN 55421).

reviewcard CHAPTER 7 SAME-SEX COUPLES AND FAMILIES

Learning Outcomes

7.1 How prevalent are homosexuality/bisexuality and same-sex households and families?

Estimates of the lesbigay population in the United States range from about 2 percent to 5 percent of the total U.S. population, with more lesbians than gay males and more bisexual than gay individuals. The 2000 census revealed that about 1 in 9 (or 594,000) unmarried-partner households in the United States involved partners of the same sex. About one-fifth (22.3 percent) of gay male couples and one-third (34.3 percent) of lesbian couples have children in the home.

7.2 Can homosexuals change their sexual orientation?

Some individuals believe that homosexual people choose their sexual orientation, and think that, through various forms of reparative therapy, homosexual people should change their sexual orientation. However, many professional organizations (including the American Psychiatric Association, the American Psychological Association, and the American Medical Association) agree that homosexuality is not a mental disorder, needs no cure, and that efforts to change sexual orientation do not work and may, in fact, be harmful. These organizations also support the notion that changing societal reaction to homosexuals is a more appropriate focus.

7.3 Who tends to have more negative views toward homosexuality?

In general, individuals who are more likely to have negative attitudes toward homosexuality and to oppose gay rights are older, attend religious services, are less educated, live in the South or Midwest, and reside in small rural towns. About half (51 percent) of U.S. adults say that homosexuality should be considered an acceptable alternative lifestyle.

7.4 How are gay and lesbian couples different from heterosexual couples?

Research suggests that gay and lesbian couples tend to be more similar to than different from heterosexual couples. Both gay and straight couples tend to value long-term monogamous relationships, experience high relationship satisfaction early in the relationship that decreases over time, disagree about the same topics, and have the same factors linked to relationship satisfaction.

However, same-sex couples' relationships are different from heterosexual relationships. Same-sex couples have more concern about when and how to disclose their relationships, are more likely to achieve a fair distribution of household labor, argue more effectively and resolve conflict in a positive way, and face prejudice and discrimination without much government support.

7.5 What forms of legal recognition of same-sex couples exist in the United States?

The link below identifies states that recognize same-sex relationships. Some work places acknowledge same-sex relationships by giving employees' partners benefits.

Key Terms

antigay bias any behavior or statement which reflects a negative attitude toward homosexuals.

biphobia (binegativity) refers to a parallel set of negative attitudes toward bisexuality and those identified as bisexual.

bisexuality emotional and sexual attraction to members of both sexes.

civil union a pair-bonded relationship given legal significance in terms of rights and privileges (more than a domestic relationship and less than a marriage). Vermont recognizes civil unions of same-sex individuals.

Defense of Marriage Act (DOMA) legislation passed by Congress denying federal recognition of homosexual marriage and allowing states to ignore same-sex marriages licensed elsewhere.

discrimination behavior that denies individuals or groups equality of treatment.

domestic partnership a relationship in which individuals who live together are emotionally and financially interdependent and are given some kind of official recognition by a city or corporation so as to receive partner benefits (for example, health insurance).

garriage term for relationship of gay individuals who are married or committed.

gay homosexual women or men.

heterosexism the denigration and stigmatization of any behavior, person, or relationship that is not heterosexual.

One resource for current information on rights for same-sex couples is the interactive gay marriage map produced by the *L.A. Times* and posted at **http://www.latimes.com/news/local/la-gmtimeline-fl,0,5345296.htmlstory**.

7.6 What does research on gay and lesbian parenting conclude?

A growing body of credible, scientific research on gay and lesbian parenting concludes that children raised by gay and lesbian parents adjust positively and their families function well. Lesbian and gay parents are as likely as heterosexual parents to provide supportive and healthy environments for their children, and the children of lesbian and gay parents are as likely as those of heterosexual parents to flourish. Some research suggests that children raised by lesbigay parents develop in less gender-stereotypical ways, are more open to homoerotic relationships, contend with the social stigma of having gay parents, and have more empathy for social diversity than children of opposite-sex parents. There is no credible social science evidence that gay parenting negatively affects the well-being of children.

7.7 In what ways are heterosexual individuals victimized by heterosexism and antigay prejudice and discrimination?

Heterosexual individuals may be victims of antigay hate crimes. Many heterosexual family members and friends of homosexuals experience concern, fear, and grief over the mistreatment of and discrimination toward their gay or lesbian friends and/or family members. Because of the antigay social climate, heterosexuals, especially males, are hindered in their own self-expression and intimacy in same-sex relationships. Homophobia has also been linked to some cases of rape and sexual assault by males who are trying to prove they are not gay. The passage of state constitutional amendments that prohibit same-sex marriage can result in denial of rights and protections to opposite-sex unmarried couples.

heterosexuality the predominance of emotional and sexual attraction to individuals of the opposite sex.

homonegativity a construct that refers to antigay responses such as negative feelings (fear, disgust, anger), thoughts ("homosexuals are HIV carriers"), and behavior ("homosexuals deserve a beating").

homophobia used to refer to negative attitudes toward homosexuality.

homosexuality the predominance of emotional and sexual attraction to individuals of the same sex.

internalized homophobia a sense of personal failure and self-hatred among lesbians and gay men resulting from social rejection and stigmatization; has been linked to increased risk for depression, substance abuse and addiction, anxiety, and suicidal thoughts.

lesbian homosexual woman.

lesbigay population collective term referring to lesbians, gays, and bisexuals.

LGBT (GLBT) refers collectively to lesbians, gays, bisexuals, and transgendered individuals.

prejudice negative attitudes toward others based on differences.

reparative therapy (conversion therapy) therapy designed to change a person's homosexual orientation to a heterosexual orientation.

second-parent adoption (also called co-parent adoption) a legal procedure that allows individuals to adopt their partner's biological or adoptive child without terminating the first parent's legal status as parent.

sexual orientation a classification of individuals as heterosexual, bisexual, or homosexual, based on their emotional, cognitive, and sexual attractions and self-identity.

sodomy oral and anal sexual acts.

transgendered individuals who express their masculinity and femininity in nontraditional ways consistent with their biological sex.

reviewcard CHAPTER 8 SEXUALITY IN RELATIONSHIPS

Learning Outcomes

8.1 What are sexual values?

Sexual values are moral guidelines for making sexual choices in nonmarital, marital, heterosexual, and homosexual relationships.

8.2 What are alternative sexual values?

Three sexual values are absolutism (rightness is defined by official code of morality), relativism (rightness depends on the situation—who does what, with whom, in what context), and hedonism ("if it feels good, do it"). Relativism is the sexual value most college students hold, with women being more relativistic than men and men being more hedonistic than women. Under the Bush administration, absolutism was taught in the public school system if the school wanted federal funds. There is no evidence that abstinence-based sex education programs are effective in stopping unmarried youth from having sex. Half of those who take the "virginity pledge" withdraw the pledge within one year. About three-fourths of college students believe that if they have oral sex, they are still virgins. About half of undergraduates reported involvement in a "friends with benefits" relationship. Women are more likely to focus on the "friendship" aspect, men on the "benefits" (sex) aspect.

8.3 What is the sexual double standard?

The sexual double standard is the view that encourages and accepts sexual expression of men more than women. For example, men may have more sexual partners than women without being stigmatized. The double standard is also reflected in movies.

8.4 What are sources of sexual values?

The sources of sexual values include one's school, family, and religion as well as technology, television, social movements, and the Internet.

8.5 What are gender differences in sexuality?

Gender differences in sexual beliefs include that men are more likely than women to believe that oral sex is not sex, that cybersex is not cheating, that men can't tell if a woman is faking orgasm, and that sex frequency drops in marriage. In regard to sexual behavior, men are more likely than females to report frequenting strip clubs, paying for sex, having anonymous sex with strangers, having casual sexual relations, and having more sexual partners. When asked what they would do in a "Vegas" context where no one would know what they did, males were more likely than females to identify a range of sexual behaviors they would engage in.

8.6 How do pheromones affect sexual behavior?

Pheromones are chemical messengers emitted from the body that activate physiological and behavioral responses. The functions of pheromones include opposite-sex attractants, same-sex repellents, and mother-infant bonding. Researchers disagree about whether pheromones do in fact influence human sociosexual behaviors. In one study, men who applied a male hormone to their aftershave lotion reported

Key Terms

absolutism belief system based on unconditional allegiance to the authority of science, law, tradition, or religion.

AIDS acquired immunodeficiency syndrome; the last stage of HIV infection, in which the immune system of a person's body is so weakened that it becomes vulnerable to disease and infection.

asceticism the belief that giving in to carnal lusts is wrong and that one must rise above the pursuit of sensual pleasure to a life of self-discipline and self-denial.

friends with benefits (FWB) a relationship between nonromantic friends who also have a sexual relationship.

hedonism belief that the ultimate value and motivation for human actions lie in the pursuit of pleasure and the avoidance of pain.

HIV human immunodeficiency virus, which attacks the immune system and can lead to AIDS.

relativism sexual value system whereby decisions are made in the context of the situation and the relationship.

satiation the state in which a stimulus loses its value with repeated exposure.

secondary virginity the conscious decision of a sexually active person to refrain from intimate encounters for a specified period of time.

sexual double standard the view that encourages and accepts sexual expression of men more than women.

sexual values moral guidelines for sexual behavior.

social script the identification of the roles in a social situation, the nature of the relationship between the roles, and the expected behaviors of those roles.

spectatoring mentally observing one's own and one's partner's sexual performance.

STI sexually transmitted infection.

significant increases in sexual intercourse and sleeping next to a partner in comparison with men who had a placebo in their aftershave lotion.

8.7 What are the sexual relationships of never-married, married, and divorced people?

Never-married and noncohabiting individuals report more sexual partners than those who are married or living with a partner. Marital sex is distinctive for its social legitimacy, declining frequency, and satisfaction (both physical and emotional). Divorced individuals have a lot of sexual partners but are the least sexually fulfilled.

8.8 How does one avoid contracting or transmitting STIs?

The best way to avoid getting an STI is to avoid sexual contact or to have contact only with partners who are not infected. This means restricting your sexual contacts to those who limit their relationships to one person. The person most likely to get an STI has sexual relations with a number of partners or with a partner who has a variety of partners. Even if you are in a mutually monogamous relationship, you may be at risk for acquiring an STI, as 30 percent of male undergraduate students and 20 percent of female undergraduate students in "monogamous" relationships reported having oral, vaginal, or anal sex with another partner outside of the monogamous relationship.

8.9 What are the prerequisites of sexual fulfillment?

Fulfilling sexual relationships involve self-knowledge, self-esteem, health, a good nonsexual relationship, open sexual communication, safer sex practices, and making love with, not to, one's partner. Other variables include realistic expectations ("my partner will not always want what I want") and not buying into sexual myths ("masturbation is sick").

Did you take the Self-Assessment on page 171?

If so, what was your score? Read on to find out scores of other students who completed the scale.

Scores of Other Students Who Completed the Scale

This scale was completed by 252 student volunteers at Valdosta State University. The mean score of the students was 40.81 (standard deviation [SD] = 13.20), reflecting that the students were virtually at the midpoint between a very negative and a very positive attitude toward premarital sex. For the 124 males and 128 females in the total sample, the mean scores were 42.06 (SD = 12.93) and 39.60 (SD = 13.39), respectively (not statistically significant). In regard to race, 59.5 percent of the sample was white and 40.5 percent was nonwhite (35.3 percent black, 2.4 percent Hispanic, 1.6 percent Asian, 0.4 percent American Indian, and 0.8 percent other). The mean scores of whites, blacks, and nonwhites were 41.64 (SD = 13.38), 38.46 (SD = 13.19), and 39.59 (SD = 12.90) (not statistically significant). Finally, regarding year in college, 8.3 percent were freshmen, 17.1 percent sophomores, 28.6 percent juniors, 43.3 percent seniors, and 2.8 percent graduate students. Freshmen and sophomores reported more positive attitudes toward premarital sex (mean = 44.81; SD = 13.39) than did juniors (mean = 40.32; SD = 12.58) or seniors and graduate students (mean = 38.91; SD = 13.10) (p = .05).

reviewcard CHAPTER 9 PLANNING TO HAVE CHILDREN

Learning Outcomes

9.1 Why do people have children?

Worldwide, 80 million unintended pregnancies occur each year (38 percent of all pregnancies). These pregnancies result in 42 million induced abortions and 34 million unintended births. Having children continues to be a major goal of most college students (women more than men). Social influences to have a child include family, friends, religion, government, favorable economic conditions, and cultural observances. The reasons people give for having children include love and companionship with one's own offspring, the desire to be personally fulfilled as an adult by having a child, and the desire to recapture one's youth. Having a child (particularly for women) reduces one's educational and career advancement. The cost for housing, food, transportation, clothing, health care, and child care for a child up to age 2 is more than $10,000 annually.

9.2 How many children do people have?

About 20 percent of women aged 40 to 44 do not have children. Whether these women will remain childfree or eventually have children is unknown. Reasons that spouses elect a childfree marriage include the freedom to spend their time and money as they choose, to enjoy their partner without interference, to pursue a career, to avoid health problems associated with pregnancy, and to avoid passing on genetic disorders to a new generation.

The most preferred type of family in the United States is the two-child family. Some of the factors in a couple's decision to have more than one child are the desire to repeat a good experience, the feeling that two children provide companionship for each other, and the desire to have a child of each sex.

9.3 Are teen mothers disadvantaged?

There are 1.7 million births annually to teenage mothers and this number is rising. The teenage birthrate in the United States, as measured by births per thousand teenagers, is nine times that of teens in the Netherlands. The reason is openness in sexuality and contraception in France, Germany, and other European countries in contrast to abstinence thinking in the United States. Teens in the United States also view motherhood as one of the only viable roles open to them. Consequences associated with teenage motherhood include poverty, alcohol or drug abuse, babies with lower birth weights, and lower academic achievement.

9.4 What are the causes of infertility?

Infertility is defined as the inability to achieve a pregnancy after at least one year of regular sexual relations without birth control, or the inability to carry a pregnancy to a live birth. Forty percent of infertility problems are attributed to the woman, 40 percent to the man, and 20 percent to both of them. The causes of infertility in women include blocked fallopian tubes, endocrine imbalance that prevents ovulation, dysfunctional ovaries, chemically hostile cervical mucus that may kill sperm, and effects of STIs. The psychological reaction to infertility is often depression over having to give up a lifetime goal. Some of the more common causes of infertility in men include low sperm production, poor semen motility, effects of STIs, and interference with passage of sperm through the genital ducts due to an enlarged prostate.

Key Terms

abortion rate the number of abortions per 1,000 women aged 15 to 44.

abortion ratio the number of abortions per 1,000 live births.

antinatalism opposition to having children.

competitive birthing pattern in which a woman will want to have the same number of children as her peers.

cryopreservation fertilized eggs are frozen and implanted at a later time.

Fertell an at-home fertility kit that allows women to measure the level of their follicle-stimulating hormone on the third day of their menstrual cycle and men to measure the concentration of motile sperm.

fertilization (conception) the fusion of the egg and sperm.

foster parent (family caregiver) a person who, either alone or with a spouse, takes care of and fosters a child taken into custody.

induced abortion the deliberate termination of a pregnancy through chemical or surgical means.

infertility the inability to achieve a pregnancy after at least one year of regular sexual relations without birth control, or the inability to carry a pregnancy to a live birth.

parental consent woman needs permission from parent to get an abortion if under a certain age, usually 18.

parental notification woman required to tell parents she is getting an abortion if she is under a certain age, usually 18; but she does not need parental permission.

pregnancy a condition that begins five to seven days after conception, when the fertilized egg is implanted (typically in the uterine wall).

procreative liberty the freedom to decide whether or not to have children.

pronatalism view that encourages having children.

spontaneous abortion (miscarriage) an unintended termination of a pregnancy.

therapeutic abortion an abortion performed to protect the life or health of a woman.

transracial adoption the practice of parents adopting children of another race.

A number of technological innovations are available to assist women and couples in becoming pregnant. These include hormonal therapy, artificial insemination, ovum transfer, in vitro fertilization, gamete intrafallopian transfer, and zygote intrafallopian transfer. Being infertile (for the woman) may have a negative lifetime effect, both personal and interpersonal (half the women in one study were separated or reported a negative effect on their sex lives).

9.5 & 9.6 Why do people adopt?

Motives for adoption include a couple's inability to have a biological child (infertility), their desire to give an otherwise unwanted child a permanent loving home, or their desire to avoid contributing to overpopulation by having more biological children. There are approximately 100,000 adoptions annually in the United States.

Whereas demographic characteristics of those who typically adopt are white, educated, and of high income, adoptees are increasingly being placed in nontraditional families including with older, gay, and single individuals; it is recognized that these individuals may also be white, educated, and of high income. Most college students are open to transracial adoption.

Some individuals seek the role of parent via foster parenting. A foster parent, also known as a family caregiver, is a person who, either alone or with a spouse, takes care of and fosters a child taken into custody in their home. A foster parent has made a contract with the state for the service, has judicial status, and is reimbursed by the state.

9.7 What are the types of and motives for an abortion?

An abortion may be either an induced abortion, which is the deliberate termination of a pregnancy through chemical or surgical means, or a spontaneous abortion (miscarriage), which is the unintended termination of a pregnancy. The most frequently cited reasons for induced abortion were that having a child would interfere with a woman's education, work, or ability to care for dependents (74 percent); that she could not afford a baby now (73 percent); and that she did not want to be a single mother or was having relationship problems (48 percent). Nearly four in ten women said they had completed their childbearing, and almost one-third of the women were not ready to have a child. Less than 1 percent said their parents' or partner's desire for them to have an abortion was the most important reason. In regard to the psychological effects of abortion, the American Psychological Association reviewed the literature and concluded that "among women who have a single, legal, first-trimester abortion of an unplanned pregnancy for nontherapeutic reasons, the relative risks of mental health problems are no greater than the risks among women who deliver an unplanned pregnancy."

Remember

There is a self-assessment card for this chapter in the self-assessment card deck. There are two tools for Chapter 9: the Attitudes towards Parenthood Scale and the Attitudes towards Transracial Adoption Scale.

reviewcard CHAPTER 10 PARENTING

Learning Outcomes

10.1 What are the basic roles of parents?

Parenting includes providing physical care for children, loving them, being an economic resource, providing guidance as a teacher or model, and protecting them from harm. One of the biggest problems confronting parents today is the societal influence on their children. These include drugs and alcohol; peer pressure; TV, Internet, and movies; and crime or gangs.

10.2 What is a choices perspective of parenting?

Although both genetic and environmental factors are at work, the choices parents make have a dramatic impact on their children. Parents who don't make a choice about parenting have already made one. The five basic choices parents make include deciding (1) whether to have a child, (2) the number of children, (3) the interval between children, (4) one's method of discipline and guidance, and (5) the degree to which one will be invested in the role of parent.

10.3 What is the transition to parenthood like for women, men, and couples?

Transition to parenthood refers to that period of time from the beginning of pregnancy through the first few months after the birth of a baby. The mother, father, and couple all undergo changes and adaptations during this period. Most mothers relish their new role; some may experience the transitory feelings of baby blues; a few report postpartum depression.

The father's involvement with his children is sometimes predicted by the quality of the parents' romantic relationship. If the father is emotionally and physically involved with the mother, he is more likely to take an active role in the child's life. In recent years, there has been a renewed cultural awareness of fatherhood.

A summary of almost 150 studies involving almost 50,000 respondents on the question of how children affect marital satisfaction revealed that parents (both women and men) reported lower marital satisfaction than nonparents. In addition, the higher the number of children, the lower the marital satisfaction; the factors that depressed marital satisfaction were conflict and loss of freedom.

Percentage of Couples Getting Divorced by Number of Children

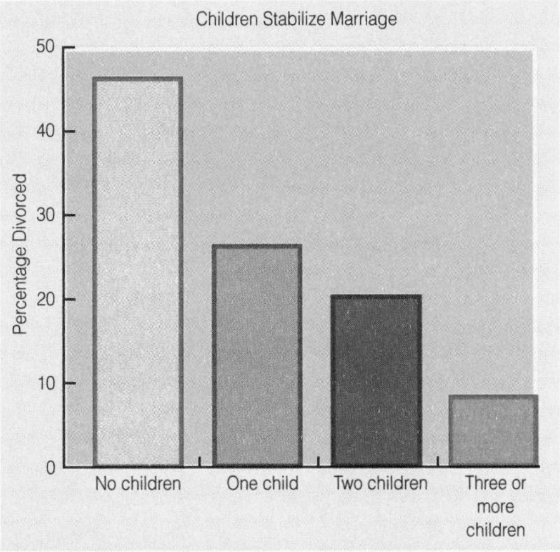

Key Terms

baby blues transitory symptoms of depression in mothers twenty-four to forty-eight hours after a baby is born.

demandingness the degree to which parents place expectations on children and use discipline to enforce the demands.

gatekeeper role refers to a mother's encouragement or criticism of the father, which influences the degree to which he is involved with his children.

menarche the time of first menstruation.

nature-deficit disorder result when children are encouraged to detach themselves from direct contact with nature.

oppositional defiant disorder a condition in which children do not comply with requests of authority figures.

overindulgence giving children too much, too soon, too long; a form of child neglect in which children are not allowed to develop their own competences.

oxytocin a hormone released from the pituitary gland during the expulsive stage of labor that has been associated with the onset of maternal behavior in lower animals.

parenting the provision by an adult or adults of physical care, emotional support, instruction, and protection from harm in an ongoing structural (home) and emotional context to one or more dependent children.

postpartum depression a reaction more severe than the "baby blues" to the birth of one's baby, characterized by crying, irritability, loss of appetite, and difficulty in sleeping.

postpartum psychosis a reaction in which a new mother wants to harm her baby.

responsiveness the extent to which parents respond to and meet the needs of their children. Refers to such qualities as warmth, reciprocity, person-centered communication, and attachment.

time-out a discipline procedure whereby a child is removed from an enjoyable context to an isolated one and left there.

transition to parenthood the period of time from the beginning of pregnancy through the first few months after the birth of a baby.

10.4 What are several facts about parenthood?

Parenthood will involve about 40 percent of the time a couple live together, parents are only one influence on their children, each child is unique, and parenting styles differ. Research suggests that an authoritarian parenting style, characterized by being both demanding and warm, is associated with positive outcomes. In addition, being emotionally connected to a child, respecting the child's individuality, and monitoring the child's behavior to encourage positive contexts have positive outcomes. Birth order effects include that firstborns are the first on the scene with parents and always have the "inside track." They want to stay that way so they are traditional and conforming to their parent's expectations. Children born later learn quickly that they entered an existing family constellation where everyone is bigger and stronger. They cannot depend on having established territory, so they must excel in ways different from the firstborn. They are open to experience, adventurousness, and trying new things because their status is not already assured.

10.5 What are some of the principles of effective parenting?

Giving time, love, praise, and encouragement; monitoring the activities of one's child; setting limits; encouraging responsibility; and providing sexuality education are aspects of effective parenting.

10.6 What are the issues of single parenting?

About 40 percent of all children will spend one-fourth of their lives in a female-headed household. The challenges of single parenthood for the parent include taking care of the emotional and physical needs of a child alone, meeting one's own adult emotional and sexual needs, money, and rearing a child without a father (the influence of whom can be positive and beneficial).

Remember

There is a self-assessment card for this chapter in the self-assessment card deck. For chapter 10, there are two tools: the Traditional Motherhood Scale and the Traditional Fatherhood Scale.

reviewcard CHAPTER 11
FAMILY AND THE ECONOMY

Learning Outcomes

11.1 How does work or money affect a couple's relationship?

Generally, the more money a partner makes, the more power that person has in the relationship. Males make considerably more money than females and generally have more power in relationships. Sixty-one percent of all U.S. wives are in the labor force. The stereotypical family consisting of a husband who earns an income and a wife who stays at home with two or more children is no longer the norm. Only 13 percent are "traditional" in the sense of a breadwinning husband, a stay-at-home wife, and their children. In contrast, most marriages may be characterized as dual-earner. Employed wives in unhappy marriages are more likely to leave the marriage than unemployed wives.

> **What do you think?**
>
> Take the self-assessment for chapter 11 in the self-assessment card deck to determine your attitudes towards maternal employment.

11.2 What is the effect of parents' work decisions on the children?

Children do not appear to suffer cognitively or emotionally from their parents' working as long as positive, consistent child-care alternatives are in place. However, less supervision of children by parents is an outcome of having two-earner parents. Leaving children to come home to an empty house is particularly problematic.

Parents view quality time as structured, planned activities, talking with their children, or just hanging out with them. Day care is typically mediocre but any negatives are offset by consistent parental attention.

11.3 What are the various strategies for balancing the demands of work and family?

Strategies used for balancing the demands of work and family include the superperson strategy, cognitive restructuring, delegation of responsibility, planning and time management, and role compartmentalization. Government and corporations have begun to respond to the family concerns of employees by implementing work-family policies and programs. These policies are typically inadequate and cosmetic.

Key Terms

absolute poverty the lack of resources that leads to hunger and physical deprivation.

dual-career marriage a marriage in which both spouses pursue careers and maintain a life together that may or may not include dependents.

dual-earner marriage both husband and wife work outside the home to provide economic support for the family.

HER/his career a wife's career is given precedence over a husband's career.

HIS/her career a husband's career is given precedence over a wife's career.

HIS/HER career a husband's and wife's careers are given equal precedence.

identity theft one person using the Social Security number, address, and bank account numbers of another, posing as that person to make purchases.

installment plan repayment plan whereby one signs a contract to pay for an item with regular installments over an agreed upon period of time.

mommy track stopping paid employment to spend time with young children.

open charge repayment plan whereby one agrees to pay the amount owed in full within the agreed upon amount of time.

opting out when professional women leave their careers and return home to care for their children.

poverty the lack of resources necessary for material well-being.

relative poverty a deficiency in material and economic resources compared with some other population.

revolving charge the repayment plan whereby you may pay the total amount you owe, any amount over the stated minimum payment due, or the minimum payment.

role compartmentalization separating the roles of work and home so that an individual does not dwell on the problems of one role while physically being at the place of the other role.

role conflict being confronted with incompatible role obligations.

shift work having one parent work during the day and the other parent work at night so that one parent can always be with the children.

supermom (superwoman) a cultural label that allows a mother who is experiencing role overload to regard herself as particularly efficient, energetic, and confident.

THEIR career a career shared by a couple who travel and work together (for example, journalists).

third shift the emotional energy expended by a spouse or parent in dealing with various family issues.

11.4 How do debt and poverty affect relationships?

The more couples are in debt, the less time they spend together, the more they argue, and the more they are unhappy. Poverty is devastating to couples and families. Those living in poverty have poorer physical and mental health, report lower personal and marital satisfaction, and die sooner. The table below reflects that a significant proportion of families in the United States continue to be affected by unemployment and low wages.

The stresses associated with low income also contribute to substance abuse, domestic violence, child abuse and neglect, and divorce. Couples with incomes of less than $25,000 are 30 percent more likely to divorce than those with $50,000 or higher.

2009 HHS Poverty Guidelines

People in Family or Household	48 Contiguous States and D.C.	Alaska	Hawaii
1	$10,830	$13,530	$12,460
2	14,570	18,210	16,760
3	18,310	22,890	21,060
4	22,050	27,570	25,360
5	25,790	32,250	29,660
6	29,530	36,930	33,960
7	33,270	41,610	38,260
8	37,010	46,290	42,560
For each additional person, add	3,740	4,680	4,300

Source: *Federal Register*, vol. 74, no. 14, pp. 4199–4201. January 23, 2009. http://aspe.hhs.gov/poverty/09poverty.shtml.

reviewcard

CHAPTER 12
VIOLENCE AND ABUSE IN RELATIONSHIPS

Learning Outcomes

12.1 What is the nature of relationship abuse?

Violence or physical abuse may be defined as the intentional infliction of physical harm by either partner on the other. Violence may be over an issue on which the partners disagree or a mechanism of control. More than four million women are victims of physical violence annually. Intimate-partner violence (IPV) is associated with posttraumatic stress disorder (PTSD) and results in social maladjustment and personal or social resource loss. Female violence is as prevalent as male violence. The difference is that female violence is often in response to male violence, whereas male violence is more often to control the partner.

Emotional abuse is designed to denigrate the partner, reduce the partner's status, and make the partner vulnerable, thereby giving the abuser more control. Stalking is unwanted following or harassment that induces fear in the target person. The stalker is most often a heterosexual male who has been rejected by someone who fails to return his advances. Stalking is typically designed either to seek revenge or to win a partner back. Cybervictimization includes being sent threatening e-mail, unsolicited obscene e-mail, computer viruses, or junk mail (spamming). It may also include flaming (online verbal abuse), and leaving improper messages on message boards designed to get back at the person. Obsessive relationship intrusion (ORI) is the interjection of a person into another's life, which makes the targeted person uncomfortable. Unlike stalkers whose goals are to harm, ORI involves hyperintimacy (telling people that they are beautiful or desirable to the point of making them uncomfortable), relentless mediated contacts (flooding people with e-mail messages, cell phone messages, or faxes), or interactional contacts (showing up at work or the gym, or joining the same volunteer groups as the pursued).

12.2 What are some explanations for violence in relationships?

Cultural explanations for violence include violence in the media, corporal punishment in childhood, gender inequality, and stress. Community explanations involve social isolation of individuals and spouses from extended family, poverty, inaccessible community services, and lack of violence prevention programs. Individual factors include psychopathology of the person (antisocial), personality (dependency or jealousy), and alcohol abuse. Family factors include child abuse by one's parents and observing parents who abuse each other.

12.3 How does sexual abuse in undergraduate relationships manifest itself?

Violence in dating relationships begins as early as grade school, is mutual, and escalates with emotional involvement. Violence occurs more often among couples in which the partners disagree about each other's level of emotional commitment and when the perpetrator has been drinking alcohol or using drugs. About 40 percent of undergraduates report being forced to have sex against their will. Acquaintance rape is defined as nonconsensual sex between adults who know each other.

One type of acquaintance rape is date rape, which refers to nonconsensual sex between people who are dating or on a date. Rohypnol, known as the date rape drug, causes profound sedation so that the person may not remember being raped. Most women do not report being raped by an acquaintance or date.

Key Terms

acquaintance rape nonconsensual sex between adults who know each other.

battered-woman syndrome general pattern of battering that a woman is subjected to, defined in terms of the frequency, severity, and injury.

corporal punishment the use of physical force with the intention of causing a child to experience pain, but not injury, for the purpose of correction or control of the child's behavior.

cybervictimization being sent unwanted e-mail, spam, viruses, or being threatened online.

date rape nonconsensual sex between two people who are dating or on a date.

emotional abuse (verbal abuse; symbolic aggression; psychological abuse) the denigration of an individual with the purpose of reducing the victim's status and increasing the victim's vulnerability so that he or she can be more easily controlled by the abuser.

entrapped stuck in an abusive relationship and unable to extricate one's self from the abusive partner.

honor crime (honor killing) killing a daughter because she brought shame to the family by having sex while not married. The killing is typically overlooked by the society. Jordan is a country where honor crimes occur.

intimate-partner violence an all-inclusive term that refers to crimes committed against current or former spouses, boyfriends, or girlfriends.

marital rape forcible rape by one's spouse.

obsessive relational intrusion the relentless pursuit of intimacy with someone who does not want it.

physical abuse intentional infliction of physical harm by either partner on the other.

red zone the first month of the first year of college when women are particularly vulnerable to unwanted sexual advances.

Rohypnol date rape drug that causes profound, prolonged sedation and short-term memory loss.

stalking unwanted following or harassment that induces fear in a target person.

uxoricide the murder of a woman by her romantic partner.

12.4 How does abuse in marriage relationships manifest itself?

Abuse in marriage is born out of the need to control the partner and may include repeated rape. About half of the women raped by an intimate partner and two-thirds of the women physically assaulted by an intimate partner have been victimized multiple times.

12.5 What are the effects of abuse?

The effects of IPV include physical harm, mental harm (depression, anxiety, low self-esteem, lost of trust in others, sexual dysfunctions), unintended pregnancy, and multiple abortions. High levels of anxiety and depression often lead to alcohol and drug abuse. Violence on pregnant women significantly increases the risk for infants of low birth weight, preterm delivery, and neonatal death. Katz and Myhr (2008) noted that 21 percent of 193 female undergraduates were experiencing verbal sexual coercion in their current relationships. The effects included feeling psychologically abused, arguing, decreased relationship satisfaction, and sexual functioning.

12.6 What is the cycle of abuse and why do people stay in an abusive relationship?

The cycle of abuse begins when a person is abused and the perpetrator feels regret, asks for forgiveness, and starts acting nice (for example, gives flowers). The victim, who perceives few options and feels guilty terminating the relationship with the partner who asks for forgiveness, feels hope for the relationship at the contriteness of the abuser and does not call the police or file charges. The couple usually experiences a period of making up or honeymooning, during which the victim feels good again about the abusing partner. However, tensions mount again and are released in the form of violence. Such violence is followed by the familiar sense of regret and pleadings for forgiveness, accompanied by being nice (a new bouquet of flowers, and so on).

The reasons people stay in abusive relationships include love, emotional dependency, commitment to the relationship, hope, view of violence as legitimate, guilt, fear, economic dependency, and isolation. The catalyst for breaking free combines the sustained aversiveness of staying, the perception that they and their children will be harmed by doing so, and the awareness of an alternative path or of help in seeking one. One must be cautious in getting out of an abusive relationship because an abuser is most likely to kill his partner when she actually leaves the relationship.

The Cycle of Abuse

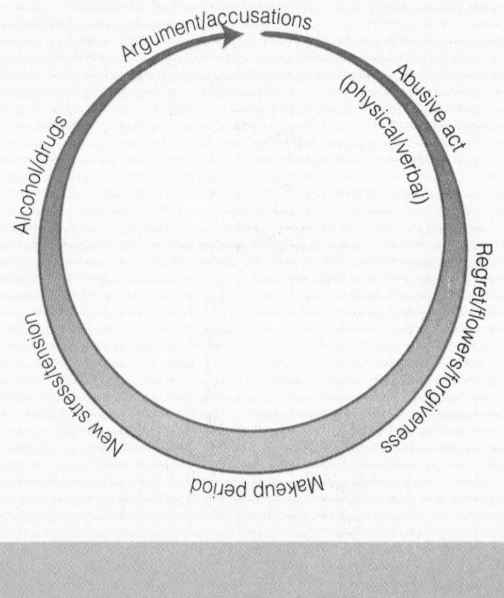

reviewcard

CHAPTER 13
STRESS AND CRISIS IN RELATIONSHIPS

Learning Outcomes

13.1 What is stress and what is a crisis event?

Stress is a reaction of the body to substantial or unusual demands (physical, environmental, or interpersonal). Stress is a process rather than a state. A crisis is a situation that requires changes in normal patterns of behavior. A family crisis is a situation that upsets the normal functioning of the family and requires a new set of responses to the stressor. Sources of stress and crises can be external (for example, hurricane, tornado, downsizing, military separation) or internal (for example, alcoholism, extramarital affair, Alzheimer's disease).

Family resilience is when family members successfully cope under adversity, which enables them to flourish with warmth, support, and cohesion. Key factors include positive outlook, spirituality, flexibility, communication, financial management, family shared recreation, routines or rituals, and support networks.

13.2 What are positive stress management strategies?

Changing one's basic values and perspective is the most helpful strategy in reacting to a crisis. Viewing ill health as a challenge, bankruptcy as an opportunity to spend time with one's family, and infidelity as an opportunity to improve communication are examples. Other positive coping strategies are exercise, adequate sleep, love, religion, friends or relatives, humor, education, and counseling. Still other strategies include intervening early in a crisis, not blaming each other, keeping destructive impulses in check, and seeking opportunities for fun.

13.4 What are harmful strategies for reacting to a crisis?

Some harmful strategies include keeping feelings inside, taking out frustrations on others, and denying or avoiding the problem.

Key Terms

Al-Anon an organization that provides support for family members and friends of alcohol abusers.

biofeedback a process in which information that is relayed back to the brain enables people to change their biological functioning.

Coolidge effect term used to describe the waning of sexual excitement and the effect of novelty and variety on sexual arousal.

crisis a sharp change for which typical patterns of coping are not adequate and new patterns must be developed.

down low African-American "heterosexual" man who has sex with men.

extradyadic (extrarelational) involvement emotional or sexual involvement between a member of a pair and someone other than the partner.

extramarital affair a spouse's sexual involvement with someone outside the marriage.

family resilience the successful coping of family members under adversity that enables them to flourish with warmth, support, and cohesion.

palliative care health care focused on the relief of pain and suffering of the individual who has a life-threatening illness and support for them and their loved ones.

resiliency the ability of a family to respond to a crisis in a positive way.

stress a nonspecific response of the body to demands made on it.

13.5 What are five of the major family crisis events?

Some of the more common crisis events that spouses and families face include physical illness, an extramarital affair, unemployment, alcohol or drug abuse, and the death of one's spouse or children. An extramarital affair is the second most stressful crisis event for a family (abuse is number one). Surviving an affair involves forgiveness on the part of the offended spouse to grant a pardon to the offending spouse, to give up feeling angry, and to relinquish the right to retaliate against the offending spouse. In exchange, an offending spouse must take responsibility for the affair, agree not to repeat the behavior, and grant the partner the right to check up on the offending partner to regain trust. The best affair prevention is a happy and fulfilling marriage as well as avoiding intimate conversations with members of the other sex and a context (for example, where alcohol is consumed and being alone), which are conducive to physical involvement. The occurrence of a "midlife crisis" is reported by less than a quarter of adults in the middle years. Those who did experience a crisis were going through a divorce.

Did you take the Attitudes towards Infidelity self-assessment in the self-assessment card deck?

Check out how other students scored on the scale. How do you compare?

Scores of Other Students Who Completed the Scale

The scale was completed by 150 male and 136 female student volunteers at Valdosta State University. The average score on the scale was 27.85 (SD = 12.02). Their ages ranged from 18 to 49, with a mean age of 23.36 (SD = 5.13). The ethnic backgrounds of the sample included 60.8 percent white, 28.3 percent African American, 2.4 percent Hispanic, 3.8 percent Asian, 0.3 percent American Indian, and 4.2 percent other. The college classification level of the sample included 11.5 percent freshmen, 18.2 percent sophomores, 20.6 percent juniors, 37.8 percent seniors, 7.7 percent graduate students, and 4.2 percent postbaccalaureate. Male participants reported more positive attitudes toward infidelity (mean = 31.53; SD =11.86) than did female participants (mean = 23.78; SD = 10.86; p = .05). White participants had more negative attitudes toward infidelity (mean = 25.36; SD = 11.17) than did nonwhite participants (mean = 31.71; SD 12.32; p = .05). There were no significant differences in regard to college classification.

Remember

Chapter 13 has another self-assessment in the self-assessment card deck. Take the Motives for Drinking self-assessment and compare your results with other students'.

reviewcard CHAPTER 14 DIVORCE AND REMARRIAGE

Learning Outcomes

14.1 What are macro factors contributing to divorce?

Macro factors contributing to divorce include increased economic independence of women (women can afford to leave), changing family functions (companionship is the only remaining function), liberal divorce laws (it's easier to leave), fewer religious sanctions (churches embrace single individuals), more divorce models (Hollywood models abound), and individualism (rather than familism as a cultural goal of happiness). Regarding individuals, 95 percent of 1,319 undergraduates at a large southeastern university disagreed with the statement, "I would not divorce my spouse for any reason"; hence, all but 5 percent would divorce under some circumstances.

14.2 What are micro factors contributing to divorce?

Micro factors include having numerous differences, falling out of love, negative behavior, lack of conflict resolution skills, satiation, value changes, and extramarital relationships. Being in one's teens at the time of marriage, having a courtship of less than two years, and having divorced parents are all associated with subsequent divorce.

14.3 How might one go about ending a relationship?

About 30 percent of undergraduates in one study reported that they were unhappy in their present relationships; another 30 percent reported that they knew they were in relationships that should end. After considering that one might improve an unhappy relationship and deciding to end a relationship, telling a partner that one needs out "for one's space" without giving a specific reason helps a partner avoid feeling obligated to stay if the other partner changes. Although some couples can remain friends, others may profit from ending the relationship completely. And recovery can take time (twelve to eighteen months).

14.4 What are gender differences in filing for divorce?

Women are more likely to file for divorce because they see that, by getting divorced, they get the husband's money (via division of property or child support), the children, and the husband out of the house. Husbands are less likely to seek divorce because they more often end up without the house, with half their money, and separated from their children. Regardless of who files, more than two-thirds of both women and men who are separated recommend that other couples who are contemplating divorce try to "work it out."

14.5 What are the consequences of divorce for spouses?

The psychological consequences for divorcing spouses depend on how unhappy the marriage was. Spouses who were miserable while in a loveless conflictual marriage often regard the divorce as a relief. Spouses who were left (for example, a spouse leaves for another partner) may be devastated and suicidal. Women tend to fare better emotionally after separation and divorce than do men. Women are more likely than men not only to have a stronger network of supportive relationships but also to profit from divorce by developing a new sense of self-esteem and confidence, because they are thrust into a more independent role.

Key Terms

binuclear a family that spans two households, often because of divorce.

blended family a family where new spouses blend their children from previous marriages.

covenant marriage type of marriage that permits divorce only under specific conditions.

December marriage a new marriage in which both spouses are elderly.

developmental task a skill that allows a family to grow as a cohesive unit.

divorce mediation process in which divorcing parties make agreements with a third party (mediator) about custody, visitation, child support, property settlement, and spousal support.

divorcism the belief that divorce is a disaster.

homogamy tendency to select someone with similar characteristics to marry.

legal custody decisional authority over major issues involving the child.

mating gradient the tendency for husbands to be more advanced than their wives with regard to age, education, and occupational success.

negative commitment individuals who remain emotionally invested in their relationship with their former spouse, despite remarriage.

no-fault divorce a divorce in which neither party is identified as the guilty party or the cause of the divorce.

parental alienation syndrome a disturbance in which children are obsessively preoccupied with deprecation or criticism of a parent.

physical custody also called "visitation," refers to distribution of parenting time following divorce.

postnuptial agreement similar to a premarital agreement, a postnuptial agreement specifies what is to be done with property and holdings at death or divorce.

satiation the state in which a stimulus loses its value with repeated exposure.

shared parenting dysfunction the set of behaviors by both parents that are focused on hurting the other parent and are counterproductive for a child's well-being.

stepfamily (step relationships) a family in which partners bring children from previous marriages into the new home, where they may also have a child of their own.

stepism the assumption that stepfamilies are inferior to biological families.

Factors associated with a quicker adjustment on the part of both spouses include mediating rather than litigating the divorce, co-parenting their children, avoiding alcohol or other drugs, reducing stress through exercise, engaging in enjoyable activities with friends, and delaying a new marriage for two years. Recovering from a broken heart may also be expedited by recalling the negative things a partner did (for example, lied, was unfaithful, and so on).

14.6 What are the effects of divorce on children?

Although researchers agree that a civil, cooperative, co-parenting relationship between ex-spouses is the greatest predictor of a positive outcome for children, researchers disagree on the long-term negative effects of divorce on children. However, there is no disagreement that most children do not experience long-term negative effects. Divorce mediation encourages civility between divorcing spouses who negotiate the issues of division of property, custody, visitation, child support, and spousal support.

14.7 What are some conditions of a "successful" divorce?

Divorce is an emotionally traumatic event for everyone involved, but there are some steps that spouses can take to minimize the pain and help each other and their children with the transition. Some of these steps include mediating the divorce, co-parenting, sharing responsibility, creating positive thoughts, avoiding drugs and alcohol, being active, releasing anger, allowing time to heal, and progressing through the psychological stages of divorce.

14.8 What are strategies to prevent divorce?

Three states (Louisiana, Arizona, and Arkansas) offer covenant marriages, in which spouses agree to divorce only for serious reasons such as imprisonment on a felony or separation of more than two years. They also agree to see a marriage counselor if problems threaten the marriage. When given the option to choose a covenant marriage, few couples do so.

14.9 What is the nature of remarriage in the United States?

Within two years, 75 percent of divorced women and 80 percent of divorced men have remarried. When comparing divorced individuals who have remarried with divorced individuals who have not remarried, those who have remarried report greater personal and relationship happiness. Most divorced individuals select someone who is divorced to remarry just as widowed individuals select someone who is also widowed to remarry.

Two years is the recommended time from the end of one marriage to the beginning of the next. Among those who have been previously remarried, living together does not seem to disadvantage the couple in terms of having a higher divorce rate. National data reflect that remarriages are more likely than first marriages to end in divorce in the early years of remarriage. After fifteen years, however, second marriages tend to be more stable and happier than first marriages. The reason for this is that remarried individuals tend not to be afraid of divorce and would divorce if unhappy in a second marriage. First-time married individuals may be fearful of divorce and stay married even though they are unhappy.

Stages of Parental Repartnering

Relationship Transition	Definition
Dating initiation	The parent begins to date.
Child introduction	The children and new dating partner meet.
Serious involvement	The parent begins to present the relationship as "serious" to the children.
Sleepover	The parent and the partner begin to spend nights together when the children are present.
Cohabitation	The parent and the partner combine households.
Breakup of a serious relationship	The relationship experiences a temporary or permanent disruption.
Pregnancy in the new relationship	A planned or unexpected pregnancy occurs.
Engagement	The parent announces plans to remarry.
Remarriage	The parent and partner create a legal or civil union.

Source: E. R. Anderson, and S. M. Greene. 2005. Transitions in parental repartnering after divorce. *Journal of Divorce & Remarriage* 43:49 (http://www.haworthpress.com/web/jdr/).

14.10 What is the nature of stepfamilies in the United States?

Stepfamilies represent the fastest-growing type of family in the United States. A blended family is one in which the spouses in a new marriage relationship are blended with the children of at least one of the spouses from a previous marriage.

There is a movement away from the use of the term *blended*, because stepfamilies really do not blend. Although a stepfamily can be created when a never-married or a widowed parent with children marries a person with or without children, most stepfamilies today are composed of spouses who were once divorced.

Stepfamilies differ from nuclear families: the children in nuclear families are biologically related to both parents, whereas the children in stepfamilies are biologically related to only one parent. Also, in nuclear families, both biological parents live with their children, whereas only one biological parent in stepfamilies lives with the children. In some cases, the children alternate living with each parent. Stepism is the assumption that stepfamilies are inferior to biological families. Stepism, like racism, heterosexism, sexism, and ageism, involves prejudice and discrimination.

Stepfamilies go through a set of stages. New remarried couples often expect instant bonding between the new members of the stepfamily. It does not often happen. The stages are fantasy (everyone will love everyone), reality (possible bitter conflict), assertiveness (parents speak their mind), strengthening pair ties (spouses nurture their relationship), and recurring change (stepfamily members know there will continue to be change). Involvement in stepfamily discussion groups such as the Stepfamily Enrichment Program provides enormous benefits.

Differences between Nuclear Families and Stepfamilies

Nuclear Families	Stepfamilies
1. Children are (usually) biologically related to both parents.	1. Children are biologically related to only one parent.
2. Both biological parents live together with children.	2. As a result of divorce or death, one biological parent does not live with the children. In the case of joint physical custody, children may live with both parents, alternating between them.
3. Beliefs and values of members tend to be similar.	3. Beliefs and values of members are more likely to be different because of different backgrounds.
4. The relationship between adults has existed longer than relationship between children and parents.	4. The relationship between children and parents has existed longer than the relationship between adults.
5. Children have one home they regard as theirs.	5. Children may have two homes they regard as theirs.
6. The family's economic resources come from within the family unit.	6. Some economic resources may come from an ex-spouse.
7. All money generated stays in the family.	7. Some money generated may leave the family in the form of alimony or child support.
8. Relationships are relatively stable.	8. Relationships are in flux: new adults adjusting to each other; children adjusting to a stepparent; a stepparent adjusting to stepchildren; stepchildren adjusting to each other.
9. No stigma is attached to nuclear family.	9. Stepfamilies are stigmatized.
10. Spouses had a childfree period.	10. Spouses had no childfree period.
11. Inheritance rights are automatic.	11. Stepchildren do not automatically inherit from stepparents.
12. Rights to custody of children are assumed if divorce occurs.	12. Rights to custody of stepchildren are usually not considered.
13. Extended family networks are smooth and comfortable.	13. Extended family networks become complex and strained.
14. Nuclear family may not have experienced loss.	14. Stepfamily has experienced loss.

14.11 What are the strengths of stepfamilies?

The strengths of stepfamilies include exposure to a variety of behavior patterns, happier parents, and greater objectivity on the part of the stepparent.

14.12 What are the developmental tasks of stepfamilies?

Developmental tasks for stepfamilies include nurturing the new marriage relationship, allowing time for partners and children to get to know each other, deciding whose money will be spent on whose children, deciding who will discipline the children and how, and supporting the children's relationship with both parents and natural grandparents. Both sets of parents and stepparents should form a parenting coalition in which they cooperate and actively participate in child rearing.

Remember

Take the chapter 14 self-assessment, called Children's Beliefs about Parental Divorce, located in your self-assessment card deck.

Learning Outcomes

15.1 What is meant by the terms age and ageism?

Age is defined chronologically (by time), physiologically (by capacity to see, hear, and so on), psychologically (by self-concept), sociologically (by social roles), and culturally (by the value placed on elderly). Ageism is the denigration of the elderly, and gerontophobia is the fear or dread of the elderly. Theories of aging range from disengagement (individuals and societies mutually disengage from each other) to continuity (the habit patterns of youth are continued in old age), as you can see in the following table. Life course is the aging theory currently in vogue.

Theories of Aging

Name of Theory	Level of Theory	Theorists	Basic Assumptions	Criticisms
Disengagement	Macro	Elaine Cumming, William Henry	The gradual and mutual withdrawal of the elderly and society from each other is a natural process. It is also necessary and functional for society that the elderly disengage so that new people can be phased in to replace them in an orderly transition.	Not all people want to disengage; some want to stay active and involved. Disengagement does not specify what happens when the elderly stay involved.
Activity	Macro	Robert Havighurst	People continue the level of activity they had in middle age into their later years. Though high levels of activity are unrelated to living longer, they are related to reporting high levels of life satisfaction.	Ill health may force people to curtail their level of activity. The older a person, the more likely the person is to curtail activity.
Conflict	Macro	Karl Marx, Max Weber	The elderly compete with youth for jobs and social resources such as government programs (Medicare).	The elderly are presented as disadvantaged. Their power to organize and mobilize political resources such as the American Association of Retired Persons is underestimated.
Age stratification	Macro	M. W. Riley	The elderly represent a powerful cohort of individuals passing through the social system that both affect and are affected by social change.	Too much emphasis is put on age, and little recognition is given to other variables within a cohort such as gender, race, and socioeconomic differences.
Modernization	Macro	Donald Cowgill	The status of the elderly is in reference to the evolution of the society toward modernization. The elderly in premodern societies have more status because what they have to offer in the form of cultural wisdom is more valued. The elderly in modern technologically advanced societies have low status because they have little to offer.	Cultural values for the elderly, not level of modernization, dictate the status of the elderly. Japan has high respect for the elderly and yet is highly technological and modernized.
Symbolic	Micro	Arlie Hochschild	The elderly socially construct meaning in their interactions with others and society. Developing social bonds with other elderly can ward off being isolated and abandoned. Meaning is in the interpretation, not in the event.	The power of the larger social system and larger social structures to affect the lives of the elderly is minimized.
Continuity	Micro	Bernice Neugarten	The earlier habit patterns, values, and attitudes of the individual are carried forward as a person ages. The only personality change that occurs with aging is the tendency to turn one's attention and interest on the self.	Other factors than one's personality affect aging outcomes. The social structure influences the life of the elderly rather than vice versa.

Key Terms

age term defined chronologically, physiologically, sociologically, and culturally.

age discrimination discriminating against a person because of age.

ageism systematic persecution and degradation of people because they are old.

blurred retirement the process of retiring gradually so that the individual works part-time before completely retiring or takes a bridge job that provides a transition between a lifelong career and full retirement.

cohousing an arrangement where elderly individuals live in group housing or shared living arrangements.

dementia the mental disorder most associated with aging, whereby the normal cognitive functions are slowly lost.

family caregiving care provided to the elderly by family members.

frail elderly person who has difficulty with at least one personal care activity or other activity related to independent living.

gerontology the study of aging.

gerontophobia fear or dread of the elderly.

Levitra, Cialis, Viagra medications taken by aging men to help them get and maintain an erection.

sandwich generation individuals who attempt to meet the needs of their children and elderly parents at the same time.

thanatology examination of the social dimensions of death, dying, and bereavement.

15.2 What is the "sandwich generation"?

Eldercare combined with child care is becoming common among the sandwich generation, adult children responsible for the needs of both their parents and their children. Two levels of eldercare include help with personal needs such as feeding, bathing, and toileting as well as instrumental care such as going to the grocery store, managing bank records, and so on. Members of the sandwich generation report feelings of exhaustion over the relentless demands, guilt over not doing enough, and resentment over feeling burdened.

Deciding whether to arrange for an elderly parent's care in a nursing home requires attention to a number of factors, including the level of care the parent needs, the philosophy and time availability of the adult child, and the resources of the adult children and other siblings. Full-time nursing care (not including medication) can be over $1,000 a week.

Elderly parents who are dying from terminal illnesses incur enormous medical bills. Some want to die and ask for help. Our society continues to wrestle with physician-assisted suicide and euthanasia. Only Oregon currently allows for physician-assisted suicide.

15.3 What issues confront the elderly?

Issues of concern to the elderly include housing, health, retirement, and sexuality. Most elderly live in their own homes, which they have paid for. Most elderly housing is adequate, although repair becomes a problem when people age. Health concerns are paramount for the elderly. Good health is the single most important factor associated with an elderly person's perceived life satisfaction.

Hearing and visual impairments, arthritis, heart conditions, and high blood pressure are all common to the elderly. Mental problems may also occur with mood disorders; depression is the most common.

Though the elderly are thought to be wealthy and living in luxury, most are not. The median household income of people over the age of 65 is less than half of what they earned in the prime of their lives. The most impoverished elderly are those who have lived the longest, who are widowed, and who live alone. Women are particularly disadvantaged because their work history may have been discontinuous, part-time, and low-paying. Social Security and private pension plans favor those with continuous, full-time work histories.

For most elderly women and men, sexuality involves lower reported interest, activity, and capacity. Fear of the inability to have an erection is the sexual problem elderly men most frequently report (Viagra, Levitra, and Cialis have helped allay this fear). The absence of a sexual partner is the sexual problem elderly women most frequently report.

15.4 What factors are associated with successful aging?

Factors associated with successful aging include not smoking (or quitting early), developing a positive view of life and life's crises, avoiding alcohol and substance abuse, maintaining healthy weight, exercising daily, continuing to educate oneself, and having a happy marriage. Indeed, those who were identified as "happy and well" were six times more likely to be in good marriages than those who were identified as "sad and sick." Success in one's career is also associated with successful aging.

15.5 What are relationships like for the elderly?

Marriages that survive into old age (beyond age 85) tend to have limited conflict, considerable companionship, and mutual supportiveness. Relationships with siblings are primarily emotional rather than functional. In regard to relationships of the elderly with their children, emotional and expressive rewards are high. Caregiving help is available but rare. Only 12 percent of one sample of adults older than 85 lived with their children.

15.6 How do the elderly face the end of life?

Thanatology is the examination of the social dimensions of death, dying, and bereavement. The end of life can involve adjusting to the death of one's spouse and to the gradual decline of one's health. Most elderly are satisfied with their lives, relationships, and health. Declines begin when people reach their eighties. Most fear the process of dying more than death itself.

The Love Attitudes Scale

This scale is designed to assess the degree to which you are romantic or realistic in your attitudes toward love. There are no right or wrong answers. After reading each sentence carefully, circle the number that best represents the degree to which you agree or disagree with the sentence.

1	2	3	4	5
Strongly Agree	Mildly Agree	Undecided	Mildly Disagree	Strongly Disagree

	SA	MA	U	MD	SD
1. Love doesn't make sense. It just is.	1	2	3	4	5
2. When you fall "head over heels" in love, it's sure to be the real thing.	1	2	3	4	5
3. To be in love with someone you would like to marry but can't is a tragedy.	1	2	3	4	5
4. When love hits, you know it.	1	2	3	4	5
5. Common interests are really unimportant; as long as each of you is truly in love, you will adjust.	1	2	3	4	5
6. It doesn't matter if you marry after you have known your partner for only a short time as long as you know you are in love.	1	2	3	4	5
7. If you are going to love a person, you will "know" after a short time.	1	2	3	4	5
8. As long as two people love each other, the educational differences they have really do not matter.	1	2	3	4	5
9. You can love someone even though you do not like any of that person's friends.	1	2	3	4	5
10. When you are in love, you are usually in a daze.	1	2	3	4	5
11. Love "at first sight" is often the deepest and most enduring type of love.	1	2	3	4	5
12. When you are in love, it really does not matter what your partner does because you will love him or her anyway.	1	2	3	4	5
13. As long as you really love a person, you will be able to solve the problems you have with the person.	1	2	3	4	5
14. Usually you can really love and be happy with only one or two people in the world.	1	2	3	4	5
15. Regardless of other factors, if you truly love another person, that is a good enough reason to marry that person.	1	2	3	4	5
16. It is necessary to be in love with the one you marry to be happy.	1	2	3	4	5
17. Love is more of a feeling than a relationship.	1	2	3	4	5
18. People should not get married unless they are in love.	1	2	3	4	5
19. Most people truly love only once during their lives.	1	2	3	4	5
20. Somewhere there is an ideal mate for most people.	1	2	3	4	5
21. In most cases, you will "know it" when you meet the right partner.	1	2	3	4	5
22. Jealousy usually varies directly with love; that is, the more you are in love, the greater your tendency to become jealous will be.	1	2	3	4	5
23. When you are in love, you are motivated by what you feel rather than by what you think.	1	2	3	4	5
24. Love is best described as an exciting rather than a calm thing.	1	2	3	4	5
25. Most divorces probably result from falling out of love rather than failing to adjust.	1	2	3	4	5
26. When you are in love, your judgment is usually not too clear.	1	2	3	4	5

		SA	MA	U	MD	SD
27.	Love comes only once in a lifetime.	1	2	3	4	5
28.	Love is often a violent and uncontrollable emotion.	1	2	3	4	5
29.	When selecting a marriage partner, differences in social class and religion are of small importance compared with love.	1	2	3	4	5
30.	No matter what anyone says, love cannot be understood.	1	2	3	4	5

Scoring

Add the numbers you circled. 1 (strongly agree) is the most romantic response and 5 (strongly disagree) is the most realistic response. The lower your total score (30 is the lowest possible score), the more romantic your attitudes toward love. The higher your total score (150 is the highest possible score), the more realistic your attitudes toward love. A score of 90 places you at the midpoint between being an extreme romantic and an extreme realist. Both men and women undergraduates typically score above 90, with men scoring closer to 90 than women.

A team of researchers (Medora et al. 2002) gave the scale to 641 young adults at three international universities in America, Turkey, and India. Female respondents in all three cultures had higher romanticism scores than male respondents (reflecting their higher value for, desire for, and thoughts about marriage). When the scores were compared by culture, American young adults were the most romantic, followed by Turkish students, with Indians having the lowest romanticism scores.

Source: Medora, N. P., J. H. Larson, N. Hortacsu, and P. Dave. 2002. Perceived attitudes towards romanticism: A cross-cultural study of American, Asian-Indian, and Turkish young adults. *Journal of Comparative Family Studies* 33:155–78.

Relationships Dynamics Scale

Please answer each of the following questions in terms of your relationship with your "mate" if married, or your "partner" if dating or engaged. We recommend that you answer these questions by yourself (not with your partner), using the ranges following for your own reflection.

Use the following three-point scale to rate how often you and your mate or partner experience the following:

1 = Almost never
2 = Once in a while
3 = Frequently

1 2 3 Little arguments escalate into ugly fights with accusations, criticisms, name-calling, or bringing up past hurts.
1 2 3 My partner criticizes or belittles my opinions, feelings, or desires.
1 2 3 My partner seems to view my words or actions more negatively than I mean them to be.
1 2 3 When we have a problem to solve, it is like we are on opposite teams.
1 2 3 I hold back from telling my partner what I really think and feel.
1 2 3 I think seriously about what it would be like to date or marry someone else.
1 2 3 I feel lonely in this relationship.
1 2 3 When we argue, one of us withdraws (that is, doesn't want to talk about it anymore; or leaves the scene).

Who tends to withdraw more when there is an argument?
Male
Female
Both Equally
Neither Tend to Withdraw

Where Are You in Your Marriage?

We devised these questions based on seventeen years of research at the University of Denver on the kinds of communication and conflict management patterns that predict whether a relationship is headed for trouble. We have recently completed a nationwide, random phone survey using these questions. The average score was 11 on this scale. Although you should not take a higher score to mean that your relationship is somehow destined to fail, higher scores can mean that your relationship may be in greater danger unless changes are made. (These ranges are based only on your individual ratings—not a couple total.)

8 to 12 "Green Light"

If you scored in the 8 to 12 range, your relationship is probably in good or even great shape *at this time*, but we emphasize *"at this time"* because relationships don't stand still. In the next twelve months, you'll either have a stronger, happier relationship, or you could head in the other direction.

To think about it another way, it's like you are traveling along and have come to a green light. There is no need to stop, but it is probably a great time to work on making your relationship all it can be.

13 to 17 "Yellow Light"

If you scored in the 13 to 17 range, it's like you are coming to a "yellow light." You need to be cautious. Although you may be happy now in your relationship, your score reveals warning signs of patterns you don't want to let get worse. You'll want to take action to protect and improve what you have. Spending time to strengthen your relationship now could be the best thing you could do for your future together.

18 to 24 "Red Light"

Finally, if you scored in the 18 to 24 range, this is like approaching a red light. Stop and think about where the two of you are headed. Your score indicates the presence of patterns that could put your relationship at significant risk. You may be heading for trouble—you may already be there. But there is ***good news***. You can stop and learn ways to improve your relationship now!

1. *We wrote these items based on understanding of many key studies in the field.* The content or themes behind the questions are based on numerous in-depth studies on how people think and act in their marriages. These kinds of dynamics have been compared with patterns on many other key variables, such as satisfaction, commitment, problem intensity, and so on. Because the kinds of methods researchers can use in their laboratories are quite complex, this actual measure is far simpler than many of the methods we and others use to study marriages over time. However, the themes are based on many solid studies. Caution is warranted in interpreting scores.
2. *The discussion of the Relationships Dynamics Scale gives rough guidelines for interpreting the meaning of the scores.* The ranges we suggest for the measure are based on results from a nationwide, random phone survey of 947 people (85 percent married) in January 1996. These ranges are meant as a rough guideline for helping couples assess the degree to which they are experiencing key danger signs in their marriages. The measure as you have it here powerfully discriminated between those doing well in their marriages or relationships and those who were not doing well on a host of other dimensions (thoughts of divorce, low satisfaction, low sense of friendship in the relationship, lower dedication, and so on). Couples scoring higher on these items are truly more likely to be experiencing problems (or, based on other research, are more likely to experience problems in the future).
3. *This measure in and of itself should not be taken as a predictor of couples who are going to fail in their marriages.* No couple should be told they will "not make it" based on a higher score. That would not be in keeping with our intention in developing this scale or with the meaning one could take from it for any one couple. Although the items are based on studies that assess such things as the likelihood of a marriage working out, we would hate for any one person to take this and assume the worst about their future based on a high score. Rather, we believe that the measure can be used to motivate high- and moderately high-scoring people to take a serious look at where their relationships are heading—and take steps to turn such negative patterns around for the better.

Source: Stanley, S. M., and H. J. Markman. 1997. *Marriage and Family: A Brief Introduction*. Reprinted with permission of PREP, Inc.

Note: For more information on constructive tools for strong marriages, or for questions about the measure and the meaning of it, please write to us at PREP, Inc., P. O. Box 102530, Denver, Colorado 80250-2530.

Supportive Communication Scale

This scale is designed to assess the degree to which partners experience supportive communication in their relationships. After reading each item, circle the number that best approximates your answer.

Strongly Disagree (SD) = 0
Disagree (D) = 1
Undecided (UN) = 2
Agree (A) = 3
Strongly Agree (SA) = 4

	SD	D	UN	A	SA
1. My partner listens to me when I need someone to talk to.	0	1	2	3	4
2. My partner helps me clarify my thoughts.	0	1	2	3	4
3. I can state my feelings without my partner getting defensive.	0	1	2	3	4
4. When it comes to having a serious discussion, it seems we have little in common.	0	1	2	3	4
5. I feel put down in a serious conversation with my partner.	0	1	2	3	4
6. I feel discussing some things with my partner is useless.	0	1	2	3	4
7. My partner and I understand each other completely.	0	1	2	3	4
8. We have an endless number of things to talk about.	0	1	2	3	4

Scoring

Look at the numbers you circled. Reverse score the numbers for questions 4, 5, and 6. For example, if you circled a 0, give yourself a 4; if you circled a 3, give yourself a 1, and so on. Add the numbers and divide by 8, the total number of items. The lowest possible score would be 0, reflecting the complete absence of supportive communication; the highest score would be 4, reflecting complete supportive communication. The average score of 94 male partners who took the scale was 3.01; the average score of 94 female partners was 3.07. Thirty-nine percent of the couples were married, 38 percent were single, and 23 percent were living together. The average age was just over 24.

Source: Sprecher, S., S. Metts, B. Burelson, E. Hatfield, and A. Thompson. 1995. Domains of expressive interaction in intimate relationships: Associations with satisfaction and commitment. *Family Relations* 44:203–10.

Involved Couple's Inventory (Excerpt)

This selection of questions from the Involved Couple's Inventory is designed to increase your knowledge of how you and your partner think and feel about a variety of issues. The full inventory (81 questions) is available online at 4ltrpress.cengage.com/M&F. Assume that you and your partner have considered getting married. Each partner should ask the other the following questions:

Partner Feelings and Issues

1. If you could change one thing about me, what would it be?
2. On a scale of 0 to 10, how well do you feel I respond to criticism or suggestions for improvement?
3. What would you like me to say or not say that would make you happier?
4. What do you think of yourself? Describe yourself with three adjectives.
5. What do you think of me? Describe me with three adjectives.
6. What do you like best about me?
7. On a scale of 0 to 10, how jealous do you think I am? How do you feel about my level of jealousy?
8. How do you feel about me emotionally?
9. To what degree do you feel we each need to develop and maintain outside relationships so as not to focus all of our interpersonal expectations on each other? Does this include other-sex individuals?
10. Do you have any history of abuse or violence, either as an abused child or adult or as the abuser in an adult relationship?

Feelings about Parents and Family

1. How do you feel about your mother? Your father? Your siblings?
2. On a ten-point scale, how close are you to your mom, dad, and each of your siblings?
3. How close were your family members to one another? On a ten-point scale, what value do you place on the opinions or values of your parents?
4. How often do you have contact with your father or mother? How often do you want to visit your parents and/or siblings? How often would you want them to visit us? Do you want to spend holidays alone or with your parents or mine?
5. What do you like and dislike most about each of your parents?
6. What do you like and dislike about my parents?
7. What is your feeling about living near our parents? How would you feel about my parents living with us? How do you feel about our parents living with us when they are old and cannot take care of themselves?
8. How do your parents get along? Rate their marriage on a scale of 0 to 10 (0 = unhappy, 10 = happy).

Social Issues, Religion, and Children

1. What are your political views? How do you feel about America being in Iraq?
2. What are your feelings about women's rights, racial equality, and homosexuality?
3. To what degree do you regard yourself as a religious or spiritual person? What do you think about religion, a Supreme Being, prayer, and life after death?
4. Do you go to religious services? Where? How often? Do you pray? How often? What do you pray about? When we are married, how often would you want to go to religious services? In what religion would you want our children to be reared? What responsibility would you take to ensure that our children had the religious training you wanted them to have?
5. How do you feel about abortion? Under what conditions, if any, do you feel abortion is justified?
6. How do you feel about children? How many do you want? When do you want the first child? At what intervals would you want to have additional children? What do you see as your responsibility in caring for the children—changing diapers,

feeding, bathing, playing with them, and taking them to lessons and activities? To what degree do you regard these responsibilities as mine?

7. Suppose I did not want to have children or couldn't have them. How would you feel? How do you feel about artificial insemination, surrogate motherhood, in vitro fertilization, and adoption?
8. To your knowledge, can you have children? Are there any genetic problems in your family history that would prevent us from having normal children? How healthy (mentally and physically) are you? How often have you seen a physician in the last three years? What medications have you taken or do you currently take? What are these medications for? Have you seen a therapist, psychologist, or psychiatrist? What for?
9. How should children be disciplined? Do you want our children to go to public or private schools?
10. How often do you think we should go out alone without our children? If we had to decide between the two of us going on a cruise to the Bahamas alone or taking the children camping for a week, what would you choose?

Sex

1. How much sexual intimacy do you feel is appropriate in casual dating, involved dating, and engagement?
2. Does "having sex" mean having sexual intercourse? If a couple has experienced oral sex only, have they "had sex"?
3. What sexual behaviors do you most and least enjoy? How often do you want to have intercourse? How do you want me to turn you down when I don't want to have sex? How do you want me to approach you for sex? How do you feel about just being physical together—hugging, rubbing, holding, but not having intercourse?
4. By what method of stimulation do you experience an orgasm most easily?
5. What do you think about masturbation, oral sex, homosexuality, sadism and masochism (S & M), and anal sex?
6. What type of contraception do you suggest? Why? If that method does not prove satisfactory, what method would you suggest next?
7. What are your values regarding extramarital sex? If I had an affair, would you want me to tell you? Why? If I told you about the affair, what would you do? Why?
8. How often do you view pornographic videos or pornography on the Internet?
9. How important is our using a condom to you?
10. Do you want me to be tested for human immunodeficiency virus (HIV)? Are you willing to be tested?

Careers and Money

1. What kind of job or career will you have? What are your feelings about working in the evening versus being home with the family? Where will your work require that we live? How often do you feel we will be moving? How much travel will your job require?
2. To what degree did your parents agree on how to deal with money? Who was in charge of spending, and who was in charge of saving? Did working or earning the bigger portion of the income connect to control over money?
3. What are your feelings about a joint versus a separate checking account? Which of us do you want to pay the bills? How much money do you think we will have left over each month? How much of this do you think we should save?
4. When we disagree over whether to buy something, how do you suggest we resolve our conflict?
5. What jobs or work experience have you had? If we end up having careers in different cities, how do you feel about being involved in a commuter marriage?
6. What is your preference for where we live? Do you want to live in an apartment or a house? What are your needs for a car, television, cable service, phone plan, entertainment devices, and so on? What are your feelings about us living in two separate places, the "living apart together" idea whereby we can have a better relationship if we give each other some space and have plenty of room?
7. How do you feel about my having a career? Do you expect me to earn an income? If so, how much annually? To what degree do you feel it is your responsibility to cook, clean, and take care of the children? How do you feel about putting young children or infants in day-care centers? When the children are sick and one of us has to stay home, who will that be?

Recreation and Leisure

1. What is your idea of the kinds of parties or social gatherings you would like for us to go to together?
2. What is your preference in terms of us hanging out with others in a group versus being alone?
3. What is your favorite recreational interest? How much time do you spend enjoying this interest? How important is it for you that I share this recreational interest with you?
4. What do you like to watch on television? How often do you watch television and for what periods of time?
5. What are the amount and frequency of your current use of alcohol and other drugs (for example, marijuana, cocaine, crack, speed)? What, if any, have been your previous alcohol and other drug behaviors and frequencies? What are your expectations of me regarding the use of alcohol and other drugs?

Relationships with Friends and Coworkers

1. How do you feel about my three closest same-sex friends?
2. How do you feel about my spending time with my friends or coworkers, such as one evening a week?
3. How do you feel about my spending time with friends of the opposite sex?
4. What do you regard as appropriate and inappropriate affectional behaviors with opposite-sex friends?

Remarriage Questions

1. How and why did your first marriage end? What are your feelings about your former spouse now? What are the feelings of your former spouse toward you? How much "trouble" do you feel your former spouse will want to cause us? What relationship do you want with your former spouse?
2. Do you want your children from a previous marriage to live with us? What are your emotional and financial expectations of me in regard to your children? What are your feelings about my children living with us? Do you want us to have additional children? How many? When?
3. When your children are with us, who will be responsible for their food preparation, care, discipline, and driving them to activities?
4. Suppose your children do not like me and vice versa. How will you handle this? Suppose they are against our getting married?
5. Suppose our respective children do not like one another. How will you handle this?

This self-assessment is intended to be thought-provoking and fun. It is not intended to be used as a clinical or diagnostic instrument. It would be unusual if you agreed with each other on all of your answers to the previous questions. You might view the differences as challenges and then find out the degree to which the differences are important for your relationship. You might need to explore ways of minimizing the negative impact of those differences on your relationship. It is not possible to have a relationship with someone in which there is total agreement. Disagreement is inevitable; the issue becomes how you and your partner manage the differences.

Attitudes toward Interracial Dating Scale

Interracial dating or marrying is the dating or marrying of two people from different races. The purpose of this survey is to gain a better understanding of what people think and feel about interracial relationships. Please read each item carefully, and in each space, score your response using the following scale. There are no right or wrong answers to any of these statements.

1	2	3	4	5	6	7
Strongly Disagree						Strongly Agree

_____ **1.** I believe that interracial couples date outside their race to get attention.
_____ **2.** I feel that interracial couples have little in common.
_____ **3.** When I see an interracial couple, I find myself evaluating them negatively.
_____ **4.** People date outside their own race because they feel inferior.
_____ **5.** Dating interracially shows a lack of respect for one's own race.
_____ **6.** I would be upset with a family member who dated outside our race.
_____ **7.** I would be upset with a close friend who dated outside our race.
_____ **8.** I feel uneasy around an interracial couple.
_____ **9.** People of different races should associate only in nondating settings.
_____ **10.** I am offended when I see an interracial couple.
_____ **11.** Interracial couples are more likely to have low self-esteem.
_____ **12.** Interracial dating interferes with my fundamental beliefs.
_____ **13.** People should date only within their race.
_____ **14.** I dislike seeing interracial couples together.
_____ **15.** I would not pursue a relationship with someone of a different race, regardless of my feelings for that person.
_____ **16.** Interracial dating interferes with my concept of cultural identity.
_____ **17.** I support dating between people with the same skin color, but not with a different skin color.
_____ **18.** I can imagine myself in a long-term relationship with someone of another race.
_____ **19.** As long as the people involved love each other, I do not have a problem with interracial dating.
_____ **20.** I think interracial dating is a good thing.

Scoring

First, reverse the scores for items 18, 19, and 20 by switching them to the opposite side of the spectrum. For example, if you selected 7 for item 18, replace it with a 1; if you selected 3, replace it with a 5, and so on. Next, add your scores and divide by 20. Possible final scores range from 1 to 7, with 1 representing the most positive attitudes toward interracial dating and 7 representing the most negative attitudes toward interracial dating.

Norms

The norming sample was based upon 113 male and 200 female students attending Valdosta State University. The participants completing the Attitudes toward Interracial Dating Scale (IRDS) received no compensation for their participation. All participants were U.S. citizens. The average age was 23.02 years (standard deviation [SD] = 5.09), and participants ranged in age from 18 to 50 years. The ethnic composition of the sample was 62.9 percent white, 32.6 percent black, 1 percent Asian, 0.6 percent Hispanic, and 2.2 percent other. The classification of the sample was 9.3 percent freshmen, 16.3 percent sophomores, 29.1 percent juniors, 37.1 percent seniors, and 2.9 percent graduate students. The average score on the IRDS was 2.88 (SD = 1.48), and scores ranged from 1.00 to 6.60, suggesting very positive views of interracial dating. Men scored an average of 2.97 (SD = 1.58), and women, 2.84 (SD = 1.42). There were no significant differences between the responses of women and men.

Source: The Attitudes toward Interracial Dating Scale. 2004. Mark Whatley, PhD, Department of Psychology, Valdosta State University, Valdosta, GA 31698. The scale is used by permission.

The Self-Report of Behavior Scale

This questionnaire is designed to examine which of the following statements most closely describes your behavior during past encounters with people you thought were homosexuals. Rate each of the following self-statements as honestly as possible using the following scale. Write each value in the provided blank.

Never = 1	Occasionally = 3	Frequently = 4
Rarely = 2		Always = 5

_____ **1.** I have spread negative talk about someone because I suspected that the person was gay.
_____ **2.** I have participated in playing jokes on someone because I suspected that the person was gay.
_____ **3.** I have changed roommates and/or rooms because I suspected my roommate was gay.
_____ **4.** I have warned people who I thought were gay and who were a little too friendly with me to keep away from me.
_____ **5.** I have attended antigay protests.
_____ **6.** I have been rude to someone because I thought that the person was gay.
_____ **7.** I have changed seat locations because I suspected the person sitting next to me was gay.
_____ **8.** I have had to force myself to keep from hitting someone because the person was gay and very near me.
_____ **9.** When someone I thought to be gay has walked toward me as if to start a conversation, I have deliberately changed directions and walked away to avoid the person.
_____ **10.** I have stared at a gay person in such a manner as to convey my disapproval of the person being too close to me.
_____ **11.** I have been with a group in which one (or more) person(s) yelled insulting comments to a gay person or group of gay people.
_____ **12.** I have changed my normal behavior in a restroom because a person I believed to be gay was in there at the same time.
_____ **13.** When a gay person has checked me out, I have verbally threatened the person.
_____ **14.** I have participated in damaging someone's property because the person was gay.
_____ **15.** I have physically hit or pushed someone I thought was gay because the person brushed against me when passing by.
_____ **16.** Within the past few months, I have told a joke that made fun of gay people.
_____ **17.** I have gotten into a physical fight with a gay person because I thought the person had been making moves on me.
_____ **18.** I have refused to work on school and/or work projects with a partner I thought was gay.
_____ **19.** I have written graffiti about gay people or homosexuality.
_____ **20.** When a gay person has been near me, I have moved away to put more distance between us.

Scoring

Determine your score by adding your points together. The lowest score is 20 points, the highest 100 points. The higher the score, the more negative the attitudes toward homosexuals.

Comparison Data

Sunita Patel (1989) originally developed the Self-Report of Behavior Scale in her thesis research in her clinical psychology master's program at East Carolina University. College men (from a university campus and from a military base) were the original participants (Patel et al. 1995). The scale was revised by Shartra Sylivant (1992), who used it with a coed high school student population, and by Tristan Roderick (1994), who involved college students to assess its psychometric properties. The scale was found to have high internal consistency. Two factors were identified: a passive avoidance of homosexuals and active or aggressive reactions.

In a study by Roderick et al. (1998), the mean score for 182 college women was 24.76. The mean score for 84 men was significantly higher, at 31.60. A similar-sex difference, although with higher (more negative) scores, was found in Sylivant's high school sample (with a mean of 33.74 for the young women, and 44.40 for the young men).

The following table provides detail for the scores of the college students in Roderick's sample (from a mid-sized state university in the southeast):

	N	Mean	Standard Deviation
Women	182	24.76	7.68
Men	84	31.60	10.36
Total	266	26.91	9.16

Sources:
Patel, S. 1989. Homophobia: Personality, emotional, and behavioral correlates. Master's thesis, East Carolina University.
Patel, S., T. E. Long, S. L. McCammon, and K. L. Wuensch. 1995. Personality and emotional correlates of self reported antigay behaviors. *Journal of Interpersonal Violence* 10:354–66.
Roderick, T. 1994. Homonegativity: An analysis of the SBS-R. Master's thesis, East Carolina University.
Roderick, T., S. L. McCammon, T. E. Long, and L. J. Allred. 1998. Behavioral aspects of homonegativity. *Journal of Homosexuality* 36:79–88.
Sylivant, S. 1992. The cognitive, affective, and behavioral components of adolescent homonegativity. Master's thesis, East Carolina University.
The SBS-R is reprinted by the permission of the students and faculty who participated in its development: S. Patel, S. L. McCammon, T. E. Long, L. J. Allred, K. Wuensch, T. Roderick, and S. Sylivant.

Student Sexual Risks Scale

Safer sex means sexual activity that reduces the risk of transmitting STIs, including the AIDS virus. Using condoms is an example of safer sex. Unsafe, risky, or unprotected sex refers to sex without a condom, or to other sexual activity that might increase the risk of transmitting the AIDS virus. For each of the following items, check the response that best characterizes your opinion.

A = Agree
U = Undecided
D = Disagree

_____ **1.** If my partner wanted me to have unprotected sex, I would probably give in.
_____ **2.** The proper use of a condom could enhance sexual pleasure.
_____ **3.** I may have had sex with someone who was at risk for HIV/AIDS.
_____ **4.** If I were going to have sex, I would take precautions to reduce my risk for HIV/AIDS.
_____ **5.** Condoms ruin the natural sex act.
_____ **6.** When I think that one of my friends might have sex on a date, I remind my friend to take a condom.
_____ **7.** I am at risk for HIV/AIDS.
_____ **8.** I try to use a condom when I have sex.
_____ **9.** Condoms interfere with romance.
_____ **10.** My friends talk a lot about safer sex.
_____ **11.** If my partner wanted me to participate in risky sex and I said that we needed to be safer, we would still probably end up having unsafe sex.
_____ **12.** Generally, I am in favor of using condoms.
_____ **13.** I would avoid using condoms if at all possible.
_____ **14.** If a friend knew that I might have sex on a date, the friend would ask me whether I was carrying a condom.
_____ **15.** There is a possibility that I have HIV/AIDS.
_____ **16.** If I had a date, I would probably not drink alcohol or use drugs.
_____ **17.** Safer sex reduces the mental pleasure of sex.
_____ **18.** If I thought that one of my friends had sex on a date, I would ask the friend if he or she used a condom.
_____ **19.** The idea of using a condom doesn't appeal to me.
_____ **20.** Safer sex is a habit for me.
_____ **21.** If a friend knew that I had sex on a date, the friend wouldn't care whether I had used a condom or not.
_____ **22.** If my partner wanted me to participate in risky sex and I suggested a lower-risk alternative, we would have the safer sex instead.
_____ **23.** The sensory aspects of condoms (smell, touch, and so on) make them unpleasant.
_____ **24.** I intend to follow "safer sex" guidelines within the next year.
_____ **25.** With condoms, you can't really give yourself over to your partner.
_____ **26.** I am determined to practice safer sex.
_____ **27.** If my partner wanted me to have unprotected sex and I made some excuse to use a condom, we would still end up having unprotected sex.
_____ **28.** If I had sex and I told my friends that I did not use a condom, they would be angry or disappointed.
_____ **29.** I think safer sex would get boring fast.
_____ **30.** My sexual experiences do not put me at risk for HIV/AIDS.
_____ **31.** Condoms are irritating.
_____ **32.** My friends and I encourage each other before dates to practice safer sex.

_____ **33.** When I socialize, I usually drink alcohol or use drugs.

_____ **34.** If I were going to have sex in the next year, I would use condoms.

_____ **35.** If a sexual partner didn't want to use condoms, we would have sex without using condoms.

_____ **36.** People can get the same pleasure from safer sex as from unprotected sex.

_____ **37.** Using condoms interrupts sex play.

_____ **38.** Using condoms is a hassle.

Scoring (to be read after completing the above scale)

Begin by giving yourself 80 points. Subtract one point for every undecided response. Subtract two points every time that you disagreed with odd-numbered items or with item number 38. Subtract two points every time you agreed with even-numbered items 2 through 36.

Interpreting Your Score

Research shows that students with higher scores on the SSRS are more likely to engage in risky sexual activities, such as having multiple sex partners and failing to consistently use condoms during sex. In contrast, students who practice safer sex tend to endorse more positive attitudes toward safer sex, and tend to have peer networks that encourage safer sexual practices. These students usually plan on making sexual activity safer, and they feel confident in their ability to negotiate safer sex, even when a dating partner may press for riskier sex. Students who practice safer sex often refrain from using alcohol or drugs, which may impede negotiation of safer sex, and often report having engaged in lower-risk activities in the past. How do you measure up?

(Below 15) Lower Risk

Congratulations! Your score on the SSRS indicates that, relative to other students, your thoughts and behaviors are more supportive of safer sex. Is there any room for improvement in your score? If so, you may want to examine items for which you lost points and try to build safer sexual strengths in those areas. You can help protect others from HIV by educating your peers about making sexual activity safer. (Of 200 students surveyed by DeHart and Berkimer, 16 percent were in this category.)

(15 to 37) Average Risk

Your score on the SSRS is about average in comparison with those of other college students. Though it is good that you don't fall into the higher-risk category, be aware that "average" people can get HIV, too. In fact, a recent study indicated that the rate of HIV among college students is ten times that in the general heterosexual population. Thus, you may want to enhance your sexual safety by figuring out where you lost points and work toward safer sexual strengths in those areas.
(Of 200 students surveyed by DeHart and Berkimer, 68 percent were in this category.)

(38 and Above) Higher Risk

Relative to other students, your score on the SSRS indicates that your thoughts and behaviors are less supportive of safer sex. Such high scores tend to be associated with greater HIV-risk behavior. Rather than simply giving in to riskier attitudes and behaviors, you may want to empower yourself and reduce your risk by critically examining areas for improvement. On which items did you lose points? Think about how you can strengthen your sexual safety in these areas. Reading more about safer sex can help, and sometimes colleges and health clinics offer courses or workshops on safer sex. You can get more information about resources in your area by contacting the Center on Disease Control's HIV/AIDS Information Line at 1-800-342-2437. (Of 200 students surveyed by DeHart and Birkimer, 16 percent were in this category.)

Source: DeHart, D. D., and J. C. Birkimer. 1997. The Student Sexual Risks Scale (modification of SRS for popular use; facilitates student self-administration, scoring, and normative interpretation). Developed specifically for this text by Dana D. DeHart, College of Social Work at the University of South Carolina, and John C. Birkimer, University of Louisville. Used by permission of Dana DeHart.

Attitudes toward Parenthood Scale

The purpose of this survey is to assess your attitudes toward the role of parenthood. Please read each item carefully and consider what you believe and feel about parenthood. There are no right or wrong answers to any of these statements. After reading each statement, select the number that best reflects your answer using the following scale:

1	2	3	4	5	6	7
Strongly Disagree						Strongly Agree

_____ **1.** Parents should be involved in their children's school.
_____ **2.** I think children add joy to a parent's life.
_____ **3.** Parents attending functions such as sporting events and recitals of their children help build their child socially.
_____ **4.** Parents are responsible for providing a healthy environment for their children.
_____ **5.** When you become a parent, your children become your top priority.
_____ **6.** Parenting is a job.
_____ **7.** I feel that spending quality time with children is an important aspect of child rearing.
_____ **8.** Mothers and fathers should share equal responsibilities in raising children.
_____ **9.** The formal education of children should begin at as early an age as possible.

Scoring

After assigning a number to each item, add the numbers and divide by 9. The higher the number (7 is the highest possible), the more positive your view and the stronger your commitment to the role of parenthood. The lower the number (1 is the lowest possible), the more negative your view and the weaker your commitment to parenthood.

Norms

Norms for the scale are based upon twenty-two male and seventy-two female students attending Valdosta State University. The scores ranged from 3.89 to 7.00, and the average was 6.36 (standard deviation [SD] = 0.65); hence, the respondents had very positive views. There was no significant difference between male participants' attitudes toward parenthood (mean [M] = 6.15; SD = 0.70) and female participants' attitudes (M = 6.42; SD = 0.63). There were also no significant differences between ethnicities.

The average age of participants completing the Attitudes toward Parenthood Scale was 22.09 years (SD = 3.85), and their ages ranged from 18 to 39. The ethnic composition of the sample was 73.4 percent white, 22.3 percent black, 2.1 percent Asian, 1.1 percent American Indian, and 1.1 percent other. The classification of the sample was 20.2 percent freshmen, 6.4 percent sophomores, 22.3 percent juniors, 47.9 percent seniors, and 3.2 percent graduate students.

Source: Mark Whatley, PhD. The Attitudes toward Parenthood Scale. 2004. Department of Psychology, Valdosta State University. Information on validity and reliability may be obtained from Dr. Whatley (mwhatley@valdosta.edu). The scale is reprinted here with permission of Dr. Whatley. Other uses of this scale by written permission only from Dr. Whatley.

Attitudes toward Transracial Adoption Scale

Transracial adoption is the adoption of children of a race other than that of the adoptive parents. Please read each item carefully and consider what you believe about each statement. There are no right or wrong answers to any of these statements, so please give your honest reaction and opinion. After reading each statement, select the number that best reflects your answer, using the following scale:

1	2	3	4	5	6	7
Strongly Disagree						Strongly Agree

_____ **1.** Transracial adoption can interfere with a child's well-being.
_____ **2.** Transracial adoption should not be allowed.
_____ **3.** I would never adopt a child of another race.
_____ **4.** I think that transracial adoption is unfair to the children.
_____ **5.** I believe that adopting parents should adopt a child within their own race.
_____ **6.** Only same-race couples should be allowed to adopt.
_____ **7.** Biracial couples are not well prepared to raise children.
_____ **8.** Transracially adopted children need to choose one culture over another.
_____ **9.** Transracially adopted children feel as though they are not part of the family they live in.
_____ **10.** Transracial adoption should occur only between certain races.
_____ **11.** I am against transracial adoption.
_____ **12.** A person has to be desperate to adopt a child of another race.
_____ **13.** Children adopted by parents of a different race have more difficulty developing socially than children adopted by foster parents of the same race.
_____ **14.** Members of multiracial families do not get along well.
_____ **15.** Transracial adoption results in "cultural genocide."

Scoring

After assigning a number to each item, add the numbers and divide by 15. The lower the score (1 is the lowest possible), the more positive one's view of transracial adoptions. The higher the score (7 is the highest possible), the more negative one's view of transracial adoptions. The norming sample was based upon thirty-four male and sixty-nine female students attending Valdosta State University. The average score was 2.27 (SD = 1.15), suggesting a generally positive view of transracial adoption by the respondents, and scores ranged from 1.00 to 6.60.

The average age of participants completing the scale was 22.22 years (SD = 4.23), and ages ranged from 18 to 48. The ethnic composition of the sample was 74.8 percent white, 20.4 percent black, 1.9 percent Asian, 1.0 percent Hispanic, 1.0 percent American Indian, and one person of nonindicated ethnicity. The classification of the sample was 15.5 percent freshmen, 6.8 percent sophomores, 32.0 percent juniors, 42.7 percent seniors, and 2.9 percent graduate students.

Source: Mark Whatley, PhD. 2004. The Attitudes toward Transracial Adoption Scale. Department of Psychology, Valdosta State University. Information on validity and reliability and permission to use this scale may be obtained from Dr. Whatley (mwhatley@valdosta.edu).

The Traditional Motherhood Scale

The purpose of this survey is to assess the degree to which students possess a traditional view of motherhood. Read each item carefully and consider what you believe. There are no right or wrong answers, so please give your honest reaction and opinion. After reading each statement, select the number that best reflects your level of agreement, using the following scale:

1	2	3	4	5	6	7
Strongly Disagree						Strongly Agree

_____ **1.** A mother has a better relationship with her children than a father does.
_____ **2.** A mother knows more about her child than a father, thereby being the better parent.
_____ **3.** Motherhood is what brings women to their fullest potential.
_____ **4.** A good mother should stay at home with her children for the first year.
_____ **5.** Mothers should stay at home with the children.
_____ **6.** Motherhood brings much joy and contentment to a woman.
_____ **7.** A mother is needed in a child's life for nurturance and growth.
_____ **8.** Motherhood is an essential part of a female's life.
_____ **9.** I feel that all women should experience motherhood in some way.
_____ **10.** Mothers are more nurturing than fathers.
_____ **11.** Mothers have a stronger emotional bond with their children than do fathers.
_____ **12.** Mothers are more sympathetic to children who have hurt themselves than are fathers.
_____ **13.** Mothers spend more time with their children than do fathers.
_____ **14.** Mothers are more lenient toward their children than are fathers.
_____ **15.** Mothers are more affectionate toward their children than are fathers.
_____ **16.** The presence of the mother is vital to the child during the formative years.
_____ **17.** Mothers play a larger role than fathers in raising children.
_____ **18.** Women instinctively know what a baby needs.

Scoring

After assigning a number from 1 (strongly disagree) to 7 (strongly agree), add the numbers and divide by 18. The higher your score (7 is the highest possible score), the stronger the traditional view of motherhood. The lower your score (1 is the lowest possible score), the less traditional the view of motherhood.

Norms

The norming sample of this self-assessment was based upon twenty male and eighty-six female students attending Valdosta State University. The average age of participants completing the scale was 21.72 years (SD = 2.98), and ages ranged from 18 to 34. The ethnic composition of the sample was 80.2 percent white, 15.1 percent black, 1.9 percent Asian, 0.9 percent American Indian, and 1.9 percent other. The classification of the sample was 16.0 percent freshmen, 15.1 percent sophomores, 27.4 percent juniors, 39.6 percent seniors, and 1.9 percent graduate students.

Participants responded to each of the eighteen items according to the 7-point scale. The most traditional score was 6.33; the score reflecting the least support for traditional motherhood was 1.78. The midpoint (average score) between the top and bottom score was 4.28 (SD = 1.04); thus, people scoring above this number tended to have a more traditional view of motherhood and people scoring below this number tended to have a less traditional view of motherhood.

There was a significant difference ($p = .05$) between female participants' scores (mean = 4.19; SD = 1.08) and male participants' scores (mean = 4.68; SD = 0.73), suggesting that males had more traditional views of motherhood than females.

Source: Mark Whatley, PhD. 2004. The Traditional Motherhood Scale. Department of Psychology, Valdosta State University. Use of this scale is permitted only by prior written permission of Dr. Whatley (mwhatley@valdosta.edu).

The Traditional Fatherhood Scale

The purpose of this survey is to assess the degree to which students have a traditional view of fatherhood. Read each item carefully and consider what you believe. There are no right or wrong answers, so please give your honest reaction and opinion. After reading each statement, select the number that best reflects your level of agreement, using the following scale:

1	2	3	4	5	6	7
Strongly Disagree						Strongly Agree

_____ **1.** Fathers do not spend much time with their children.
_____ **2.** Fathers should be the disciplinarians in the family.
_____ **3.** Fathers should never stay at home with the children while the mother works.
_____ **4.** The father's main contribution to his family is giving financially.
_____ **5.** Fathers are less nurturing than mothers.
_____ **6.** Fathers expect more from children than their mothers do.
_____ **7.** Most men make horrible fathers.
_____ **8.** Fathers punish children more than mothers do.
_____ **9.** Fathers do not take a highly active role in their children's lives.
_____ **10.** Fathers are very controlling.

Scoring

After assigning a number from 1 (strongly disagree) to 7 (strongly agree), add the numbers and divide by 10. The higher your score (7 is the highest possible score), the stronger the traditional view of fatherhood. The lower your score (1 is the lowest possible score), the less traditional the view of fatherhood.

Norms

The norming sample was based upon twenty-four male and sixty-nine female students attending Valdosta State University. The average age of participants completing the Traditional Fatherhood Scale was 22.15 years (SD = 4.23), and ages ranged from 18 to 47. The ethnic composition of the sample was 77.4 percent white, 19.4 percent black, 1.1 percent Hispanic, and 2.2 percent other. The classification of the sample was 16.1 percent freshmen, 11.8 percent sophomores, 23.7 percent juniors, 46.2 percent seniors, and 2.2 percent graduate students.

Participants responded to each of the ten items on the 7-point scale. The most traditional score was 5.50; the score representing the least support for traditional fatherhood was 1.00. The average score was 3.33 (SD = 1.03), suggesting a less-than-traditional view.

There was a significant difference ($p = .05$) between female participants' attitudes (mean = 3.20; SD = 1.01) and male participants' attitudes toward fatherhood (mean = 3.69; SD = 1.01), suggesting that males had more traditional views of fatherhood than females. There were no significant differences between ethnicities.

Source: Mark Whatley, PhD. 2004. The Traditional Fatherhood Scale. Department of Psychology, Valdosta State University. Use of this scale is permitted only by prior written permission of Dr. Whatley (mwhatley@valdosta.edu).

Maternal Employment Scale

Using the following scale, please mark a number on the blank next to each statement to indicate how strongly you agree or disagree.

1	2	3	4	5	6
Disagree Very Strongly	Disagree Strongly	Disagree Slightly	Agree Slightly	Agree Strongly	Agree Very Strongly

_____ **1.** Children are less likely to form a warm and secure relationship with a mother who works full-time outside the home.

_____ **2.** Children whose mothers work are more independent and able to do things for themselves.

_____ **3.** Working mothers are more likely to have children with psychological problems than mothers who do not work outside the home.

_____ **4.** Teenagers get into less trouble with the law if their mothers do not work full-time outside the home.

_____ **5.** For young children, working mothers are good role models for leading busy and productive lives.

_____ **6.** Boys whose mothers work are more likely to develop respect for women.

_____ **7.** Young children learn more if their mothers stay at home with them.

_____ **8.** Children whose mothers work learn valuable lessons about other people they can rely on.

_____ **9.** Girls whose mothers work full-time outside the home develop stronger motivation to do well in school.

_____ **10.** Daughters of working mothers are better prepared to combine work and motherhood if they choose to do both.

_____ **11.** Children whose mothers work are more likely to be left alone and exposed to dangerous situations.

_____ **12.** Children whose mothers work are more likely to pitch in and do tasks around the house.

_____ **13.** Children do better in school if their mothers are not working full-time outside the home.

_____ **14.** Children whose mothers work full-time outside the home develop more regard for women's intelligence and competence.

_____ **15.** Children of working mothers are less well-nourished and don't eat the way they should.

_____ **16.** Children whose mothers work are more likely to understand and appreciate the value of a dollar.

_____ **17.** Children whose mothers work suffer because their mothers are not there when they need them.

_____ **18.** Children of working mothers grow up to be less competent parents than other children because they have not had adequate parental role models.

_____ **19.** Sons of working mothers are better prepared to cooperate with a wife who wants both to work and have children.

_____ **20.** Children of mothers who work develop lower self-esteem because they think they are not worth devoting attention to.

_____ **21.** Children whose mothers work are more likely to learn the importance of teamwork and cooperation among family members.

_____ **22.** Children of working mothers are more likely than other children to experiment with alcohol, other drugs, and sex at an early age.

_____ **23.** Children whose mothers work develop less stereotyped views about men's and women's roles.

_____ **24.** Children whose mothers work full-time outside the home are more adaptable; they cope better with the unexpected and with changes in plans.

Scoring

Items 1, 3, 4, 7, 11, 13, 15, 17, 18, 20, and 22 refer to "costs" of maternal employment for children and yield a Costs Subscale score. High scores on the Costs Subscale reflect strong beliefs that maternal employment is costly to children. Items 2, 5, 6, 8, 9, 10, 12, 14, 16, 19, 21, 23, and 24 refer to "benefits" of maternal employment for children and yield a Benefits Subscale score. To obtain a Total Score, reverse the score of all items in the Benefits Subscale so that 1 = 6, 2 = 5, 3 = 4, 4 = 3, 5 = 2, and 6 = 1. The higher one's Total Score, the more one believes that maternal employment has negative consequences for children.

Source: Greenberger, E., W. A. Goldberg, T. J. Crawford, and J. Granger. 1988. Beliefs about the consequences of maternal employment for children. *Psychology of Women* 12:35–59. Used by permission of Blackwell Publishing.

Attitudes toward Infidelity Scale

The purpose of this survey is to gain a better understanding of what people think and feel about issues associated with infidelity. There are no right or wrong answers to any of these statements; we are interested in your honest reactions and opinions. Please read each statement carefully, and respond by using the following scale:

1	2	3	4	5	6	7
Strongly Disagree						Strongly Agree

_____ **1.** Being unfaithful never hurt anyone.
_____ **2.** Infidelity in a marital relationship is grounds for divorce.
_____ **3.** Infidelity is acceptable for retaliation of infidelity.
_____ **4.** It is natural for people to be unfaithful.
_____ **5.** Online/Internet behavior (for example, visiting sex chat rooms, porn sites) is an act of infidelity.
_____ **6.** Infidelity is morally wrong in all circumstances, regardless of the situation.
_____ **7.** Being unfaithful in a relationship is one of the most dishonorable things a person can do.
_____ **8.** Infidelity is unacceptable under any circumstances if the couple is married.
_____ **9.** I would not mind if my significant other had an affair as long as I did not know about it.
_____ **10.** It would be acceptable for me to have an affair, but not my significant other.
_____ **11.** I would have an affair if I knew my significant other would never find out.
_____ **12.** If I knew my significant other was guilty of infidelity, I would confront him/her.

Scoring

Selecting a 1 reflects the least acceptance of infidelity; selecting a 7 reflects the greatest acceptance of infidelity. Before adding the numbers you selected, reverse the scores for item numbers 2, 5, 6, 7, 8, and 12. For example, if you responded to item 2 with a "6," change this number to a "2"; if you responded with a "3," change this number to "5," and so on. After making these changes, add the numbers. The lower your total score (12 is the lowest possible), the less accepting you are of infidelity; the higher your total score (84 is the highest possible), the greater your acceptance of infidelity. A score of 48 places you at the midpoint between being very disapproving and very accepting of infidelity.

Scores of Other Students Who Completed the Scale

The scale was completed by 150 male and 136 female student volunteers at Valdosta State University. The average score on the scale was 27.85 (SD = 12.02). Their ages ranged from 18 to 49, with a mean age of 23.36 (SD = 5.13). The ethnic backgrounds of the sample included 60.8 percent white, 28.3 percent African American, 2.4 percent Hispanic, 3.8 percent Asian, 0.3 percent American Indian, and 4.2 percent other. The college classification level of the sample included 11.5 percent freshmen, 18.2 percent sophomores, 20.6 percent juniors, 37.8 percent seniors, 7.7 percent graduate students, and 4.2 percent postbaccalaureate. Male participants reported more positive attitudes toward infidelity (mean = 31.53; SD = 11.86) than did female participants (mean = 23.78; SD = 10.86; $p = .05$). White participants had more negative attitudes toward infidelity (mean = 25.36; SD = 11.17) than did nonwhite participants (mean = 31.71; SD = 12.32; $p = .05$). There were no significant differences in regard to college classification.

Source: Mark Whatley, PhD. 2006. Attitudes toward Infidelity Scale. Department of Psychology, Valdosta State University. Used by permission. Other uses of this scale by written permission of Dr. Whatley only. Information on the reliability and validity of this scale is available from Dr. Whatley (mwhatley@valdosta.edu).

Motives for Drinking Alcohol Scale

Read the list of reasons people sometimes give for drinking alcohol. Thinking of all the times you drink, how often would you say you drink for each of the following reasons?

1 = Almost never/never
2 = Some of the time
3 = Half of the time
4 = Most of the time
5 = Almost always

_____ **1.** Because it helps me to enjoy a party
_____ **2.** To be sociable
_____ **3.** To make social gatherings more fun
_____ **4.** To improve parties and celebrations
_____ **5.** To celebrate a special occasion with friends
_____ **6.** To forget my worries
_____ **7.** To help my depression or nervousness
_____ **8.** To cheer me up
_____ **9.** To make me feel more self-confident
_____ **10.** To help me forget about problems
_____ **11.** Because the effects of alcohol feel good
_____ **12.** Because it is exciting
_____ **13.** To get high
_____ **14.** Because it gives me a pleasant feeling
_____ **15.** Because it is fun
_____ **16.** Because of pressure from friends
_____ **17.** To avoid disapproval for not drinking
_____ **18.** To fit in with the group
_____ **19.** To be liked
_____ **20.** To avoid feeling left out

Scoring

The four basic drinking motives are social, coping, enhancement, and conformity. The lowest score reflecting each motive would be 1 = never; the highest score reflecting each motive would be 5 = always. The most frequent motive for women and men is to be sociable. The least frequent motive for women and men is to conform. To compare your score with other respondents, add the numbers you circled for each of the following social, coping, enhancement, and conformity reasons identified. For example, to ascertain the degree to which your motivation for drinking alcohol is to be sociable, add the numbers you circled for items 1 through 5.

Social Score	Coping Score	Enhancement Score	Conformity Score
Add Items 1–5	Add Items 6–10	Add Items 11–15	Add Items 16–20

Average scores for 1,243 respondents

	Social	Coping	Enhancement	Conformity
Female:	2.29	1.61	1.99	1.34
Male:	2.63	1.59	2.33	1.43

Source: Adapted from M. Lynne Cooper. 1994. Motivations for alcohol use among adolescents: Development and validation of a four-factor model. *Psychological Assessment* 6:117–28. Used by permission of the American Psychological Association.

Children's Beliefs about Parental Divorce Scale

The following are some statements about children and their separated parents. Some of the statements are ***true*** about how you think and feel, so you will want to check ***yes***. Some are ***not true*** about how you think or feel, so you will want to check ***no***. There are no right or wrong answers. Your answers will just indicate some of the things you are thinking now about your parents' separation.

1. It would upset me if other kids asked a lot of questions about my parents. ___ Yes ___ No
2. It was usually my father's fault when my parents had a fight. ___ Yes ___ No
3. I sometimes worry that both my parents will want to live without me. ___ Yes ___ No
4. When my family was unhappy, it was usually because of my mother. ___ Yes ___ No
5. My parents will always live apart. ___ Yes ___ No
6. My parents often argue with each other after I misbehave. ___ Yes ___ No
7. I like talking to my friends as much now as I used to. ___ Yes ___ No
8. My father is usually a nice person. ___ Yes ___ No
9. It's possible that both my parents will never want to see me again. ___ Yes ___ No
10. My mother is usually a nice person. ___ Yes ___ No
11. If I behave better, I might be able to bring my family back together. ___ Yes ___ No
12. My parents would probably be happier if I were never born. ___ Yes ___ No
13. I like playing with my friends as much now as I used to. ___ Yes ___ No
14. When my family was unhappy, it was usually because of something my father said or did. ___ Yes ___ No
15. I sometimes worry that I'll be left all alone. ___ Yes ___ No
16. Often I have a bad time when I'm with my mother. ___ Yes ___ No
17. My family will probably do things together just like before. ___ Yes ___ No
18. My parents probably argue more when I'm with them than when I'm gone. ___ Yes ___ No
19. I'd rather be alone than play with other kids. ___ Yes ___ No
20. My father caused most of the trouble in my family. ___ Yes ___ No
21. I feel that my parents still love me. ___ Yes ___ No
22. My mother caused most of the trouble in my family. ___ Yes ___ No
23. My parents will probably see that they have made a mistake and get back together again. ___ Yes ___ No
24. My parents are happier when I'm with them than when I'm not. ___ Yes ___ No
25. My friends and I do many things together. ___ Yes ___ No
26. There are a lot of things I like about my father. ___ Yes ___ No
27. I sometimes think that one day I may have to go live with a friend or relative. ___ Yes ___ No
28. My mother is more good than bad. ___ Yes ___ No
29. I sometimes think that my parents will one day live together again. ___ Yes ___ No
30. I can make my parents unhappy with each other by what I say or do. ___ Yes ___ No
31. My friends understand how I feel about my parents. ___ Yes ___ No
32. My father is more good than bad. ___ Yes ___ No
33. I feel my parents still like me. ___ Yes ___ No
34. There are a lot of things about my mother I like. ___ Yes ___ No
35. I sometimes think that my parents will live together again once they realize how much I want them to. ___ Yes ___ No
36. My parents would probably still be living together if it weren't for me. ___ Yes ___ No

Scoring

The Children's Beliefs about Parental Divorce Scale (CBAPS) identifies problematic responding. A ***yes*** response on items 1, 2, 3, 4, 6, 9, 11, 12, 14–20, 22, 23, 27, 29, 30, 35, and 36, and a ***no*** response on items 5, 7, 8, 10, 13, 21, 24–26, 28, and 31–34 indicate a problematic reaction to one's parents divorcing. A total score is derived by adding the number of problematic beliefs across all items, with a total score of 36. The higher the score, the more problematic the beliefs about parental divorce.

Norms:

A total of 170 schoolchildren whose parents were divorced completed the scale; of the children, 84 were boys and 86 were girls, with a mean age of 11. The mean for the total score was 8.20, with a standard deviation of 4.98.

Source:
L. A. Kurdek, and B. Berg. 1987. Children's Beliefs about Parental Divorce Scale: Psychometric characteristics and concurrent validity. *Journal of Consulting and Clinical Psychology* 55:712–18.